AF784753

Voltage Regulators for Next Generation Microprocessors

Toni López · Reinhold Elferich · Eduard Alarcón

Voltage Regulators for Next Generation Microprocessors

Springer

Toni López
Philips Research
Weisshausstrasse 2
D-52066 Aachen
Germany
toni.lopez@philips.com

Reinhold Elferich
Philips Research
Weisshausstrasse 2
D-52066 Aachen
Germany
reinhold.elferich@philips.com

Eduard Alarcón
Technical University of Catalunya
Department of Electronics Engineering
Telecom Engineering Office 105
c/Gran Capità s/n
08034 Barcelona
Spain

ISBN 978-1-4419-7559-1 e-ISBN 978-1-4419-7560-7
DOI 10.1007/978-1-4419-7560-7
Springer New York Dordrecht Heidelberg London

© Springer Science+Business Media, LLC 2011
All rights reserved. This work may not be translated or copied in whole or in part without the written permission of the publisher (Springer Science+Business Media, LLC, 233 Spring Street, New York, NY 10013, USA), except for brief excerpts in connection with reviews or scholarly analysis. Use in connection with any form of information storage and retrieval, electronic adaptation, computer software, or by similar or dissimilar methodology now known or hereafter developed is forbidden.
The use in this publication of trade names, trademarks, service marks, and similar terms, even if they are not identified as such, is not to be taken as an expression of opinion as to whether or not they are subject to proprietary rights.

Printed on acid-free paper

Springer is part of Springer Science+Business Media (www.springer.com)

To our families

Preface

With the rapid advances in computer processing speed and consequent increasing demands in energy consumption of CPUs, soon it became obvious that breakthroughs may need to be realized to enable compliant power units featuring compact size and high efficiency. It was then, around year 2000, when a group of scientists led by Prof. Dr. Thomas Dürbaum decided to start it all up at the Philips research labs from Aachen, Germany.

In the power electronics community, researchers were already wondering about what could be done to drive such stringent loads. The Center of Power Electronics Systems (CPES) from the VirginiaTech institute, among others, had already started exploring numerous converter topologies as alternatives to the most widely adopted hard-switching scheme: The buck converter.

The strategy in Aachen was far more conservative and yet not less ambitious. The key question to start up with was: What makes the MOSFETs so hot? At that time, many specialists believed that the ON state resistance and the Miller capacitance were the major causes of losses in the converter. As MOSFET technologists effusively exploited the simple and extensively used $Q_{GD} \cdot R_{DSon}$ expression as baseline for their technology developments, soon it was realized that behind the apparent simplicity of the buck converter there were numerously hidden, generally unwanted elements that made fast switching operation a far more complex combination of dynamic effects than initially expected. The added difficulty was such that switching phenomena may only be accurately describable with an awfully large nonlinear impedance network. "*That's fun!!*" Or that is at least what I thought back then when I first glanced at a first attempt to model the converter with more than 30 lumped elements: "*Parasitic elements come always for free*," Thomas used to say. "*Well...*" I used to think, "*if it was so, MOSFETs would be a lot cheaper.*"

The complexity of the equivalent network that accurately described switching transients had to be implemented in a circuit simulator. The Aachen team had thus the task to build a dedicated macro MOSFET model for SPICE. Combined with an extensive device characterization for the calibration of the model parameters, the numerical calculations should enlighten the foundations of switching phenomena, which will then lead the team to the right answer to the initially formulated question. This answer would be of great value to Philips Semiconductors (now NXP Semiconductors), as they could already experience that their figure of merits started to be highly inaccurate and even misleading. Therefore, collaborations between the two organizations started.

Shortly after everything was in place and rolling, I joined the Aachen team, formed by Reinhold Elferich, Dr. Tobias Tolle, Thomas, and myself. The team from Philips Semiconductors was led by Dr. Phil Rutter, from Hazel Grove, UK. Colleagues like Steven T. Peake and Nick Koper were crucial to provide us with model data from device physics calculations and other relevant information to investigate the performance benefits of solutions such as power multichip modules, which were emerging at that time as potential alternatives to overcome a number of limitations of existing discrete solutions.

Through this intensive communication the so-called *virtual design loop* concept emerged as a methodology for Hazel Grove to receive direct feedback from circuit simulations regarding the performance of their novel *virtual* MOSFET structures, which they could refine accordingly before their actual technological implementation.

Unfortunately for us, Thomas joined the University of Erlangen (Germany) and later, Tobias moved to Philips Lighting, leaving Reinhold and myself as the only members of the Aachen team.

The disentanglement of Philips Semiconductors that gave birth to NXP Semiconductor changed the course of events quite considerably. Results were transferred to NXP so that Phil's team could carry on with their relentless technology developments employing the new methods and modeling tools, whereas I continued a parallel work in collaboration with Prof. Dr. Eduard Alarcón, from the Polytechnic University of Catalunya, Barcelona, Spain, who gave me all the support I have ever needed to turn this industrially oriented project into an academic thesis.

As the reader will notice, this treatise goes beyond finding the answer to the initial formulated question. Essential to meeting the target requirements has been an overview of all critical aspects of the converter by means of active simulations. Thus, Chap. 2 through Chap. 5 extensively analyzes a number of circuit and device models mainly devoted to MOSFET switching analysis. Each modeling approach is thoroughly described in separate chapters in terms of implementation, functionality, computation requirements, and performance predictions. The latter are extensively supported with experimental data.

The potential use of every model will be highlighted in each corresponding chapter by means of analyzing particular aspects of the converter circuit. Therefore, in Chap. 2, an in-depth analysis of power MOSFETs' switching behavior and related loss mechanisms will be provided based on a MOSFET model for circuit simulations. In Chap. 3, a piecewise linear model will be developed to study switching phenomena at a fundamental level. Different case examples will be presented to illustrate the need to move towards integrating power solutions. In Chap. 4, a power loss model will be used to breakdown the losses of a multichip module and quantify important loss mechanisms under fast switching operation.

Chap. 5 focuses on design guidelines for converter optimization. Relevant elements of the converter will be considered, including the passive filter elements and the gate drivers. Special emphasis will be given to the design of the power MOSFETs and the selection of the switching frequency and number of phases.

Chap. 6 exploits the developed models to determine the performance limitations of the converter topology and proposes measures to enhance its performance. Based on the guidelines of Chap. 5, the power density and efficiency limits of the converter based on present and future technologies will be identified. Finally, roadmap targets and improvement options for next generation power solutions will be presented and discussed.

Conclusions and future work will be exposed in Chap. 7.

The appendix sections offer technical details that, for the sake of completeness, allow the reader to dwell on some of the central discussions with extensive ancillary information.

Because this dissertation would not have been possible without them, thanks go to Dr. Phil Rutter, Prof. Dr. Thomas Duerbaum, and Prof. Dr. Eduard Alarcón. My thanks go as well with great appreciation to Reinhold Elferich, Dr. Tobias Tolle, and Nick Koper for their incommensurable guidance. Special thanks go to Steven T. Peak, Jim Parkin, Steven Hodgskiss, Victor Guijarro, and Mark Gadja from NXP Semiconductors for their priceless support.

Aachen, Germany Toni López

Abstract

Prospects of Voltage Regulators for Next Generation Computer Microprocessors

Synchronous buck converter based multiphase architectures are evaluated to determine whether or not the most widespread voltage regulator topology can meet the power delivery requirements of next generation computer microprocessors. According to the prognostications, the load current will rise to 200 A along with the decrease of the supply voltage to 0.5 V and staggering tight dynamic and static load line tolerances. In view of these demands, researchers face serious challenges to bring forth compliant solutions that can further offer acceptable conversion efficiencies and minimum mainboard area occupancy.

Among the most prominent investigation fronts are those surveying fundamental technology improvements aiming at making power semiconductor devices more effective at high switching frequency. The latter is of critical importance as the increase of the switching frequency is fundamentally recognized as the way forward to enhance power density conversion. Provided that switching losses must be kept low to enable the miniaturization of the filter components, one primary goal is to cope with semiconductor and system integration technologies enabling fast dynamic operation of ultralow ON resistance power switches.

This justifies the main focus of this thesis work, centered around a comprehensive analysis of the MOSFET switching behavior in the synchronous buck converter.

The MOSFET's dynamic operation, far from being well describable with the traditional clamped inductive hard-switching mode, is strongly influenced by a number of frequently ignored linear and nonlinear parasitic elements that must be taken into account in order to fully predict real switching waveforms, understand their dynamics, and most importantly, identify and quantify the related mechanisms leading to heat generation. This will be revealed from in-depth investigations of the switched converter under fast switching speeds and heavy load.

Recognizing the key relevance of appropriate modeling tools that support this task, the second focal point of the thesis aims at developing a number of suitable models for the switching analysis of power MOSFETs.

Combined with a series of design guidelines and optimization procedures, these models form the basis of a proposed methodological approach, where numerical computations replace the usually enormous experimental effort to elucidate the most effective pathways towards reducing power losses. This gives rise to the concept referred to as *virtual design loop*, which is successfully applied to the development of a new power MOSFET technology offering outstanding dynamic and static performance characteristics. From a system perspective, the limits of the power density conversion will be explored for this and other emerging technologies that promise to open up a new paradigm in power integration capabilities.

Contents

Abbreviations

A	Magnetic potential (V·s/m)
ABD	Avalanche breakdown
AC	Alternating current
ACPI	Advanced configuration and power interface
A_{die}, $A_{die(s)}$, $A_{die(c)}$	Die area, SyncFET A_{die}, CtrlFET A_{die} (m^2)
$A_{die(s)opt}$, $A_{die(c)opt}$	Optimum SyncFET and CtrlFET die areas (m^2)
AVP	Adaptive voltage positioning
B	Magnetic flux density (T)
B_{Sat}	Magnetic flux density saturation of a magnetic material (T)
BV_{DSS}	MOSFET's avalanche rating voltage (V)
CCM	Continuous conduction mode
C_{DS}, $C_{DS(s)}$, $C_{DS(c)}$	Drain–source capacitance, SyncFET C_{DS}, and CtrlFET C_{DS} (F)
C_{DSsp}	Specific drain–source capacitance (F)
C_{GD}, $C_{GD(s)}$, $C_{GD(c)}$	Gate–drain capacitance, SyncFET C_{GD}, and CtrlFET C_{GD} (F)
C_{GDsp}	Specific gate–drain capacitance (F/m^2)
C_{GS}, $C_{GS(s)}$, $C_{GS(c)}$	Gate–source capacitance, SyncFET C_{GS}, and CtrlFET C_{GS} (F)
C_{GSsp}	Specific gate–source capacitance (F/m^2)
CHS	Current hard-switching time interval
C_{iss}, $C_{iss(s)}$, $C_{iss(c)}$	Input capacitance for $v_{DS} = 0$, SyncFET C_{iss}, and CtrlFET C_{iss}
CMOS	Complementary metal oxide semiconductor
C_{oss}, $C_{oss(s)}$, $C_{oss(c)}$	Output capacitance for $v_{GS} = 0$, SyncFET C_{oss}, and CtrlFET C_{oss} (F)
C_{rss}, $C_{rss(s)}$, $C_{rss(c)}$	Reverse transfer capacitance, SyncFET C_{rss}, and CtrlFET C_{rss} (F)

CPU	Computer processor unit
CtrlFET	Control MOSFET
d	Converter duty cycle (output voltage to input voltage ratio)
DC	Direct current
DCM	Discontinuous conduction mode
DPA	Distributed power architecture
DrMOS	Driver plus MOSFET
DT	Dead time interval
DUT	Device under test
E	Electric field (V/m)
EMC	Electromagnetic compatibility
EVRD	Enterprise voltage regulator down
FE	Finite element
FE	Switched-node falling edge transition
FEM	Finite element method
FoM	Figure of merit(s)
FoM_t	Sum of $\mathrm{FoM}_{(s)}$ and $\mathrm{FoM}_{(c)}$ (J)
$\mathrm{FoM}_{(s)}$, $\mathrm{FoM}_{(c)}$	SyncFET and CtrlFET figure of merits defined as the loss per cycle in a particular application (J)
F_s	Switching frequency (Hz)
GB	Gate bounce
$\mathrm{GBS}_{\mathrm{GSO(s)}}$	SyncFET capacitive gate bounce susceptibility
$\mathrm{GBS}_{\mathrm{GSOx(s)}}$	SyncFET gate bounce susceptibility
GC	Gate charging loss mechanism
g_{m}, $g_{\mathrm{m(s)}}$, $g_{\mathrm{m(c)}}$	MOSFET Transconductance (S), SyncFET g_{m} and CtrlFET g_{m}
GPU	Graphic processor unit
H	Magnetic field strength (A/m)
H_c	Coercivity of ferromagnetic material (A/m)
HBC	Half-bridge charging loss mechanism
HBC(L)	Inductive half-bridge charging loss mechanism
HS	Hard-switching
i_{AB}, $i_{\mathrm{AB(s)}}$, $i_{\mathrm{AB(c)}}$	Avalanche breakdown current, SyncFET i_{AB} and CtrlFET i_{AB} (A)
IBA	Intermediate bus architecture
i_{cn}, $i_{\mathrm{cn(s)}}$, $i_{\mathrm{cn(c)}}$	Channel current, SyncFET i_{cn} and CtrlFET i_{cn} (A)
I_{D}, i_{D}, $i_{\mathrm{D(s)}}$, $i_{\mathrm{D(c)}}$	Drain current, SyncFET i_{D} and CtrlFET i_{D} (A)
I_{D0}, $I_{\mathrm{D(s)0}}$, $I_{\mathrm{D(c)0}}$	Initial drain current condition, SyncFET I_{D0} and CtrlFET I_{D0} (A)
i_{Dcap}	Drain capacitive MOSFET's current (A)
i_{diff}	Diode diffusion current (A)
i_{dio}, $i_{\mathrm{dio(s)}}$, $i_{\mathrm{dio(c)}}$	Body diode current, SyncFET i_{dio} and CtrlFET i_{dio} (A)
$i_{\mathrm{dio\_dc}}$	Static diode current characteristics (A)

i_{Dnc}	Noncapacitive drain current (A)
I_G, i_G, $i_{G(s)}$, $i_{G(c)}$	Gate current, SyncFET i_G and CtrlFET i_G (A)
I_{G0}, $I_{G(s)0}$, $I_{G(c)0}$	Initial gate current condition, SyncFET I_{G0} and CtrlFET I_{G0} (A)
i_{Lo}	Current through output filter inductor (A)
$i_{Lo(max)}$	Maximum peak current through output filter inductor (A)
$i_{Lo(min)}$	Minimum peak current through output filter inductor (A)
IMVP	Intel® mobile voltage positioning
I_o, $I_{o(av)}$, i_o	Load current (A)
$i_{o(max)}$	Maximum load current (A)
IPM	Integrated power module
$I_{rr(s)}$	SyncFET reverse recovery peak current (A)
i_S, $i_{S(s)}$, $i_{S(c)}$	Source current, SyncFET i_S and CtrlFET i_S
I_{S0}, $I_{S(s)0}$, $I_{S(c)0}$	Initial source current condition, SyncFET I_{S0} and CtrlFET I_{S0} (A)
J	Current density through a conductor (A/m^2)
J_e	Current density impressed by an external current source (A/m^2)
J–A	Jiles–Atherton magnetic hysteresis model
K_{id}, $K_{id(s)}$, $K_{id(c)}$	Time derivative of the drain current, SyncFET K_{id}, CtrlFET K_{id} (A/s)
K_v, $K_{v(s)}$, $K_{v(c)}$	Time derivative of the drain–source voltage, SyncFET K_v, CtrlFET K_v (V/s)
L_D, $L_{D(s)}$, $L_{D(c)}$	Drain inductance, SyncFET L_D, and CtrlFET L_D (H)
L_{drv}, $L_{drv(s)}$, $L_{drv(c)}$	Gate driver inductance, SyncFET L_{drv}, and CtrlFET L_{drv} (H)
LE	Switched node leading edge transition
L_G, $L_{G(s)}$, $L_{G(c)}$	Gate inductance, SyncFET L_G, and CtrlFET L_G (H)
L_{HB}	Half-bridge loop inductance (H)
L_{in}	Converter input filter inductance (H)
L_o	Converter output filter inductance (H)
L_S, $L_{S(s)}$, $L_{S(c)}$	Source inductance, SyncFET L_S, and CtrlFET L_S (H)
L_{sh}	ESL *of s*hunt resistor for current sensing (H)
M	Material's magnetization (A/m)
M_{an}	Anhysteretic magnetization (A/m)
MCM	Multichip module
M_{irr}	Irreversible magnetization (A/m)
MLCC	Multilayer ceramic capacitors
MOSFET	Metal oxide semiconductor field effect transistor
M_{rev}	Reversible magnetization (A/m)
N_{C1}	Number of output filter bulk capacitors
$N_{C1(d)}$	Minimum number of output filter bulk capacitors to avoid undue output voltage span during load transient
N_{C2}	Number of output filter ceramic capacitors

$N_{C2(sr)}$	Minimum number of output filter ceramic capacitors to avoid undue output voltage span due to load transient slew-rate
$N_{C2(ss)}$	Minimum number of output filter ceramic capacitors to avoid undue output voltage steady-state ripple
N_{Ci}	Number of input filter capacitors
N_{Co}	Number of output filter capacitors
N_{p}	Number of interleaved phases
N_{typ}	Number of different capacitor types used as the output filter components
OLGA	Organic land grid array
OnClo	ON conduction loss mechanism
PCB	Printed circuit board
PDN	Power delivery network
PEEC	Partial element equivalent circuit
PFC	Power factor correction
PFM	Pulse frequency modulation
PGN	Power ground network
PLA	Piecewise linear analytical
PoL	Point of load
PSiP	Power supply in package
PWM	Pulse width modulation
PwrSoC	Power supply on chip
Q	Quality factor of resonators
Q_{DS}	MOSFET's drain–source charge under specified switching conditions (C)
Q_{DSsp}	Specific MOSFET's drain–source charge under specified switching conditions (C/m^2)
Q_{GD}	MOSFET's gate–drain charge under particular switching conditions (C)
Q_{GDsp}	MOSFET's gate–drain charge under particular switching conditions (C/m^2)
Q_{GS}	MOSFET's gate–source charge under particular switching conditions (C)
Q_{GSsp}	Specific MOSFET's gate–source charge under particular switching conditions (C/m^2)
Q_{Gt}	MOSFET's input gate charge under particular switching conditions (C)
Q_{iss}, $Q_{iss(s)}$, $Q_{iss(c)}$	MOSFET's input charge with $v_{DS} = 0$ and specified gate–source voltage transient, SyncFET Q_{iss}, and CtrlFET Q_{iss} (C)
Q_{oss}, $Q_{oss(s)}$, $Q_{oss(c)}$	MOSFET's output charge with $v_{GS} = 0$ and specified drain–source voltage transient, SyncFET Q_{oss}, and CtrlFET Q_{oss} (C)

Q_r, $Q_{r(s)}$	Reverse recovery charge, SyncFET Q_r (C)
QSW	Quasisquare wave
$R_{ac(Co)}$	AC ESR of output filter capacitors
$R_{ac(Ci)}$	AC ESR of input filter capacitors
RAM	Random access memory
RESURF	Reduced surface field
R_D, $R_{D(s)}$, $R_{D(c)}$	Gate driver resistance, SyncFET R_{drv}, and CtrlFET R_{drv} (Ω)
R_{drv}, $R_{drv(s)}$, $R_{drv(c)}$	Gate driver resistance, SyncFET R_{drv}, and CtrlFET R_{drv} (Ω)
R_{DSon} (also $R_{DS(on)}$), $R_{DSon(s)}$, $R_{DSon(c)}$	MOSFET's drain–source ON state resistance, SyncFET R_{DSon}, CtrlFET R_{DSon} (Ω)
$R_{DSon(s)ac}$, $R_{DSon(c)ac}$	SyncFET and CtrlFET MOSFET's drain–source AC ON state resistance (Ω)
$R_{DS(on)sp}$	MOSFET's specific drain–source ON state resistance (Ω mm^2)
R_{DSpak}	MOSFET's package resistance (Ω)
R_{HB}	Total half-bridge loop resistance (Ω)
R_{LL}	Load line resistance (Ω)
RMS	Root mean square
R_G, $R_{G(s)}$, $R_{G(c)}$	Gate resistance, SyncFET R_G, and CtrlFET R_G (Ω)
R_{GM}	Polysilicon resistance of MOSFET's gate (Ω)
R_{Lo}	ESR of output filter inductor (Ω)
RR	Reverse recovery
R_S, $R_{S(s)}$, $R_{S(c)}$	Source resistance, SyncFET R_S, and CtrlFET R_S (Ω)
R_{sh}	Shunt resistor for current sensing (Ω)
RSO	RESURF stepped oxide
R_{Sub}	Substrate resistance (Ω)
SHS	Snubbed hard-switching
SiP	System in package
SMPS	Switched mode power supply
SoC	System on chip
SR	Slew rate
SRBC	Synchronous buck converter
SyncFET	Synchronous rectifier MOSFET
T_c	Curie temperature of magnetic materials (°C)
T_j	Junction temperature (°C)
Trench4, Trench6	Fourth and sixth technology generation of power trench MOSFETs from NXP Semiconductors
t_{rr}, $t_{rr(s)}$	Reverse recovery time, SyncFET t_{rr} (s)
t_{rr0}	Reverse recovery time interval between the zero current crossing and the reverse current peak (s)
T_S	Switching period (s)
t_{swFE}	CtrlFET hard-switching turn-on time (s)

t_{swLE}	CtrlFET hard-switching turn-off time (s)
$t_{\beta rr}$	Reverse recovery time parameter of diode reverse recovery model (s)
V_{AB}, $V_{AB(s)}$, $V_{AB(c)}$	Avalanche breakdown voltage, SyncFET $V_{AB(s)}$, CtrlFET $V_{AB(c)}$ (V)
V_{DRV}, $V_{DRV(s)}$, $V_{DRV(c)}$	ON state gate driver voltage, SyncFET V_{DRV}, CtrlFET V_{DRV} (V)
v_{drv}, $v_{drv(s)}$, $v_{drv(c)}$	Time dependent gate driver voltage, SyncFET v_{drv}, CtrlFET v_{drv} (V)
V_{DRVi}, $V_{DRV(s)i}$, $V_{DRV(c)i}$	ON state gate driver voltage of gate drive switches, SyncFET V_{DRVi}, CtrlFET V_{DRVi} (V)
V_{drvn}	OFF state gate driving voltage level of power switch (V)
V_{drvni}	OFF state gate driving voltage level of gate driver switches (V)
V_{drvp}	ON state gate driving voltage level of power switch (V)
V_{drvpi}	ON state gate driving voltage level of gate driver switches (V)
V_{DS}, v_{DS}, $v_{DS(s)}$, $v_{DS(c)}$	Drain–source voltage, SyncFET v_{DS}, CtrlFET v_{DS} (V)
V_{DS0}	Initial condition in v_{DS} (V)
V_{GS}, v_{GS}, $v_{GS(s)}$, $v_{GS(c)}$	Gate-source voltage, SyncFET v_{GS}, CtrlFET v_{GS} (V)
V_{GS0}	Initial condition in v_{GS} (V)
VHS	Voltage hard-switching time interval
VID	Voltage identification digital
V_{in}	Converter input voltage (V)
VNA	Vector network analyzer
V_o	Converter output voltage (V)
VR	Voltage regulator
VRD	Voltage regulator down
VRM	Voltage regulator module
V_s	Supply voltage (V)
V_{sh}	Voltage across shunt resistor for current sensing
V_{TH}, $V_{TH(s)}$, $V_{TH(c)}$	MOSFET's threshold voltage, SyncFET V_{TH} and CtrlFET V_{TH} (V)
v_x	Half-bridge switched node voltage (V)
$W_{cHBC(c)}$, $W_{cHBC(s)}$	Capacitive half-bridge charging loss of CtrlFET and SyncFET, respectively (J)
$W_{DRV(c)}$, $W_{DRV(s)}$	CtrlFET and SyncFET gate driver loss, excluding idle operation (J)
$W_{DRVi(c)}$, $W_{DRVi(s)}$	CtrlFET and SyncFET gate driver loss in idle operation (J)
W_{iHBC}	Inductive half-bridge charging loss (J)
$W_{ioHS(c)}$	CtrlFET output current hard-switching loss (J)
$W_{ioON(c)}$, $W_{ioON(s)}$	Output current ON conduction loss of CtrlFET and SyncFET, respectively (J)

ZCS	Zero current switching
ZVS	Zero voltage switching
β_{rr}	Reverse recovery overshoot parameter of diode reverse recovery model
Γ	Contribution weight of a given current or voltage in the circuit network
Δi_{Lo}	Output coil current ripple amplitude (A)
$\Delta v_{os(max)}$	Steady-state output voltage tolerance window (V)
$\Delta v_{ot(max)}$	Maximum transient output voltage overshoot (V)
δ	Skin depth (m)
ζ	Damping factor of second order systems
μ	Material's magnetic permeability (H/m)
μ_o	Magnetic permeability in vacuum (H/m)
μ_r	Relative magnetic permeability
ρ	Material's electrical resistivity (Ω·m)
σ	Material's electrical conductivity (S/m)
τ_{rr}, $\tau_{rr(s)}$	Reverse recovery single exponential decay time constant, SyncFET τ_{rr} (s)
ω_n	Natural angular frequency of second order systems (rad/s)

Chapter 1
Introduction

The challenges of powering modern computer microprocessors are so relevant that major advances in the power electronics industry are motivated by the evolution of the energy consumption of CPUs. So much so, that recent innovations in low voltage semiconductor technology, packaging, integration, control schemes and system architectures have been vastly conditioned by the energy needs of the microprocessor and related solutions for conversion and transfer of power.

From the power supply standpoint, high-end computer microprocessors are one of the most demanding electronic loads, for a combination of numerous stringent performance requirements have to be simultaneously met. Namely, CPU power delivery solutions must critically comply with large conversion ratios, high current operation, low output voltage ripple, high power density, high efficiency and wide bandwidth response to rapid load changes all in one and the same design.

The level of sophistication demanded to energize power hungry CPUs have rapidly increased over the past decades. This chapter outlines the evolution of the microprocessor load from their early days to the present time. Focusing on server, workstation and high-end desktop computer applications, future perspectives on microprocessor energy requirements are presented and analyzed. The predictions justify the current efforts in research and development to seek for effective power delivery solutions for next generation computer microprocessors. A constructive overview of the state of the art on this subject is provided, which will lead to the objectives of the thesis and a precise definition of its contributions. The last sections of the chapter describe the adopted methodological approach for addressing the challenges set forth, which constitute the central focus of the dissertation.

1.1 The Microprocessor Load

Ever since the release of the first microprocessor back in 1971, the power consumption of CPUs has increased more than two orders of magnitude, exceeding 130 W in modern high-performance computers. With the staggering advances in integration technology, the minimum feature size of transistors in state of the art CPU microstructures has decreased from 10 μm to 45 nm in 40 years, and appears to continue

T. López et al., *Voltage Regulators for Next Generation Microprocessors*,
DOI 10.1007/978-1-4419-7560-7_1, © Springer Science+Business Media, LLC 2011

to shrink down to the 32 nm scale by year 2010 [1–7]. This miniaturization trend combined with Moore's empirical observation and the increasing operating clock frequency has resulted in a progressive increase of the power density, presently approaching 100 W/cm^2 existing high-performance microprocessors. Putting it into perspective, such concentration of generated heat greatly exceeds that of the heating coil of a kitchen hot plate [8]. With the power density of CPUs ascending to even higher levels, severe thermal management limitations for heat removal are expected to become the single greatest jeopardy to further advances in computer performance in the coming years.

In view of the apparent thermal dissipation constraints, hardware solutions have been found and applied to mitigate heat generation while maintaining or even increasing computational speed. Some relevant measures have been the implementation of the CMOS technology in the 1980s, the constant improvements on gate oxide materials for leakage mitigation, the reduction of the supply voltage, the standardization of *advanced configuration and power interface* (ACPI) specifications, and the more recent introduction of multi-core architectures, among others [9–11].

These solutions have in turn leveled off the power scaling curve in the last years as reflected in the evolution chart of the high-performance Intel® family of Fig. 1.1. The power consumption of modern microprocessors has well exceeded 100 W, reaching over 150 W in some overclocked CPUs. For the next 5 years, the typical consumption is expected to rise to an average of 180 W[1] [5–8, 12–30].

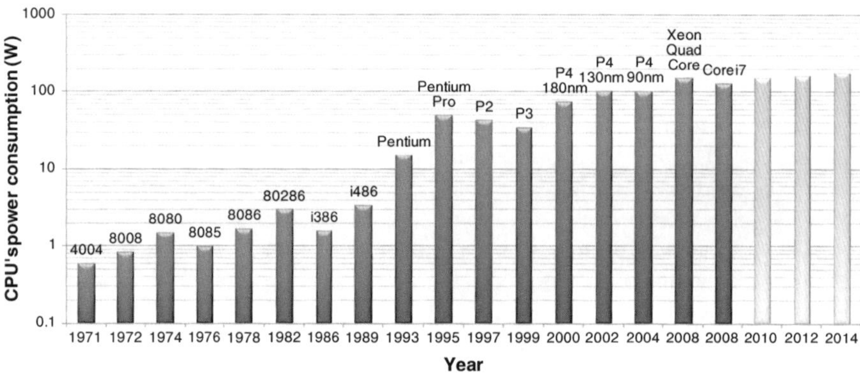

Fig. 1.1 Average power consumption evolution of Intel® microprocessors

[1]These power levels are far from deserving careless attention of their environmental and economical impact when considering the massive expansion of computer systems, particularly, server farms and supercomputers. A number of governmental agencies, industry consortiums and conservation organizations have taken consequent action to implement standards and regulations that enable computer power savings and promote the advances of energy efficiency in data centers and business computing ecosystems [31–38].

Because of its square proportionality to switching losses, the reduction of supply voltage has undoubtedly been one of the key measures to slow down the increase of heat generation despite significant developments in computation speed. The reduction of the supply voltage further helps to mitigate reliability issues concerning short-channel effects like punchthrough, oxide breakdown, and injection into the oxide [39], which become increasingly relevant as processing geometry shrinks and processor speed increases. As depicted in Fig.1.2, the supply voltage has dropped by more than 5 Volts in the last 20 years. For reasons of signal integrity and device performance, however, significant technical advances are required to keep on reducing the voltage level, which according to the present slow scaling pace, is not expected to drop below 0.5 V in the next 5 years.

The continuous rise of transistor count and clock frequency has generally been accompanied by the demand of higher current operation. In the last decade alone, the current supply has progressively increased by one order of magnitude, as depicted in Fig. 1.3. The forecast indicates a monotonic increase in the next

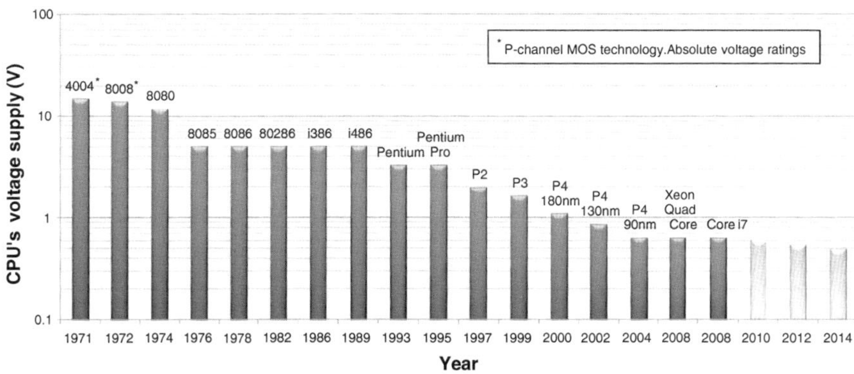

Fig. 1.2 Voltage supply evolution of Intel® microprocessors (specified at full load) [5–8, 12–30]

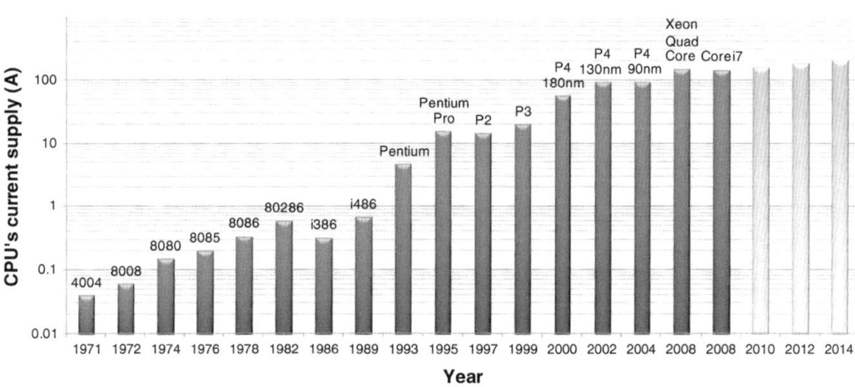

Fig. 1.3 Current supply evolution of Intel® microprocessors [5–8, 12–30]

years, with the possibility to reach 200 A peaks by 2014. Negative implications associated with the distribution of high current throughout the CPU chip area include heat generation, metal migration, and system malfunction due to voltage fluctuations across the power distribution grid (also called *power delivery network* PDN) and interconnections to the voltage supply [40, 41].

Consumption, distribution, and dissipation of power are not however the only ongoing difficulties that can potentially deteriorate the performance and reliability of computer systems. Power generation is also among the forefront challenges to endeavor [11]. Of critical importance are the static and dynamic forms in which the energy must be provided to the microprocessor in order to guarantee proper operation. The quality of the DC voltage level is one essential steady-state condition for the load. Namely, the DC voltage supply must remain within a specified margin defined by a tolerance band, which takes into consideration deviations in components spread, temperature ranges, aging factors, and ripple voltage. Figure 1.4 illustrates the evolution of the maximum voltage deviation in case of the Intel® CPU family. The tolerance band is presently around ± 20 mV for modern microprocessors, typically implying a static voltage ripple from the power supply of no more than ± 10 mV. These values are expected to lower in the next 5 years in proportion with the supply voltage so as to maintain a relative low error margin.

A second relevant static aspect of the microprocessor load is the defined linear dependence of the supply voltage and the load current, known as the load line. Such relation helps to improve the dynamic response of the power supply to sudden load changes. The associated regulation technique that deliberately adds a slope to the output characteristic is referred to as *adaptive voltage positioning* (AVP), and was widely introduced at the beginning of the second millennium [42–44]. With the AVP, the average DC output voltage is skewed to a higher value at light loads and to a lower value at heavy loads. Thus, when a load current step-up transient generates a voltage decay transition, the biased voltage produces enough headroom for this transient before reaching the lower regulation limit. Conversely, the

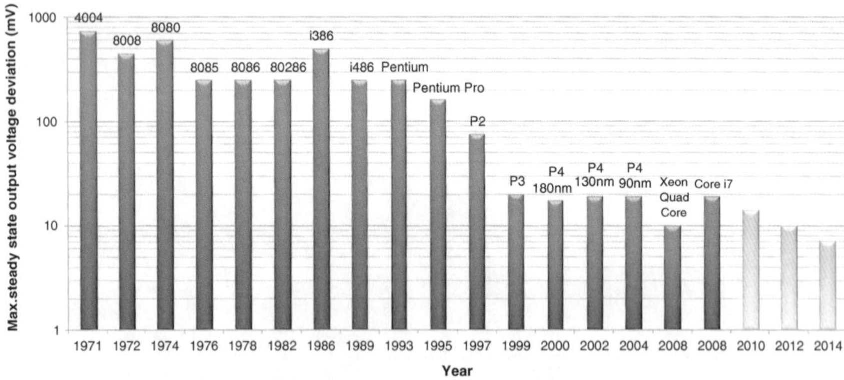

Fig. 1.4 Evolution of maximum steady-state output voltage deviation of Intel® microprocessors [5–7, 12–30]

positive-going transient after a sudden reduction in load current is accommodated with sufficient amount of headroom to avoid upper limit control saturation.

The load line is fully defined with two parameters: The voltage at zero current operation, and the resistance determining the voltage decay rate as a function of the load current. Such a load line resistance has been decreasing since the introduction of Pentium® IV processors from the 2 mΩ range down to present values slightly below 1 mΩ, as illustrated in Fig. 1.5. Expectations situate it around 0.5 mΩ in 5 years as a consequence of the predictions of higher current and lower voltage operation.

Steady-state characteristics aside, critical dynamic aspects of the microprocessor load include current slew rate, maximum voltage overshoot, and settling time. The current slew rate essentially defines the maximum speed at which the load current can change as a consequence of sudden variations in CPU usage. A limited slew rate is necessary due to the parasitic inductances in the power distribution grid and output filter capacitors. During the current ramp, the induced voltage across the parasitic inductances produces spikes across the supply terminals of the CPU that may yield circuit malfunctions and other system reliability problems. These voltage spikes are produced from what is known as the *di/dt problem* since the magnitude of the voltage ripple is affected by the instantaneous change of current with respect to time. Current fluctuations are primarily derived from dynamic resource utilization fluctuations, which are heavily influenced by architectural power-saving events such as clock and power supply gating and idle or sleep modes. The industry

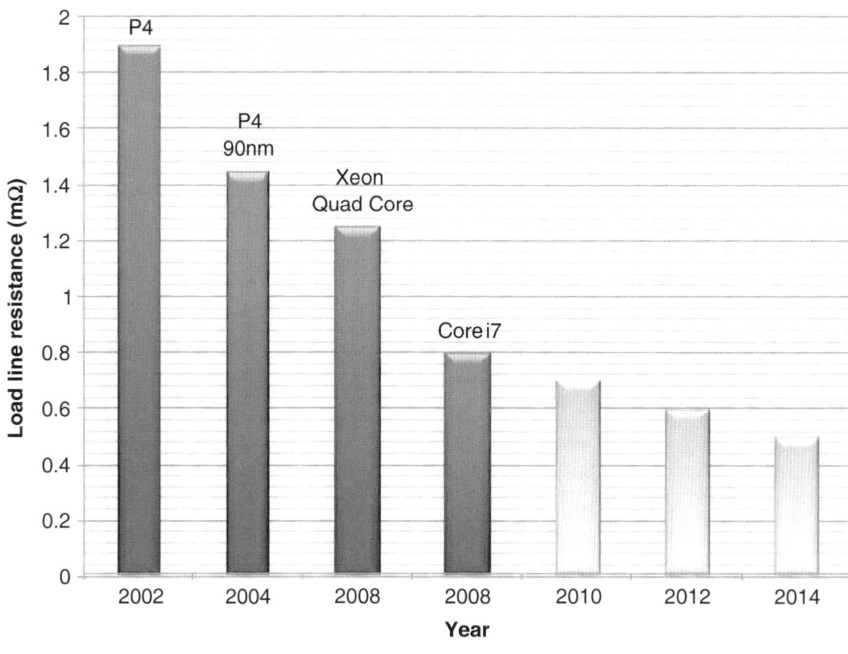

Fig. 1.5 Load line evolution of Intel® microprocessors [5–7, 12–30]

trend toward aggressive power management and voltage scaling in multi-core designs makes it increasingly important to understand the susceptibility to voltage fluctuations across the PDN at both on-chip and PCB levels. This is being addressed by the use of distributed power delivery models, such as the partial element equivalent circuit (PEEC) and related lumped power ground network (PGN) models that are designed to analyze both PCB and local on-chip voltage variations [45–48].

The determination and specification of the maximum current slew rate of the CPU is important for the design of the power supply, in particular, the output filter size and control bandwidth. The decoupling capacitors of the output filter play an important role in the PDN as they act as charge reservoirs for the switching circuits, and thus must be considered in the design of PDN target impedances suitable to provide decoupling in a wide frequency range [49, 50].

As a consequence of persistent higher operating clock frequencies and growing current demands, the current slew rate at the package terminals of the CPU has dramatically increased by almost two orders of magnitude, going from 24 mA/ns to 1.2 A/ns in the last 10 years, as shown in Fig. 1.6. Extrapolations for the next 5 years result in slew rates in the order of 2.5 A/ns.

Another key parameter of the load from the power supply design standpoint is the absolute maximum voltage overshoot that may occur when the microprocessor goes from full to light load. The overshoot level is presently specified at 50 mV over

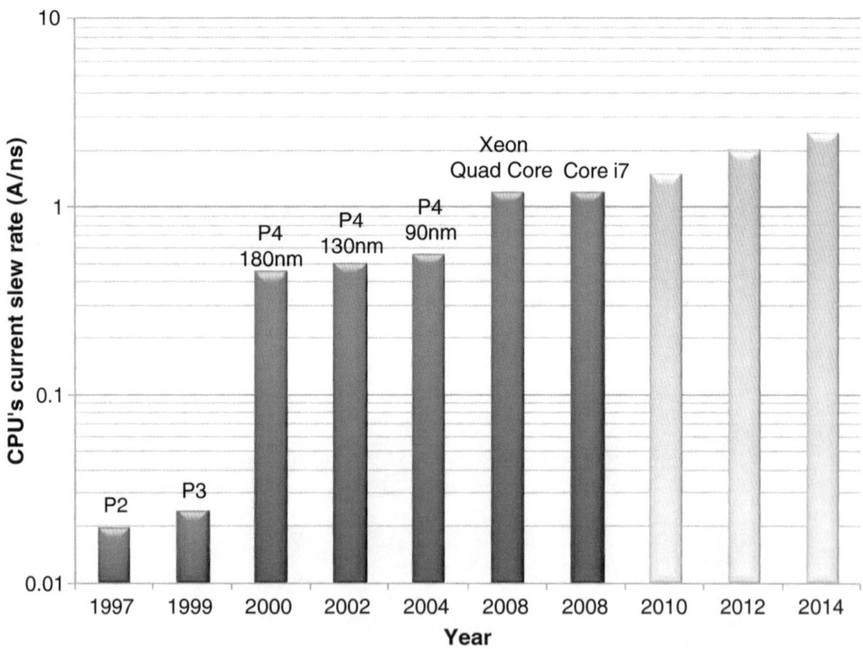

Fig. 1.6 Current slew rate evolution of Intel® microprocessor (package terminals) [5–7, 12–30]

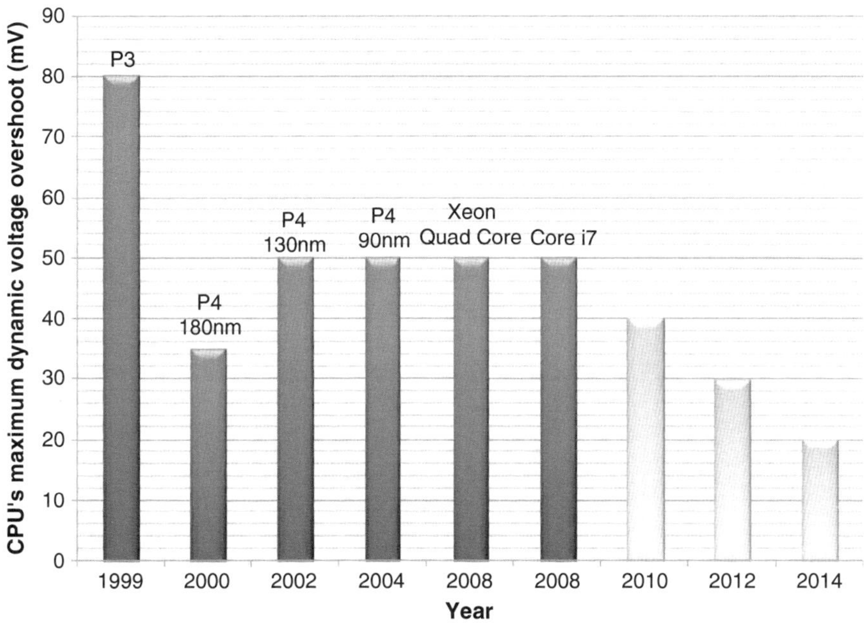

Fig. 1.7 Maximum voltage overshoot evolution of Intel® microprocessors [5–7, 12–30]

the maximum DC value, as Fig. 1.7 illustrates for the case of the Intel® products. Although this boundary has been maintained in the last years, a worst-case scenario contemplates 30 mV decay in the next 5 years.

Although it may occur repetitively, the microprocessor can only be exposed to voltage overshoots of limited time length, which is defined by the settling time parameter. The maximum settling time specified to avoid malfunctions and fast deterioration of modern CPUs is in the order of a few tenths of microseconds, as indicated in Fig. 1.8. Reducing the settling time in next generation microprocessors as expected will significantly impact both the output filter and control loop requirements of the power supply.

The evolution of high-performance microprocessors shows that the continuous enhancements in processing speed and functionality of every new process technology tends to result in more stringent electrical specifications, which are required to maintain proper and reliable system operation. The progression in this group of high-end microprocessors has clearly led to power consumption levels that are inadmissible for mobile computer applications. Thus, a second group of microprocessors emerged a decade ago to satisfy the rapid growing customer demands for portability [49–54]. The design of this new category of mobile CPUs is based on exclusive energy efficient circuit techniques and architectures [55, 56], the technology of which has enabled the commercialization of laptops featuring long battery autonomy, slim form factors, and light weight. Despite the reduced energy consumption, other important constraints such as size, heat generation, and energy

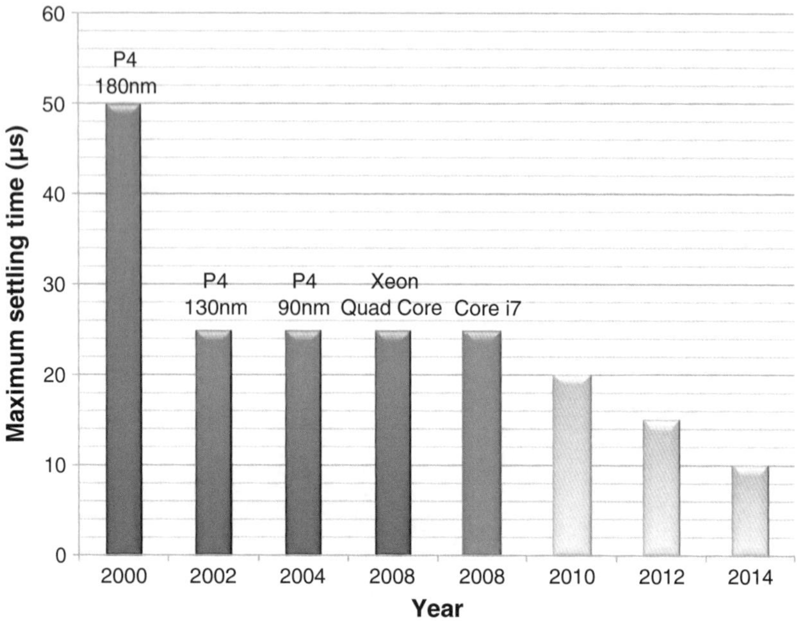

Fig. 1.8 Evolution of maximum settling time to load transients of Intel® microprocessors [5–7, 12–30]

Table 1.1 Past, present, and future of high-performance Intel® CPUs [5–7, 12–30]

Specification	Past (2002)	Present (2008)	Future (2014)
Microprocessor's name	Pentium® IV	Core™ i7	TBD
Maximum power consumption[a] (W)	103	130	180
Minimum supply voltage[b] (V)	0.87	0.65	0.5
Maximum supply current[c] (A)	90	145	200
Tolerance band (mV)	±19	±19	±7
Load line (mΩ)	1.9	0.8	0.5
Current slew rate (A/ns)	0.5	1.2	2.5
Dynamic voltage overshoot (mV)	50	50	20
Settling time (μs)	25	25	10

[a]Total contribution of the multiple voltage supplies
[b]Core supply voltage
[c]Core current consumption

conversion can make the realization of power supplies for mobile CPUs as challenging as those for servers and workstations.

A summary of past, present, and future CPU load characteristics is given in Table 1.1 and Table 1.2 for the Intel® family groups of high-performance and mobile microprocessors, respectively. The values contained in these tables are the primary basis for the analysis and discussions of the following chapters, particularly Chaps. 5 and 6.

Table 1.2 Past, present, and future of mobile Intel® CPUs [5–7, 49–54]

Specification	Past (2004)	Present (2008)	Future (2014)
Mobile microprocessor's name	Pentium® M	Core™ 2 Extreme	*TBD*
Maximum power consumption[a] (W)	30	55	65
Minimum supply voltage[b] (V)	1.24	0.9	0.6
Maximum supply current[c] (A)	21	64	80
Tolerance band (mV)	±25	±20	±10
Load line (mΩ)	3	2.1	1
Current slew rate (A/ns)	0.5	0.6	1
Maximum voltage overshoot (mV)	230	150	50

[a]Total contribution of the multiple voltage supplies
[b]Core supply voltage
[c]Core current consumption

1.2 The Microprocessor Power Supply

Most generally, an electric power supply is a power electronic system principally concerned with processing energy. It converts electrical energy from the form supplied by a source to the form required by an electric load. The input source of dedicated microprocessor power supplies may typically be of the form of a DC voltage (around 9 V to 19 V for laptops and 12 V for servers, workstations, and desktops), which must be converted to lower DC voltage levels. Because of its primary ability to maintain the output voltage compliant to the load requirements, these power supplies commonly receive the name of voltage regulators (VRs).

The VR function is broadly used nearly in every other electronic system, which may explain the wide variety of existing solutions for its implementation [57, 58]. VRs especially designed for modern microprocessors are perhaps the most sophisticated of all solutions encountered today [59, 60].

The first microprocessors, however, consumed low power and did not necessarily require dedicated power supplies since the demanded voltage and current levels were common in many parts of the computer system. Thus, inexpensive linear VR based solutions [61] were the primary choice until high power consumption microprocessors such as the Pentium® entered the scene and brought along the introduction of more sophisticated energy efficient alternatives.

Soon the power consumption became not just one of the greatest concerns of computer system designers. The relentless demands for excellence in bandwidth response, output ripple, and power density triggered a call to the power electronics community to quickly tackle the new challenges and bring forth innovative solutions [62–66].

1.2.1 Voltage Regulator (VR) Specifications

In order to aid the development of advanced VRs, microprocessor manufacturers began to publish design guidelines intended to provide a detailed description of the electrical performance of the load, requirements of the associated power supply, conditions of operation, and recommended implementation practices [67–81]. Intel® VR guidelines consider two general approaches commonly applied to modern computer systems: The *voltage regulator module* (VRM) and the *voltage regulator down* (VRD), the latter also referred to as *enterprise voltage regulator down* (EVRD) when dedicated to high-performance microprocessors. A VRM is a VR that is plugged into a baseboard (or motherboard) via a connector or soldered in with signal and power leads (see Sect. 1.2.9). On the other hand, a VRD or EVRD is a VR that is permanently embedded on the mainboard. Some major aspects of each VR category are highlighted in Table 1.3.

Note that despite of the modular and thermal advantages that VRM may offer, VRDs have become the most typical solution adopted in desktop applications primarily due to cost reasons; and so is the case in laptops and slim server computers, where tight height constraints further restrict the use of VRMs.

The main VR characteristics covered by Intel® specifications are represented in the diagram of Fig. 1.9 and briefly described as follows:

(1) *Power-in* aspects are concerned with the form of the energy provided by a separate power source at the input of the VR, which involves the specification of the corresponding voltage and current waveforms. The input voltage consists typically of a 12 V DC level with a limited specified tolerance margin. On the other hand, the input current is characterized by a smooth shape, which results from the use of a bypass filter that limits the slew rate. The input filter must be located on the VRM or on the baseboard. Recommendations on the inductor and bulk decoupling capacitors that define the filter size are typically provided.
(2) *Power-out* specifications refer to the form in which the output energy must be delivered to the load. The most relevant aspects have already been presented in section 1.1. In addition to those, other specifications related to processor power-up sequencing and shut-down response are provided.
(3) *Control signals* are available for the microprocessor to manage the output voltage of the VR according to its individual requirements. This provides the

Table 1.3 VR considerations of VRD and VRM based solutions [83]

VR considerations	VRD	VRM
Motherboard space utilization	Mainly determined by inductors and heatsink	Defined by connector size. VRM must fit within system height constraints
Environmental conditions	Ineffective cooling airflow. Motherboard temperature strongly affected by VR	Efficient air cooling of VRM board and components. Thick PCB traces
Cost performance	Cost effective at low and moderate current loads	It might be more cost and space effective than VRDs for high-performance CPUs
Replaceability and maintenance	Not applicable in most cases	Possibility of replacements for repair, expand or upgrade
Testing and evaluation	Difficult	Module can be independently tested for performance evaluation

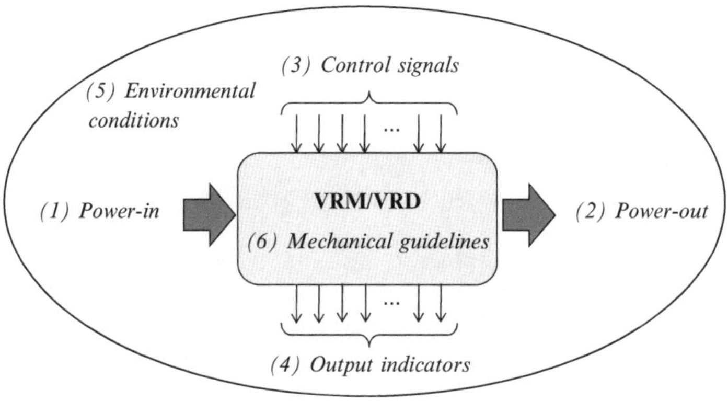

Fig. 1.9 Microprocessor VR system diagram representing five main categories of aspects addressed by Intel® guidelines

flexibility of employing common platforms for different CPUs as well as the possibility to minimize power consumption during idle periods of operation by means of reducing the supply voltage. Two digital inputs are accessible for such functionality: The *voltage identification digital* (VID), and the load line select. The VID is used by the microprocessor to generate an output reference voltage with a precision of up to 7 bits. The load line select consists of a 3-bit code to adjust the output voltage droop. Additional input signals allow disabling the VR function whenever needed as well as sensing voltage deviations during load transients. In notebook applications, control signals may be available to support heavy–light load efficiency by means of alternating between different phase operating modes.

(4) *Output indicators* provide relevant information to the microprocessor such as the finalization of the start-up sequence, VR overheating and load line regulation status.

(5) *Environmental conditions* must be compliant with the specified requirements of operating temperature, humidity, electrostatic discharge, shock, vibrations, electromagnetic compatibility (EMC), reliability, and safety regulations. Of relevant importance is to limit the maximum temperature operation for reasons of reliability and lifetime. In case of VRDs, board temperature is presently restricted to a maximum of 105°C when using low-cost (PCB) materials. For VRMs board temperature shall not exceed 90°C. These conditions include an ambient temperature range of 0°C to 45°C with minimum airflow of 400 LFM (Linear Feet per Minute) or 2 m/s, which frequently corresponds to the residual airflow from the CPU fan as no dedicated air cooling conditions are deployed.

(6) *Mechanical guidelines* impose constraints involving the physical dimensions of the VR. Although these are generally conditioned to traditional tower mount desktop or workstations, smaller form factors are emerging and gaining market share such as the 1U rack mount server, where the total height is only 1.75 in. This solution restricts the use of low cost bulk capacitors on the baseboard and

other components with heights greater than 1 inch, which include the standard 2.5 inch height VRMs.

Intel® VR guidelines are subject to change as innate consequence of technology advances. A number of revisions have been released ever since the publication of the first version more than 10 years ago [67–82]. Within each new revision, added or modified specifications and other mandatory features often significantly impact the power electronics industry in general, and motherboard manufacturers in particular. The importance of these changes is such that they usually dictate the rate of innovation and development efforts of new technologies and form the basis of new reference designs. Monitoring the evolution of these guidelines thus provides good insights as to where microprocessor power management is heading to.

1.2.1.1 VR Guidelines for Servers and Workstations

One of Intel®'s most recent VR guidelines revisions, the VRM/EVRD 11.0 [80], focuses on requirements for high-end CPU based server/workstation platforms. Some specifications described in that document are summarized in Table 1.4. Excluding *Power-out*, the parameters of all other system categories have not significantly changed with respect to revisions 10.x [74–77], and may remain so for the next couple of years.

1.2.1.2 VR Guidelines for Notebooks

With the release of the mobile Pentium® III microprocessor, Intel® introduced dedicated power delivery guidelines known as *Intel mobile voltage positioning* (IMVP) specifications [84, 85]. IMVP is primarily based on the AVP concept (see Sect. 1.1) to effectively reduce power dissipation and boost battery lifetime in notebook computers. IMVP emerges as a smart voltage regulation technology to further minimize leakage-related losses, thereby enhancing the CPU consumption

Table 1.4 Relevant aspects covered by Intel® VRM/EVRD 11.0 guidelines [80]

System category	Parameter(s)	Specification
Power-in	Voltage	12 Vdc +5%/-8%
	Current slew rate	0.5 A/μs
Power-out	Dynamic/static voltage and current	See Sect. 1.1
Control signals	Voltage identification VID	7-bit resolution
	Load line select	3-bit resolution
Environmental conditions	Ambient temperature	0–45°C
	VRM board temperature	90°C at connector
	Minimum airflow	400LFM
VRM Mechanical guidelines	Board height	66.34 mm
	Board length	96.52 mm
	Connector current rating	130 Adc

during the periods of inactivity. Energy loss savings from idle operation can be substantial since the times of inactivity are generally the longest in mobile computers. Systems supporting IMVP can experience power savings up to 75% depending on the statistical distribution of idle operation.

In addition to the already introduced VID control input, Intel®'s latest IMVP revision (IMVP-6) [86] includes the definition of three VR signals to specify different CPU operation modes: High load, low load, and idle load (the later referred to as deeper sleep mode). These signals are used by the VR control to optimize the conversion performance as well as the voltage supply level to the CPU.

Other particular VR requirements for mobile CPUs include higher input voltage (up to 19 V), tight dimension constraints (commonly less than 15 mm) and, in contrast to high-performance microprocessors, moderate output power demands [87] (compare Table 1.1 with Table 1.2).

1.2.1.3 The DrMOS Specification

The technological pathways of VRs are adeptly converging toward the sophistication of functionally integrated hardware solutions. Such proliferation is the industry attempt to undertake the microprocessor demands for high power density energy conversion. Some of these recent innovation efforts have resulted in power semiconductor integrated modules referred to as DrMOS [82]. The different requirements needed to enable DrMOS have been specified by Intel® for effective implementation in PC platforms. The aim of this specification is to enable features that are necessary to develop an integrated system such that inter-operability between various devices and controllers is feasible. Further details of DrMOS will be provided in Sect. 1.2.8.

1.2.2 Basic Circuit Topology

Because of the imperative need of high efficient energy conversion, modern microprocessor VRs are *switched mode power supply* (SMPS) based [88]. The most extended SMPS topology used to perform the required DC–DC step-down conversion function is the synchronous rectifier buck converter (SRBC) of Fig. 1.10. The half-bridge configuration comprises power metal oxide semiconductor field effect transistors[2] (MOSFETs) [89]. The upper and lower devices are named *control MOSFET* (CtrlFET) and *synchronous MOSFET* (SyncFET), respectively.

The SRBC is an advanced version of the basic buck converter topology, where the rectifying diode is replaced by an active device (SyncFET) which is controlled to conduct or block current in synchronism with the CtrlFET [90, 91].

[2]The use of MOSFET devices will be justified in Sect. 1.2.4.

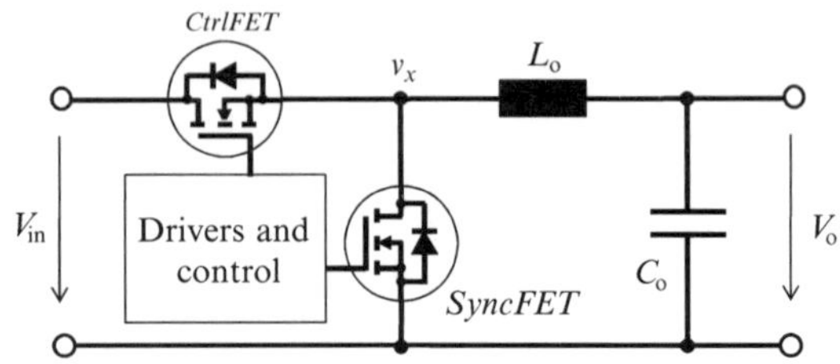

Fig. 1.10 Synchronous rectifier buck converter (SRBC) topology

This additional switch and associated control complexity represents the most viable and effective alternative to mitigate the principal efficiency limitation of the diode's forward voltage in low-voltage applications.

In the SRBC, both active devices operate sequentially in switched-mode to chop the DC input voltage, thereby generating a unipolar square waveform at switch node voltage v_x. The basic characteristics of the v_x signal are readily defined by input voltage V_{in}, switching frequency F_s and duty cycle d, the latter being determined by the CtrlFET ON conduction time over one switching period (T_s). The ideally lossless low pass filter formed by L_o and C_o is such that mainly the DC component of v_x is extracted at the output (voltage V_o).

The switching frequency dictates the cut-off frequency of the output filter and thus the physical dimensions of its elements. That is, for a given fixed output voltage ripple, the increase of switching frequency yields a higher cut-off frequency and hence a smaller filter size.

In addition to enabling higher power density, the increase of the switching frequency further allows a faster load transient response and a reduced dynamic voltage overshoot since the stored energy in the output filter becomes smaller. This argument alone justifies the continuous efforts to come up with practical ways to boost F_s, which is central in this thesis work. Typical switching frequencies in existing commercial solutions go from 300 kHz to 1 MHz.

The challenges of increasing F_s are directly related to switching losses, which predominantly occur in the power MOSFETs. Attempts to mitigate them usually compromise other loss mechanisms, such as the conduction losses [92]. One main source of switching related losses is the hard-switching operation of the CtrlFET. In this mode, the load current commutes during the time in which the voltage across the device is still high (i.e., around V_{in}). The simultaneous presence of high current through the device and high voltage across it produces high stress and energy dissipation in every switching cycle. The effect can be reduced by applying soft-switching techniques, some of which give rise to other types of converter topologies: Soft-switching converters [93, 94]. Soft-switching is characterized by ideally lossless switching events, implying that either current or voltage is zero during commutation, that is, zero current switching (ZCS) or zero voltage switching (ZVS).

Alternative soft-switching topologies known as resonant converters have been proposed and analyzed for microprocessor VRs due to their presumable suitability to efficiently operate at high F_s. Most of these resonant SMPS [95] feature reduced switching losses by virtue of ZVS attained in the power switches. Although, as it will be shown, ZVS can also be fully realized in the SRBC resonant topologies can effectively exploit the use of transformers to perform energy transformations that minimize the additional current swing (i.e., resonant current) required to achieve ZVS, thereby mitigating the associated losses that such RMS current increase may produce. The use of transformers however adds design complexity and troublesome miniaturization challenges.

Nonetheless, despite of these alternative solutions and the hindrances of switching losses, the SRBC continues to be the topology of choice for microprocessor VRs. The combination of extreme simplicity, cost-effectiveness, maturity, extensive use, and potential qualities for integration and compactness makes the SRBC topology as of today irreplaceable.

What appears to be uncertain is whether the SRBC topology will still be feasible for next generation VRs, and whether its practical use will be subject to key conditionals, in which case the determination of these will be essential to lay out the development of potential enabling technologies.

The SRBC can operate in a number of control modes and modulation schemes [62], [88, 91, 96–98] some of which include pulse width modulation (PWM), pulse frequency modulation (PFM), discontinuous conduction mode (DCM), continuous conduction mode (CCM), and quasi-square wave (QSW). Combining them may maximize the performance over the entire load range.

1.2.3 System Architecture

The evolution of VRs is highly influenced by the power delivery architecture of the computer system, which has conversely evolved from centralized to distributed designs to supply energy not just to the CPU, but also to other power hungry loads such as RAM memory modules, video cards, and chipset controllers. A centralized power system consists of a single power supply residing in a particular location and powering all system's elements through a network of cables and PCB buses. With the demands of high current operation and power quality, this conceptual approach has become obsolete since the parasitic elements of the interconnections can excessively increase power loss and compromise the signal integrity of the intended power delivery form.

Distributed power architectures (DPA) were created to address these issues by essentially bringing power processing closer to where energy is used [99–105]. Such concept gave rise to a special class of power supplies known as point-of-load (PoL) converters [106]. A PoL converter is in essence a dedicated VR (high- or low-power) that supplies power locally to a complex load according to its particular requirements. PoL has proven to effectively boost power conversion density while improving efficiency, steady-state, and transient regulation. The PoL approach is in line with present system design development trends, forming the basis for the power supply-in-package (PSiP) and power supply-on-chip (PwrSoC) concepts that have recently gained much attention in the power supply industry [107]. Front-end PoL is

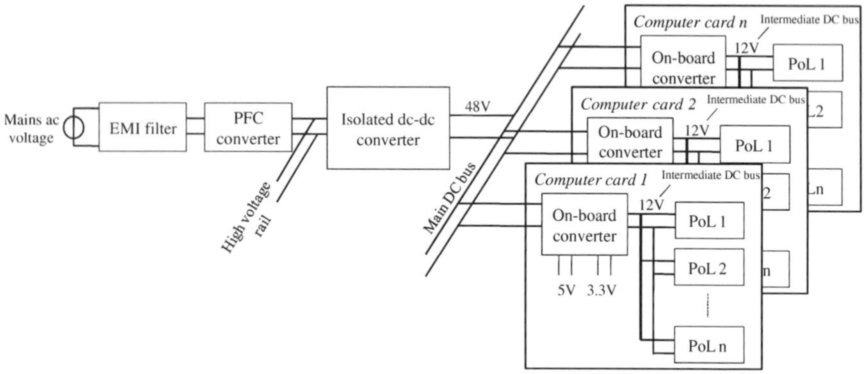

Fig. 1.11 Diagram of a distributed power architecture (DPA) in a computer server

thus a central topic of this thesis in the context of power supply miniaturization capabilities.

Generally, DPA based on an intermediate bus architecture (IBA) may consist of a series of cascaded power stages with intermediate buses or power rails as illustrated in Fig. 1.11, [108–111]. In complex computer systems like servers, the structure may be composed by exchangeable functional cards or processing modules powered by on-board power stages. In order to power a large number of system card loads, a main DC bus of 48 V may be adopted. Each on-board central power stage then converts this voltage to lower levels, thus providing an intermediate bus voltage of, for instance, 12 V. This secondary power rails distribute the energy throughout the computer card. In the final stage, individual PoL converters power the different loads, which may include multiple high-performance CPUs or GPUs.

The subsystem of high-power PoL converters comprises a number of parallely connected SRBCs to drive a common load, forming what it is known as a *multiphase architecture* [112, 113]. When the load is a microprocessor, the associated PoL converter becomes part of a VRD or VRM, which may basically consist of a large number of paralleled SRBCs, also called phases or unit cells. Some commercially available high-performance workstations already employ VRDs with 16 phases.

Multiphase cells operate at a common frequency (i.e., F_s), but are phase shifted so that switching occurs at regular intervals lower than T_s. Namely, a phase controller staggers the switching time of each SRBC in an interleaved fashion such that the phase angle of consecutive phases is $360°/n$, n being the number of unit cells. As a result, the effective ripple frequency at the output of the parallel connection is $n{\cdot}F_s$, thereby improving both dynamic and static performance.

Another significant advantage of the multiphase architecture is that, as the load current distributes among the different phases, so does the generated heat, hence reducing the power demands of each individual converter and improving the thermal design. This, in turn, may be translated into lower costs (i.e., simpler packaging and reduced chip size area) and higher power density (i.e., smaller heatsinks).

Although system complexity is increased (by the need for instance of including current sharing monitoring and control), the multiphase architecture represents today the only practical approach to drive high-performance computer microprocessors. Primarily, all proposed PoL subsystem structures from recent publications based on the existing multiphase approach. A central focus of attention is currently on multiphase control algorithms to further improve conversion efficiency, particularly at light loads, by means of methods such as phase shedding or PFM (see Sect. 1.2.7), which have already been succesfully applied to battery operated systems.

One relevant subsystem level solution that has recently been proposed consists of a PoL featuring two cascaded power stages [114], thus forming a local intermediate bus voltage of closer value to V_o. The arrangement allows significant reduction of power losses in the second stage due to reduced voltage switching. The high currents at such energy conversion level imply though that the first stage must also consist of multiple parallel converters. Consequently, the benefits of the approach in terms of overall efficiency and power density do not always seem significant when compared (under optimum conditions) to the conventional multiphase architecture for the same number of total phases [115]. Nevertheless, the solution may be considered as alternative to avoid the need of resolving extreme low duty cycles, which may impose a practical limit in the conventional topology as microprocessor voltage continues to scale down.

1.2.4 Semiconductor Power Devices

As mentioned earlier, power switched devices are key elements of the VR since their associated power losses usually account for a significant portion of the overall generated heat [116]. The loss contribution is such that it often represents the main barrier to achieve higher system efficiency and F_s operation. Thus, meeting forthcoming VR specifications highly relies on performance improvements in the semiconductor device.

The SRBC was introduced in Sect. 1.2.2 along with the power MOSFET as switching device. Among all transistor structures, the power MOSFET is the preferred implementation because it combines a series of basic characteristics that makes it the most suitable semiconductor device for the low voltage range of interest [117]. The power MOSFET is compatible with the demands of high-switching frequency and high load current due to its inherent fast switching speed capabilities, low ON conduction resistance (R_{DSon}) and high input gate impedance. The unique advantages are such that other basic silicon-based structures have hardly ever been considered for microprocessor VRs [118].

The variety of existing process capabilities enables a number of different MOSFET types. The most extended of all is the *n-channel enhancement-mode* MOSFET, for it offers the best performance and desired functionality [89].

On the other hand, the arguments for which to decide how the structure of this device type can be best adapted to the targeted application are subject to continuous

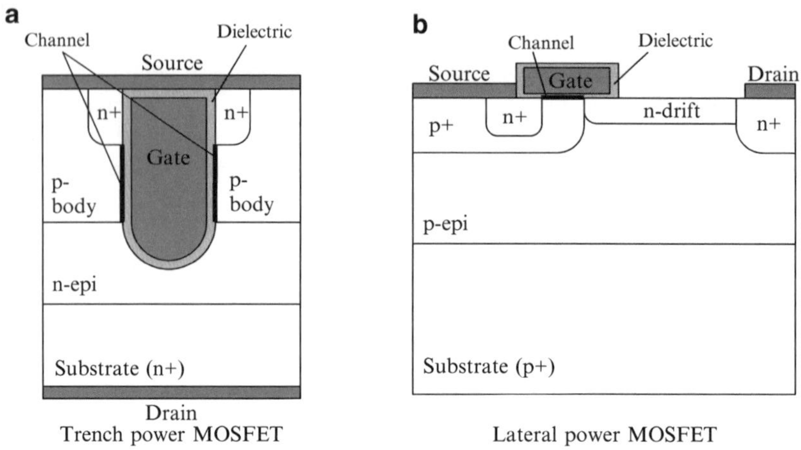

Fig. 1.12 Cross-sectional representation of two kinds of power MOSFET structures: (a) Vertical trench power MOSFET; (b) Lateral power MOSFET

debate. The two basic structures under consideration are the lateral power MOSFET and the vertical trench power MOSFET [119], whose cell cross-sectional representations are depicted in Fig. 1.12. As it can be observed, the fundamental difference between the two devices lays on the fact that, unlike the planar arrangement, the vertical structure employs the substrate material to form the drain contact.

To be able to carry high currents, many of these individual unit cells are connected in parallel and arranged in regular spaces defined by a cell pitch, which determines the density of cells per unit area [89]. Therefore, the required current level capability, the cell pitch, and the unit cell size basically determine the total semiconductor area. Minimizing it is crucial to reduce manufacturing costs, which is one of the main driving aspects to decide between one and the other solution.

There are two major reasons why the planar approach is unsatisfactory with regard to area occupancy. First, the drain-source spacing has to be increased in comparison to low power MOSFETs in order to obtain a high voltage blocking capacity. And second, the three electrode connections have to be made from the upper surface (the one opposite to the substrate). While this facilitates the monolithic integration of components, it complicates the interconnection in a single device package. Both effects reduce the effective area of utilization to form the active region.

Vertical trench power MOSFETs (or trench MOSFETs) mitigate this problem by forming the MOS channel on the vertical walls of a trench, which is etched into the semiconductor material and filled with dielectric and a gate electrode, thus allowing closer cell spacing and increased current density.

Consequently, trench MOSFETs offer low specific R_{DSon} ($R_{DS(on)sp}$) in detriment of parasitic gate charge, which has been traditionally higher than in lateral

MOSFETs. The latter has been the main argument (together with the monolithic integration quality) to encourage the use of lateral MOSFETs for high-switching frequency despite of the cost penalty associated to the low utilization factor of the semiconductor area [120].

For its cost-effectiveness however, the majority of semiconductor manufacturers have progressively adopted the use of trench MOSFETs, which have accordingly evolved at a much faster pace than any other structure [121]. As a result of the remarkable improvements achieved in every technology generation, trench MOSFETs have quickly caught up with lateral devices in terms of switching capabilities while maintaining a continuous decay in $R_{DS(on)sp}$ [122, 123].

In the last decade alone, silicon-based trench MOSFET technologies have tremendously improved by as much as 90% in $R_{DS(on)sp}$, as shown in the evolution chart of Fig. 1.13. The data refer to 30 V power MOSFETs, which is the typical device voltage rating for the 12 V conversion level. The trend line clearly reveals a progressive saturation of $R_{DS(on)sp}$ as it approaches the fundamental 1D limit of silicon [124], which describes the theoretical relationship between the drift region resistance and the OFF state blocking voltage (BV_{DSS}). Having reached the 1D limit suggests that Si is approaching maturity, and thus a continuous cost-effective innovation rate will be difficult to maintain in the near future.

A number of variations of the basic trench MOSFET structure have been developed to further improve the performance of the device in the switched-converter. The main driver has been primarily directed at producing the lowest possible $R_{DS(on)sp}$ for an acceptable switching performance. To meet this challenge, solutions based on the charge-balanced concept such as *superjunctions* and the isolated field-plates along the drift region have been demonstrated to achieve $R_{DS(on)sp}$ values below the fundamental 1D limit [125–128], as Fig. 1.13 shows. The charge-balanced concept

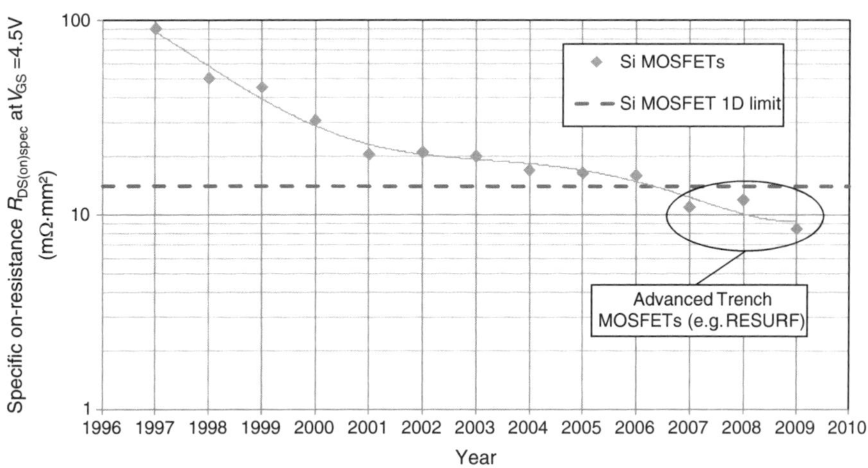

Fig. 1.13 Specific ON resistance evolution of 30 V power MOSFETs. Sources: [122, 123, 125–128] and NXP internal competitor analysis

exploits the so-called reduced surface field (RESURF) effect to enable the increase of conductivity in the drift region without compromising BV_{DSS}. Unconventional device structures such as the RESURF stepped oxide (RSO) and the split-gate RSO have proven to be very effective especially for high voltage applications. Their industrialization [129] however suffers from significant cost premiums due to troublesome technological issues such as tight silicon specification requirements, additional mask layers, and stringent process tolerances.

In order to achieve the best trade-off between BV_{DSS} and $R_{DS(on)sp}$, other alternative cost-effective trench structures [130] employ a relatively thick gate oxide at the bottom of the trench and reduced drift region, thereby allowing the reduction of both gate-drain charge Q_{GD} and R_{DSon} for the same BV_{DSS} capability. Some of these trench structures have proven to produce $R_{DS(on)sp}$ values around 12 mΩ·mm^2, which compare competitively with the more complex RESURF devices [133, 134].

Another alternative device structure known as the NexFET™ [131–132] adopts a hybrid approach between lateral and trench MOSFETs. This solution allows achieving lower $R_{DS(on)sp}$ than the lateral devices and better switching performance than the trench MOSFETs, thereby making it suitable for applications demanding high current and high-switching frequency operation.

Besides the innovations on device structures, investigations are further devoted to advances in alternative semiconductor materials that can potentially offer superior physical and electrical properties than Si. Compound semiconductors like SiC, GaAs, and GaN provide greater thermal conductivity, electric field breakdown strength, and current density capabilities [138]. Although these materials were discovered long ago, their progress has lagged behind silicon due to the difficulties of growing good quality crystals. In recent years though, the advances on these semiconductor compounds have been tremendous, allowing the fabrication of reliable power devices exhibiting significant advantages over silicon. The manufacturing cost and complexity of the required technology platforms are still high, which has restricted their use to particular applications in high voltage, ultra high frequency or optoelectronics.

Foreseeing the window of possibilities in power electronics, great effort has been put into the development of cost-effective GaN-based technologies, which have been recently claimed by International Rectifier Corp. [139–141]. The main breakthrough regarding this has been to overcome the great difficulties of successfully growing GaN on Si substrates. Combined with high electron mobility transistor (HEMT) structures [142], the result has been a GaN technology platform based solution viable for a wide range of power electronics applications, including microprocessor VRs. According to the projected evolution of this GaN technology, a dramatic reduction of both parasitic charges and R_{DSon} will be attained in the coming years, thus creating a new paradigm of opportunities for extreme high power density conversion. Figure 1.14 depicts the evolution of various FoM (see section 1.2.4.1 for a definition) from 30 V Si power MOSFETs and presents the projected expectations of the low voltage GaN HEMT devices. The curves suggest GaN as a potential candidate to replace Si and achieve breakthrough performance improvements of one order of magnitude in less than 5 years.

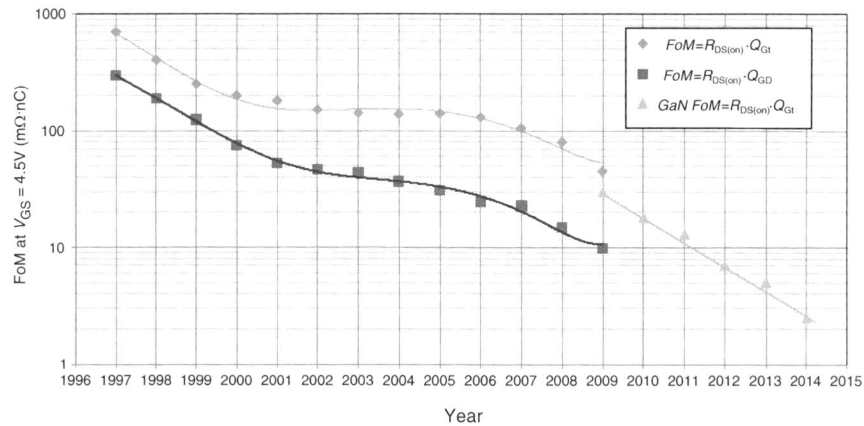

Fig. 1.14 Evolution of various figure of merits (FoM) for 30 V Si power MOSFETs and projected expectations for low voltage GaN devices. Sources: [139–141] and NXP internal competitor analysis of Si devices

These tremendous improvements may have to be accompanied by novel parasitic free packages that improve the existing LFPAK [143], DirectMOS [144], or PolarPak [145], as well as new driving solutions, as it shall be discussed in the following sections.

1.2.4.1 Device Modeling and Optimization Tools

The analysis and design of these novel materials and device structures require sophisticated modeling tools to provide sufficiently accurate performance predictions in the application. The most extended tools are multidimensional device physics simulators based on numerical methods such as MEDICI [135]. These software packages can produce outstanding accurate results but generally demand significant computation power and long simulation times, which restrict their use to rather simple device/circuit simulations in mixed-mode.

For the sake of simplicity and reduced computations, device physicists rely on figure of merits (FoM) for device optimization and technology comparisons [136, 137]. These FoM characterize the performance of the MOSFETs in the switched-converter by way of a power loss index deduced under optimum circuit conditions. Due to differences in operation, individual FoM are defined for the CtrlFET and SyncFET. Typically, the SyncFET FoM is defined by the product $R_{DSon}{\cdot}Q_{Gt}$, where Q_{Gt} is the total gate (input) parasitic charge. The CtrlFET FoM differs from the previous expression in that it only considers the portion of Q_{Gt} associated to the transfer capacitance, hence resulting in the product $R_{DSon}{\cdot}Q_{GD}$.

Both FoM derive from simplified loss calculations that neglect a number of circuit elements under the assumption of low or moderate switching speeds. Upon fast switching though, the expressions become inaccurate because some of those

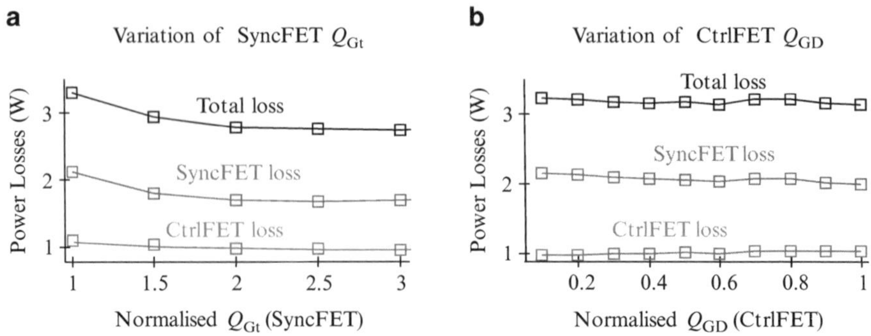

Fig. 1.15 Power MOSFETs loss dependence on (a) SyncFET gate charge, and (b) CtrlFET transfer charge. Simulation results of a typical SRBC for VRs

disregarded elements start influencing dramatically the dynamic behavior of the switches. This is illustrated in the loss graphs of Fig. 1.15 where the increase of the parasitic MOSFETs charges leads to a reduction of switching losses. Figure 1.15(a) shows that increasing by factor of 3 the SyncFET charge Q_{Gt} causes a 17% loss reduction. The example of Fig. 1.15(b) indicates that the total MOSFETs losses may be reduced by more than 3% after an increase of CtrlFET charge Q_{GD}. To explain such discrepancies with the predictions given by common FoM, a thorough analysis of switching phenomena needs to be attained.

Further relevant circuit parameters of fast switching are the parasitic inductance of the device package, the half-bridge loop inductance and the gate resistance among others. In order for these to be considered, more complex circuit models and new FoM may need to be established and evaluated as F_s moves into the MHz range. The new tools may need to be compatible with a co-design approach, where the two perspectives of circuit and device physics are brought together in the development process flow. Also, for the level of accuracy required, the issues related to high computation power and long simulation times may have to be addressed so as to generate stable simulated results within reasonable times.

1.2.5 Gate Driving Schemes

Gate drivers provide a means to energize power MOSFETs so that they can turn on and off according to the commands of the control system [146]. Driving a MOSFET is essentially equivalent to charging a capacitance, and as such the driving process may involve significant charge transfer related losses, particularly at high-switching frequency, high current applications.

When driving, the charging energy must be high enough to drive the power MOSFET from a blocking to a conducting voltage level and vice-versa. For convenience, in *n-channel enhancement-mode* device types, the blocking level is usually zero volts with respect to the source terminal. This usually gives a sufficient tolerance margin of two to three volts deviation from threshold voltage V_{TH}, thereby preventing the device from any undesired spurious turn-on caused by coupled noise. Regardless of a tolerance margin, the ON state voltage level, referred to as driver voltage V_{DRV}, must correspond to the ohmic region of the device. Determining the optimum V_{DRV} for both SyncFET and CtrlFET is per se an important design consideration that influences conversion efficiency. A trade-off between ON conduction and switching loss arises as a consequence of the inverse proportionality of V_{DRV} over R_{DSon} and its square dependence on the driving loss. Commonly, voltage V_{DRV} may be set anywhere between 4.5 V and 12 V depending on the converter specifications and availability of the DC voltage levels.

The ability of gate drivers to produce fast switching transients is perhaps the most important feature to allow the reduction of power MOSFET switching losses and hence the increase of F_s. The form in which the energy is supplied to or absorbed from the power MOSFET is the key driving aspect that determines the switching speed. Fast switching can be achieved by means of high current driving pulses so that charge can be quickly transferred from and to the device in the shortest time possible. It is fundamentally indispensable to minimize the parasitic elements of the gate circuit in order to enable these high current driving pulses. It implies the reduction of parasitic elements in the gate circuit, such as the package inductances, the polysilicon gate resistance of the power MOSFET, and the ON resistance of the gate switches. While the parasitic inductances may be addressed by the integration of the driver and the switch (i.e., DrMOS concept), the gate resistance can be reduced by improving the contribution of the spreading resistance corresponding to the gate busbar structure that distributes the current throughout the chip area of the device [147–149]. Other significant improvement options include the use of gate switches that can sustain higher current levels (e.g. 5 A or more), thereby avoiding their saturation during the periods of charge and discharge. Generally however, the increase in current capabilities is in detriment of an increase of the parasitic charge of the gate switches, which also contributes to switching loss. Selecting the right type of gate switches is therefore another design task in the optimization of the driver circuit.

It is important that the gate resistance reduction is carried out in proportion with the gate inductance in order to avoid excessive ringing during commutation. This ringing is a resonant oscillation between the gate inductance and the input capacitance of the MOSFET that, when not sufficiently damped, it can lead to circuit malfunction and additional EMI issues.

Figure 1.16 illustrates a simplified representation of a conventional gate driver scheme, which comprises a half-bridge of low power MOSFETs driven by a synchronization circuitry. The upper and lower driver MOSFETs operate complementary. To avoid crossover conduction, a dead time is programmed during which both devices are off prior to a switching event of the power MOSFET.

The driving conditions of SyncFET and CtrlFET may differ considerably. The SyncFET operates in ZVS, and as such the input capacitance can be considered practically linear and lower than in hard-switching. Driving the SyncFET however may be lossier than driving the CtrlFET since the die size of the first is significantly

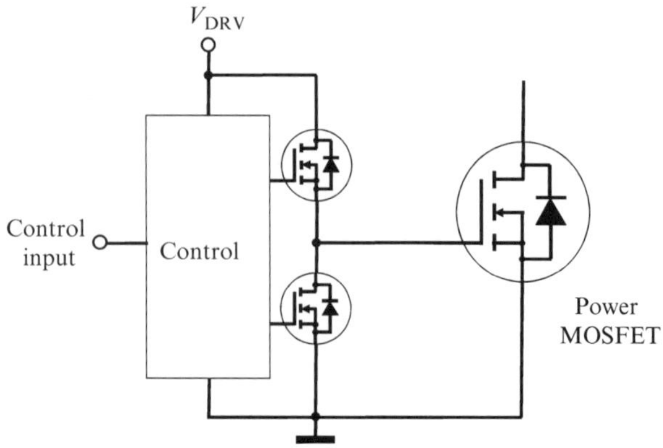

Fig. 1.16 Conventional gate driver based on gate MOSFET switches

larger (e.g. as much as four times, depending on the application), that is, higher input capacitance. The size unbalance is the consequence of the low duty cycle operation that makes the SyncFET carry the load current for most of the time.

The large size of the SyncFET input capacitance may have more negative implications other than gate charge losses. First, without substantial reduction in gate resistance, long switching times will critically limit the minimum duty cycle operation and hence the maximum F_s. And second, the ON resistance during the switching times is higher than during the ON time (i.e., time when the gate voltage equals V_{DRV}). Consequently, slowing the switching transients will result in higher ON conduction losses.

The CtrlFET is a significantly smaller device and therefore the input capacitance and associated gate losses are generally lower. Nonetheless, the CtrlFET operates in hard-switching, and as such reducing the switching times is particularly critical so as to effectively mitigate switching loss in the power device. Under hard-switching, the device undergoes a potentially loss Miller and *di*/*dt* plateau transitions, which are the most critical to reduce. One effective measure to shorten the *di*/*dt* transition is by mitigating the source inductance of the power device. Packages like the DirectFET [144] and the PolarPAK® [145] are specially designed for that purpose, featuring a source inductance in the order of just 100 pH.

Although this and other solutions are well recognized under the assumption of ideal clamped inductive switching mode, their effectiveness becomes uncertain at high switching frequency due to the influence of neglected parasitic elements. Their implications on switching losses and gate driving requirements thus need to be addressed in order to identify adequate solutions. Some studies based on more accurate models have already been initiated, revealing indeed the importance that certain parasitic elements have on switching phenomena.

A different gate driving concept is that of resonant gate drivers. For years, resonant gate drivers have been presented as alternatives to conventional solutions to drive power MOSFETs with reduced energy loss. The conceptual idea of

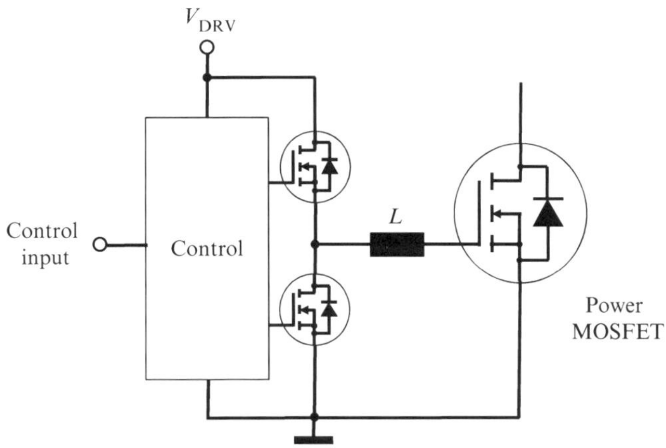

Fig. 1.17 Basic resonant gate driver

resonant gate drivers consists of replacing the resistance of the path through which gate charge flows in the conventional approach by an inductance, as shown in the basic diagram of Fig. 1.17. With the ideal absence of dissipative elements, the full energy required to change the state of the power device can be controllably recovered and returned to the power supply, or alternatively stored in auxiliary tanks for reuse. Several basic circuit topologies have been proposed to most effectively implement this concept [150, 151].

Commonly, the inductance of the resonant gate drivers determines the switching speed of the power MOSFET as the gate resistance does in the conventional way. Thus, in high-switching frequency applications, the use of low loss resonant gate drivers featuring low inductance may be of high interest, for size and switching times need also be reduced.

On the other hand, reducing the gate inductance negatively impacts the gate losses as soon as parasitic resistances cannot be neglected. In fact, the gate path circuit is not exempt from dissipative elements as described above. Therefore, whenever the inductance of the resonant gate driver is to be minimized, both the parasitic charge and resistance of the gate switches may have to be considered in the circuit optimization. This strong loss dependency on the switching speed in resonant gate drivers contrasts with the behavior of conventional solutions.

Despite of the long list of publications about resonant gate drivers, there are no conclusive results known to the author demonstrating that resonant gate drive topologies may be beneficial for microprocessor VRs.

1.2.6 Filters

An inherent disadvantage of SMPS is that they represent a powerful source of high frequency harmonics that can generate a great deal of conducted and radiated EMI

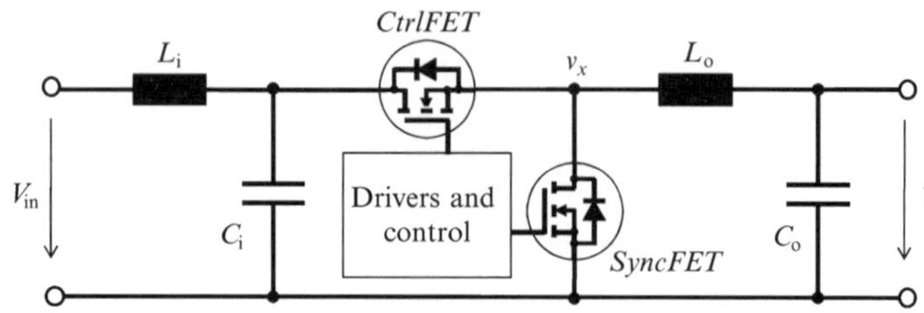

Fig. 1.18 SRBC topology with a low-pass input filter

as well as high frequency ringing at the input/output(s) of the converter stage. Existing guidelines and regulations [152] impose the use of proper layout designs and filters that minimize their harmful effects. In the SRBC, this translates in the need of adding a low-pass filter at the input of the converter, which may result in a circuit like the one shown in Fig. 1.18.

High efficiency demands the use of reactive elements, these are, capacitors and inductors for smoothing the output ripple. The effectiveness in carrying out this function lays on the energy storage capacity of the filter elements, which is proportional to their dimensions. Thus, the filter size increases with the smoothing requirements and, as mentioned earlier, decreases with F_s. Since the filter components largely dictate the size of the converter, the current trend is to keep on increasing F_s in detriment of the efficiency, which, on the other hand, should not excessively lower to avoid the use of bulky heatsinks that could counteract the filter size reduction. Switching frequency F_s must therefore be adjusted to best compromise between size and efficiency.

1.2.6.1 Input Filter

Filtering the current drawn by the SRBC such that it is compliant with current Intel® specifications (see *power-in* specs in Table 1.4 for the enterprise case) can be accomplished with low equivalent series resistance (ESR) capacitors and a power choke of approximately 1 μH capable of carrying 10 A [153].

The input capacitance size depends on the duty cycle and F_s. Its value has to be high enough so that the voltage drop during CtrlFET conduction is limited. For instance, a load of 200 A will discharge an input capacitance of 47 μF by 350 mV when converting 12 V to 1 V at 1 MHz switching frequency. The ESR of 2 × 22μF can be as high as 2 mΩ at that frequency, thus producing an additional voltage drop of 400 mV. The losses caused by such ESR need to be sufficiently low to guarantee the required capacitors' lifetime and converter efficiency. Thus, multiple parallel capacitors may be arranged to improve heat distribution and reduce filter losses.

It turns out that, in order to perform the decoupling function effectively, the input filter capacitor must be physically located as close as possible to the power MOSFETs, thereby minimizing the half-bridge loop inductance. The stored energy in the loop is responsible for overvoltage stress, ringing, and additional losses in the power switches under fast switching conditions. Thus, the use of low *equivalent series inductance* (ESL) *surface-mount technology* (SMT) components such as *multilayer ceramic capacitors* (MLCC) is probably the best currently available choice. Other advanced solutions consider the possibility of integrating the input capacitor together with the power MOSFETs, thereby reducing considerably the half-bridge inductance to just a few hundreds of picohenries, as reported in [154].

In desktop VRs, this input filter inductor may consist of a low-cost iron-powder core wound with a single strand of magnetic wire on a vertically mounted toroid. This type of inductor is cost-effective and occupies limited board space. However, restrictions in component height to allow the location of the inductors close to the processor and under the overhanging CPU heatsink may make vertically mounted toroids unfeasible. Given that the current drawn from the input voltage must have little AC-ripple, a possible low-cost alternative may be a ferrite rod-core inductor comprising a cylindrical ferrite rod with a coil over it. The use of ferrite materials allow for lower core losses but require the use of air-gaps to avoid saturation. In the rod-core approach, the air-gap corresponds to the volume around the inductor. Therefore, unlike the toroid inductor, the magnetic energy is not confined within the volume of the component. While permissible as input filter, the rod-core inductor may be prohibitive for the output since it must tolerate higher ripple currents, and thus any uncontained significant AC field would produce EMI and Eddy current losses in nearby conductors.

1.2.6.2 Output Filter

The advances in output filter components have experienced significant progress driven by the demands of fast dynamic output response, reduced component count, increased power density, and high-switching frequency.

The output inductor has evolved in both structure and magnetic material characteristics. Traditional vertically mounted toroids have progressively given way to SMT power cube packages offering reduced volume, improved heat dissipation, and easy-to-assembly advantages. At the same time, fundamental research on magnetism has resulted in a variety of commercial and under development core materials suitable for microprocessor VRs.

Essential for high-switching frequency operation is to mitigate the three known contributions to core losses [155]: The classical Eddy current loss, the hysteretic loss, and the excess (Eddy current) loss. Among the techniques to tackle this difficult task include the formation of thin film-based core structures from small granular magnetic materials featuring low coercivity and reduced hysteresis area. Materials' anisotropy is also frequently exploited in combination with a proper flux orientation arrangement to effectively avoid domain growth, a phenomena which is responsible for excess Eddy current loss [156].

Furthermore, in order to cope with high temperature and high-current density operation, suitable magnetic materials must feature high magnetic flux saturation (B_{Sat}), and a high Curie temperature (T_c).

Table 1.5 Classification of soft magnetic materials and typical characteristic values

Material	Permeability μ_r	Resistivity r (Ω·m)	Curie temp. T_C (°C)	Coercivity H_c (A/m)	Saturation B_{Sat} (T)
Steel/Si Steel	2,000	$0.8{\cdot}10^{-6}$	700	40	1.9
Powdered iron	300	$\sim10^{-6}$	700	280	1.2
Powdered alloy	10^4	$0.15{\cdot}10^{-6}$	500	80	1
Ferrite MnZn	1,000	10	300	60	0.34
Ferrite NiZn	300	10^5	260	140	0.3
Amorphous alloy	$6{\cdot}10^5$	$1.3{\cdot}10^{-6}$	400	8	1.6
Nanocrytallines	10^5	$1.2{\cdot}10^{-6}$	600	0.5	1.2

Sources: From [157–163]

Table 1.5 presents a classification of relevant soft magnetic materials and compares their main characteristics. Unlike the hard ones, soft magnetic materials are easily magnetized and demagnetized. Depending on the type of material used, these can be split into steels (or Si Steels), powdered iron, powdered alloys (such as permalloys), ferrites, amorphous Fe-based, Co-based, and nanocrystallines [157–163]. As one can clearly see, the primary advantage of ferrites over the rest of soft metal based magnetic materials is ferrite's high electrical resistivity, which keeps Eddy current losses low at high frequency. This property makes ferrites the most commonly used material for high-switching frequency converters. On the contrary, ferrites have low relative magnetic flux saturation B_{Sat}, and thus are unsuitable for high power density operation without the use of airgaps.

The field of soft magnetic materials has undergone a rapid progress in the last 30 years. The search has been focused on new materials with low coercive force and high value of magnetic permeability and flux saturation, which has led to the discovery of new amorphous alloys and nanocrystalline structures. Because of their superior magnetic characteristics, these emerging materials have found rapid use in various power electronics applications. Nonetheless, the high electrical conductivity of these and other metal based composites constitutes a fundamental barrier for high frequency operation. This problem has been circumvented by core formation techniques based on stacking multiple electrically isolated thin layers of magnetic material. Because of the resulting drastically increased effective resistivity, these thin film based cores can operate at higher frequencies with reduced Eddy current losses compared to bulk cores of same dimensions and material characteristics. This approach heavily relies on existing methods for film deposition such as electroplating, sputtering, and others, which can generate multilayer stacks of just a few tenths of micrometers each. Although still under development, thin film techniques have been successfully applied for on-chip integration of low power coils [164–167].

With regard to loss performance, Fig. 1.19 shows how emerging electroplated metal and granular film alloys are already competitive with ferrite materials for high frequency [168–171]. The results invite to optimism regarding the potential advantages of these emerging materials for microprocessor VRs in the near future.

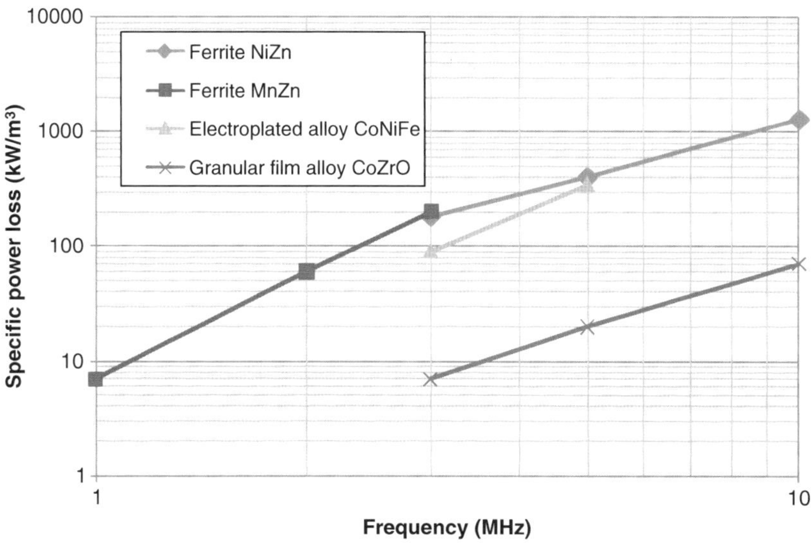

Fig. 1.19 Core loss density comparison of various soft magnetic materials at a peak flux density of 15 mT. Sources [168–171]

Nonetheless, Eddy current loss in both magnetic core and copper winding continues to be a main barrier for high-switching frequency operation, particularly in operating modes implying high current ripples, such as QSW. The selection of the right control scheme strongly influences the output filter size and ultimately the power density of the converter.

Historically, the first output filter inductors employed in high-end computer microprocessors were iron powdered vertically mounted toroids that featured low material cost, high inductance, excellent heat dissipation, superior shielding properties and limited board space.

As transient requirements increased and multiphase architectures were introduced, iron powdered cores became impractical due to excessive Eddy current and hysteretic losses. Magnetic materials suitable for higher frequencies soon replaced the powdered iron material. One broadly adopted cost-effective material solution has been the ferrites due to their high resistivity. The inconvenient low B_{Sat} of ferrites though has to be compensated with the use of airgaps, thus limiting the maximum inductance value. Nevertheless, typical targeted inductances in multiphase converters are sufficiently low, e.g., below 200 nH, which makes low effective permeability airgaped ferrites suitable for high current and high frequency.

The inductor size is determined by a large number of factors, some of which include the current level, the magnetic flux saturation, and the inductance value. Taking them into account, designers came up with an optimized core package design known as the *power cube*. While offering similar benefits compared to the toroid core in terms of shielding capabilities and heat dissipation, the power cube also provides compactness, reduced volume (both height and footprint), and easy-to-assemble (SMT based) capabilities. The manufacturing techniques applied for the power cube further guarantees a tight control of the inductance and ESR spread, which enable the use of the inductor as an attractive lossless means to monitor the output current flow. The latter is necessary in multiphase architectures for current sharing control, as pointed out in Sect. 1.2.3.

The power cube also has advantages over the toroid shape with regard to the winding. Namely, the copper losses can be reduced by more than 25% due to the fact that the coil mean-length-turn is considerably smaller with the coil inside the core. This is important at high-switching frequencies to minimize additional power losses associated to skin and proximity effects.

The power cube is in line with integration trends, where at very high frequencies the inductor can be simply reduced to a short copper stripe surrounded by a core material, thus forming a miniaturized power inductor of just a few tenths of nanohenries [172].

These numerous advantages over the traditional toroid core are causing the power cube to progressively become the most utilized core package in microprocessor VRs. Inductor manufacturers currently offer these molded low profile SMT inductors in a variety of sizes and inductance values, most of which are manufactured with powdered iron materials for frequencies below 1 MHz. Smaller ferrite based high frequency power cube inductors are also provided by a number of manufacturers for currents up to 100 A.

Overall, the power cube inductor outperforms the traditional toroid inductor for high power density VRs.

Besides common output inductors, a special class of coupled inductors for multiphase architectures [173] was initially presented by year 2000, and rapidly introduced in the market [174]. Several studies have shown that, by means of magnetically coupling the inductor of each of the phases, output ripple cancellation, and transient response can further be enhanced. Presently, significant work is being carried out to analyze and quantify the performance benefits of this approach. Recent results demonstrate that certain coupling arrangements can improve the output dynamic response, yet in detriment of the steady-state output ripple [175]. Another disadvantage of couple inductors may be the inconvenience to implement phase shedding control to maximize converter efficiency at light load.

Despite of the bulky appearance of the output power inductor, the output filter capacitance of computer microprocessor VRs usually occupies more PCB area than the inductance does. This is because the capacitance realization consists of a large bank of parallel connected capacitors. The primary reason of paralleling small capacitors as opposed to using just one of higher capacitance is the need to reduce the equivalent intrinsic ESR and ESL seen by the load. These parasitic elements are responsible for steady-state output voltage ripple and voltage spikes during load transients, both of which need to be kept below certain limits. Since capacitor paralleling add costs and area occupancy, it is essential to employ low profile capacitors with reduced ESR and ESL as well as minimize the output current ripple. For a given switching frequency, the latter can be achieved by a proper selection of output inductors and number of interleaved phases.

The right balance between these various aspects needs to be attained by means of an optimization process based on relevant adopted criteria. For years, the most important design criterion has been around improving transient response, for what increasing the switching frequency has become the most essential choice to reduce the inductor size without increasing the ripple current per phase. Combined with multiphase architectures, the switching frequency boost from 300 to 600 kHz and above has in the last years allowed the elimination of major bulk capacitors, thereby reducing volume, PCB space, and costs. Moreover, the output ripple voltage has been reduced because the bulky capacitors could be replaced by others with low profile and reduced ESR and ESL. In fact, the increase of F_s may become ineffective in terms of output voltage ripple reduction if the parasitic elements are not mitigated along with the capacitance. This may occur, for instance, if the corresponding capacitance reduction does not lead to any further improvements in package size or integration capabilities.

Advances in capacitors technology have resulted in higher performance miniaturized SMT polymer aluminum electrolytic capacitors (OS-CON). This type of components has been quickly adopted by motherboard designers, who tend to replace the traditional bulky leaded electrolytic capacitors. The short height of these new capacitors makes them as well appropriate for laptop computers, although in these applications other SMT chip polymer aluminum electrolytic or tantalum capacitors may be the preferred choice for their extreme flatness and smaller footprint [176].

Regardless of their reduced capacity, MLCC devices are key elements of the output filter because of their reduced ESR and ESL. Presently, MLCC of up to 100 μF capacity are commercially available in a 3.2×2.5 mm case with typical ESL and ESR of less than 1 nH and 2 mΩ, respectively. One trend is to make exclusive use of MLCC to build the output filter capacitance, possibly requiring a reduced portion of PCB area such as that of the central cavity of the CPU socket LGA 775 from Intel® [177]. The maximum switching frequency operation may primarily dictate the size of the output capacitor bank, and ultimately, the power density of the VR.

1.2.7 Control Systems

Computer microprocessor VRs must unavoidably incorporate a number of advanced monitoring and control schemes that are essential to fulfill the troublesome tasks of maximizing performance, avoiding instabilities, and guaranteeing proper operation despite of disturbances, uncertainties, and variation tolerances in the source, load, and circuit components [178]. Table 1.6 lists relevant control and monitoring functions typically encountered in modern VRD and VRM. From a system level perspective, several of these schemes coexist with the primary purpose of protecting the system from excess of temperature and current operation, as well as providing proper start-up functionality and power delivery conditions.

Table 1.6 Monitoring and control aspects in microprocessor VRs

Control level	Functionality
VRD/VRM system	Over temperature protection Over current protection Start-up sequence Load line (AVP) and transient response
Multiphase PoL	Current sharing Transient response enhancement Phase shedding
Single phase	Multimode switching modulation
Gate driver	Dead time optimization Driving voltage optimization Synchronous rectification ON/OFF

Classified as control subsystems, on the other hand, multiphase inner loops are employed to maintain balanced current sharing, enhance transient response, and improve converter efficiency. In addition, within each individual phase, local control loops and monitoring schemes may be implemented to enable multimode operation, accurate switching synchronization, and optimum gate driving. The following subsections highlight the principles and techniques of these fundamental regulation systems.

1.2.7.1 Load Line Regulation

As mentioned in section 1.1, the load line regulation is based on the AVP scheme. Its implementation is usually integrated in one of the control architectures of Table 1.7.

Traditionally, the voltage mode control [179] (also categorized as active droop control) and current mode control [180, 181] have been the most widespread architectures for VR applications. In the voltage mode control, the difference between the set-point and the feedback signal is compared to an artificially generated sawtooth ramp to control the duty cycle. In one possible arrangement, the set-point is composed by the output voltage plus a droop voltage of the load line, which is emulated by a sensed output current signal and a resistor. Such AVP implementation is known as AVP- because the droop voltage is fed to the negative input of the error amplifier. Alternatively, the droop voltage may be combined with a reference potential to form a current dependent set-point signal, which is then compared in the error amplifier with the output voltage of the converter. This configuration is named AVP+ because the droop voltage is fed to the positive input of the error amplifier.

Current mode control is a multiloop approach where the inductor current is the main control variable as opposed to the duty cycle in the voltage mode control. Thus, the duty cycle conceptually becomes an auxiliary variable. Current mode control results in an inner loop that regulates either the peak current, the valley current, or the average current. In either case, an outer control loop of the output voltage specifies a target value for the current control. Such information is

Table 1.7 Classification of relevant control architectures for the VR load line

Voltage mode (active droop control)	AVP+
	AVP−
Current mode	Current peak
	Current valley
	Current average
Current mode with feedforward loop	Input current feedforward
	Output current feedforward
V^2 scheme	
Hysteretic	Voltage mode
	Current mode

combined with the droop voltage and input into the nested loop to generate the right set-point for the load current.

The fact that the current mode control approximately transforms the output inductor into a current source offers significant advantages over the voltage mode control, some of which include easy compensation, proper current sharing regulation, line rejection, and optimal control in both CCM and DCM. On the contrary, there are certain disadvantages such as subharmonic instabilities and low audio susceptibility that make the voltage mode control a better choice in some cases [182, 183].

Other more advanced control methods have been developed with the purpose of providing a faster transient response capability than the voltage and current mode controls. Feedforward control loops offer dynamic advantages by virtue of measuring the load current changes and feeding them to the process for corrective action before they disturb the control parameters. Feedforward control is used in conjunction with feedback control to provide multiple-input-single-output control. Several implementations combining current mode control with load current feedforward have been analyzed in [184]. By summing the error signal of the output voltage control (outer loop) and the load current signal, and feeding the sum to the current mode controller, the inductor current can automatically, and with virtually no delay, follow the load current variations even without reaction of the voltage regulation loop. The concept can be best exploited when the current command signal and the inductor current are closely related. The V^2 control emerges as an attempt to achieve this in a simple and yet effective implementation approach [185–189]. In the V^2 architecture, the feedforward signal corresponds to the converter output voltage, which inherently contains the information of the inductor current generated by the ESR of the output capacitors. The feedforward signal is then used as reference signal to the modulator as opposed to an artificially generated sawtooth like in previous control schemes. A change in load current will have an effect on the output voltage, thus altering the ramp signal. This causes a rapid change in the modulator output to correct the duty cycle of the switches without changing the error signal of the feedback loop. Such low frequency loop provides DC accuracy and improves noise immunity. The simplicity and effectiveness of the V^2 control has been successfully implemented by several semiconductor manufactures offering to the industry a variety of IC controllers for high-performance VR applications [190].

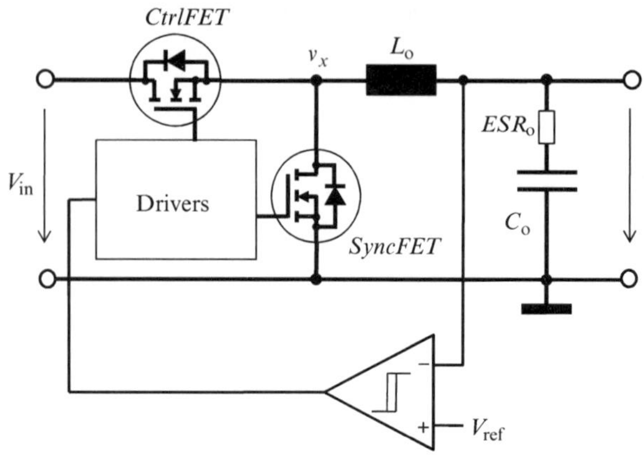

Fig. 1.20 Basic representation of the hysteretic controller in a SRBC

Boundary control techniques with linear switching surface such as hysteresis and sliding-mode control arise as alternatives to PWM controllers. The nonlinear hysteretic control is the most popular in microprocessor VRs for its simple implementation and extreme fast transient response [191]. As shown in Fig. 1.20, a basic hysteretic controller uses only a comparator and a Schmitt trigger to modulate the state of the switches as function of the difference between the reference and feedback signals. Thus, as in the V^2 control, no reference sawtooth generator or timing function whatsoever is fundamentally required.

In voltage mode control, the feedback signal corresponds to the output voltage, whereas in current mode control, the inductor current is used instead. In either case, accurate sensing of the signal's AC component is crucial for the hysteretic control to function correctly. With proper signal sensing, no feedback loop compensation is necessary to guarantee stability. The absence of a low frequency bandwidth compensator makes the hysteretic control one of the fastest existing close loop architectures, with potential for nearly instantaneous response to load transients. Nonetheless, propagation delays exist, which affect the output ripple voltage and the switching frequency operation. In addition, these parameters are inherently dependent on the parasitic ESR and ESL of the output filter. The switching frequency, although being very predictable from simple equations, may further vary substantially with the input and output voltages, load conditions, and components spread. This may lead to undesirable chattering phenomena, EMI issues, and difficulties in the optimization of the switches and filter components. These drawbacks have been addressed in the last years, resulting in several solution approaches to essentially achieve nearly constant switching frequency operation [192]. Among the proposed solutions, the variable width hysteresis comparator approach [193], and the constant ON time or OFF time [194] and adaptive ON time modulation [195] methods prove to be very effective. These modulation techniques, summarized in Table 1.8, correspond to PFM and can be applied to other control strategies, particularly for operation in DCM, as it shall be described in section 1.2.7.3.

Table 1.8 Classification of relevant modulation techniques

Constant frequency	Pulse width modulation (PWM)	
Variable frequency	Pulse frequency modulation (PFM)	Constant ON time
		Constant OFF time
		Adaptive constant ON time

For the AVP implementation both current and voltage need be simultaneously measured. In the hysteretic voltage mode control, for instance, the droop compensation is realized by subtracting the sensed current signal from the reference signal at the input of the comparator [196].

1.2.7.2 Multiphase Regulation

One of the most critical aspects of multiphase architectures is the control mechanism by which all individual phases equally share the output power. A stable and even load distribution improves performance and reliability without the need of overdimensioned components due to reduced peak thermal stresses.

Current sharing techniques are brought up to achieve this goal. Load sharing information is most commonly generated and distributed over a single interconnection among the cells. Typically, the interconnection circuit is designed so that a current sharing error signal is produced from the output current of one cell and the average current of all cells. The error signal is then utilized to locally readjust the compensation signal of the feedback loop, as shown in Fig. 1.21 [197].

Other approaches exist where current sharing information is communicated implicitly through the converter output, and no additional interconnections among converter cells are required [198].

Current sharing in hysteretic control architectures may also be accomplished with an alternative approach of phase selection. The algorithm by which it is ruled controls that the active phase delivering the power at any given time corresponds to that one carrying the lowest instantaneous current [199].

A means for accurately sensing the current of the individual phases is primordial in the majority of current sharing methods. Table 1.9 summarizes the most commonly used current sensing methods. For the sake of maximizing efficiency, the lossless approaches predominate, most of which may exploit the R_{DSon} of the power switches or the ESR of the passive filter components. In any case, and since the voltage across the ESR cannot be measured directly, the use of compensation filters is required to subtract the useful information from the measured quantities.

Added control functionality to multiphase architectures has been provided to leverage the transient response of conventional trailing and leading edge modulations. In these basic topologies, the phases are synchronized with a fixed clock and shifted phase such that the system may react to load steps with lags of T_s/n, n being the number of phases, and with a corrective action just consisting of a duty cycle readjustment.

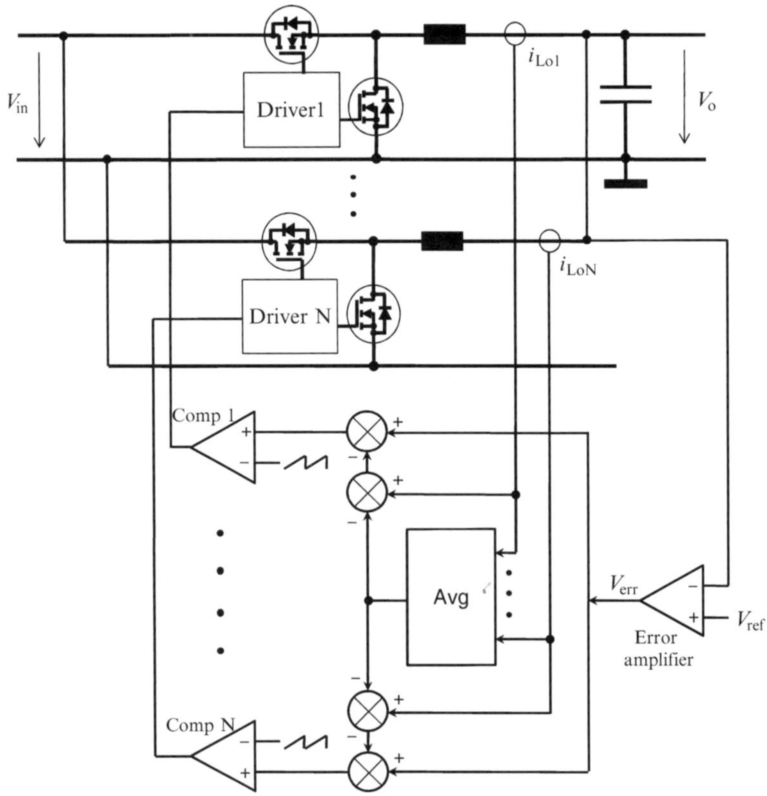

Fig. 1.21 Voltage modulation with current balance function

Table 1.9 Current sensing approaches

External resistor	
Output capacitor ESR	
MOSFET R_{DSon}	
Output inductor ESR	With passive filter
	With active filter
MOSFET R_{DSon} & output inductor ESR	

Sources: From [185, 198, 202, 203]

The proposed dual-edge PWM control [200] combines both leading and trailing edge methods without a fixed clock. This enables the controller to react by turning on more than one phase at once whenever a load step occurs. The transient response is thus improved compared with conventional modulators because the phase activation frequency is not restricted to multiples of F_s. Furthermore, in case of a transient step-up, paralleling multiple active phases allows faster current buildup in the converter.

In order to maximize efficiency in the entire load range, multiphase architectures may incorporate phase shedding control. At light load, conduction loss is low compared with switching loss. Thus, in this case, the overall losses can be effectively reduced by disabling phases. Phase shedding may however jeopardize the benefits of interleaving with regard to output capacitance size, since the reduction of phases may cause an increase of the output ripple voltage.

The optimum selection of the phase number as a function of the load current has been recently analyzed in [201]. The results suggest that the current range between phase shedding thresholds may be approximately constant, which simplifies the implementation of the control.

1.2.7.3 Multimode Switching Modulation

Multimode switching modulation represents an additional approach to phase shedding for maximizing efficiency over a wide load range, particularly, at light loads. The concept arises based on the fact that the switching modulation technique leading to minimum power losses depends on the load current.

Switching modulation techniques have been classified earlier in Table 1.8 as constant and variable frequency. Constant frequency PWM is effective at high loads, where the converter runs in CCM. At light load, PWM suffers from excessive switching losses. PFM thus becomes more effective since the switching frequency can be drastically reduced without increasing the output ripple, that is, at low currents the output capacitors can maintain the voltage longer during the OFF cycle.

Multimode switching modulation has been introduced in commercial products to offer significant efficiency gains at light load. Controllers featuring *hyper light load*^{TM} or *deep sleep* functionality are based on this concept and can increase the system efficiency by more than 50% at very low currents [60]. Other commercial parts use burst-mode control, which produces trains of fixed-frequency pulses followed by OFF periods [204]. At some lighter load the controller may reach the minimum duty cycle capability. What happens is that the energy being pushed into the converter is in excess of the output power. Therefore, the controller may typically respond by trying to omit cycles. This is commonly known as the variable frequency pulse skipping mode, which is available in various commercial controller ICs [205]. All these control techniques may be implemented both in analog- and digital-based circuits [206, 207].

The ability to switch between modulation modes as a function of the load current has commonly been a feature of VRs for portable devices. For higher power applications this technique is less frequently used, yet it is becoming increasingly relevant due to the paramount importance that light load efficiency is acquiring.

1.2.7.4 Gate Driving Control

The simplest SRBC gate driver controllers have fixed delay times of sufficient duration to insure that the ON state of the power switches do not overlap under any conditions, thereby preventing shoot-through (or cross-conduction) related losses. Such dead times must account for components spreads, temperature dependencies, and operating conditions, which usually result in excessive times incurring SyncFET body diode conduction. Body diode conduction is undesirable for it produces high ON conduction losses as well as additional ringing and switching losses associated to reverse recovery. Consequently, control techniques are required to adapt the switching time with enough delay so as to concurrently shun both shoot-through and body diode conduction.

One proposed technique is the adaptive dead time control, which allows on the fly delay readjustments by means of monitoring the gate voltages [208]. When the SyncFET gate voltage goes below a certain threshold, it is assumed that the device is off and thus the CtrlFET can be turned on. Similarly, the switched node falling edge dead time undergoes a converse sequence of events. Additionally, the SyncFET turn-on may also rely on the switch node voltage and a zero crossing detector to determine the condition for device activation. Although this control strategy can substantially reduce excessive dead times, body diode conduction may not be fully prevented due to the propagation delays of the controller and drivers.

Solutions have been introduced to overcome these propagation delays by predictive gate driving. The technique works on the premise that the delay time required for the next switching cycle can be self-adjusted by corrective action of the chosen dead time from previous cycles. The concept employs a feedback control that adaptively detects and minimizes the body diode conduction cycle by cycle. Regardless of the inherent propagation delays, the control ends up producing the precise timing signals necessary to effectively operate near the threshold of cross-conduction, thus virtually preventing body diode conduction and reverse recovery losses. Predictive gate driving was initially proposed and implemented by Texas Instruments [210–212].

An alternative approach to optimize the dead time is based on the idea of minimizing some measure of the power loss. A number of works studying this concept have been presented. In [213], the dead times are stepped to measure the resulting change in the converter input current, which is related to the efficiency. The dead time is then adjusted in the direction of increasing efficiency. A similar method proposed in [214] adjusts the dead times so that the duty cycle is minimized, corresponding to maximized efficiency.

At light loads, the SyncFET may need to be deactivated in order to emulate diode behavior and enable DCM. Many controller ICs implement this function by detecting a negative inductor current [215]. Other solutions simply turn off the SyncFET whenever a command of light load is received by the VR from the host microprocessor [216]. To mitigate the need of high bandwidth sensing due to propagation delays, adaptive and predictive techniques may also be suitable to switch off the SyncFET at the right zero current crossing times.

The gate drive voltage of the power switches are other variables of adjustment in addition to the dead times. Namely, the optimum gate voltages depend on the load current. At high current, the gate voltage may require to be rather high so as to minimize the R_{DSon} of the switches. On the contrary, low gate voltages at light load may be more beneficial since gate charging losses dominate over conduction losses. Accordingly, an adaptive adjustment of the gate voltages may lead to further efficiency improvements in the entire load range.

1.2.8 Packaging and Integration

The market demand for increased power density conversion cannot be met with conventional device packaging and interconnect technologies. Limitations in interconnect density, thermal management, bandwidth, and signal integrity need to be addressed with advanced thermally efficient encapsulation and integration techniques.

Inferior thermal dissipation imposes the most serious bottleneck for the realization of superior performance power management solutions. Not only do the thermal path characteristics and heat flux density dictate the maximum temperature rise, but the limits in reliability, quality, and yield as well. Therefore, it is of primary importance to create a thermo-mechanical environment that does not hinder the advances of new device technology generations.

The industry faces a difficult trade-off with popular semiconductor packaging options. Traditional baseplate D-Pak and D^2-Pak solutions offer good thermal dissipation capabilities but rather large footprints. In contrast, popular SOIC outlines feature space efficient choices but poor thermal performances.

Innovation in power MOSFET packaging has followed a trend in SMT technologies to enable the reduction in footprint while maintaining levels of thermal resistance and reliability performance that compare to traditional larger packages. Solutions like the DirectFET [144] offer double-sided cooling to achieve excellent heat removal capabilities around 1 K/W with the size similar to SOIC packages. The elimination of wire bonding, molding, and lead frames further allows significant reduction of parasitic elements. Other cost-effective solutions showing competitive performance are the LFPak [143] and the PolarPak [145]. These devices also feature double-sided cooling, thus providing low thermal resistance in extremely compact housings.

Further miniaturization has been achieved through the ability of semiconductor companies to deliver advanced processing and functional integration in the form of power management *system in package* SiP and *system on chip* SoC. The rapid proliferation of functionally integrated hardware solutions is seen as an inflection point in the power supply industry, which is causing a dramatic move away from traditional discrete solutions to an increasing emphasis on power supply products deriving directly from semiconductor products and microelectronics technologies.

The success of integrated power systems lies on the significant performance advantages of placing components physically close in a single chip or module. Compactness allows not just miniaturization, but also a great deal of parasitic elements reduction, which is fundamental to enable fast switching and hence high frequency operation. Furthermore, the better parameter controllability of integrated systems reduces tolerances and facilitates the introduction of key power saving functionalities, such as predictive dead time control.

The decision whether to adopt SoC or SiP to best exploit the benefits of system integration depends on a number of considerations. On the one hand, SoC implies the multi-functional integration of essential and ancillary system components into a single semiconductor integrated circuit. This inherently offers superior miniaturization advantages, high interconnection density, and a virtual suppression of relevant parasitic inductances. The performance capabilities are, however, generally bounded to the selected chip technology. This can be a major performance limitation drawback since the optimum technology for achieving the best possible result is not unique but varies depending on the function to be implemented.

Serving the same purpose, SiP [217] provides integration by way of assembling multiple chips in a single unit, package, or module. Thus, SiP is frequently referred to as *multichip module* (MCM) [123, 218, 219]. By proper interconnection of these chips, the module can perform the desired function associated with a system or subsystem. Bond wires are most frequently used to make the interconnections, which may add generally low but noticeable parasitic inductances and resistances. In contrast to SoC, the SiP approach fundamentally allows the use of multiple front end technologies for discrete and integrated components to be combined in one package, thus offering the best practical alternative for the implementation of each function. This advantage has conditioned the main adoption of SiP in high current, low voltage VRs.

SiP quickly emerged in the market segment of computer power management around year 2000, and has progressively evolved ever since. Commercial MCM products integrate the power MOSFETs, driver, controls, and, in some cases, the input capacitor filter of the SRBC in a single package. While the power devices may typically be based on trench MOSFET technology, drivers and control may be conventional CMOS based, whereas the input filter may consist of MLCCs. Practically all these *integrated power module* (IPM) products further feature automatic dead time reduction control, internal thermal shutdown, power sequencing functions, compatibility for multiphase operation and extremely low profile, surface mounted package with a thermal resistances typically below 5 K/W (e.g., 8 mm × 8 mm × 0.85 mm of PIP212 from NXP Semiconductors) [220].

The MCM approach was adopted by Intel® in 2004 as reference for the DrMOS specification introduced in Sect. 1.2.1.3. The importance of such specification has made most competitor products to be DrMOS compatible.

Future MCM developments may aim at, 1: Continuing the reduction of parasitic elements, in particular, the half-bridge loop inductance [221–224]; 2: Adding further power-saving functionalities, like the control modes for light load described in Sect. 1.2.7.3., and 3: Explore advanced cost-effective packaging techniques, such as for instance, embedded technologies [225, 226].

Ultimate approaches aiming at the full power supply integration such as the PSiP, PwrSoC (see Sect. 1.2.3) [107, 227, 228] or the combination of CPU and power supply on the same *organic land grid array* (OLGA) [229], are still major challenges even for low power portable applications since switching frequencies cannot reach the levels where the size of the passive components makes their integration feasible [107, 230]. Major breakthroughs like the GaN power switch devices (see Sect. 1.2.4) may first need to be introduced.

1.2.9 Commercial Voltage Regulators

In the last years, most commercial motherboards with embedded VRDs have been employing multiphase architectures formed by 3–5 interleaved SRBCs, depending on the characteristics of the CPU. Figures 1.22 and 1.23 show one VRD design usually encountered in motherboards back in 2005. In such arrangement, the maximum height of the VRD is determined by the leaded aluminum electrolytic capacitors, followed by the toroid chokes. Power MOSFETs are packaged in D-Pak, which do not require heatsink, yet occupy a substantial PCB area comparable to that of electrolytic capacitors. The high inductance of this package is one of the main limiting factors preventing higher-switching frequency operation. Three discrete MOSFET parts may be used per phase, two of which are parallelly connected to constitute the SyncFET, since it carries the output current during most of the time. A rod-core inductor may be frequently employed as input filter for the reasons given in Sect. 1.2.6.1. Note that the overall VRD PCB area is significantly larger than the CPU socket area.

More recent high-performance motherboards start to utilize VRDs with larger number of phases as a response to the higher power demands of CPUs. Figure 1.24 shows an example of a 16 phases VRD design to drive overclocked Intel® Extreme Quad Core™ processors. As it can be seen from a first glance, the cooling system

Fig. 1.22 VRD module and CPU socket of Asus A8N-VM CSM motherboard for AMD's Athlon™ 64-bit processors

Fig. 1.23 Three phases low switching frequency VRD of Asus A8N-VM CSM motherboard for AMD's Athlon™ 64bits processors

Fig. 1.24 Sixteen phases VRD of Asus P5Q-E motherboard for Intel® Extreme Quad Core™ processors

around the CPU is far from basic. It features a large bridge heatsink, which is connected via a single copper heatpipe to another series of heatsinks, designed to cool the board's power MOSFETs. Together with the heatsink fan, this cooling system dictates most of the VRD volume, which is significantly higher than in previous motherboard designs. The large number of phases also leads to higher PCB area occupancy, even with the use of modern passives, such as the power cube inductors and the low profile OS-CON capacitors shown in detail in Fig. 1.25. The reduced height of these components, however, allows their placement close to the processor and under the overhanging CPU heatsink. The increase of cost, volume, and area of this extreme design offers in turn high efficiency at full loads, thus generating the least amount of heat to enhance overclocking capabilities and reliable operation.

Fig. 1.25 Sixteen phases VRD of Asus P6T Deluxe Intel® X58 motherboard for Intel® Core™ i7 processors

Fig. 1.26 Sixteen phases VRD without heatsinks of Asus P6T Deluxe Intel® X58 motherboard for Intel® Core™ i7 processors

A look at the 16 phases' solution of Fig. 1.26 without heatsink reveals the type of power MOSFETs and their arrangement on the motherboard. As depicted in Fig. 1.27, each phase consists of two LFPak surface mounting based power switches. In contrast to other VRD designs with fewer phases, paralleling is not necessary since the reduced current per phase can be reliably handled by modern ultra low ON resistance power MOSFETs. Furthermore, output ripple cancellation

Fig. 1.27 Six phases DrMOS VRD of MSI X58 Pro motherboard for Intel® Core™ i7 processors

is remarkably enhanced by the high number of interleaved phases, which reduces the OS-CON count and the high frequency decoupling MLCCs confined in the CPU socket cavity area.

Another generation of VRDs is based on DrMOS compatible MCM products. An example of that is shown in Fig. 1.28, corresponding to a state-of-the-art motherboard for Intel® Core™ i7 processors. Providing current from six phases, this VRD employs some of the most modern components available in the industry. Besides the MCMs, the output filter is formed by state-of-the-art ferrite-based power cube inductors, and low profile polymer electrolytic capacitors. The VRD also uses a rather voluminous heatsink for proper heat dissipation of the MCMs. This heatsink primarily dictates the size of the VRD. The PCB area, however, is comparable to that of VRD for lower power microprocessors. A top view of the power supply without heatsink is shown in Fig. 1.29. Comparing it to that of Fig. 1.22 and Fig. 1.23 gives a good sense of how much power density in VRs has improved in the last 5 years.

The large dimensions of most heatsinks in computer workstations are prohibitive in laptops and some server computers. Such volume restrictions imply achieving higher efficiencies at full load. This is currently possible in laptops since the power consumption of the microprocessor is drastically lower, thus simplifying the VRD structure to typically two phases, as shown in the example of Fig. 1.28. The use of low profile OS-CON is possible but unusual in ultra slim laptops, where the utilization of polymer electrolytic or tantalum capacitors offers superior form factors.

Figure 1.30 shows a commercial VRM solution in a standard board size. The total PCB area occupied by VRMs is usually smaller than VRDs because the output capacitors are not included in the module. Nevertheless, VRMs frequently employ rather small heatsinks for its vertical arrangement allows for better heat dissipation. Thus, overall, the volume of the VRM may be usually smaller than that of VRDs.

Fig. 1.28 Six phases DrMOS VRD without heatsink of MSI X58 Pro motherboard for Intel® Core™ i7 processors

Fig. 1.29 Example of two phases VRD (left hand side of the image) for laptop computers

1.2.10 Survey on Power MOSFET Models for Circuit Simulations

As pointed out throughout this review chapter, one key challenge of the power delivery system is to reduce the switching losses of the power MOSFETs. To address this most effectively, the analysis of the switch device in the converter circuit is fundamental so as to understand switching phenomena, identify relevant loss mechanisms, and find measures to mitigate them. A significant amount of

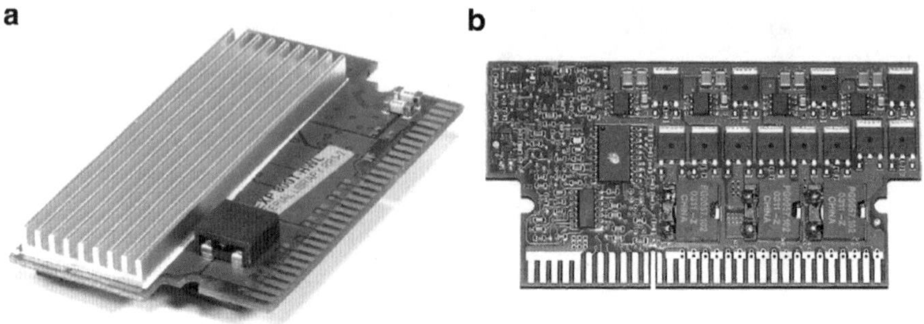

Fig. 1.30 Four phases VRM for PowerEdge 4600 processor, Dell 9T434. (a) Heatsink side, (b) Opposite side

Table 1.10 Power MOSFET model approaches for circuit simulations

	Semiconductor device physics		Empirical	
Description method	Analytical lumped	Numerical distributed	Analytical lumped	Numerical lumped
Basic representation	Approximated solution functions	Partial differential equations (PDEs)	Fitting functions	Look-up tables
Implementation	Structural macro-model circuit	PDEs/spatial discretization FD, FE	Structural macro-model circuit	
Solver type	Impedance network	Mixed-mode	Impedance network	
Device variables involved	Material related or circuit related	Material related	Fitting parameters	

modeling approaches has been developed to assist these tasks. However, there appears to be only a scarce number of effective methods for high-switching frequency when investigations demand accurate loss quantifications and the assessment of improvement options at both device and circuit levels.

To provide a sense of it, this section overviews existing power MOSFET models for circuit simulation aiming at the device performance evaluation.

In the literature of power MOSFET models for circuit simulations a number of distinguished approaches can be classified and summarized as shown in Table 1.10. Essentially, two main categories exist that conceptually differentiate between theoretical semiconductor device physics and empirical models. At this point, the suitability of the latter ones may apparently be restricted to merely circuit analysis, for the approach may not provide insights of the internal device behavior (i.e., the black box concept). One should, however, become aware that the representation, implementation, and performance of the model will eventually dictate the applicability and potential use of the model upon consideration.

Both models' main categories commonly present methods based on analytical and numerical solutions.

Models based on device physics closed-form expressions usually derive from simplifications of the spatial distribution of the device structure into a network of lumped elements, each of which represents an intrinsic functional (desirable or parasitic) discrete electrical entity [231, 232]. The resulting equivalent sub-circuit network constitutes a structural macro-model, the parameters of which can be device material or circuit related.

The main differences between existing macro-models lay on the network topology and implementation approach [233–236], which may significantly influence the simulation performance (e.g., convergency errors). Nevertheless, fast computation times can be regarded as a common feature in most implementation approaches of this kind. On the other hand, a major drawback of physical based macro-models may be the limited accuracy caused by the underlying approximations.

As an alternative approach, the partial differential equations (PDEs) governing the device physics of a multidimensional spatial distribution may be solved by numerical methods such as finite differences (FD) and finite elements (FE). These methods can further be coupled to a lumped impedance network solver and thus simultaneously resolve the PDEs and associated circuit network in what are known as mixed-mode simulations.

Mixed-mode simulations allow the study and optimization of carrier dynamics and device geometry in the actual switched-circuit. PDEs solvers deliver outstanding accuracy when the physical and geometric parameters are well-known. On the contrary, mixed-mode simulations usually require long computation times and their use is presently restricted to rather simplified circuit descriptions [237–241].

By means of device characterization techniques, empirical models aim at reproducing the device behavior at its terminals without required knowledge of the underlying physics. Nonetheless, the majority of advanced empirical models are, in part, intrinsically physics based in the sense that their internal structure conceptually resembles that of device physics macro-models. Thus, the word "empirical" usually refers to the way by which the individual lumped elements that constitute the macro-structure are represented. This approach significantly simplifies the device representation, and most importantly, it extends the scope of applicability beyond that of pure device black box models, as it allows exploring the interaction between the different internal characteristics as well as their contribution to the overall device behavior.

Applying the right testing conditions, individual device characteristics can effectively be decoupled by measurements [242], or alternatively by accurate device physics simulations [243].

The representation of these characteristics by lumped elements is usually based on fitting functions [244] or look-up tables linked to voltage or current dependent sources in the macro-model sub-circuit [236, 245]. Analytical expressions derived from simpler MOSFET structures (e.g., standard low-power devices) are often potential fitting functions. Their use is though purely empirical since the dependent variables may lose their physical meaning [246].

Empirical models generally provide high accuracy with reasonable short computation times. This is though restricted to the operating regions covered by the characterization data. Beyond these regions extrapolation techniques may yield inaccurate simulations. Also, unlike device physics models, empirical models do not generally offer direct correlations between device performance in the switched-circuit and the physical parameters of the device.

1.3 Objectives of the Thesis

In view of the current state of the art, the obvious question arises as to whether the present basic VR topology will be suitable to fulfill the requirements of next generation microprocessors. A rigorous analysis on fundamental limitations of the SRBC needs to be carried out to estimate and compare the ultimate power density levels attained with existing and forecasted technologies. The identification of crucial circuit and device parameter dependencies on power losses are essential to establish the conditions upon which certain system specification requirements can be met. Such parameter correlations form the basis to lay out roadmap targets of future VRs, which, based on the discussions thus far, may be potentially constituted by improvement aspects related to control strategies, system integration, and power MOSFET performance. Accordingly, solutions may only be considered within the framework of SRBC based multiphase architectures. Nonetheless, clear indications as to whether alternative circuit topologies may or not be better suited shall be revealed.

Upon such premises, the following particular VR aspects shall be addressed:

- Guidelines for the design and optimization of multiphase SRBC. It includes the optimum size of power MOSFETs, gate driving voltage, optimum number of phases, and filters size.
- Provided that the focus is on maximizing efficiency at high-switching frequency, special attention is given to analyzing the performance of power MOSFET technology trends in close correlation with the operating conditions of the switched converter. Combined with a comprehensive examination of switching phenomena and related power losses, this study comprises the core of the thesis work.
- With regard to gate driving, investigations on control parameters, topologies, and operation modes leading to both fast switching and reduced losses shall be carried out. A procedure to derive the optimum driving conditions for a given topology and converter specifications will be proposed.
- Concerning the output filter size, the advantages of going to higher-switching frequency operation shall be quantified. In particular, an optimum switching frequency may be determined based on a series of design criteria for reducing both converter size and component count. The size and heat dissipation of the output filter inductor will strongly depend not only on the load current, but also on the converter operation mode as well as the switching frequency. Thus, the analysis of AC inductor losses will be essential to determine the right design

parameters that compromises between power density conversion and power losses.
- Control strategies offering loss saving functionality will be explored to maximize system efficiency in a wide load range. Techniques such as phase shedding will be brought in to enable fine dynamic adjustments of the number of phases as function of the load demands. The advantages of dead time control algorithms shall also be examined paying rigorous attention to efficiency gains at high-switching frequency.
- The move toward integrated solutions will mainly be motivated from performance benefits on power density, switching loss savings, and higher-switching speed capabilities. Arguments will be centered around demonstrating the importance of further pursuing the reduction of critical layout parasitics as the switching frequency goes above 1 MHz.

The information resulting from these investigations will merely be referent to technologists and physicists, serving as guidelines to develop optimized technologies that can be most effectively applied to fulfill the outlined roadmap targets. Improvement options may be manifold, and so may the combinations of them leading to similar performance results. While the objective of this work is to provide possible scenarios of compliant system operation, it will be the technologists' decision to select among them the most appropriate one as basis for their development strategy. The best solution to be adopted may be strongly dependent on many factors, such as technology trends, available expertise, time to market, and cost. Making the right choice at the right time will then be essential to reach excellence in program delivery.

To this end, modeling tools are developed to form the main pillars of a methodological approach for the analysis of switching behavior, conversion efficiency predictions, and optimization. As it will be described in the following section, the proposed methodology consists of a combination of circuit and device models with a series of analysis procedures, the applicability of which extends beyond the particular purposes of this thesis work. Thus, emphasis is given to provide a detailed description of the methods applied such that the reader can identify from a much broader scope the potential advantages and benefits of the approach to aid the development process of advanced power delivery solutions.

1.4 Methodological Approach

Advanced microprocessor VRs pose a number of practical challenges to their designers, who must tackle the task to analyze complex switching phenomena and estimate their associated power losses [247]. For that matter, existing design models [248, 249] may become exceedingly inaccurate because the simplifications they are based upon may not be applicable in the framework of high-current, high-switching frequency operation. Perhaps the most critical simplification is the omission of parasitic inductances, which turn out to play a critical role at high-switching frequency. Ignoring them may thus result in misleading prognostications and improper optimization designs. As such, there is the great need to develop dedicated design oriented loss models for modern VRs. These models should allow sufficiently accurate performance estimations in a wide range of the entire design space and still be practical for circuit designs. Crucial goals are to derive design guidelines and FoM for loss minimization based on functional relations between circuit design parameters and performance related factors.

These modeling requirements are targeted in this thesis work. Following a systematic methodology, the SRBC is comprehensively studied in the context of computer microprocessor VRs, going from the fundamentals of switching phenomena, to the optimization design and roadmap targets identification.

Circuit modeling is central in the proposed methodological approach, which comprises four study phases. In every one of them, different modeling levels are suitably employed according to specific analysis demands. Figure 1.31 shows the scheme of the adopted procedure with a description of the primary purposes in each course of action.

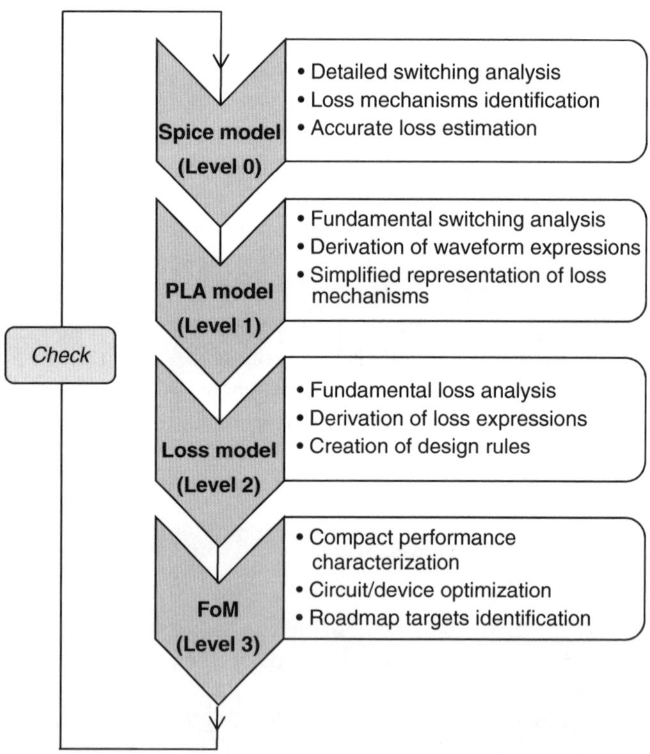

Fig. 1.31 VR power efficiency study workflow consisting of four modeling phases

In the initial study phase, switching phenomena is rigorously analyzed for the purposes of understanding fast-switching operation, identifying loss mechanisms, and quantifying their effects.

Although featuring outstanding simulation accuracy, the proposed level 0 model (or SPICE circuit model) is typically associated to both high-computational power and long simulation times. Furthermore, the simulation analysis may be rather cumbersome when requiring extensive parameter sweep calculations, thereby often making level 0 impractical as circuit design tool.

In that respect, analytical modeling [250–254] may be a more convenient approach as it aims at simplifying complexity and concisely describing the essential and most relevant parameter correlations. Additionally, simulations become simpler and faster, thus reducing the need of high computation power and long processing times. Analytical models frequently neglect second-order effects and restrict their applicability to narrower operating regimes of interest. The art for deriving practical analytical models lays on the ability to make the right simplifications, in particular, those which can provide a compact mathematical description of the significant aspects of the circuit behavior while maintaining sufficient accuracy.

Accordingly, a piecewise linear analytical (PLA) model (or level 1 model) is developed for the second study phase. As it will be shown, the SRBC operation is divided in switching intervals and represented with various linear circuit models from which waveforms are analytically defined. The ensuing equations elucidate the foundations of switching phenomena without significant loss of generality and simulation accuracy.

The focus of the third and fourth study phases is on the description and analysis of power losses. The time dependent equations from the level 1 model aid the derivation of closed-form expressions of individual switching loss quantities. As it will be described, the adopted simplifications restrict the applicability of the loss model to certain boundary conditions, which turn out to constitute a series of design rules necessary to avoid critical loss mechanisms. The resulting loss equations, though being less accurate than the numerical simulations of level 0 and level 1 models, form the basis to define FoM in the last study phase. These FoM compactly characterize the performance of the switches in the converter circuit and thus help designers optimize their parameters, compare technologies, and identify roadmap targets. The final design can eventually be checked with the more accurate level 0 or level 1 model.

The proposed multilevel modeling structure is motivated by the adopted co-design approach in which the perspectives of device behavior, circuit topology, and control aspects are brought together in the analysis process. It is therefore intended to provide sufficient levels of accuracy, comprehensiveness, and rigor in describing the performance limitations of the prescribed VR solution.

A matrix of correlation between the addressed power efficiency aspects and the model levels is represented in Fig. 1.32. As indicated, model level 0 is mainly devoted to the behavioral analysis of the power switches in the converter circuit, which allows the identification and understanding of switching phenomena. Once the relevant switching aspects have been recognized and assessed, simplifications

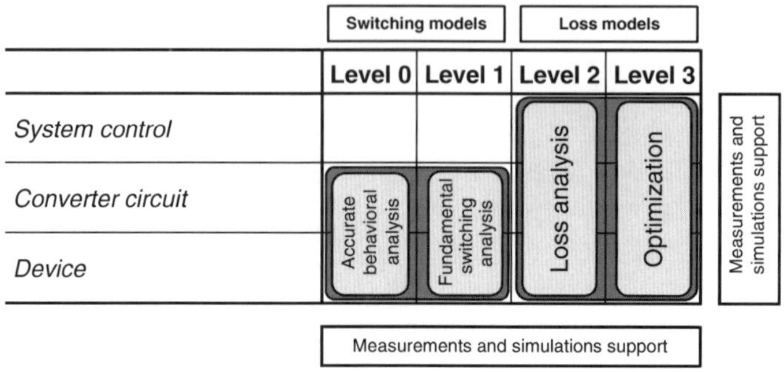

Fig. 1.32 Next generation VR development process flow

are deduced to build model level 1, which concisely describes the essentials of the converter's switching operation while neglecting secondary order effects.

The study of control aspects by means of model levels 0 and 1 is of little convenience due to the generally required high-computation power. In contrast, the underlying approximations of model level 2 allow proper assessments of system control techniques to improve power efficiency. Model level 3 may also be employed for loss savings related control analysis, although it is dedicatedly developed for optimization purposes.

All these models are supported at all levels with lab experiments and accurate FE and device simulations. The latter complements the validation of the proposed model levels in those operating regimes where measurements become inaccurate or difficult to obtain.

These modeling tools are essential in the development process of next generation VRs presented in this treatise. The process workflow of the proposed method is illustrated in Fig. 1.33. It starts with the characterization of current state-of-the-art VRs. A set of measurements from which to characterize the existing system performance is required to determine the parameters of the circuit models. The accuracy of the measurements will be influential to generate meaningful simulations that can reproduce the performance of current state-of-the-art solutions. Based on the market input (i.e., CPU and computer specifications), such circuit simulations will primarily consist of parameter variations and sensitivity analysis to translate the load requirements into clear roadmap targets for the VR. The identification of roadmap targets encompasses a precise definition of the system characteristics and parameter values that elements of the VR (such as the MOSFETs and power coil) may need to feature in order to achieve the targeting performance. Any modifications in the VR that allow meeting the roadmap targets will be defined as improvement options.

The key aspect of the proposed process flow is to iteratively identify the most appropriated improvement options by means of active simulations. This gives rise to the concept called *virtual design loop*, where numerical computations replace the usually enormous experimental/empirical effort to elucidate the pathway toward the most effective solutions. Thus, the ability of the models to accurately predict performance trends as function of control, circuit, and device parameters is of critical importance for a successful technology development.

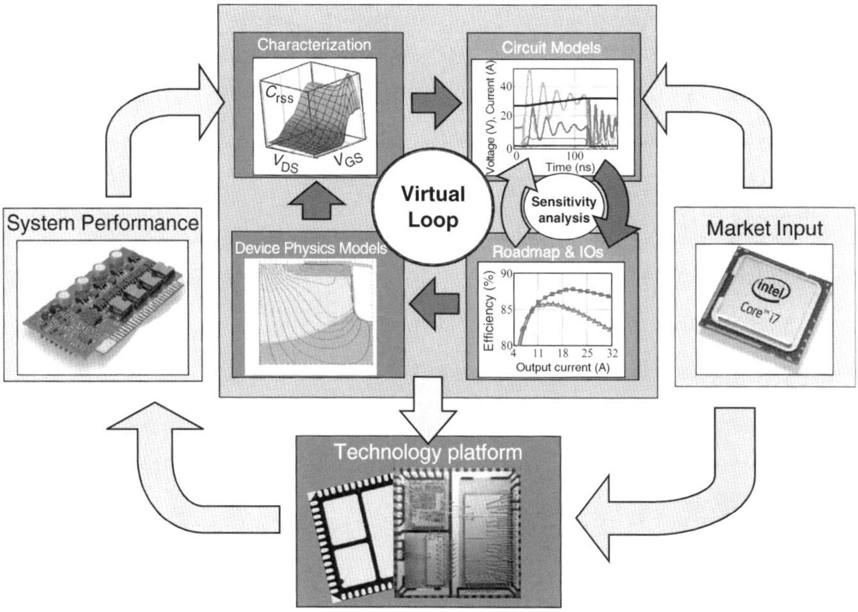

Fig. 1.33 Matrix correlation of the proposed co-design approach

The virtual loop exploits the use of device physics models to accurately predict the impact of new structures and/or materials in the application. Therefore, it is important that the characterization process be compatible with both experimental (real devices) and simulated data (virtual devices).

It is only then, after several virtual loop iterations have been run and the most suitable solution chosen, when the development of the technology platform and eventually the prototyping of the VR are worked out. The selection of the right technology platform may typically result from trade-offs between technology requirements (e.g., process capabilities and development effort) and performance benefits.

The process flow begins once again after the new system is available and the performance can be tested, hence validating the modeling predictions.

Note that the proposed development process flow can be generalized and applied to other engineering disciplines.

1.5 Thesis Outline

The next four chapters of the thesis are dedicated to analyzing the above introduced model levels. Each modeling approach is thoroughly described in separate chapters in terms of implementation, functionality, computation requirements, and performance predictions. The latter are extensively supported with experimental data, particularly for the case of the most sophisticated model level 0. The characterization work to determine model parameters from both experimental and simulation data will be common to all model levels. However, it will only be presented and illustrated in Chap. 2 for the case of model level 0.

The potential use of each model will be highlighted in each corresponding chapter by means of analyzing particular aspects of the VR. Therefore, Chap. 2 provides an in depth analysis of power MOSFETs' switching behavior based on model level 0. In Chap. 3, the use of model level 1 will serve to study the critical aspects of the SRBC topology. Different case examples will be presented to illustrate the need to move toward integrating solutions. In Chap. 4, model level 2 will predict the loss breakdown of a MCM solution to reveal the relevant loss mechanisms under fast-switching operation as well as effective control methods to mitigate them. Chap. 5 focuses the attention on design guidelines for optimizing the VR for a particular application. All elements of the VR are considered, including the passive filter elements and the gate drivers. Special emphasis is given to the design of the power MOSFETs, the selection of the switching frequency, and the optimum number of phases.

Chap. 6 exploits the developed models to examine SRBC based multiphase architectures in order to determine the limits of operation and propose measures for boosting the overall performance of the VR. Based on the guidelines of Chap. 5, the power density and efficiency limits of the converter based on present and future technologies will be identified. Finally, roadmap targets and improvement options for next generation VRs will be presented and discussed. A discussion of the findings in Chap. 7 will conclude the work.

References

1. Intel® (2004) Technology and research at Intel, architectural innovation for the future. www.intel.com/technology
2. Intel® (2006) Designing for power, Intel leadership in power efficient silicon and system design. www.intel.com/technology
3. Intel® (2006) Intel silicon innovation, fueling new solutions for the digital planet. www.intel.com/technology/silicon
4. Intel®, Moore's law 40th anniversary. Intel® technology innovation trends, http://www.intel.com/cd/corporate/techtrends/emea/eng/209729.htm
5. International Technology Roadmap for Semiconductors (2008) update, overview, http://public.itrs.net/reports.html
6. Ed Stanford (2004) Power technology roadmap for microprocessor voltage regulators, Intel® presentation at the Applied Power Electronics Conference, February
7. Chickamenahalli S (2008) Power electronics research efforts at Intel. Special presentation at the Applied Power Electronics Conference and Exposition, February
8. Kursun V, Friedman E (2008) Multi-voltage CMOS circuit design, Wiley, New York
9. Zagacki P, Ponnala V (2008) Power Improvements on 2008 desktop platforms, Intel Technology Journal 12(3), October ISSN 1535-864X
10. Advanced Configuration and Power Interface (ACPI), http://www.acpi.info/

11. Ed Stanford (2006) Power delivery challenges in computer platforms, Special presentation at the Applied Power Electronics Conference and Exposition, February 2006
12. Intel® 4004 single chip 4-bit p-channel microprocessor. http://www.applelogic.org/AL/4004Data.pdf, www.4004.com
13. Intel® (1972) 8008 8-bit parallel central processor unit. users manual, November
14. Intel® (1986) 8080A/8080A-1/8080A-2 8-bit n-channel microprocessor, datasheet, November, order number: 231453-001
15. Intel®, 8085AH/8085AH-2/8085AH-1 8-bit HMOS microprocessors, datasheet
16. Intel® (1990) 8086 16-bit HMOS microprocessor, datasheet, September, order number 231455-005
17. Intel® (1988) 80286 high performance microprocessor with memory management and protection, datasheet, September, order number 210253-013
18. Intel® (1998) Intel386™ EX embedded microprocessor, datasheet, October, order number 272420-007
19. Intel® (2004) Embedded Intel486™ SX processor, datasheet, August, order number 272769-004
20. Intel® (1997) Pentium® processor, datasheet, June, order number 241997-010
21. Intel® (1998) Pentium® Pro processor with 1MB L2 cache at 200MHz, datasheet, January, order number 243570-001
22. Intel® (1998) Pentium® II processor at 233MHz, 266MHz, 300MHz and 333MHz, datasheet, January, order number 243335-003
23. Intel® (2002) Intel® Pentium® III processor for the SC242 at 450MHz to 1.0GHz, datasheet, July, document number 244452-009
24. Intel® (2001) Intel® Pentium® 4 processor in the 423-pin package at 1.30, 1.40, 1.50, 1.60,1.70, 1.80, 1.90 and 2GHz, datasheet, August, order number 249198-005
25. Intel® (2004) Intel® Pentium® 4 processor with 512KB L2 cache on 0.13 micron process and Intel® Pentium® 4 processor extreme edition supporting hyper-threading technology, datasheet, February, document number 298643-012
26. Intel®, (2005) Intel® Pentium® 4 processor on 90nm process, datasheet, February, document number 300561-003
27. Intel®, (2005) Intel® Pentium® 4 processor 6xx sequence and Intel® Pentium® 4 processor extreme edition, datasheet on 90nm process in the 775 land LGA package and supporting Intel® extended memory 64 technology, and supporting Intel® virtualization technology, November, document number: 306382-003
28. Intel®, (2009) Intel® Core™ 2 extreme processor QX9000 series, Intel® Core™ 2 quad processor Q9000, Q9000S, Q8000 and Q8000S series, datasheet on 45nm process in the 775 land package, January, document number 318726-007
29. Intel®, (2009) Intel® Core™ i7 processor extreme edition series and Intel® Core™ i7 processor, specification update, January, document number 320836-003
30. International Technology Roadmap for Semiconductors, ITRS edition reports and ordering, years 2000 through 2007, executive summary, http://public.itrs.net/reports.html
31. Edward Stanford (2008) Environmental trends and opportunities for computer system power delivery, Proceeding of the 20th International Symposium on Power Semiconductor Devices & IC's, May 18–22, 2008 Orlando, FL
32. Climate savers computing initiative, http://www.climatesaverscomputing.org/
33. Energy Star, http://www.energystar.gov/
34. The green grid, http://www.thegreengrid.org/
35. Green computing impact organization, Inc., http://www.gcio.org/
36. Green electronics council, http://www.greenelectronicscouncil.org/
37. Advanced configuration and power interface, http://www.acpi.info/
38. Intel® (2009) Whitepaper ENERGY STAR Version 5.0 System Implementation, published by Intel® with technical collaboration from the U.S. Environmental Protection Agency, February 2009, document number: 321556-001

39. Pizano JE (1995) Low voltage microprocessors, the inevitable future, Digital Avionics Systems Conference, 1995., 14th DASC Volume, Issue, 5–9 Nov 1995 pp 169–172
40. Amelifard B, Pedram M (2007) Design of an efficient power delivery network in an SoC to enable dynamic power management. Proceedings of the 2007 international symposium on Low power electronics and design, pp 328–333, 2007
41. Schmitt MA, Lam K, Mosely LE, Choksi G, Bhattacharyya K (1988) Current Distribution on Power and Ground Planes of a Multilayer Pin Grid Package. Proceedings International Electronics Packaging Society, 1988, pp 467–475
42. Sheehan R (1999) Active voltage positioning reduces output capacitors. Linear Technology, design solutions 10, November 1999, http://www.linear.com/
43. Zhan D (2000) Using dynamic voltage positioning to reduce the number of output capacitors in microprocessor power supplies. National Semiconductor, application note 1145, July 2000
44. Lenk R (2000) Understanding droop and programmable active droop™. Fairchild Semiconductor, application bulletin AB-24 rev. 0.4, September 2000, http://www.fairchildsemi.com/an/AB/AB-24.pdf
45. Gupta MS, Oatley JL, Joseph R, Wei G-Y, Brooks DM (2007) Understanding voltage variations in chip multiprocessors using a distributed power-delivery network. Design, Automation & Test in Europe Conference & Exhibition, April 2007, pp 1–6
46. Wang T-Y, Chen CC-P (2002)Optimization of the power/ground network wire-sizing and spacing based on sequential network simplex algorithm. Symposium on Quality Electronic Design, proceedings 2002, international volume, pp 157–162
47. Ruehli AE (1972) Inductance calculations in a complex integrated circuit environment. IBM J Res Develop 16(5):470–481
48. Altera® (2009) Power delivery network (PDN) tool user guide. version 2.0, March 2009, http://www.altera.com
49. Rozman AF, Fellhoelter KJ (1995) Circuit considerations for fast, sensitive, low-voltage loads in a distributed power system. IEEE Applied Power Electronics Conference and Exposition, APEC 1:34–42
50. Zhang MT, Jovanovic MM, Lee fC (1996) Design considerations for low voltage on-board DC/DC modules for next generations of data processing circuits. IEEE Transactions on Power Electronics 11(2):328–337
51. Intel® (2002) Mobile Intel® Pentium® 4 processor-M and Intel® 845MP/845MHz chipset platform, datasheet, April 2002, order number 250688-002
52. Intel® (2003) Mobile Intel® Pentium® 4 processor-M, datasheet, June 2003, order number 250686-007
53. Intel® (2006) Intel® Pentium® M processor on 90 nm process with 2-MB L2 Cache, datasheet, January 2006, document number 302189-008
54. Intel® (2009) Intel® Core™ 2 extreme quad Core™ mobile processor and Intel® Core™ 2 quad mobile processor on 45-nm process, datasheet, January 2009, document number 320390-002
55. Gochman S, Ronen R, Anati I, Berkovits A, Kurts T, Naveh A, Saeed A, Sperber Z, Valentine RC (2003) The Intel® Pentium® M processor: microarchitecture and performance. Intel Technol J 07:02
56. Genossar D, Shanmir N (2003) Intel® Pentium® M processor power estimation, budgeting optimization and validation. Intel Technol J 07 (02), ISSN 1535-864X
57. ON Semiconductor™ (2002) Linear & switching voltage regulator handbook. Technical note, HB206/D, rev.4, February 2002
58. Perez R (2000) A comparative assessment between linear and switching power supplies in portable electronic devices. IEEE Int Symp Electromagn Compatibility 2:839–843
59. Sippola M, Sepponen R DC/DC converter technology for distributed telecom and microprocessor power systems – a literature review. http://www.hut.fi/Yksikot/Elektroniikka/reports
60. Huang B Beyond LDOs and switching regulators. Bodo's Power Systems, October 2007

61. Simpson C Linear and switching voltage regulator fundamentals. National Semiconductor, power management application note, http://www.national.com/appinfo/power/files/f4.pdf
62. Lee FC,Zhou X (1999) Power management issues for future generation microprocessors. International Symposium on Power Semiconductor Devices and ICs, ISPSD 1999, pp 27–33
63. Elbanhawy A The road to 200 Ampere at one volt VRM. Fairchild Semiconductors, www.fairchildsemi.com/an/AN/AN-7016.pdf
64. Lidow Alex, Sheridan Gene (2003) Defining the future for microprocessor power delivery. IEEE Applied Power Electronics Conference and Exposition, APEC vol 1, pp 3–9
65. Schuellein G (2003) VRM design optimization for varying system requirements. Powersystems world, November 2003, www.irf.com/technical-info/whitepaper/pswus03vrmdesign.pdf
66. Stanford E (2002) New processors will require new powering technologies. Power Electronics Technology, February 2002, www.powerelectronics.com
67. Intel® (1999) Pentium® III processor power distribution guidelines. Application note, April 1999, order number 245085-001
68. Intel® (1999) VRM 8.3 DC-DC converter design guidelines. March 1999, order number 243870-002
69. Intel® (2001) VRM 8.4 DC-DC converter design guidelines. October 2001, order number 245335-006
70. Intel® (2002) VRM 8.5 DC-DC converter design guidelines. March 2002, order number 249659-002
71. Intel® (2002) VRM 9.0 DC-DC converter design guidelines. April 2002, order number 249205-004
72. Intel® (2002) VRM 9.1 DC-DC converter design guidelines. January 2002, order number 295646-001
73. Intel® (2003) Dual Intel® Xeon® processor voltage regulator down (VRD) design guidelines. July 2003, order number 298644-003
74. Intel® (2004) Voltage regulator down (VRD) 10.0. design guide for desktop socket 478, February 2004, document number 252885-003
75. Intel® (2005) Voltage regulator module (VRM) and enterprise voltage regulator down (EVRD) 10.0. design guidelines, July 2005, document number 302731-002
76. Intel® (2005) Voltage regulator down (VRD) 10.1. design guide for desktop LGA775 socket, April 2005, document number 302356-004
77. Intel® (2005) Voltage regulator module (VRM) 10.2L. design guidelines, March 2005, document number: 306761-001
78. Intel® (2006) Voltage regulator down (VRD) 11.0. Processor power delivery design guidelines for desktop LGA775 socket, November 2006
79. Intel® (2007) Embedded voltage regulator-down (EmVRD) 11.0. design guidelines for embedded implementations supporting PGA478, January 2007, order number: 311395-005
80. Intel® (2008) Voltage regulator module (VRM) and enterprise voltage regulator down (EVRD) 11.0. design guidelines, April 2008, reference number: 315889-002
81. Intel® (2009) Voltage regulator-down (VRD) 11.1, processor power delivery design guidelines. September 2009, document number: 322171-001
82. Intel® (2004) DrMOS Specifications. Revision 1.0, November 2004, www.intel.com
83. Davis S (2004) Basics of design microprocessor power management. a supplement to electronic design, September 13, 2004
84. Dendinger B (2001) Advancing Intel® Mobile Voltage Positioning (IMVP) Technology. Intel Developer Forum, Spring 2001
85. Analog Devices (2002) 2-Phase IMVP-II & IMVP-III core controller for mobile CPUs. preliminary technical data ADP3203, 2002, www.analogdevices.com
86. Intersil® (2006) Two-phase core regulator for IMVP-6 mobile CPUs, datasheet ISL6262, May 15, 2006, document FN9199.2, www.intersil.com

87. AMD (1999) Mobile AMD-K6® processor, power supply design. application note, publication 22495, rev.C, May 1999
88. Kassakian JG, Schlecht MF, Verghese GC (1991) Principles of power electronics, Addison Wesley, Reading, MA, ISBN: 0201096897, Chap. 6, pp 103–138
89. Grant DA, Gowar J (1989) Power MOSFETs theory and applications, Wiley, New York
90. Erickson RW, Maksimovic D (2001) Fundamentals of power electronics, 2nd edn, Springer, Netherlands, February 2001, ISBN-10: 0792372700, Sect. 4.1.5, p 73
91. Luo FL, Ye H (2005) Synchronous and resonant DC/DC conversion technology, energy factor and mathematical modeling, CRC, Boca Raton, ISBN-10: 0849372377, Chaps. 1 & 8
92. Huang AQ, Sun NX, Bo Z, Zhou X, Lee FC (1998) Low voltage power devices for future VRM. Proceedings of 1998 International Symposium on Power Semicondcutor Devices & ICs, Kyoto
93. Cheron Y (1992) Soft commutation, Kluwer , Dordrecht, ISBN-10: 041239510X
94. Mohan N (1995) Power electronics: converters, applications and design, 2nd edn, Wiley, New York, ISBN: 0471584088, Chaps 9 & 25
95. Kazimierczuk MK, Czarkowski D (1995) Resonant power converters, Wiley, New York, ISBN-10: 0471047066
96. Erickson RW, Maksimovic D (2001) Fundamentals of power electronics. 2nd edn, Springer Netherlands, ISBN-10: 0792372700, Chap. 5, pp 107–130
97. Song J-H, Kim Y-D, Choy I, Choi J-Y (1998) A pulse frequency modulation control method for single-switch three-phase buck rectifiers. Telecommunications Energy Conference, 1998. INTELEC. Twentieth International 1998, pp 231–236
98. Chen S,Ng WT (2003) High-efficiency operation of high-frequency DC/DC conversion for next-generation microprocessors. The 29th annual conference of the IEEE industrial electronics society, IECON, vol. 1, pp 30–35, November 2003
99. Mammano B (1993) Distributed power systems, Texas Instruments, April 1993, http://ac.aua.am/erik_arevshatyan/Public/Thesis/slup099.pdf
100. Greenland P (2004) Trends in distributed-power architecture, Three elements of architectural change represent a new piece in the distributed-power puzzle. application brief, August 2004, www.national.com
101. Liu KH (1994) A cost-effective desktop computer power management architecture for the energy star computer program. IEEE Power Electronics Specialists Conference, PESC Record, 2:1337–1341
102. Lee FC, Ming Xu, Wang S (2007) High density approaches of AC to DC converter of distributed power systems (DPS) for telecom and computers. Power Conversion Conference - Nagoya. PCC '07, 2–5 April 2007, pp 1236–1243
103. Amelifard B, Pedram M (2007) Optimal selection of voltage regulator modules in a power delivery network. Design Automation Conference, 2007. DAC '07. 44th ACM/IEEE, June 2007
104. Intel® (1997) Slot 1 processor power distribution guidelines. application note AP-587, May 1997, order number: 243332-001
105. Intel® (2007) Power supply. design guide for desktop platform form factors, revision 1.1, March 2007
106. Greenland P (2009) Developments in point of load regulation. Bodo's Power Systems, February 2009
107. Alderman A (2008) PSMA power supply in a package PSiP and power supply on a chip PwrSoC project – phase I & II. Invited talk at the first International Workshop on Power Supply on Chip, September 2008
108. White RV (2003) Emerging on-board power architectures. Annual IEEE Applied Power Electronics Conference and Exposition, APEC 2:799–804
109. Marshall Miles F (2005) An alternative power architecture for next generation systems. International Power Electronics and Motion Control Conference, IPEMC vol. 1, pp 67–72

110. Miftakhutdinov R (2008) Power distribution architecture for tele- and data communications system based on new generation intermediate bus converter. IEEE International Telecommuncations Energy Conference, INTELEC 2008, pp 1–8
111. Brown J Addressing the topologies, converters, and switching devices for intermediate bus architectures. European Conference on Power Electronics and Applications, EPE 2005, pp P.1–P.9
112. Miwa BA Interleaved conversion techniques for high density power supplies, Ph.D. dissertation, Elect. Eng. And Comp. Sci. Dept., Massachusetts Inst. of Technol., Cambridge, 1992
113. Chang C, Knights MA (1995) Interleaving technique in distributed power conversion systems. IEEE Transactions on Circuits and Systems I: Fundamental Theory and Applications vol. 42:no. 5, May 1995
114. Wei J (2004) High frequency high efficiency voltage regulators for future microprocessors. Ph.D. dissertation Electrical Engineering Virginia Polytechnic Institute and State University, Blacksburg, Virginia, USA, September 2004
115. Nguyen H (2004) Design, analysis and implementation of multiphase synchronous buck DC-DC converter for transportable processor, Master Science in Electrical Engineering Virginia Polytechnic Institute and State University, Blacksburg, VA, April 2004
116. Lidow A, Kinzer D, Sheridan G, Tam D (2001) The semiconductor roadmap for power management in the new millennium. Proc IEEE 89(6)803–812
117. Sze SM (2002) Semiconductor devices, physics and technology 2nd edn, Wiley, New York, Part II: Semiconductor Devices, ISBN: 0471333727
118. Yu W, Bo T, Huang H, Dongqing H, Sin JKO, Kang B (2008) Evaluation of trench-gate bipolar-mode JFETs used as high-side transistors in low-voltage buck converters. Proceedings of the 20th International Symposium on Power Semiconductor Devices & ICs, pp 127–130
119. Xiong Y, Cheng X, Okada D, John Shen Z (2007) Comparative study of lateral and trench power MOSFETs in multi-MHz buck converter applications. IEEE Power Electronics Specialists Conference, PESC, June 2007, pp 2175–2181
120. John Shen Z, Okada DN, Lin F, Anderson S,Cheng X (2006) Lateral power MOSFET for Megahertz-frequency, high-density DC/DC converters. IEEE Transactions on Power Electronics 21(1):11–17
121. Cheng X, Xiong Y, Wang X, Kumar P, Shen ZJ (2007) Performance analysis of trench power MOSFETs in synchronous buck converter applications. IEEE Applied Power Electronics Conference, APEC 2007, pp 332–338
122. Peake ST, Kelly B,Grover R (2003) A novel high side FET with reduced switching losses. Proceedings of the International Symposium on Power Semiconductor Devices & ICs, April 2003, pp 362–365
123. Rutter P (2008) Design considerations for integrated powertrains. Invited talk at the first International Workshop on Power Supply on Chip, September 2008
124. Zingg RP (2001) New benchmark for RESURF, SOI, and super-junction power devices. Proceedings of the International Symposium on Power Semiconductor Devices & ICs, 2001, pp 343–346
125. Deboy G, März M, Stengl J-P, Strack H, Tihanyi J, Weber H (1998) A new generation of high voltage MOSFETs breaks the limit line of silicon. International Electron Devices Meeting, IEDM Technical Digest, Dec 1998, pp 683–685
126. Ludikhuize AW (2000) A review of RESURF technology. Proceedings of the International Symposium on Power Semiconductor Devices & ICs, May 2000, pp 11–18
127. Goarin P, Koops GEJ, van Dalen R, Cam Cle, Saby J (2007) split-gate resurf stepped oxide (RSO) MOSFETs for 25V applications with record low gate-to-drain charge. Proceedings of the 19th International Symposium on Power Semiconductor Devices & ICs, May 2007, pp 61–64

128. Koops GEJ, Hijzen EA, Hueting RJE, Zandt MAA (2004) Resurf stepped oxide (RSO) MOSFET for 85V having a record-low specific on-resistance. Proceedings of the International Symposium on Power Semiconductor Devices & ICs, 2004, pp 185–188
129. Gadja MA, Hodgskiss SW, Mounfield LA, Irwin NT (2006) Industrialisation of resurf stepped oxide technology for power transistors. Proceedings of the 18th International Symposium on Power Semiconductor Devices & ICs, June 2006, pp 1–4
130. Darwish M (2003) Next-generation semiconductors for DC-DC converters. Proceedings of the Bipolar/BiCMOS Circuits and Technology Meeting, September 2003, pp 15–21
131. Shuming X, Korec J, Jauregui D, Kocon C, Molly S, Lin H, Daum G, Perelli S, Barry K, Pearce C, Lopez O,Herbsommer J (2009) NexFET a new power device. International Electron Devices Meeting, IEDM 2009, pp 145–148
132. N-channel NexFET™ Power MOSFETs, CSD16325Q5, August 2010. www.ti.com
133. Ma L, Amali A, Kiyawat S, Mirchandani A, He D, Thapar N, Sodhi R, Spring K, Kinzer Dan (2003) New trench MOSFET technology for DC-DC converter applications. Proceedings of the International Symposium on Power Semiconductor Devices & ICs, April 2003, pp 354–357
134. Peake ST, Rutter P, Hodgskiss S, Gadja M, Irwin N (2008) A fully realized 'field balanced' trenchMOS technology. Proceedings of the 20th International Symposium on Power Semiconductor Devices & ICs, May 2008, pp 28–31
135. MEDICI 2D and TSUPREM4 TMA Avant Software
136. Baliga BJ (1989) Power semiconductor device sure of merit for high-frequency applications. IEEE Electron Dev Lett 10(10):455–457
137. Huang AQ (2004) New unipolar switching power device figure of merit,. IEEE Electron Dev Lett 25(5):298–301
138. Adachi S (1992) Physical Properties of III-V Semiconductor Compounds. InP, InAs, GaAs, GaP, InGaAs, and InGaAsP: InP, InAs, GaAs, GaP, InGaAs and InGaAsP, Wiley, New York, 1992, ISBN-10: 047573299
139. Briere MA (2009) GaN-based power device platform. Power systems magazine, Jan 2009, http://ftp.powersystemsdesign.com
140. Briere MA (2008) GaN based power devices: cost-effective revolutionary performance. Power Electron Europe Magazine 7, http://www.power-mag.com
141. Briere MA (2008) High frequency GaN-based power conversion stages. Invited talk at the first International Workshop on Power Supply on Chip, September, 2008
142. Liu W (1999) Fundamentals of III-V devices: Hbts, Mesfets and Hfets/Hemts. Wiley, New York, April 1999, ISBN-10: 0471297003
143. NXP Semiconductors (2004) MOSFETs in LFPak. Brochure, November 2004, document number: 939775913927
144. Sawle A, Blake C, Mariae D (2002) Novel power MOSFET packaging technology doubles power density in synchronous buck converters for next generation microprocessors. International Rectifier white paper, http://www.irf.com/technical-info/whitepaper/mospacapec02.pdf
145. Siliconix V (2005) New Vishay Siliconix PolarPAK® power MOSFETs with double-sided cooling offer on-resistance down to 1.9milliohms maximum, current handling up to 60A in 5mm by 6mm footprint. Announcement December, 2005, http://www.vishay.com
146. López T (2002) High frequency resonant gate drivers for MOSFETs with gate charge recovery Master Science in Electrical Engineering at the Polytechnic University of Catalunya, School of Telecommunication Engineering, Barcelona, Spain, June 2002
147. Shenay K (1990) Effect of gate resistance on high-frequency power switching efficiencies of advanced power MOSFET's. IEEE J Solid-State Circuits 25(2):595–601
148. Kim L-S, Dutton W (1989) Modeling of the distributed gate RC effect in MOSFET's. IEEE Trans Comp Aid Design 8(12):1365–1367
149. Chimento F, Musumeci S, Privitera F, Raciti A, Frisina F, Magri A, Melito M (2006) Modeling and simulation of low-voltage MOSFETs accounting for the effect of the gate

prarasitic-RC distribution. IAS Annual Meeting, Conference Record of the IEEE Industry Applications Conference , vol 3, pp 1443–1449
150. Strydom JT, de Rooij MA, van Wyk JD (2004) A comparison of fundamental gate driver topologies for high frequency applications. IEEE Applied Power Electronics Conference APEC, 2004, pp 1045–1052
151. Dwane P, O'Sullivan D, Egan MG (2005) An assessment of resonant gate drive techniques for use in modern low power DC-DC converters. IEEE Applied Power Electronics Conference APEC 2005, pp 1572–1580
152. Murata Manufacturing Co. (2006) EMI regulations. outline of major noise regulation standards, http://www.murata.com/products/emc/standard/index.html
153. Fairchild Semiconductor™ (2006) Designing the input filter for a Pentium® II processor converter. Appl Bull AB-16, rev.1.0
154. International Rectifier (2008) High frequency synchronous buck optimized LGA power stage, datasheet iP2005APbF, August, 2008
155. Bertotti G (1998) Hysteresis in magnetism for physicists, materials scientists and engineers, Academic Press, New York ISBN-10: 0120932709, Chap. 12: Eddy currents, pp 391–432
156. Bertotti G (1992) Physical interpretation of induction and frequency dependence of power losses in soft magnetic materials. IEEE Trans Magn 28(5):2787–2789
157. Buschow KHJ, De Boer FR (2003) Physics of magnetism and magnetic materials, Kluwer Academic Publishers, Dordrecht, ISBN: 0306474212, Chap. 14: Soft-magnetic materials, pp 147–164
158. Bossche AV, Valchev VC (2005) Inductors and transformers for power electronics. Taylor & Francis, London, ISBN: 1574446797, Chap. 3: Soft magnetic materials
159. Osaka T, Takai M, Hayashi K, Ohashi K, Saito M, Yamada K (1998) A soft magnetic CoNiFe film with high saturation magnetic flux density and low coercivity. Nature 392:796–798
160. Lyons BJ, Hayes JG, Egan MG (2007) Magnetic material comparisons for high-current inductors in low-medium frequency DC-DC converters. Twenty Second Annual IEEE Applied Power Electronics Conference, APEC 2007, pp 71–77
161. Rylko MS, Lyons BJ, Hartnett KJ, Hayes JG, Egan MG (2008) Magnetic material comparisons for high-current gapped and gapless foil wound inductors in high frequency DC-DC converters. 13th Power Electronics and Motion Control Conference, EPE-PEMC 2008, pp 1249–1256
162. Prabhakaran S, Sullivan CR, Venkatachalam K (2003) Measured electrical performance of v-groove inductors for microprocessor power delivery. IEEE Trans Magn 39(5):EA-06
163. Dhagat P, Prabhakaran S, Sullivan CR (2004) Comparison of magnetic materials for v-groove inductors in optimized high-frequency DC-DC converters. IEEE Trans Magn 40 (4):2008–2010
164. Sullivan CR, Sanders SR (1996) Design of microfabricated transformers and inductors for high-frequency power conversion. IEEE Trans Power Electron 11(2):228–238
165. Kowase I, Sato T, Yamasawa K, Miura Y (2005) A planar inductor using Mn-Zn ferrite/ polymide composite thick film for low-voltage and large-current DC-DC converter. IEEE Trans Mag 41(10):3991–3993
166. Donnell TO, Wang N, Meere R, Rhen F, Roy S, O'Sullivan D,O'Mathuna C (2008) Microfabricated inductors for 20MHz DC-DC converters. Twenty-Third Annual IEEE Applied Power Electronics Conference and Exposition, APEC, February 2008, pp 689–693
167. Duffy M, Collins C, Rhen FMF, McCloskey P, Roy S (2008) High current inductor design for MHz switching. IEEE Power Electronics Specialists Conference, PESC, June 2008, pp 2672–2677
168. Ferroxcube (2002) 4F1 Material specification, datasheet, February 2002
169. Ferroxcube (2002) 3F5 preliminary material specification, datasheet, February 2002
170. Li Weidong, Sun Yuqin, Sullivan Charles R (2005) High-frequency resistivity of soft magnetic granular films. IEEE Trans Magn 41(10):3283–3285

171. Kelly S, Collins C, Duffy M, Rhen FMF, Roy S (2007) Core materials for high frequency VRM inductors. IEEE Power Electronics Specialists Conference, PESC, June 2007, pp 1767–1772
172. Coilcraft (2008) SMT power inductors – SLC7530 Series, datasheet, document 366-1, April 2008
173. Wong P-L, Peng X, Bo Y, Lee FC (2001) Performance improvements of interleaving VRMs with coupling inductors. IEEE Transact Power Electron 16(4):499–507
174. Li J, Sullivan CR (2004) Using coupled inductors to enhance transient performance of multi-phase buck converters. Nineteenth Annual IEEE Applied Power Electronics Conference and Exposition, APEC, vol 2, pp 1289–1293
175. Zhu G, McDonald B, Wang K (2009) Modelling and analysis of coupled inductors in power converters. 24th Annual IEEE Applied Power Electronics Conference and Exposition, APEC 2009, pp 83–89
176. Panasonic, Specialty polymer aluminum electrolytic capacitors, SP-Cap. technical guide, http://industrial.panasonic.com/www-ctlg/tech/tABE0000_WW.html
177. Intel® (2008) Voltage regulator module (VRM) and enterprise voltage regulator down (EVRD) 11.0. design guidelines, April 2008, reference number: 315889-002, Appendix A4, pp 53–56
178. Kassakian JG, Schlecht MF, Verghese GC (1991) Principles of power electronics, Part II: Dynamics and control, Addison Wesley, Reading, MA, pp 251–402, ISBN: 0201096897
179. Fairchild Semiconductor® (2002) Voltage mode control IC. Application note AN4128, rev.1.0.1, 2002
180. Deish CW (1978) Simple switching control method changes power converter into a current source. IEEE Power Electronics Specialists Conference, 1978 Record, pp 300–306
181. Dixon L (1998) Average current mode control of switching power supplies. Unitrode application note 1998, http://server.oersted.dtu.dk/ftp/database/Data_CDs/Component_-data/Unitrode_1998/catalog/apps/u140.pdf
182. Mammano R (1994) Switching power supply topology voltage mode vs. current mode, Unitrode design note, DN-62, October 1994
183. Lee M, Chen D, Liu C-W, Huang K, Tseng E, Tai B (2006) Comparison of three control schemes for adaptive voltage position (AVP) droop for VRMs applications. IEEE Power Electronics and Motion Control Conference, EPE-PEMC, 2006, 12th International, August 2006, pp 206–211
184. Redl R, Erisman BP Zansky Z (1998) Optimizing the load transient response of buck converter. Proceedings of the Thirteenth Annual IEEE Applied Power Electronics Conference and Exposition, vol 1, February 1998, pp 170–176
185. Huang W, (2001) A new control for multi-phase buck converter with fast transient response, Proceedings IEEE Applied Power Electronics Conference and Exposition, Sixteenth Annual, vol 1, February 2001, pp 273–279
186. Jianping Xu, Cao Xiaohong, Luo Qianchao (1999) The effects of control techniques on the transient response of switching DC-DC converters. IEEE 1999 International Conference on Power Electronics and Drive Systems, PEDS, July 1999, pp 794–796
187. Goder D, Pelletier WR (1996) V^2 architecture provides ultra-fast transient response in switch mode power supplies. Proceeding HFPC power conversion, September, 1996
188. Schuellein G (1998) Current sharing of redundant synchronous buck regulators powering high performance microprocessors using the V^2 control method. IEEE Thirteenth Annual Proceedings Applied Power Electronics Conference and Exposition, APEC, vol 2, February 1998, pp 853–859
189. Veerachary M (2003) V2 control of interleaved buck converters. IEEE Proceedings of the 2003 International Symposium on Circuits and Systems, ISCAS, volume 3, May 2003, pp III-344–III-346
190. On Semiconductor, (2002) Four-phase VRM 9.0 buck controller, datasheet, July 2002, rev.8, publication number: CS5307/D

191. Skelton DJ, Miftakhutdinov RK (2000) Hysteretic regulator and control method having switching frequency independent from output filter. US patent 6147478, November 14, 2000
192. Tso C-H, Jiin-Chuan W (2003) A ripple control buck regulator with fixed output frequency. IEEE Power Electronics Letters 1(2)61–63
193. Yan WT, Chung HSH, Keith TKA, Ho ClNM (2008) Fixed-frequency boundary control of buck converters with second-order switching surface. IEEE Power Electronics Specialists Conference, PESC 2008, pp 629–635
194. Varga C How a constant on-time voltage regulator reduces ripple. Power Design EE Times-India, http://www.eetimes.com/design/power-management-design
195. Ni C, Tetsuo T (2007) Adaptive constant on-time (D-CAP™) control study in notebook applications. Texas Instruments application report, SLVA281B, July 2007
196. Texas Instruments (1999) Designing fast response synchronous buck regulators using the TPS5210. Application report SLVA044, March 1999
197. Intersil (2002) Current sharing technique for VRMs. Technical brief, July 2002, TB385.1
198. Zhang X, Huang AQ (2004) Investigation of VRM controllers. Proceedings of the International Symposium on Power Semiconductor Devices & ICs, 2004, pp 51–54
199. Abu-Qahouq J, Mao H, Batarseh I (2004) Multiphase voltage-mode hysteretic controlled DC-DC converter with novel current sharing. IEEE Trans Power Electron 19(6)1397–1407
200. Qiu W, Miller G (2007) Dual-edge PWM improves multiphase regulators. Power Electronics Technology, July 2007, www.powerelectronics.com
201. Qiu W, Cheung C, Xiao S, Miller G (2009) Power loss analyses for dynamic phase number control in multiphase voltage regulators. IEEE Proceedings Applied Power Electronics Conference and Exposition, APEC 2009, pp 102–108
202. Forghani-zadeh HP, Rincon-More GA (2002) Current-sensing techniques for DC-DC converters. The 45th Midwest Symposium on Circuit and Systems, 2002, vol 2, pp II-577–II-580
203. Zho X, Xu P (2000) A novel current-sharing control technique for low-voltage high-current voltage regulator module applications. IEEE Trans Industrial Electron 47:1249–1252
204. Edgar D, Pietkiewicz S (1994) Applications of the LT1300 and LT1301 micropower DC/DC converters. application note 59, Linear Technology, January 1994
205. Linear Technology, Synchronous step-down switching regulator controller, datasheet LTC3851, http://www.linear.com
206. Burd TD, Pering TA, Stratakos AJ, Brodersen RW (2000) A dynamic voltage scaled microprocessor system. IEEE JSSC 35(11):1571–1580, November. http://www.linear.com/
207. Stratakos AJ, Sanders SR, Brodersen RW (1994) A low-voltage CMOS DC-DC converter for a portable battery-operated system. Proceedings IEEE Power Electronics and Specialists Conference vol 1, pp 619–626
208. Intersil®, High voltage synchronous rectified buck MOSFET driver, datasheet ISL6205, http://www.intersil.com/cda/home
209. Intersil®, Advanced synchronous rectified buck MOSFET drivers with Pre-POR OVP, datasheet ISL6612B and ISL6613B, http://www.intersil.com/cda/home
210. Bridge CD (2002) Control method to reduce body diode conduction and reverse recovery losses. Texas Instruments US patent, US6396250B1, May 28, 2002
211. Texas Instruments (2001) High-efficiency predictive synchronous buck driver, datasheet UCC27221 and UCC27222, SLUS486A, August 2001
212. Texas Instruments (2002) 12V to 1.8V, 20A high-efficiency synchronous buck converter using the UCC272222 with predictive gate drive. user's guide, 2002
213. Abu-Qahouq JA, Mao H, Al-Atrash HJ, Batarseh I (2004) Maximum efficiency point tracking (MEPT) method and dead time control. Proceedings IEEE Power Electronics Specialists Conference, vol 5 pp 3700–3706
214. Yousefzadeh V, Maksimovic D (2005) Sensorless optimization of dead times in DC-DC converters with synchronous recitfication. Proceedings IEEE Applied Power Electronics Conference and Exposition, pp 911–917, 2005

215. Fairchild Semiconductor (2003) FAN5093: Two phase interleaved synchronous buck converter for VRM 9.x applications, datasheet, March 2003
216. Chinn G, Desai S, DiStefano E, Ravichandran K, Thakkar S (2007) Mobile PC platforms enabled with Intel® Centrino™ mobile technology. Intel Technol J 7(2):6–15
217. International Technology Roadmap for Semiconductors, The next step in assembly and packaging: system level integration in the package (SiP). SiP white paper, http://www.itrs.net/Links/2007ITRS/LinkedFiles/AP/AP_Paper.pdf. Accessed 23 May 2008
218. de Groot LA (2000) SAPFET-2: a power module for power converters. Power Conversion and Intelligent Motion, 2000
219. Rutter P (2004) Challenges of integrated power trains. Intel Technology Symposium 2004
220. NXP Semiconductors (2006) DC-DC converter powertrain, datasheet PIP212-12M, 2006, www.nxp.com
221. Hashimoto T, Uno T, Satou Y, Shiraishi M, Kawashima T, Matsuura N (2007) Advanced power SiP with wireless bonding for voltage regulators. Proceedings of the 19th international symposium on power semiconductor devices & ICs, ISPSD 2007, pp 125–128
222. Shiraishi M, Iwasaki T, Akiyama N, Kawashima T, Matsuura Nk Chiba S (2005) Low loss and small SiP for DC-DC converters. Proceedings of the 17th international symposium on power semiconductor devices & ICs, ISPSD 2005, pp 175–178
223. Hashimoto T, Kawashima T, Uno T, Satou Y, Matsuura N (2008) System in package with mounted capacitor for reduced parasitic inductance in voltage regulators. Proceedings of the 20th International Symposium on Power Semiconductor Devices & IC's, ISPSD 2008, pp 187–191
224. Kawaguchi Y, Kawano T, Takei H, Ono S, Nakagawa A (2005) Multi chip module with minimum parasitic inductance for next generation voltage regulator. Proceedings of the 17th international symposium on power semiconductor devices & ICs, ISPSD 2005, pp 371–374
225. Chin R, Chia TS, Wan K, Houng T, Peels W (2008) Development of flex-based embedded actives packages. Electronic Circuit World Convention 11, March 2008
226. Peels W, Heyes D, Kengen M (2009) Embedded die technology, next generation packaging for discrete semiconductors, SEMICON Europa, October 2009, Dresden, Germany
227. Schrom G, Hazucha P, Hahn J, Gardner DS, Bloechel BA, Dermer G, Narendra SG, Karnik T, De Vivek (2004) A 480-MHz, multi-phase interleaved buck DC-DC converter with hysteretic control. Proceedings of the 35th Annual IEEE Power Electronics Specialists Conference, 2004, vol 6, pp 4702–4707
228. Schrom GH, Paillet P, Rennie F, Moon DJ, Gardner ST, Kamik DS, Sun T, Nguyen P, Hill TT, Radhakrishnan MJ, Memioglu K, T (2007) A 100MHz Eight-Phase Buck Converter Delivering 12A in 25mm2 Using Air-Core Inductors. Applied Power Electronics Conference, APEC 2007 - Twenty Second Annual IEEE, February 2007, pp 727–730
229. Zhang X, Huang AQ, Sun NX (2003) Microprocessor power management integration by VRB-CPU approach. 34th Annual IEEE Power Electronics Specialists Conference, 2003, vol 2, pp 875–880
230. Lee FC (2008) Survey of trends for integrated point-of-load converters, Plenary talk at the first International Workshop on Power Supply on Chip, September 2008
231. Budihardjo I, Lauritzen PO (1995) The lumped charge power MOSFET model, including parameter extraction. IEEE Trans Power Electron 10(3):379–387
232. Chung Y (2000) LADISPICE-1.2: a nonplanar-drift lateral DMOS transistor model and its application to power IC TCAD, IEE Proceedings Circuits, Devices and Systems, August, 2000, pp 219–227
233. Shenai K (1991) A circuit simulation model for high-frequency power MOSFETs. IEEE Trans Power Electron 6(3)539–547
234. Moncoqut D, Farenc D, Rossel P, Charitat G, Tranduc H, Victory J, Pages I (1996) LDMOS transistor for smart power circuits: modelling and design, IEEE Proceedings on Bipolar/BiCMOS Circuits and Technology Meeting, September, 1996, pp 216–219

235. Pierce DG (1991) A temperature dependent SPICE macro-model for power MOSFETs, Proceedings of the 34th Midwest Symposium on Circuits and Systems, vol 2, May 1991, pp 592–596
236. Cotorogea M (1998) Implementation of mathematical models of power devices for circuit simulation in PSpice. Sixth Workshop on Computers in Power Electronics, July 1998, pp 17–22
237. Shenai K (1994) Mixed-mode circuit simulation: An emerging CAD tool for the design and optimisation of power semiconductor devices and circuits. Workshop on Computers in Power Electronics, 1994, pp 1–5
238. John Shen Z, Xiong Yali, Cheng Xu, Yue Fu, Kumar Pavan (2006) Power MOSFET switching loss analysis: A new insight. IEEE Ind Appl Conf 3:1438–1442
239. Shenai K, Cavallaro C, Musumeci S, Pagano R, Raciti A (2003) Modeling low-voltage Power MOSFETs as synchronous rectifiers in buck converter applications. Conference Record of the Industry Applications Conference, 2003, vol 3, pp 1794–1801
240. Kocon C, Gladish J, Challa A (2008) Advanced physics based modelling of power MOSFET devices. Proceedings Power Electronics Intelligent Motion Power Quality Europe 2008
241. Cheng X, Xiong Y, Wang X, Kumar P, Shen ZJ (2007) Performance analysis of trench power MOSFETs in synchronous buck converter applications. IEEE Appl Power Electron Conf 332–338
242. Leonardi C, Frisina F, Letor R, Raciti A (1998) Power MOSFET macromodel accounting for temperature dependence: parameter extraction and simulation. IEEE 6th Workshop on Computers in Power Electronics, July 1998, pp.37–47
243. Varadarajan KR, Sinkar A, Chow TP (2006) A circuit simulation model of a novel silicon lateral trench power MOSFET for high frequency switching applications. IEEE Compel workshop, July 2006, pp 306–309
244. Shousha AHM, Aboulwafa M (1993) A generalised Tanh law MOSFET model and its applications to CMOS inverters. IEEE J Solid-State Circuits 28(2):176–179
245. Maxim A, Maxim G (1999) A high accuracy power MOSFET SPICE behavioural macro-model including the device self-heating and safe operating area simulation. IEEE Appl Power Electron Conf Exp 1:177–183
246. Scott RS, Franz GA, Johnson JL (1991) An accurate model for power DMOSFET's including interelectrode capacitances. IEEE Trans Power Electron 6(2):192–198
247. Panv Y, Jovanovic MM (2001) Design considerations for 12V/1.5V, 50A voltage regulator modules. IEEE Trans Power Electron 16(6):39–46
248. Spaziana L (1996) A study of MOSFET performance in processor targeted buck and synchronous rectifier buck converters. High Frequency Power Conversion Conference, HFPC 1996, pp 123–137
249. Deuty S (2000) Optimizing transistor performance in synchronous rectifier, buck converters. IEEE Appl Power Electron Conf Exp, APEC 2:675–678
250. Xiao Y, Shah H, Chow TP, Gutmann RJ (2004) Analytical modeling and experimental evaluation of interconnect parasitic inductance on MOSFET switching characteristics. IEEE Appl Power Electron Conf Exp, APEC 1:516–521
251. Ren Y, Ming X, Zhou J, Lee FC (2006) Analytical loss model of power MOSFET. IEEE transactions on power electronics 21(2):310–319
252. Eberle W, Zhang Z, Liu Y-F, Sen PC (2008) A simple analytical switching loss model for buck voltage regulators. IEEE Applied Power Electronics Conference and Exposition, APEC 2008, pp 36–42
253. Meade T, O'Sullivan D, Foley R, Achimescu C, Egan M, MacCloskey P (2008) Parasitic inductance effect on switching losses for a high frequency DC-DC converter. IEEE Applied Power Electronics Conference and Exposition, APEC 2008, pp 3–9
254. Kawaguchi Y, Yamaguchi Y, Kanie S, Baba A, Nakagawa A (2008) Proposal of the method for high efficiency DC-DC converters and the efficiency limit restricted by silicon properties. IEEE Power Electronics Specialists Conference, PESC 2008, pp 147–152

Chapter 2
Model Level 0: Switching Behavior of Power MOSFETs

This chapter describes the model level 0 introduced in the previous chapter. One main goal is to provide the reader with sufficient knowledge on the adopted modeling approach, characteristics, basic structure, advantages, and drawbacks of its use and associated analysis methods.

Within the scope of the investigations, the chapter includes simulation results corresponding to state-of-the-art SRBCs for VRs, which are compared to extensive experimental data. The validation assesses the accuracy of the model to reproduce switching waveforms and estimate power losses.

The purposes of the model are thus to accurately represent the behavior of the power MOSFETs in the switched converter, understand fast switching phenomena, identify mechanisms of power losses, and quantify their effects. To meet all these requirements, model level 0 is based around an accurate behavioral model of power MOSFETs.

Furthermore, those circuit components of the switched converter that may potentially influence the operation of the power switches are properly modeled according to the analysis demands. The model description of these circuit elements is provided in the different sections of the chapter.

For its complexity and relevant importance in the converter circuit though, most of the chapter is devoted to the modeling of the power MOSFETs. The proposed MOSFET model consists of a proper combination of individually identified device interterminal elements,[1] such as the capacitances, channel and body diode, each of which are characterized and represented by means of lumped elements. The superposition of the electrical effects produced by these elements effectively reproduces the behavior at the terminals of the device. One of the benefits of this implementation is that it provides consistent simulations of the MOSFETs internal operation corresponding to each of the extracted device characteristics. Such ability aids the purpose of elucidating switching phenomena and their impact on the converter performance.

Besides the basic structure, composition and adopted approach, the description of the converter components includes the derivation of model parameters, which

[1]All circuit elements considered herein are assumed time invariant, unless otherwise specified.

T. López et al., *Voltage Regulators for Next Generation Microprocessors*,
DOI 10.1007/978-1-4419-7560-7_2, © Springer Science+Business Media, LLC 2011

result from experimental characterization and, alternatively, accurate device physics simulations. Both methods will be briefly outlined and their use compared and justified for each particular case.

The resulting converter model combines power switches, gate drivers, input/output filters and layout parasitic elements to form a complex lumped network structure that may be solved with circuit simulators such as SPICE. In addition to the power circuit, a PWM control and gate drive logic with dead time adjustments are implemented in the simulator to enable switching synchronization and power balance control. The control module further incorporates an algorithm for the estimation of initial values of the state variables so as to simulate switching cycles in steady-state conditions. This allows skipping the transient time of a usually large number of switching periods, which in turn significantly reduces simulation times.

The analysis of simulated waveforms will not only be the basis to understand fast switching operation but also to identify and quantify associated loss mechanisms. Special emphasis is given to a proposed systematic procedure to perform the loss breakdown by separation of loss components. Although the losses not related to the power MOSFETs are not addressed in detail, the results of the loss extraction yield quantities associated to several identified loss mechanisms which happen to be critical to identify key parameters influencing the performance of the overall converter.

The proposed loss breakdown and analysis methods will be illustrated with various case examples that demonstrate the benefits of the approach to help investigate switching phenomena, identify major loss contributors, and propose measures to improve converter performance.

2.1 Power MOSFET Model for Circuit Simulations

Given the need to effectively combine device behavior and circuit analysis in one single simulation environment, the goal is to develop an accurate power MOSFET model for circuit simulators that features device analysis capabilities. As highlighted in Chap. 1, empirical models appear to be the right choice for this purpose as they offer a proper balance trade-off between required computation power and simulation accuracy, which cannot be met with device physics simulations in mixed mode.

Thus, the following is a description of an adopted behavioral modeling approach. The proposed model aims at reproducing the device behavior at its terminals by proper combination of individual device characteristics. By means of the right testing conditions, such characteristics can be derived from experimental measurements resulting in DC output characteristics in first and third quadrant, avalanche breakdown (ABD), capacitances, gate resistance, package impedances, and body diode reverse recovery (RR). Numerical device simulation is principally an alternative method to obtain one or more of these characteristics by means of rather simple simulation setups, as it shall be shown later on. Once available, the data may be represented in terms of look-up tables, fitting functions, and differential equations. These can then be combined in a simulation model that reproduces the actual device behavior at its terminals, to which the original testing conditions have been applied. The resulting empirical model is meant for circuit simulators like SPICE in order to cope with rather complex circuit environments.

The accuracy of the loss predictions of the complete circuit and individual loss contributors is determined by the employed characteristics and can be tailored to the analysis demands. In this regard, both characterization methods (i.e., measurements and device physics simulations) play different roles. Within experimental limits, measurements allow validating device simulations. The latter in turn is not only useful to replace cumbersome measurements; it also enables to cover characterization ranges that go beyond experimental limits and hence to complete the desired characteristics. In cases where no experimental data are available, like for instance in the process development of a novel technology, device physics simulations may be the only means to generate accurate characterization data.

The following sections outline implementation issues of the behavioral model, describe methods for comprehensive device characterization, which is the crucial part of the approach, and explain analysis opportunities by means of simulation examples.

2.1.1 Model Structure and Implementation

The dynamic behavior of the power MOSFET may be represented with the lumped network structure of Fig. 2.1. Note that, for simplicity, all parasitic lumped elements associated to the package, substrate, and polysilicon gate are at this point omitted. As shown in the circuit diagram, three current sources are parallel connected to individually model the channel, body diode, and ABD current contributions. These are i_{cn}, i_{dio}, and i_{AB}, respectively. Added to these current paths, the model further considers the three equivalent parasitic terminal capacitances of the MOSFET, namely C_{GS}, C_{GD}, and C_{DS}. In order to properly represent the behavior of each individual characteristic, the elements of the model may be dependent on interelectrode voltages v_{DS} and v_{GS}. In particular, diode current i_{dio} may additionally

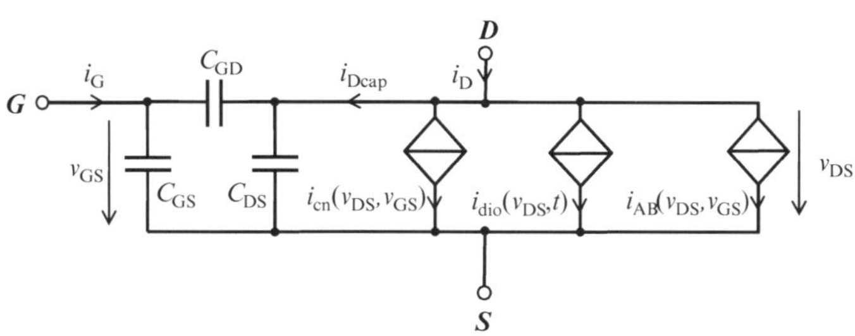

Fig. 2.1 Circuit diagram of the power MOSFET model for circuit simulations. For simplicity, package-related parasitic elements and gate and substrate resistances are omitted

be time dependent so as to represent the dynamics of RR, as it shall be described in detail later.

Establishing appropriate criteria to de-embed the terminal currents into the defined current contributions is a primary aspect of the modeling approach. This is extensively discussed in the next section in the context of model parameter extraction. This section focuses on the model implementation in circuit simulators.

Regarding DC output characteristics, the channel current is defined in both first and third quadrants, i.e., both directions of drain current flow. The channel behavior considerably varies in each quadrant, for which different empirical functions are utilized. In the first quadrant it follows that,

$$i_{q1}(v_{DS}, v_{GS}) = A_{die} \cdot c_0(v_{GS}) \cdot \begin{bmatrix} \tanh\left(\dfrac{v_{DS} \cdot c_2(v_{GS})}{1 + v_{DS} \cdot e^{c_1(v_{GS})}}\right) + \cdots \\ + v_{DS} \cdot c_3(v_{GS}) + c_4(v_{GS}) \cdot (e^{\kappa_e \cdot v_{DS}} - 1) \end{bmatrix}. \quad (2.1)$$

Coefficients c_0–c_4 are dependent on v_{GS}. Together with constant κ_e, these coefficients are determined from fitting procedures of the original MOSFET characteristics. The resulting relations with v_{GS} are implemented with look-up tables. Parameter A_{die} is the effective active chip area of the power device, which is introduced in the equation to enable die area scaling. This feature will be extended throughout the rest of the model components and frequently employed in subsequent analysis.

Third quadrant operation is highly influenced by the so-called body-effect, which describes the changes in the threshold voltage as function of a negative v_{DS} [1]. The analysis of this effect in high-density trench power MOSFETs is provided in Appendices A and B.

An empirical equation that reproduces the channel DC behavior under the influence of the body-effect is as follows:

$$i_{q3}(v_{DS}, v_{GS}) = A_{die} \cdot v_{DS} \cdot e^{c_7(v_{GS})} \cdot [1 - \tanh(c_5(v_{GS}) \cdot (v_{DS} - c_6(v_{GS})))]. \quad (2.2)$$

Similar to (2.1), c_5–c_7 correspond to fitting coefficients. Expressions (2.1) and (2.2) fully describe the DC characteristics of the channel. Combining them yields the expression for the channel current

$$i_{cn}(v_{DS}, v_{GS}) = i_{q1}(v_{DS}, v_{GS}) \cdot \max(v_{DS}, 0) + i_{q3}(v_{DS}, v_{GS}) \cdot \min(v_{DS}, 0). \quad (2.3)$$

Note that functions max and min restricts the use of the functions to their respective boundary limits.

Besides the channel current, the third quadrant DC output characteristics are further composed by the body diode current. Its DC component is assumed to be uniquely dependent on v_{DS} since the drain current dependence on the gate voltage is entirely attributed to the channel. This is justified in the Appendices A and B by means of extensive simulations and experimental analysis. Thus, in DC operation it follows that

$$i_{\mathrm{dio\_dc}}(v_{\mathrm{DS}}) = i_{\mathrm{dio}}(v_{\mathrm{DS}}, t \to \infty). \tag{2.4}$$

The right-hand side of (2.4) represents the body diode's response to a step function. The implementation of the DC diode current is based on (2.2) at a given gate voltage, that is,

$$i_{\mathrm{dio\_dc}}(v_{\mathrm{DS}}) = i_{q3}(v_{\mathrm{DS}}, V_{\mathrm{GSd}}), \tag{2.5}$$

where V_{GSd} is usually well below zero. As it shall be illustrated later in this chapter, condition $V_{\mathrm{GSd}} \leq -5V$ is sufficient in the devices characterized so far.

Similarly, the ABD is implemented by a diode function in the first quadrant as follows:

$$i_{\mathrm{AB}}(v_{\mathrm{DS}}, v_{\mathrm{GS}}) = -i_{\mathrm{dio\_dc}}(\min(-v_{\mathrm{DS}} + V_{\mathrm{AB0}} - \kappa_{\mathrm{AB}} \cdot v_{\mathrm{GS}}, 0)). \tag{2.6}$$

Constants V_{AB0} and κ_{AB} allow adjusting the avalanche voltage level as well as its dependence on v_{GS}.

In the next section, the use of (2.1)–(2.6) will be illustrated to fit the characteristics of a particular MOSFET technology. The equations are suitable for most of the vertical MOSFET structures studied thus far. However, certain device technologies may involve the definition of alternative empirical functions.

The latter is also the case for the dynamic model of the body diode, which is defined as an extension of its DC characteristics by means of the following integral equation

$$i_{\mathrm{dio}}(v_{\mathrm{DS}}, t) = \beta_{\mathrm{rr}} \cdot i_{\mathrm{dio\_dc}}(v_{\mathrm{DS}}) + \frac{1}{t_{\beta\mathrm{rr}}} \int_0^t (i_{\mathrm{dio\_dc}}(v_{\mathrm{DS}}) - i_{\mathrm{dio}}(v_{\mathrm{DS}}, t))\, \mathrm{d}t + I_{\mathrm{dio0}}. \tag{2.7}$$

Empirical equation (2.7) compactly represents the dynamics of RR, which may be adjusted by parameters β_{rr} and $t_{\beta\mathrm{rr}}$. The empirical model is an adaptation of the simple diode model from Lauritzen [2]. A detailed description of the model performance and comparison to other implementation approaches are provided in Appendix C. Note that forward recovery is neglected since it has not been observed in the studied low-voltage devices.

The diode capacitance is implicitly described by the model element C_{DS}. The implementation of this and the other nonlinear capacitances in circuit simulators is generally nontrivial as it frequently involves a rather complex custom design by means of basic built-in functions and components. The realization method of interest is that one which best compromises simulation accuracy, speed, and convergence. Nonetheless, one of the most common approaches is the use of equivalent voltage controlled current sources, as described in [3, 4]. In such configuration, the interconnection of nodes D–S and G–S is established solely by variable current sources, which may be linked to look-up tables[2] and interpolation functions to define the voltage-dependent capacitances. Consequently, any series connection of inductances to these nodes is potentially susceptible to generate convergence problems because the piece-wise approximations yield time-discontinuous derivatives of the drain and gate currents.

In this regard, a more topologically stable configuration implies the use of current-controlled voltage sources for the capacitance implementation. To achieve this, the capacitance network from the model of Fig. 2.1 needs to be transformed to an alternative equivalent circuit. Let the capacitance currents of the MOSFET be expressed in vector form as follows:

$$\begin{pmatrix} i_{\mathrm{G}} \\ i_{\mathrm{Dcap}} \end{pmatrix} = C \cdot \begin{pmatrix} \frac{\mathrm{d}v_{\mathrm{GS}}}{\mathrm{d}t} \\ \frac{\mathrm{d}v_{\mathrm{DS}}}{\mathrm{d}t} \end{pmatrix}, \tag{2.8}$$

where,

$$\boldsymbol{C} = \boldsymbol{C}(v_{\mathrm{GS}}, v_{\mathrm{DS}}) = \begin{pmatrix} C_{\mathrm{iss}} & -C_{\mathrm{rrs}} \\ -C'_{\mathrm{rss}} & C_{\mathrm{oss}} \end{pmatrix}, \tag{2.9}$$

$$C_{\mathrm{iss}} = C_{\mathrm{iss}}(v_{\mathrm{GS}}, v_{\mathrm{DS}}) = C_{\mathrm{GS}}(v_{\mathrm{GS}}, v_{\mathrm{DS}}) + C_{\mathrm{GD}}(v_{\mathrm{GS}}, v_{\mathrm{DS}}), \tag{2.10}$$

$$C_{\mathrm{oss}} = C_{\mathrm{oss}}(v_{\mathrm{GS}}, v_{\mathrm{DS}}) = C_{\mathrm{GD}}(v_{\mathrm{GS}}, v_{\mathrm{DS}}) + C_{\mathrm{DS}}(v_{\mathrm{GS}}, v_{\mathrm{DS}}), \tag{2.11}$$

$$C_{\mathrm{rss}} = C_{\mathrm{GD}}, \tag{2.12}$$

$$C'_{\mathrm{rss}} \cong C_{\mathrm{rss}}. \tag{2.13}$$

The latter expresses the reciprocity property [5] of the capacitance matrix from (2.9), which has been observed in all devices analyzed thus far.

The capacitances may be conveniently specified in terms of relative values as follows:

$$\begin{pmatrix} C_{\mathrm{GS}} \\ C_{\mathrm{GD}} \\ C_{\mathrm{DS}} \end{pmatrix} = A_{\mathrm{die}} \cdot \begin{pmatrix} C_{\mathrm{GSsp}} \\ C_{\mathrm{GDsp}} \\ C_{\mathrm{DSsp}} \end{pmatrix}. \tag{2.14}$$

[2]Unlike in the case of the channel and body diode, look-up tables are employed for the full representation of the capacitances since no suitable empirical function was found. Generally, simple fitting functions are preferred over look-up tables for better scaling capabilities, compactness, and computation speed, particularly when multivariable dependencies need to be implemented and high accuracy is required.

The vector to the right of (2.14) contains specific capacitance values given in Farads per unit area.

The implementation of the nonlinear capacitances as voltage sources implies expressing (2.8) in integral form, that is,

$$\begin{pmatrix} v_{GS} \\ v_{DS} \end{pmatrix} = \int_0^t C^{-1} \cdot \begin{pmatrix} i_G \\ i_{Dcap} \end{pmatrix} \cdot dt + \begin{pmatrix} V_{GS0} \\ V_{DS0} \end{pmatrix}, \tag{2.15}$$

where V_{GS0} and V_{DS0} are the initial conditions of state variables v_{GS} and v_{DS}.

Based on the above transformation, the general model of Fig. 2.1 may be equivalently represented as shown in Fig. 2.2. This structure differs from the previously exposed in that the MOSFET capacitances are implemented by means of two current-dependent voltage sources, which offers three major advantages. First, current discontinuities through the terminals of the structure can no longer occur due to the integral form of the impressed voltage across terminals D–S and G–S. Second, initial conditions can be readily defined, which contrasts with the implementation based on current sources. And third, implementation is simplified since only two voltage sources are required to represent three capacitances.

The proposed implementation can be further simplified by just combining terms and expressing capacitive current i_{Dcap} as follows:

$$i_{Dcap}(v_{DS}, v_{GS}, t) = i_D - i_{Dnc}(v_{DS}, v_{GS}, t), \tag{2.16}$$

where i_{Dnc} defines the noncapacitive term of the drain current, that is,

$$i_{Dnc}(v_{DS}, v_{GS}, t) = i_{cn}(v_{DS}, v_{GS}) + i_{dio}(v_{DS}, t) + i_{AB}(v_{DS}, v_{GS}). \tag{2.17}$$

Thus, Fig. 2.3 shows the resulting simplified power MOSFET model structure composed by just two dependent voltage sources governed by (2.1)–(2.17). The model can be readily implemented in circuit simulators such as PSpice, which allows the declaration of all required nonlinear functions, integral operations, and use of 2D look-up tables for the definition of the voltage-dependent capacitances.

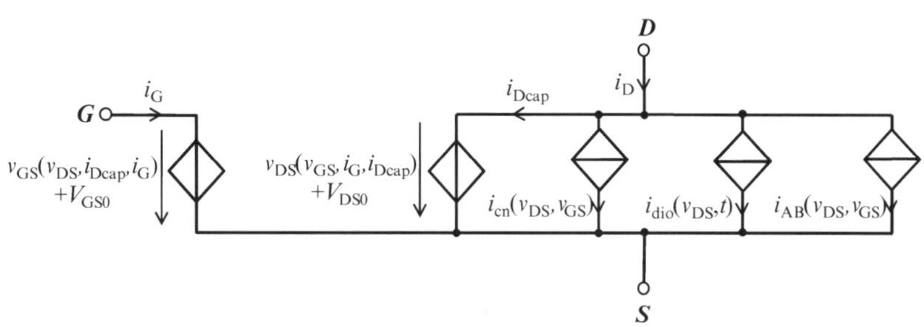

Fig. 2.2 Model of Fig. 2.1 with nonlinear capacitances represented as voltage-dependent sources

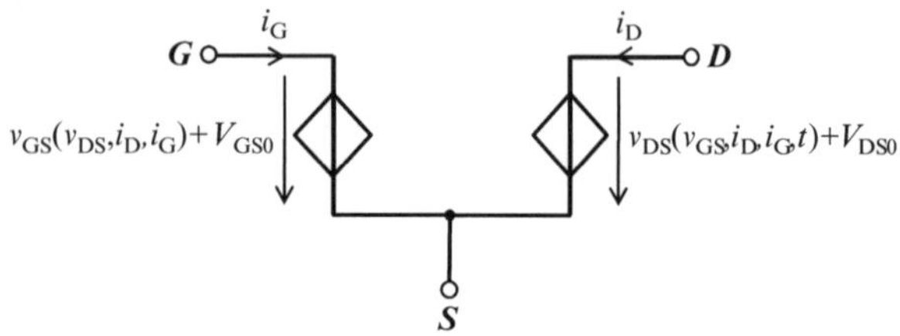

Fig. 2.3 Simplified circuit implementation of proposed power MOSFET model for circuit simulations

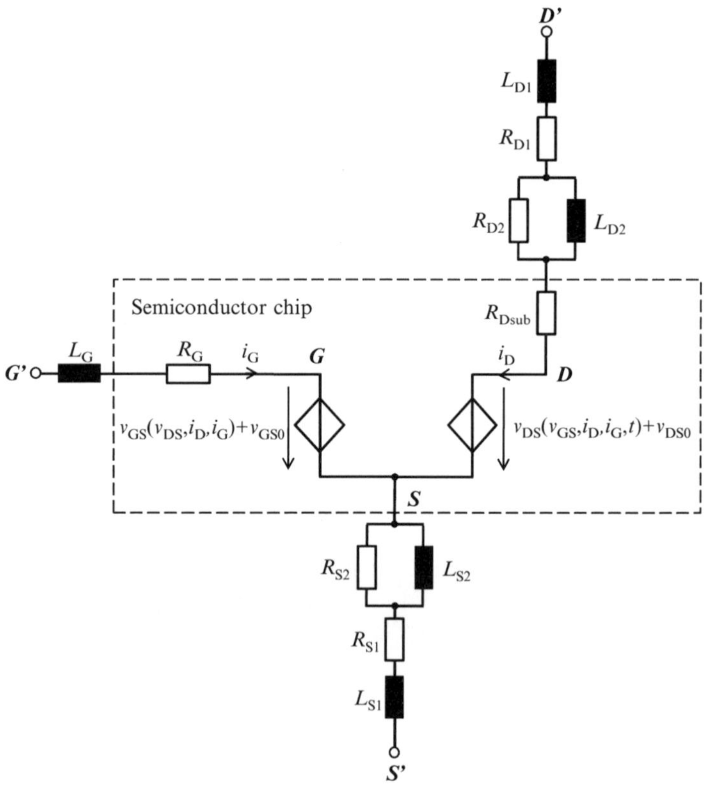

Fig. 2.4 Proposed power MOSFET model for circuit simulations, including package parasitic elements, and both gate and substrate resistances

The complete dynamic power MOSFET model including all parasitic resistances and inductances from chip and package is depicted in Fig. 2.4. The highlighted area corresponds to the portion of the semiconductor chip, which includes substrate resistance R_{Dsub} and polysilicon gate resistance R_{G} in addition to the elements considered in Fig. 2.1. While substrate resistance R_{Dsub} may scale with A_{die}, gate resistance R_{G} strongly depends on the gate busbar layout design, as it shall be seen later in this section.

The rest of the model representation corresponds to the package and electrode equivalent parasitic elements. Therefore, electrodes labeled as G', D', and S' correspond to the terminals of the MOSFET package. The network formed by elements L_{D2}, R_{D2} and L_{S2}, R_{S2} represents the skin losses and proximity effects of the package interconnections, e.g., bond wires. At high-switching frequency, package loss dissipation due to these magnetic effects may be substantial, and thus need to be considered. In the gate terminal, such effects are frequently neglected since the polysilicon gate resistance dominates.

2.1.2 Model Data Acquisition

Although the required model data may be obtained either by measurements or by device physics simulations, there are differences in the ranges that can be covered with the two approaches. Table 2.1 summarizes the required characteristics and methods for their acquisition. The resolution and extension of the gathered data may be adapted to the modeling needs which, for the present purpose, go far beyond the information usually contained in datasheets.

The characterization techniques may involve postprocessing algorithms to ultimately subtract the model parameters from the raw data (see (2.1)–(2.17)). The extraction procedures of each device characteristic are described in the following subsections.

2.1.2.1 Gate Resistance and Package Impedances

Parasitic inductances of device package and polysilicon gate resistance are experimentally derived from scattering parameters (or S-parameters) measurements, which describe the electrical behavior of impedance networks when undergoing various steady-state stimuli by small signals. Assuming an equivalent impedance network at the terminals of the MOSFET, the measured S-parameters can be processed to

Table 2.1 MOSFET device characterization by measurements and device physics simulations

MOSFET characteristics	Measurements	Device physics simulations
Gate resistance Package impedances	S-parameter extraction with vector network analyzer	N/A[a]
DC output characteristics (channel, body diode, ABD, and substrate resistance)	Curve tracer based on voltage ramps or voltage pulses	DC parameter sweep simulations
Interelectrode capacitances	Impedance analyzer measurements for $V_{GS} < V_{TH}$ Measurement of currents resulting from applied voltage ramps	Transient simulations of current response to applied voltage ramps
Body diode RR	Half-bridge in clamped inductive switching mode	Transient simulations of resistive or inductive turn-off

[a]device physics simulations are presented in Appendix E to estimate package impedances

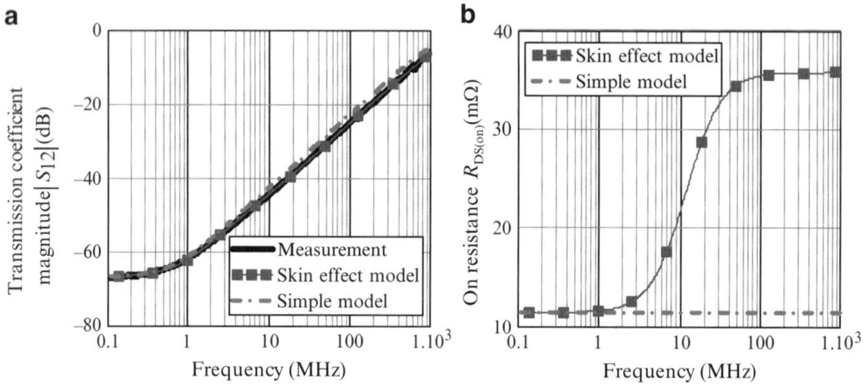

Fig. 2.5 (a) Drain-source transmission coefficient magnitude of a MOSFET in on-conduction, (b) extracted on resistance. Experiment performed at $v_{GS} = 6$ V. Measurement and fitting results. DUT: PHD77NQ03T from NXP Semiconductors (DPak encapsulation). Fitting results refer to model of Fig. 2.6: $L_1 = 2.13$ nH, $R_1 = 11$ mΩ, $L_2 = 0.33$ nH, $R_2 = 24$ mΩ

quantitatively determine the elements constituting such a network under different bias conditions.

S-parameter measurements are performed with *a vector network analyzer* (VNA). The principle of operation of VNAs enables to effectively measure the terminal impedance of the device under test (DUT) without the influence of interconnection parasitic impedances. With a well-calibrated setup, the VNA allows characterization of low impedance packages at frequencies of interest, which may extend into the GHz range. Proper interconnections of the DUT to the VNA are essential to achieve consistent measurements at high frequency. The setup employs RF cables to connect the VNA ports to a test fixture for coplanar microstrip lines, to which the DUT is suitably attached.

Figure 2.5a illustrates the magnitude of the transmission coefficient corresponding to the terminal drain–source impedance of an ON state power MOSFET mounted on DPak. The transmission coefficient of the series R–L lumped structure (i.e., referred to as "simple model" in Fig. 2.5) cannot finely match the experimental results due to the presence of skin effects at the package terminals, which translate into an effective inductance reduction as frequency increases. The lumped skin effect model of Fig. 2.6 allows a simple and yet more accurate representation of the measured response, as Fig. 2.5a illustrates[3] [6, 7]. A look at the real part of the estimated impedance reveals the strong resistance dependency on frequency, which clearly manifests already in the MHz range, as Fig. 2.5b depicts. The curve

[3]The phase of the impedance is not considered for parameter fitting since the distributed nature of the underlying skin effect cannot be accurately represented with the proposed models. A far more complex ladder network may be required for that purpose. The simple lumped-based skin effect model aims at representing the fundamentals of such magnetic phenomena by first-order approximations. Estimations of the loss contribution of this effect will dictate whether the approximations are satisfactory enough or, on the contrary, further model refinements are demanded.

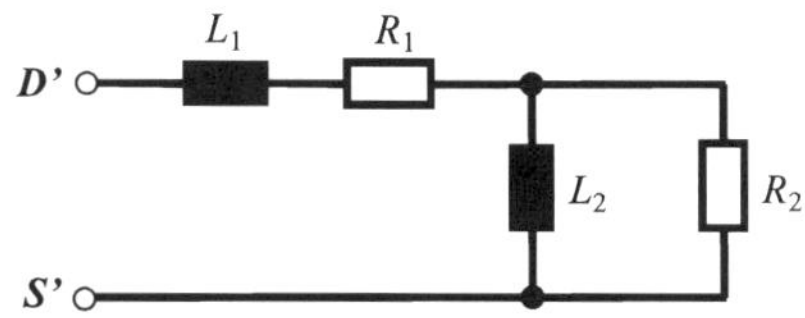

Fig. 2.6 Terminal drain–source equivalent impedance network of a MOSFET in ON conduction considering package skin effect

is characterized by two values, R_1 and R_2, which define the resistance of the network at low and high frequency, respectively. A transition between the two values occurs at intermediate frequencies. The transition interval is given by the inverse time constants R_1/L_1 and R_2/L_2.

The skin effect may be mostly attributed to the source terminal since the drain is contacted through the mounting base, whose equivalent inductance may be negligible.

The results of the characterization suggest limiting the DPak to applications where switching frequencies are well below 1 MHz. At sufficient low switching frequencies, the CtrlFET package loss contribution as a consequence of the output inductor ripple current may be approximately estimated from the following expression,

$$P_{\mathrm{DPak(ac)}}\big|_{\mathrm{CtrlFET}} \cong (R_2)^{-1} \cdot \left(L_2 \frac{V_{\mathrm{in}} - V_{\mathrm{o}}}{L_{\mathrm{o}}}\right)^2 \cdot \left(\frac{V_{\mathrm{o}}}{V_{\mathrm{in}}}\right). \qquad (2.18)$$

According to the proposed skin effect model, the square term in (2.18) equals the voltage across R_2 during the ON state of the CtrlFET when assuming steady-state operation. A similar expression may be deduced for the SyncFET. Given the extracted values of Fig. 2.5, $P_{\mathrm{DPak(ac)}}$ of both devices result in less than 1 mW for typical applications (i.e., down conversion of 12–1 V). Expression (2.18) may no longer be accurate at high-switching frequencies as the ON times approach time constant L_2/R_2. Nonetheless, the simplified skin effect model is helpful to qualitatively represent the increase of ON resistance with the switching frequency.

A far more sophisticated skin effect model may be obtained through the use of FE software tools that enable to numerically solve multidimensional Eddy current problems. Appendix E describes how this is achieved to study the parasitic inductance and dynamic resistance of MOSFETS packages and PCB layouts. A quasi-static formulation of Maxwell's equations can be further developed for the performance estimation of the output inductor, as described in Appendix E.

Figure 2.7 illustrates some of the results derived from FE harmonic analysis, where a MOSFETs package impedance is evaluated in combination with a portion of a PCB track, as representative arrangement from a realistic application. Details of the layout arrangement are provided in Appendix E. The double layer PCB with return ground plane allows to effectively neglect the PCB inductance.

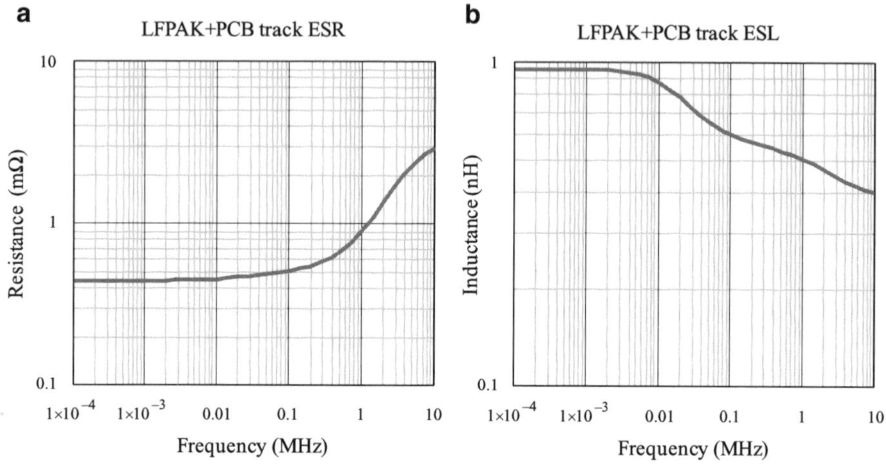

Fig. 2.7 Frequency dependence of the equivalent series resistance (a), and inductance (b) (from drain to source) of the LFPAK plus PCB section with ground plane. Simulation results (see Appendix E for details)

Table 2.2 Measurement extraction of parasitic elements from various MOSFET packages

Package type	Package inductance neglecting skin effects (nH)	
D2Pak	Drain	$L_D = 0$
	Source	$L_S = 5$
	Gate	$L_G = 7$
DPak	Drain	$L_D = 0$
	Source	$L_S = 3.5$
	Gate	$L_G = 4.5$
LFPak	Drain	$L_D = 0$
	Gate	$L_G = 1.1$
	Source	$L_S = 0.5$

As such, the curve from Fig. 2.7b mainly accounts for the MOSFETs ESL corresponding to the package leadframe of the source contact. The extracted value varies around 500 pH in the high-frequency range. The ESR increases rapidly above 200 kHz, reaching five times the DC values at 5 MHz. Unlike the case of the ESL, the ESR contribution of PCB layout cannot be neglected, as shown in Appendix E.

Table 2.2 summarizes the results from VNA measurements of three different power MOSFET packages. To simplify the comparison, skin effects are neglected and so every terminal parasitic inductance can be simply represented with a single lumped inductor. Note that the source inductance of the LFPak matches the simulation results from Fig. 2.7a at 1 MHz.

Besides the extraction of model parameters, S-parameter measurements can further be exploited for device design and failure diagnosis related to manufacturing processes. Effects such as spreading resistance of bounding connections and split gate [8] can be detected and assessed with no influence of large series parasitic inductances.

A parameter that has an effect on the gate resistance is the gate mask structure, i.e., the gate busbars that distribute the gate current over the area of polysilicon. Figure 2.8 shows some layout examples of possible gate busbar designs. A design criterion for layout optimization may be based on the best compromise between occupied semiconductor area and uniform current distribution. Regarding the later, S-parameter measurements may be used to determine whether or not a design is adequate for current distribution by analyzing the transmission coefficients. Ideally, in case of a uniform current density all cells switch simultaneously. This should be translated into a frequency response equivalent to a simple series R–L–C network, as shown in the example of Fig. 2.9 for the case of an optimum gate busbar design. Note that the value at the resonant frequency yields the gate resistance of the device.

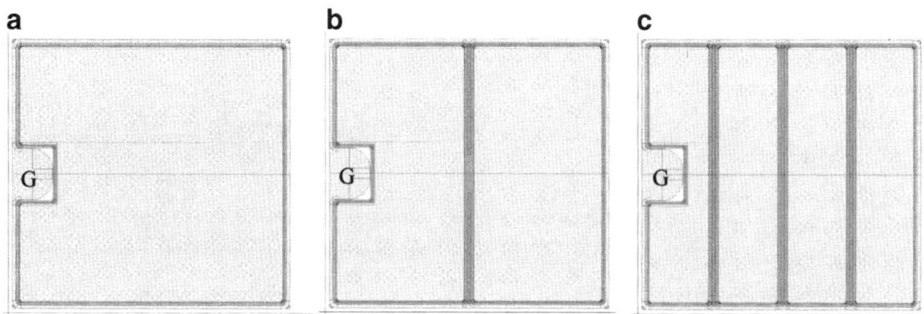

Fig. 2.8 Example of different gate mask structures; (**a**) metal contact along the chip periphery only, (**b**) with one additional center busbar, (**c**) with three additional uniformly distributed parallel busbars

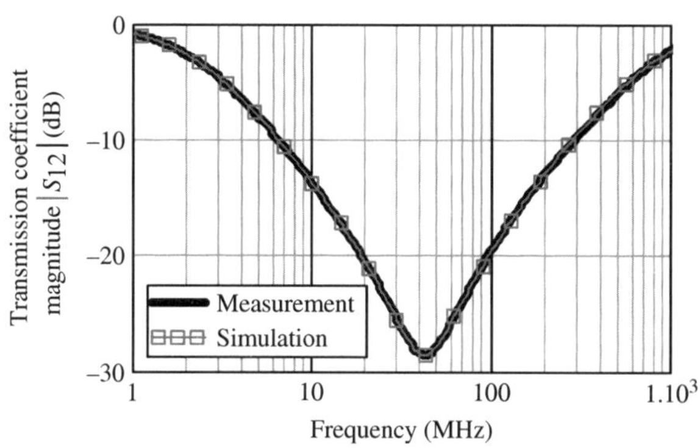

Fig. 2.9 Magnitude of the transmission coefficients of an optimized gate mask layout. Gate-drain terminal measurement and fitting results. The following parameters correspond to a series R–L–C network: $L = 5.18$ nH, $R = 0.97\ \Omega$, $C = 2.76$ nF

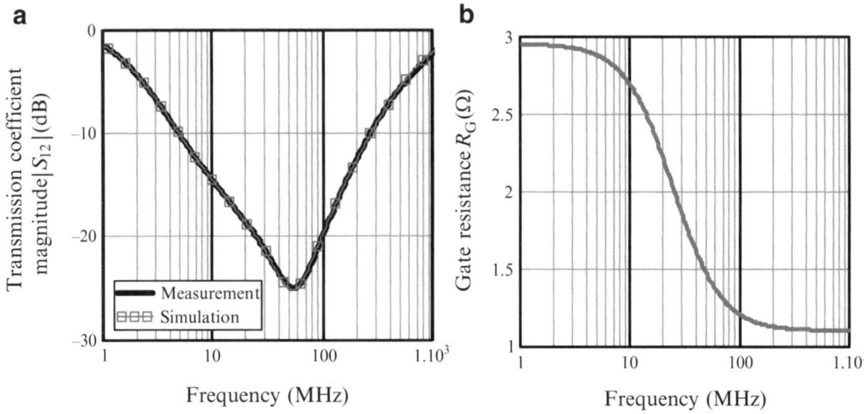

Fig. 2.10 (a) Gate-drain transmission coefficient magnitude of a MOSFET with nonoptimized gate mask. (b) Extracted gate resistance. Terminal gate-drain measurement and fitting results. The following parameters refer to model of Fig. 2.11: $L_1 = 5.18\,\text{nH}$, $R_1 = 1.1\,\Omega$, $R_2 = 6.85\,\Omega$, $C_1 = 1.8\,\text{nF}$, $C_2 = 1.95\,\text{nF}$

Fig. 2.11 Terminal gate-drain impedance equivalent network corresponding to a MOSFET with split gate effect, as illustrated in Fig. 2.10

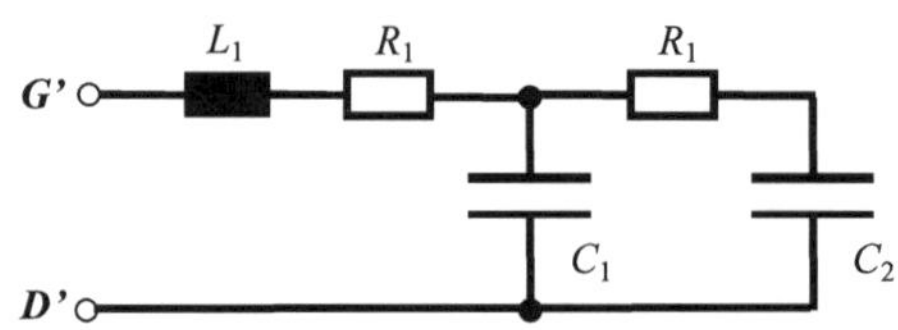

Transmission measurements could, however, result in a rather strong frequency dependence of the real part of the gate-source impedance, which indicates a nonuniform current distribution. This is illustrated in the case example of Fig. 2.10 corresponding to a device with nonoptimized gate mask layout. The frequency-dependent gate resistance of Fig. 2.10b results from the impedance network of Fig. 2.11, the parameters of which are determined to match the measured transmission coefficients of the terminal gate-drain impedance, as Fig. 2.10a shows. The model represents two portions of the device that commute asynchronously in the application. This is the split gate effect, which is undesirable in most applications since it leads to increased switching losses [8].

2.1.2.2 DC Output Characteristics

Static characteristics include channel, forward body diode, and reverse ABD currents. These can be derived from either measurements or device physics simulations. The later may be necessary when the devices are not physically available (e.g., not yet fabricated or prototyped) or the accuracy of the measurement method in certain regions of operation becomes too low.

Regarding experimental characterization, the accurate measurement of the DC output characteristics is particularly a challenging issue because of device self-heating [9]. Dedicated curve tracers have been developed that improve the accuracy of some commercially available solutions. The proposed measurement approach presented in [10] is proven to mitigate self-heating while covering the full dynamic range of operation in both first and third quadrants.

ABD characteristics are separately obtained with a dedicated current pulse driver. The approach can capture ABD levels up to 50 V with pulse widths of less than 2 μs. Since the shape of the ABD is irrelevant for the purpose of this work, only a few points of the characteristics in the low current regime are measured, thereby avoiding the need of stressing the DUT. Note that in the measurements, gate voltage and junction temperature dependencies are taken into account.

Regarding device physics simulations, DC output characteristics are readily obtained from parameter sweep bias point calculations. With proper calibration of the device physics model, the simulations are accurate and representative of real behavior, as Fig. 2.12 illustrates.

Once the DC output characteristics are obtained, model parameters need to be calculated so that the fitting functions proposed in Sect. 2.1.1 match the raw data. In the first quadrant, the curves describing the channel characteristics (i.e., data below avalanche voltage) are used to determine by fitting procedures the parameters of (2.1). Figure 2.13 shows the results of a typical match. In the high v_{DS} range (Fig. 2.13a), the empirical function reproduces effects such as channel length modulation (or short-channel effect) and early avalanche, known from the physics of the device [11]. Furthermore, when fitting experimental raw data, the best agreement in the low v_{DS} range (Fig. 2.13b) is typically obtained when assuming a series

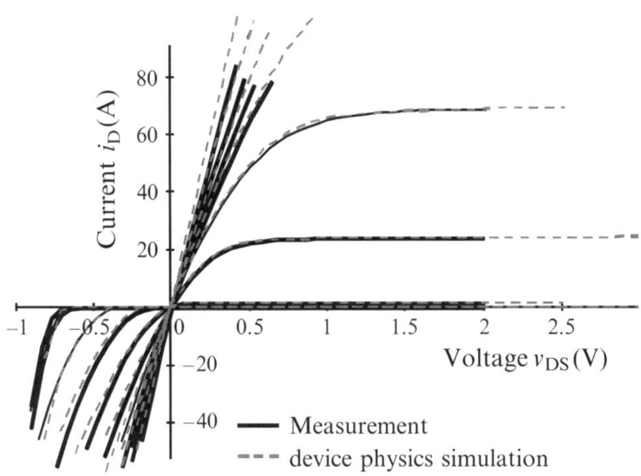

Fig. 2.12 First and third quadrant DC output characteristics ($i_{cn} + i_{dio}$) resulting from device physics simulations and measurements of a trench power MOSFET at 25°C (PHB96NQ03LT from NXP Semiconductors). Package resistance is considered in the comparison

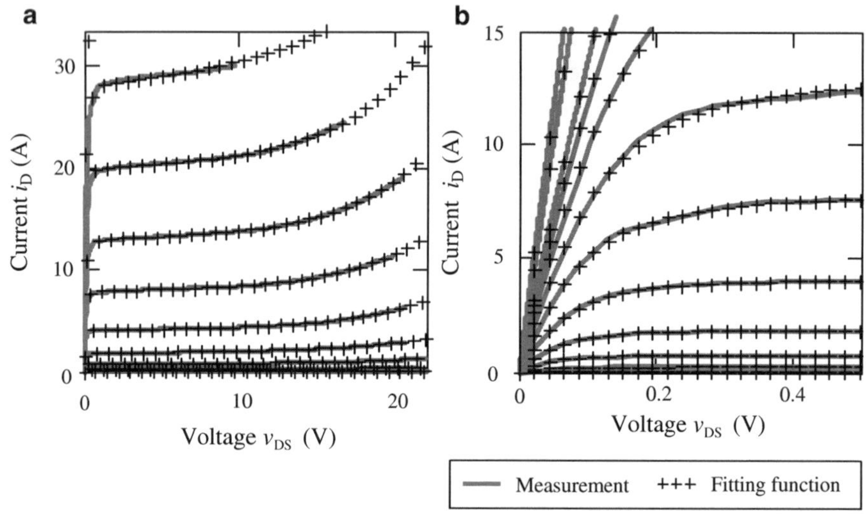

Fig. 2.13 First quadrant output characteristics (channel current) of a commercial device (PH3330 from NXP Semiconductors). at 25°C junction temperature. Comparison between measurements and fitting function (2.1) after calibration of model parameters. (a) Drain-source voltage range 0–20V. (b) Drain-source voltage range 0–0.5V

linear resistor to the channel. This extracted quantity is associated to the substrate resistance (R_{Sub} from Fig. 2.4).

Model parameter adjustments for the third quadrant undergo the de-embedding of the drain current into the channel and body diode contributions. For consistency with the MOSFETs body-effect theory (see Appendix A for details), the body diode characteristics are assumed to be independent on the gate voltage. The criterion to extract the body diode current from the sensed drain current is to assume that the channel current is negligible for gate voltages well below zero. In this operating region, any existing drain current is attributed to minority carriers in the body region. Figure 2.14 shows the result of de-embedding the drain current into the two current components of a simulated trench MOSFET structure. As it can be observed, the channel current can be a significant portion of the total current for subthreshold levels above zero volts. This will have a major impact on the switching dynamics, as it shall be described later.

Similarly to the first quadrant channel current, the parameters of the fitting functions for the third quadrant are calculated to accurately match the corresponding extracted current components. This completes the full static model calibration since the adjustments of the body diode fitting function are basically the same as for the ABD characteristics (see (2.6)).

2.1.2.3 Interelectrode Capacitances

Hard-switching applications particularly require accurate modeling of the nonlinear interelectrode capacitance [12] introduced in Sect. 2.1.1. This is because switching losses depend primordially on the value and shape of these capacitances, which are

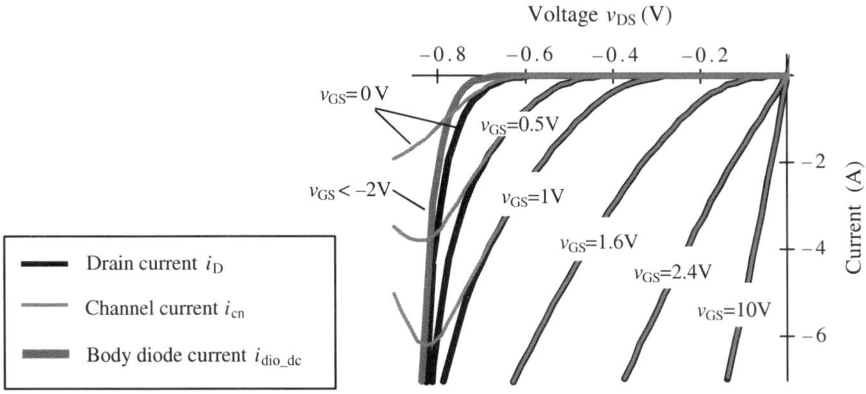

Fig. 2.14 Typical third quadrant output characteristics of high power trench power MOSFETs. Drain current de-embedding into channel and body diode currents. Data derived from 2D device physics simulations of a trench MOSFET cell with scale factor corresponding to a 1 mm^2 active area

generally dependent on both gate and drain voltages. Such dependence makes favorable the identification of operating regions of capacitive influence, thus enabling the possibility of finely defining the critical areas while simplifying others outside the ranges of interest. Important areas of operation are determined from the switching trajectories of the MOSFETs, as illustrated in the state plane diagram of Fig. 2.15. The highlighted areas correspond to the blocking and active regions of the devices. While the former may be experimentally covered by means of small signal impedance analyzers, measurements in the active region essentially involve signal excitation by power sources. Among the various alternatives, the proposed approach measures the current response to large signal voltage ramps, which are generated by means of a linear capacitor and a constant current source as charging pump. In addition, SMT capacitors are employed as high frequency shorts that allow the simplification of the capacitance extraction. These capacitors are connected in close proximity to the DUT so as to mitigate the effect of interconnect and package parasitic inductances. This is important because the ramp slope is limited by the disturbances caused by these parasitic elements. Note that steep ramps are desirable to increase the sensitivity of the measurement and reduce self-heating. Figure 2.16 shows the simplified diagram of the measurement setup, which is valid for both impedance analyzer and voltage ramps. The terminal voltage of the DUT and switches S_i and S_o are controlled by the measurement instrument, which can be automated to cover the entire capacitance range. In order to achieve this, the measurement instrument must be capable of providing not just the excitation signals and measurement means but also the DC bias voltage sources. Two of them are required for small signal measurements, whereas only one may be necessary in combination with the voltage ramp approach.

The measurement instrument is configured such that it allows various contact configurations between the DUT, excitation signal, and bias voltages. The full capacitance range is obtained by means of six different measurements, the configurations

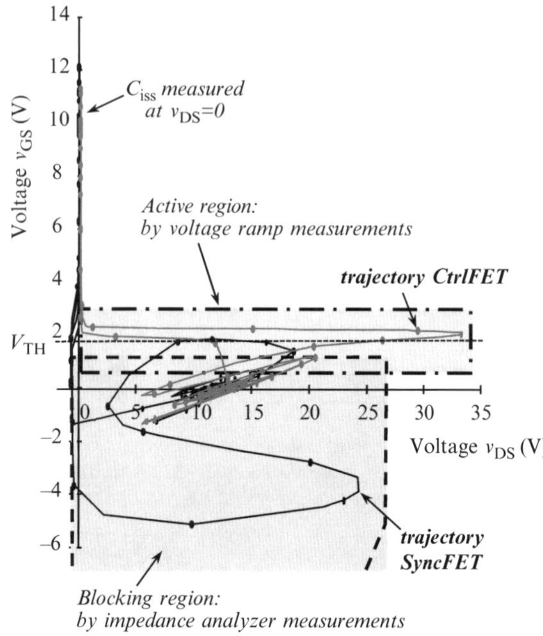

Fig. 2.15 State space diagram of MOSFET voltages. Trajectories result from simulations of typical switching events in a VR application

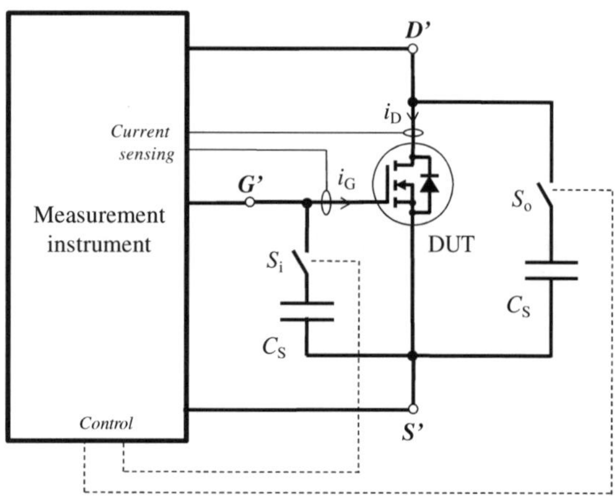

Fig. 2.16 Setup for extraction of MOSFET capacitances. Measurement instrument may comprise an impedance analyzer or a voltage ramp generator with current sensing means

of which are summarized in Table 2.3. Measurements C_a to C_c are carried out with an impedance analyzer and two DC bias voltages for v_{DS} and v_{GS}. The combination of these three measurements allows the extraction of three capacitances. Other measurement combinations are possible. The measurement covers the entire blocking region well below V_{TH} (typically for $v_{GS} < 1$ V). In addition, input capacitance C_{iss} can be determined for the particular case of $v_{DS} = 0$, which is important during ON conduction. Besides the advantage of not requiring high frequency capacitance

Table 2.3 Summary of methods for MOSFET capacitance extraction in both blocking (shaded rows) and active regions of device operation

Reference capacitance	Excitation	Measurement	Switch states		Extracted capacitance
			S_i	S_o	
C_a	Small AC signal G–S	Impedance Z_{GS}	0	0	$C_{GS} + \left(\frac{1}{C_{GD}} + \frac{1}{C_{DS}}\right)^{-1}$
C_b	Small AC signal D–S	Impedance Z_{DS}	0	0	$C_{DS} + \left(\frac{1}{C_{GD}} + \frac{1}{C_{GS}}\right)^{-1}$
C_c	Small AC signal G–D	Impedance Z_{GD}	0	0	$C_{GD} + \left(\frac{1}{C_{GS}} + \frac{1}{C_{DS}}\right)^{-1}$
C_d	Ramp voltage G–S	Gate current i_G	0	1	$C_{iss} = C_{GS} + C_{GD}$
C_e	Ramp voltage D–S	Gate current i_G	1	0	$C_{rss} = C_{GD}$
C_f	Ramp voltage D–S	Drain current i_D	1	0	$C_{oss} = C_{DS} + C_{GD}$

shorts, these three independent measurements relate to the inverse of the capacitance matrix of (2.9) with the following simple equations:

$$C^{-1} = \begin{pmatrix} m_2 & m_0 \\ m_0 & m_1 \end{pmatrix}, \tag{2.19}$$

$$\begin{pmatrix} m_0 \\ m_1 \\ m_2 \end{pmatrix} \equiv \begin{pmatrix} m_0(v_{DS}, v_{GS}) \\ m_1(v_{DS}, v_{GS}) \\ m_2(v_{DS}, v_{GS}) \end{pmatrix} = \begin{pmatrix} -2 & 1 & 1 \\ 0 & 1 & 0 \\ 0 & 0 & 1 \end{pmatrix} \cdot \begin{pmatrix} C_c^{-1}(v_{DS}, v_{GS}) \\ C_b^{-1}(v_{DS}, v_{GS}) \\ C_a^{1}(v_{DS}, v_{GS}) \end{pmatrix}. \tag{2.20}$$

The coefficients in matrix $\boldsymbol{C}^{-1}$ determined from voltage ramps, i.e., measurements C_d to C_f, involve more complex relations, as it can be deduced from (2.9). This may lead to a lower accuracy of the calculated matrix coefficients as a consequence of inherent noise in the raw data. Furthermore, measurement C_f in the active region can only be accurate as long as the channel current is comparatively low with respect to the capacitive current. This can be seen from the following expression for the output capacitance calculation:

$$C_f \equiv C_{oss}(v_{DS}, v_{GS}) = \left.\frac{i_D(v_{DS}, v_{GS}) - i_{cn}(v_{DS}, v_{GS})}{\frac{dv_{DS}}{dt}}\right|_{v_{GS}=\text{const.}} . \tag{2.21}$$

The above expression clearly evidences that any uncertainty in the prediction of the channel current will lead to an error of C_{oss} that increases in proportion to the subtracted magnitude relative to the capacitive current. Therefore, in practice, and due to the limited voltage ramp slope, the maximum gate voltage is typically limited slightly above V_{TH}. On the other hand, the drain voltage must be maintained below the ABD voltage level, and above the body diode conduction for negative values of v_{GS}. Note that the body diode stored charge in forward conduction is assumed to be dominated by minority charges in the quasi-neutral regions of the device. The effects of these charges are analyzed and modeled separately in the next section.

In case of C_{iss} and C_{rss}, the covered ranges are typically wider than for C_{oss} since the measured gate current is not subject to the subtraction of large quantities, as the following equations suggest:

$$C_d \equiv C_{\text{iss}}(v_{\text{DS}}, v_{\text{GS}}) = \left.\frac{i_{\text{G}}(v_{\text{DS}}, v_{\text{GS}})}{\frac{\text{d}v_{\text{GS}}}{\text{d}t}}\right|_{v_{\text{DS}}=\text{const.}}, \tag{2.22}$$

$$C_e \equiv C_{\text{rss}}(v_{\text{DS}}, v_{\text{GS}}) = \left.\frac{i_{\text{G}}(v_{\text{DS}}, v_{\text{GS}})}{\frac{\text{d}v_{\text{DS}}}{\text{d}t}}\right|_{v_{\text{GS}}=\text{const.}}. \tag{2.23}$$

Both measurements C_e and C_f require DC voltage bias for the gate-source, whereas measurement C_d needs it for the drain–source. The ability to keep constant values depends on the ramp slope, the value of C_{S} and the parasitic elements between the DUT and C_{S}.

Figure 2.17 shows an example of extracted reverse and input capacitance of a measured device in both blocking and active regions. Consistent experiments demonstrate that the voltage ramp approach allows extending the capacitance range into areas corresponding to switching trajectories of up to 30 A loads. The voltage ramp measurements partially cover the blocking region in order to compare the results of the two methods. Further data validation is possible by comparing

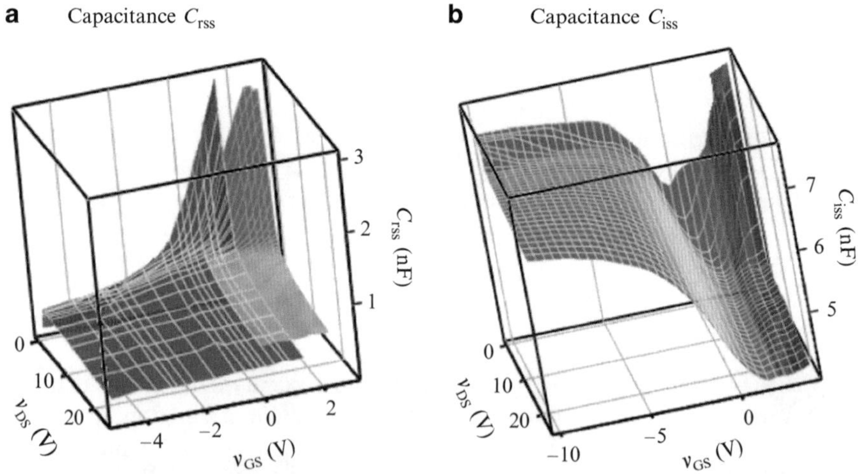

Fig. 2.17 MOSFET capacitance extraction in both blocking and active regions of device operation resulting from combining impedance analyzer and voltage ramps measurement approaches. Data extended into the active region (i.e. gate-source voltage higher than 2V are derived from voltage ramp measurements. Experimental results of device PH3330 from NXP Semiconductors at 25°C (a) Transfer capacitance, (b) input capacitance

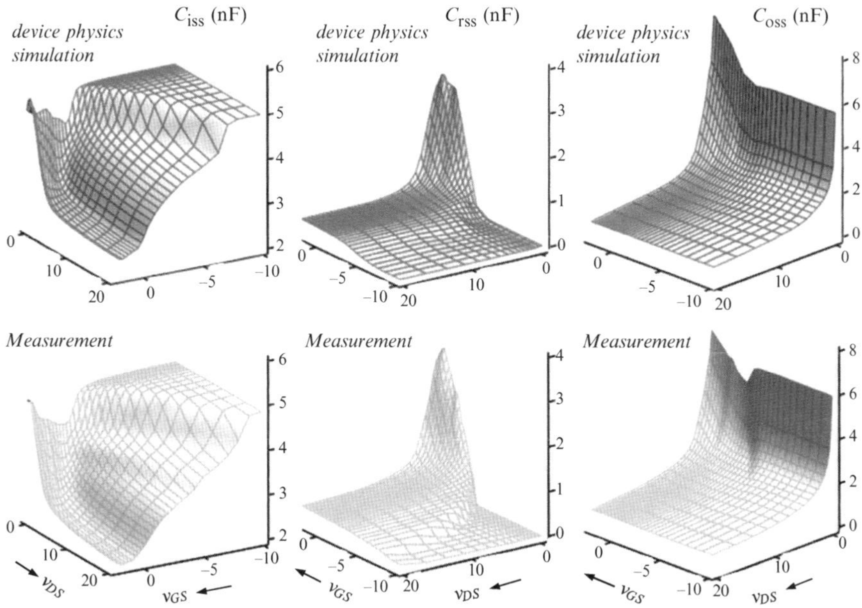

Fig. 2.18 Comparison of capacitance extraction results from device physics simulations and measurements of a trench MOSFET at 25°C

the shape of the capacitance curves with device physics simulations, as shown in Fig. 2.18. In order to extract the capacitance from the device physics model, transient simulations are performed based on the same principles applied for the measurements in the active region. As Fig. 2.18 shows, the results are comparable both in magnitude and shape. Because of this close correlation, device physics simulations can be further employed to extend the data beyond the experimental limits of the measurement approach. This includes the output capacitance in the active region.

2.1.2.4 Body Diode Reverse Recovery

The proposed dynamic lumped diode model defined in (2.7) is suitable for the representation of RR characteristics of low-voltage trench MOSFETs. As it has been shown, the model has a very simple structure, it is easy to implement in a circuit simulator and, as it will be shown in this section, it predicts with virtually no error transient responses of both device physics simulations and measurements.

The model derives from the lumped charge model proposed by Lauritzen and Ma [2], as described in Appendices B and C. It employs parameters β_{rr} and $t_{\beta rr}$, both of which need to be adjusted by empirical methods. To carry out this task, practically all other MOSFET characteristics are demanded in advance. It includes the capacitance characteristics, channel conduction in the third quadrant, DC body diode forward characteristics and package parasitic elements (the latter only in case of experimental characterization). This implies that both β_{rr} and $t_{\beta rr}$ are usually the last parameters of the MOSFET model to be derived. These parameters are adjusted by curve fitting of a reference RR transient, which may be derived from either device physics simulations or measurements. In either case, curve fitting may involve a procedure to de-embed the body diode diffusion current from the total drain current.

RR may be produced by resistive or inductive loads. Both modes are equally valid for the extraction of the model parameters. For convenience, parameter extraction by device physics simulations is illustrated with a resistive recovery setup, whereas inductive RR will be analyzed experimentally.

Figure 2.19 shows the circuit diagram of a resistive RR setup, which can readily be arranged for device physics transient simulations. Note that parasitic elements are ignored to facilitate the characterization. Thus, $v_{DS'} = v_{DS}$ and $v_{GS'} = v_{GS}$.

Simulation examples of the proposed circuit arrangement are shown in Fig. 2.20. The transient curves result from both device physics simulations and SPICE simulations of the behavioral model with properly adjusted RR parameters. In all cases, the body diode is forward biased with the same initial current. The plots in the left column show the impact of the di/dt drain current variation while keeping v_{GS} constant. On the other hand, the plots in the right column illustrate variations of v_{GS} at constant di/dt. Regarding voltage and current waveforms, the simple RR model together with the capacitance and static characteristics agree with the device physics simulations. The variations of di/dt and gate voltage do not imply the modification of the RR parameters. However, the RR current at zero v_{GS} is already lower than at negative v_{GS}. At v_{GS}= 1 V, the body diode current (as it appears from the device terminal) vanishes completely. These effects are studied in detail in Appendix B.

Calculation of RR current from experimental waveforms involves more steps than from device physics simulations, because of transient voltage drops at the package impedances of the real device. Namely, the source inductance may cause a voltage drop affecting v_{GS}, which cannot be avoided since the gate impedance makes impressing this "internal" voltage impossible. This means that the gate current cannot be kept zero and all capacitance characteristics are involved in the experiment. Here, the measured $v_{DS'}$ is applied to the model in order to compare drain and gate currents resulting from measurements and SPICE simulations of the behavioral model. The simulated v_{GS} trace in Fig. 2.21 indicates that, depending on gate voltage fluctuations due to capacitance characteristics and package inductances, measurements can only be performed up to a certain value of $v_{GS'}$ that is far below threshold. Hence, Fig. 2.21 illustrates the limits of experimental RR

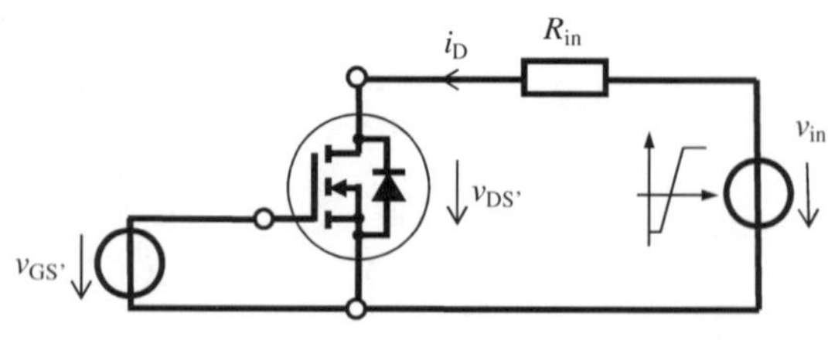

Fig. 2.19 Setup for resistive reverse recovery (RR)

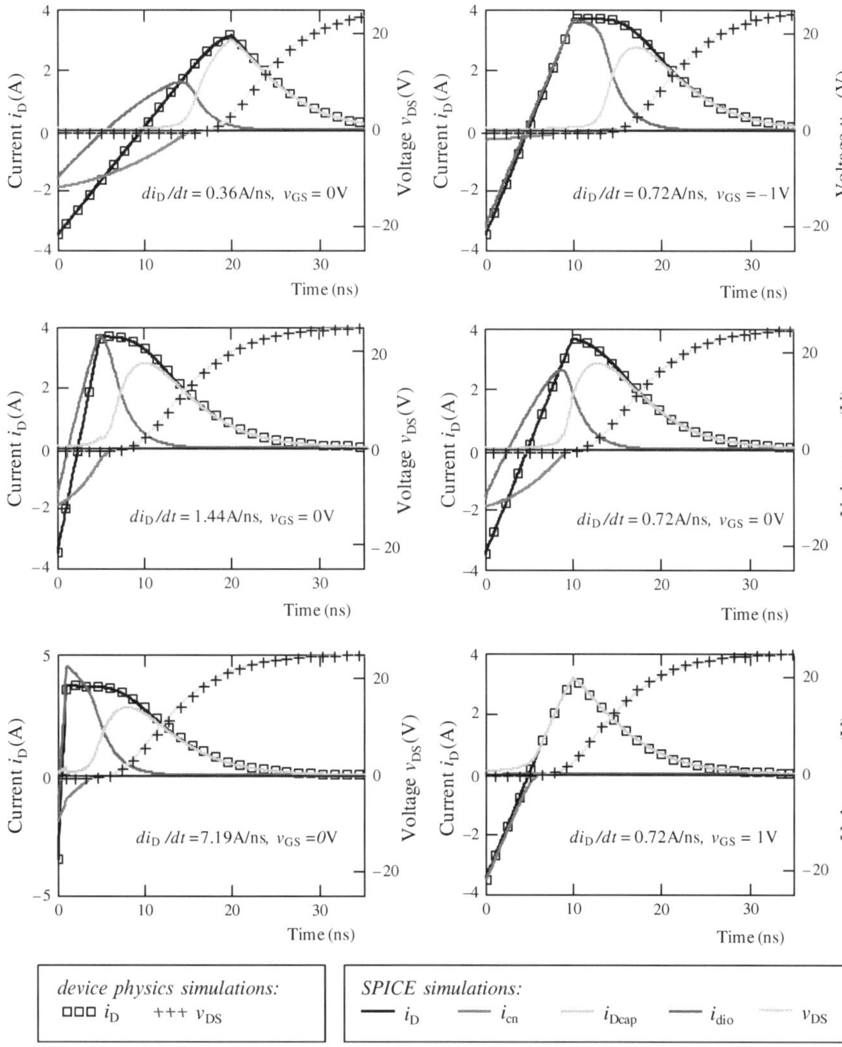

Fig. 2.20 Comparison of resistive RR resulting from device physics simulations and SPICE simulations of the proposed behavioral model. Left column: di/dt variations; right column: v_{GS} variations. Data refers to a 25 V, 5 mΩ logic level n-channel trench MOSFET at 25°C. Extracted model parameters: $t_{\beta rr} = 2$ ns, $\beta_{rr} = 7.8$

characterization. However, within those limits, i.e., at sufficiently low gate voltage $v_{GS'}$, measurements show good agreement with device simulations in terms of RR parameters β_{rr} and $t_{\beta rr}$. Since these parameters as resulting from device simulations show no dependence on gate voltage, measurements at a sufficiently negative $v_{GS'}$ may yield all required data.

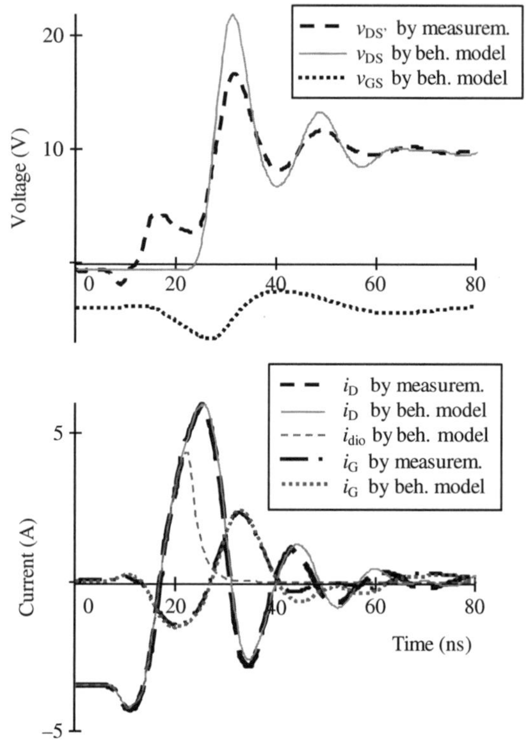

Fig. 2.21 Inductive RR transient. Comparison between measured and SPICE simulations of behavioral model at $v_{GS'} = -4$ V and $di_D/dt = 1$ A/ns at time = 15 ns. MOSFET type: PHB96NQ03LT from NXP Semiconductors. Extracted model parameters: $t_{\beta rr} = 2$ ns, $\beta_{rr} = 8$

2.2 Switched Converter Test Board

MOSFET performance must be evaluated under the conditions established by the application at hand. Particularly, the main focus is on the analysis of the half-bridge of SRBCs for VRs. All the elements of the switched circuit that may influence the behavior of the MOSFETs have to be considered. These include the gate drivers, dead time control, PCB layout parasitic elements, input filter capacitors, and output inductor. In order to investigate the interaction of these elements with the MOSFETs, a number of SRBC test boards are prototyped that experimentally aid the analysis of various power MOSFET discrete solutions. These prototype boards are taken as reference to build a SRBC lumped model for circuit simulators that integrates the proposed MOSFET model of Sect. 2.1. To accomplish this, relevant components comprising the test circuit are individually characterized and modeled according to the analysis demands. Combining them properly in a simulator environment will result in a circuit model setup that enables to meet two goals: First, a thorough validation of the MOSFET model by means of experimental data; and second, a realistic representation of the switched converter for investigations on switching behavior. Further details of the test boards are described in Appendix G with a summary of their main features and specifications.

2.2.1 Gate Driver

The gate circuit and control for driving the power MOSFETs are key elements influencing switching performance. Thus, fine adjustments of their related parameters will be critical in the design phase of the switched converter. Regardless of the circuit topology, important aspects to be considered in the gate drive design are the gate impedance, maximum gate current, ON and OFF state voltage levels, and driving losses. These driver characteristics are considered in the circuit diagram of Fig. 2.22, corresponding to a conventional gate drive topology. It consists of a half-bridge configuration that enables the commutation of the power device by driving its gate between two voltage levels, V_{drvp} and V_{drvn}. For convenience, V_{drvn} usually corresponds to zero volts, whereas V_{drvp} strongly depends on the driving demands and requires to be optimized accordingly. These two voltage levels are provided from what are considered ideal voltage sources. Nonetheless, parasitic inductances are included in the half-bridge, which model the interconnection impedance of these usually external voltage sources to the drive circuit.

Like in most high speed gate drivers, MOSFET devices (M_p and M_n) are the preferred transistor type for the implementation of the half-bridge switches. These drive MOSFETs are driven by buffers connected to the control circuitry. The driving voltages of these buffers (V_{drvpi} and V_{drvni}) are relevant for the estimation of the generated losses [13].

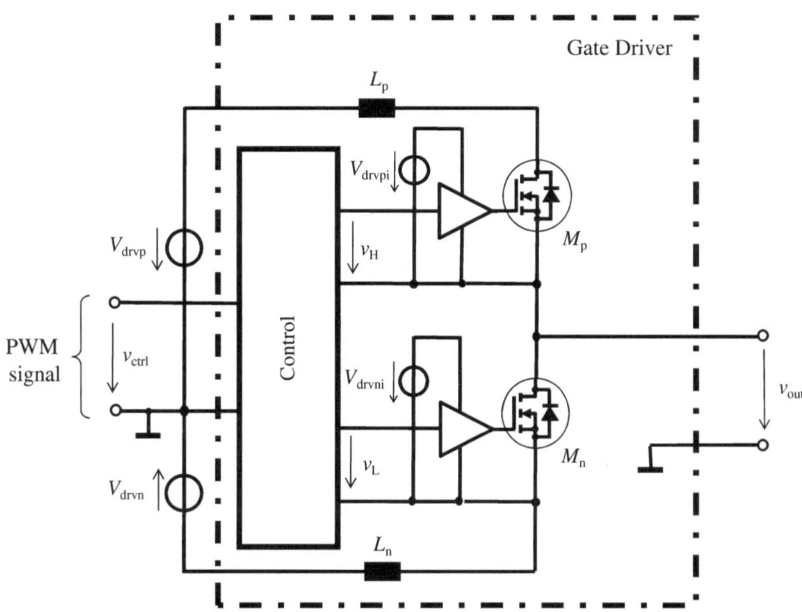

Fig. 2.22 Circuit diagram of MOSFET gate driver

Special emphasis is given to the modeling of M_p and M_n, for they may limit the gate driving current, thus representing a main barrier to the increase of switching frequency. Furthermore, the gate switches can be major sources of losses in the gate circuit. One reason of this is that, even though the gate switches can reach ZCS operation, switching losses may still be significant as consequence of the parasitic input and output charges. The representations of these charges as linear capacitors are frequently sufficiently good approximations to determine the total charge related losses. In addition, the channel of these gate devices has to be modeled so that the behavior in the first quadrant is properly represented with the ohmic, active, and blocking regions.

Similarly as described in Sect. 2.1.1, but to a higher level of simplification, the gate drive circuit implementation of Fig. 2.22 results in the circuit diagram of Fig. 2.23. As it can seen, each driver MOSFET in the half-bridge circuit is modeled with an output capacitor, current-dependent source (i.e., channel current) and a series resistance. The gate of these MOSFETs consists of a series connection of a voltage-dependent source, a gate resistor and a linear capacitor representing the input capacitance of the device. The voltage source is managed by input control signal v_{ctrl} and the gate voltage of its associated device from the half-bridge. The condition for the turn-on is such that shoot-through is avoided, as expressions (2.24) and (2.25) reflect.

Expression (2.26) is a simplified version of (2.1) to represent the DC output characteristics of a MOSFET in the first quadrant. Parameters I_S and R_{on} are defined to conveniently adjust the levels and shape of the curves. Figure 2.24 shows that the characteristics represented by (2.26) realistically reproduce the saturation effects observed in conventional gate drivers. Computing the derivative of (2.26) with respect to v_{DS} yields the dynamic resistance, which depends on the applied current as illustrated in Fig. 2.25.

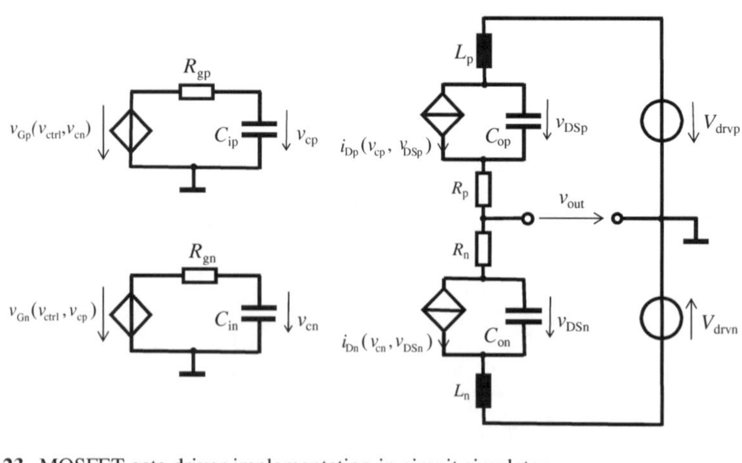

Fig. 2.23 MOSFET gate driver implementation in circuit simulator

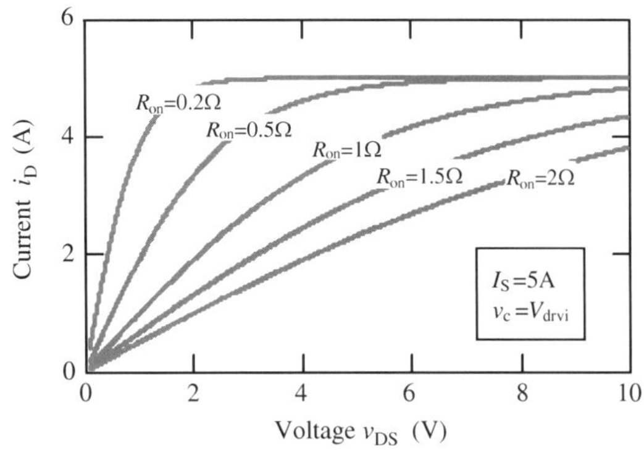

Fig. 2.24 DC static characteristics of driver MOSFETs based on (2.26)

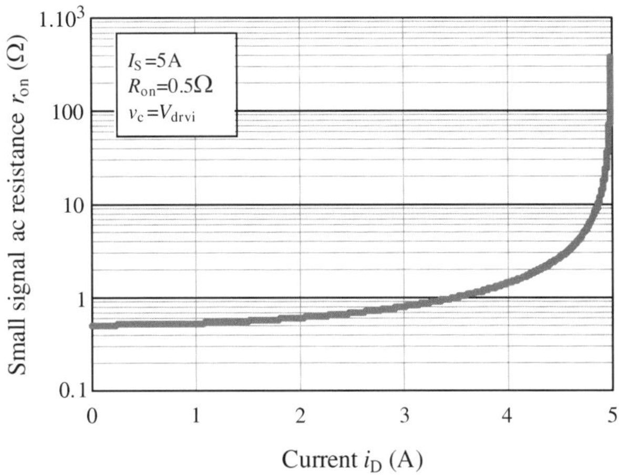

Fig. 2.25 ON state resistance of drive MOSFETs defined as the derivative of (2.26) with respect to v_{DS}

$$v_{Gn}(v_{ctrl}, v_{cp}) = \begin{vmatrix} V_{drvni} & \text{if } (v_{ctrl} = 0 \;\&\; v_{cp} < V_{THp}), \\ 0 & \text{otherwise}, \end{vmatrix} \tag{2.24}$$

$$v_{Gp}(v_{ctrl}, v_{cn}) = \begin{vmatrix} V_{drvpi} & \text{if } (v_{ctrl} = 1 \;\&\; v_{cn} < V_{THn}), \\ 0 & \text{otherwise}, \end{vmatrix} \tag{2.25}$$

$$i_D(v_c, v_{DS}) = I_S \cdot \frac{v_c}{V_{drvi}} \tanh\left(\frac{v_{DS}}{I_S \cdot R_{on}}\right). \tag{2.26}$$

Linear input and output capacitance values for M_p and M_n are calculated based on the following average expression:

$$C_{av} = \frac{1}{|V_f - V_i|} \int_{V_i}^{V_f} C_x(v)\, dv. \tag{2.27}$$

Thus, $C_{av} \cdot |V_f - V_i|$ corresponds to the total charge stored in or removed from nonlinear capacitor C_x. The value of C_{av}, like the rest of parameters of the driver model, can be usually derived from the datasheet of the employed driver, since it contains all required information referent to the characteristics of the gate switches. The datasheet information may be complemented with specific measurement tests to extract additional driver characteristics such as idle losses, as described in [14].

The proposed model, as such, allows the study of interactions between the elements of the driver and power MOSFET in the circuit simulator, as it shall be described later in this chapter.

2.2.2 Input/Output Filters

Of key relevance to the switching performance of the MOSFETs are the input filter capacitors, mainly by virtue of their contributions to the parasitic inductances of the half-bridge loop. The input filter of the test boards (see Appendix G) is used as illustration in the following discussion. Such a filter comprises a capacitor bank of three different types of parallel decoupling capacitors; the OSCON is the one type of capacitors used to supply most of the DC energy content during the ON state period of the CtrlFET. The other two types are ceramic capacitors of various dimensions featuring low ESR and ESL to provide high frequency decoupling capabilities. The equivalent circuit of the impedance network seen at the terminals of the filter is represented in Fig. 2.26. The parameters of the network are determined from curve fitting of the frequency response measured with a VNA. Figure 2.27 compares the results of the measured and simulated transmission coefficient from the scattering matrix. The resonances at 4.5 and 16 MHz correspond to SMD ceramic capacitors, which effectively contribute to significantly lower the impedance well above the resonance of the OSCON capacitors (i.e., ≈100 kHz).

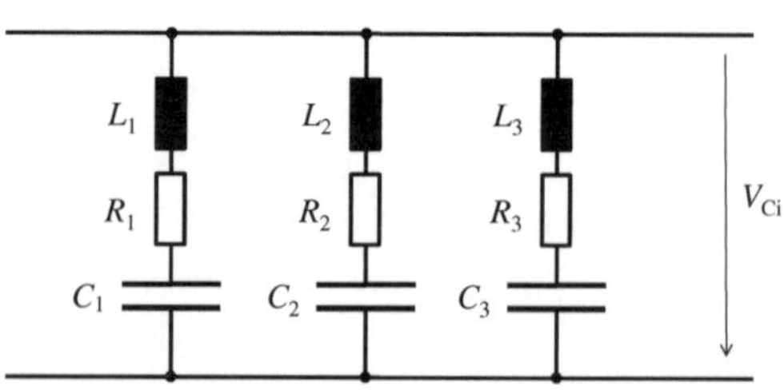

Fig. 2.26 Equivalent input filter capacitor representation of test board (see Appendix G)

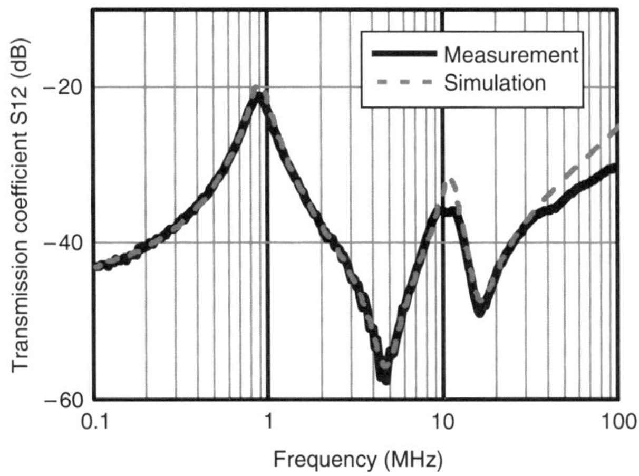

Fig. 2.27 Frequency response of input filter capacitor. Measurement response derives from VNA. Simulations correspond to circuit of Fig. 2.26. Values: $L_1 = 150$ pH, $L_2 = 217$ pH, $L_3 = 4.9$ nH, $R_1 = 4.5$ mΩ, $R_2 = 1.6$ mΩ, $R_3 = 6.5$ mΩ, $C_1 = 620$ nF, $C_2 = 5{,}243$ nF, $C_3 = 8{,}400$ nF. Capacitances C_2 and C_3 are derived from impedance analyzer measurements due to the limited frequency range of the VNA

The output filter can be highly simplified for the study of MOSFET performance. The most important simplification is the replacement of the output filter capacitors by an ideal voltage source, thereby neglecting any influence that the output ripple may have on MOSFET switching. The current ripple, however, must be taken into account for the estimation of both ON conduction and switching losses. For this it is sufficient to model the output choke as an ideal inductance. A series resistor may be simply added to cope with the resistance of the track, contacts, and winding. A more accurate model of the output inductor will be considered latter for the estimations of the overall converter losses.

2.2.3 PCB Layout Impedance Characterization

One of the main concerns of hard-switching SRBCs is the energy loss produced by the parasitic inductances in the half-bridge. It is therefore essential to determine the magnitude of such parasitic elements. The main purpose of this section is to obtain an equivalent lumped network of the half-bridge PCB layout from the test boards. The main difficulty of this task is to discern the impedance contribution of the track portions constituting the half-bridge circuit layout. This is accomplished by the time domain measurement method described in [15]. The technique basically consists of generating high amplitude oscillations at resonant frequencies close to those excited in the application. To achieve this, the power MOSFETs are replaced by accurately characterized linear capacitors of low ESR and ESL. A high power driver is then used to inject pulses from the side of the input filter capacitor, thereby generating oscillations between the parasitic layout inductance and the reference capacitors. Track inductances can therefore be determined by measuring sections of the PCB with differential probes in such a way that disturbances are minimized (e.g., reduction of coupling effects in the sensing probes). The estimated inductances derive from fitting procedures of the measured signals. The resulting PCB electrical model consists of a series of equivalent lumped inductors related to certain areas of the layout. Figure 2.28 illustrates the results of a characterized test board, where portions of the half-bridge and gate circuit are represented. Reference voltages $v_{drv(s)}$, $v_{drv(c)}$, and $v_{DSx(c)}$ are probing points for switching waveform measurements, whereas V_{Ci} corresponds to the input filter capacitor voltage, as indicated in Fig. 2.26. Resistance R_{sh} is incorporated for current sensing means. Note that when measuring the voltage across this shunt resistor, the influence of its ESL (i.e., inductor L_{sh}) must be taken into account for determining the current waveform. Further details of the PCB layout design are provided in Appendix G.

The derived circuit and MOSFET models can be experimentally validated by measuring V_{Ci}, $v_{drv(s)}$, $v_{drv(c)}$, v_{sh}, and $v_{DSx(s)}$. The resulting waveforms from V_{Ci}, $v_{drv(s)}$, and $v_{drv(c)}$ are then used in the simulator as time-dependent voltage excitation

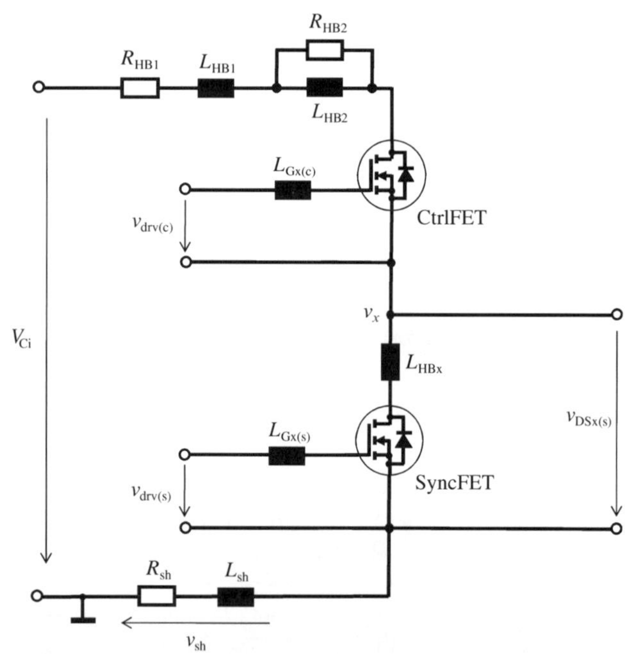

Fig. 2.28 Lumped model representation of half-bridge and gate drive layout parasitic elements from switching test board (see Appendix G, Sect. G.1). Package parasitic elements of MOSFETs are not represented. Extracted values: $R_{HB1} = 3.5$ mΩ, $L_{HB1} = 1.5$ nH, $R_{HB2} = 75$ m, $L_{HB2} = 500$ pH, $L_{Gx1} = 250$ pH, $L_{Gx2} = 250$ pH, $L_{HBx} = 600$ pH, $L_{sh} = 170$ pH, $R_{sh} = 100$ mΩ

sources. Thus, predictions of v_{sh} and $v_{DSx(s)}$ are produced and compared to the corresponding measured data in order to assess the accuracy of the models. Such validation procedure will be discussed in more detail in Sect. 2.4.

2.3 Switched Converter Simulation Setup

This section presents the final circuit representation of the SRBC based on the previously introduced models of the power MOSFETs, PCB layout, gate drivers, and input/output filters. Control circuitry is additionally included to allow automatic adjustments of the dead times as well as regulation of steady-state conditions that effectively enhance simulation performance.

Since the main purpose of the converter model is to investigate the performance of the power MOSFETs in the application, special emphasis is given to properly represent the elements of the circuit influencing the behavior of the power devices, such as the half-bridge impedance, gate drivers, and output choke. The rest of the elements from the switched circuit are simplified or completely neglected so as to improve tractability and explicitness. Regarding control aspects, disturbances such as the load transient and other similar dynamic events are not taken into consideration. Namely, the analysis is devoted to the nominal operating conditions, which involve periodic steady-state waveforms.

Figure 2.29 shows the proposed simulation setup of the SRBC. The block diagrams of the input filter capacitors and half-bridge PCB layout contain the models of Sects. 2.2.2 and 2.2.3, respectively. The MOSFET subcircuit model is based on the diagram of Fig. 2.4.

The gate driver and control system of Fig. 2.29 is depicted in more detail in the diagram of Fig. 2.30. Two main blocks are identified. The MOSFET driver control regulates the state of the half-bridge switches according to a PWM input signal. Furthermore, the dead times of the switched node leading and falling edges (FEs) can be programmed to predefined values. The controller thus keeps the body diode conduction times of the SyncFET constant and independent on the circuit conditions. The operation is as follows: the simulation starts in the first switching cycle with an initial guess for the delay times between switching actions of the gate signals. In the same switching cycle the controller monitors the internal drain– source voltage of the SyncFET (i.e., $v_{DS(s)}$ of Fig. 2.4) to determine the body diode conduction times, (e.g., time during which $v_{DS(s)}$ less than -0.6 V). In the next switching cycle the delay time is readjusted to correct for the difference between the programmed and sensed body diode conduction times. This is done individually for both the leading and FEs of the PWM signal. Thus, each transition edge can be independently adjusted with different dead time values.

The control signal outputs are connected to buffers that drive the power MOSFETs. The implementation of these gate buffers is as described in Sect. 2.2.1. Additionally, $L_{drv(c)}$, $L_{drv(s)}$, $R_{drv(c)}$, and $R_{drv(s)}$ may represent PCB layout

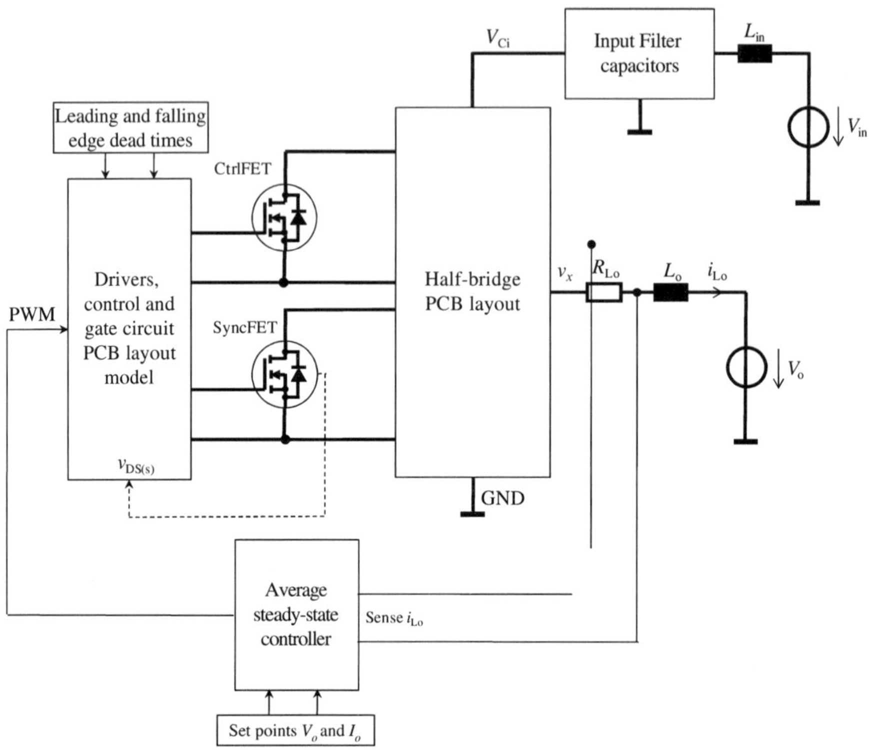

Fig. 2.29 Simulation setup diagram of the SRBC

parasitic elements or external components added for current sensing and other test purposes.

The main purpose of the average steady-state controller is to provide a PWM signal that leads to the specified steady-state conditions in the least number of switching cycles, thereby minimizing simulation times. The steady-state conditions occur when the following two conditions are met,

$$\frac{1}{T_s}\int_{t_x}^{t_x+T_s} i_{Lo}\cdot dt = I_{Lo} = I_o, \tag{2.28}$$

$$\frac{1}{T_s}\int_{t_x}^{t_x+T_s} v_{Lo}\cdot dt = V_{Lo} = 0, \tag{2.29}$$

where v_{Lo} is the voltage across output inductor L_o and I_o is the load current. Two control variables are required to satisfy the above two conditions. On the one hand, the average current through the output choke is regulated by the switching period of the PWM signal, which must equal to the target value once steady-state has been reached. On the other hand, the average voltage across the inductor is controlled by means of the duty cycle ON time, or in other words, CtrlFET ON time.

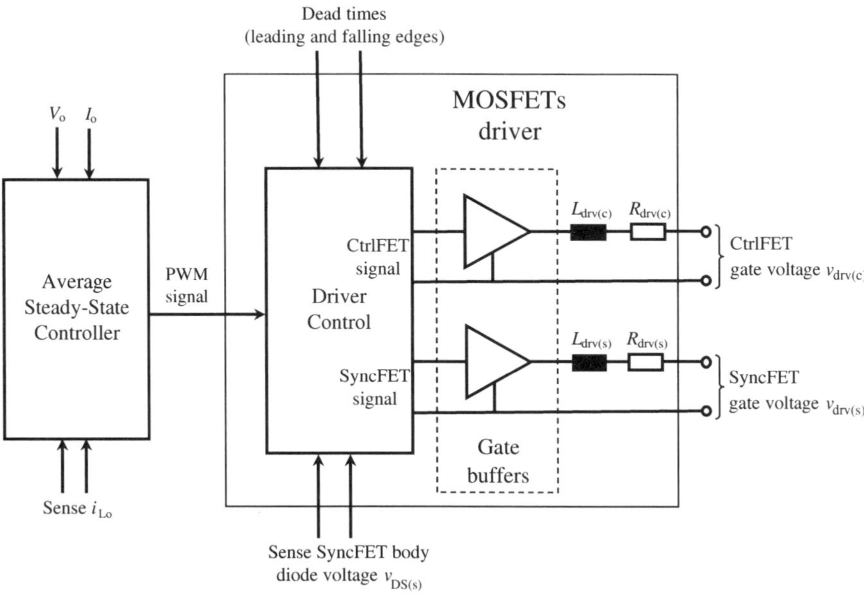

Fig. 2.30 Block diagram of driver and average steady-state controller. Implementation in a circuit simulator for transient analysis

The operation of the average steady-state controller is illustrated in Fig. 2.31. The simulation starts at the beginning of the switching period, which is defined as the leading edge (LE) of the PWM signal. The initial guess for the output coil current ($i_{Lo0(G)}$) at that time is as follows:

$$i_{Lo0(G)} = I_o - \frac{I_{o(pp)}}{2} + \frac{V_o + I_o \cdot R_{sp}}{L_{sp}} .t_{dCr}, \tag{2.30}$$

where $I_{o(pp)}$ is the peak-to-peak output current ripple, R_{sp} and L_{sp} are the total resistance and inductance of the SyncFET and output paths, respectively (i.e., including all parasitic components). Time t_{dCr} is the delay existing between the PWM and the CtrlFET signal at the LE transition. The initial guess for the CtrlFET ON time is as follows:

$$t_{on(G)} = \frac{V_0}{V_{in}} \cdot T_s + t_{dCr} - t_{dCf}. \tag{2.31}$$

Time t_{dCf} is the delay time between the PWM and the CtrlFET signal at the FE transition.

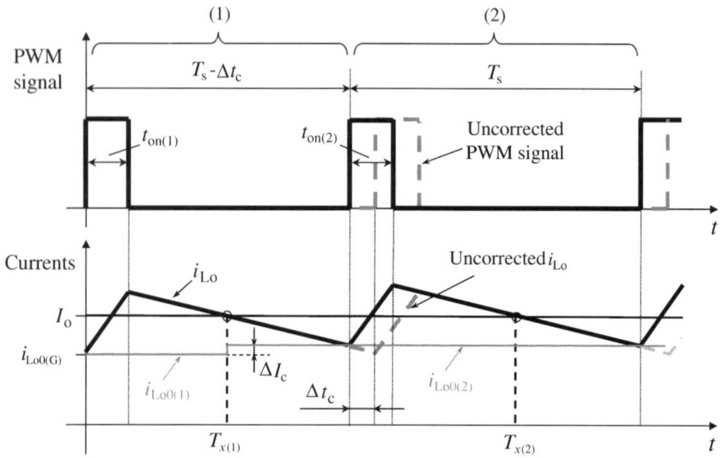

Fig. 2.31 Example of operation of the average steady-state controller

The controller computes the average values of i_{Lo} and v_{Lo} from the actual switching cycle and estimates both CtrlFET ON time and minimum output inductor current for the next period as follows:

$$t_{on(2)} = t_{on(1)} - \frac{V_{Lo(1)}}{V_{in}} \cdot T_s, \tag{2.32}$$

$$i_{Lo0(2)} = i_{Lo0(1)} + I_o + \frac{V_{Lo(1)}}{L_o} \cdot T_s - I_{Lo(1)}. \tag{2.33}$$

Note that subindexes (1) and (2) refer to the past and present switching cycles, respectively. Conditions $t_{on(2)} = t_{on(1)}$ and $i_{Lo0(2)} = i_{Lo0(1)}$ are given when steady state is reached, that is to say, when (2.28) and (2.29) are both satisfied.

Equations (2.32) and (2.33) are calculated by the controller at particular time instants corresponding to those at which the inductor current equals the load current during the SyncFET ON times. This is indicated in Fig. 2.31 with times $T_{x(1)}$ and $T_{x(2)}$. Thus, the averaged state variables of the output coil at the actual switching cycle can be calculated as follows:

$$I_{Lo(1)} = \frac{1}{T_s} \cdot \left(\int_0^{T_{x(1)}} i_{Lo}(t) \cdot dt + \int_{T_{x(1)}}^{T_s} i_{Lo}(t) \cdot dt \right). \tag{2.34}$$

The first integral in (2.34) is numerically calculated during the simulation run. The calculation makes use of the simulated current waveform data that are available until that time instant. The second integral corresponds to a section of the ON time of the synchronous MOSFET where no switching events occur. It is therefore possible to calculate it analytically as follows:

$$\int_{T_{x(1)}}^{T_s} i_{\mathrm{Lo}}(t)\cdot \mathrm{d}t = \frac{V_o}{R_{\mathrm{sp}}}(T_s - T_{x(1)}) - \frac{L_{\mathrm{sp}}}{R_{\mathrm{sp}}}\left(I_o + \frac{V_o}{R_{\mathrm{sp}}}\right)\left(\mathrm{e}^{-(R_{\mathrm{sp}}/L_{\mathrm{sp}})\cdot(T_s - T_{x(1)})} - 1\right). \quad (2.35)$$

Equation (2.35) defines the integral current through a series RL circuit with constant voltage source V_o and initial current condition I_o. Similarly for $V_{\mathrm{Lo}(1)}$,

$$V_{\mathrm{Lo}(1)} = \frac{1}{T_s}\cdot\left(\int_0^{T_{x(1)}} v_{\mathrm{Lo}}(t)\cdot \mathrm{d}t + \int_{T_{x(1)}}^{T_s} v_{\mathrm{Lo}}(t)\cdot \mathrm{d}t\right), \quad (2.36)$$

$$\int_{T_{x(1)}}^{T_s} v_{\mathrm{Lo}}(t)\cdot \mathrm{d}t = L_o\left(I_o + \frac{V_o}{R_{\mathrm{sp}}}\right)\left(\mathrm{e}^{-(R_{\mathrm{sp}}/L_{\mathrm{sp}})\cdot(T_s - T_{x(1)})} - 1\right). \quad (2.37)$$

In the illustration of Fig. 2.31 it is assumed that the average output coil current in the first cycle is lower than I_o, whereas $V_{\mathrm{Lo}(1)}$ is zero. At time $T_{x(1)}$, current ΔI_c, corresponding to the difference between I_o and the average current through L_o, is added to $i_{\mathrm{Lo}0(1)}$ as correction for the next cycle. The next cycle starts as soon as i_{Lo} reaches the new $i_{\mathrm{Lo}0(2)}$ value. Note that in order to start the next cycle with a higher current, the period of the first cycle must be lower than T_s. The controller reaches its target in the second switching cycle since ΔI_c is zero and thus the switching period must equal T_s.

2.4 Model Validation

This section evaluates the accuracy of the proposed MOSFET model by comparing switching waveform and power loss simulations with experiments performed on a synchronous buck converter test board. The prototype circuit is specially designed to carry out studies on switching behavior of MOSFETs mounted on LFPak or other compatible discrete packages (for details of the test board, see Appendix G). A number of different test conditions are rigorously examined, which involve variations of the load current, input voltage, and driver parameters such as the ON state voltage and dead times.

Figure 2.32 represents the equivalent circuit of the test board, the parameters of which derive from the characterization techniques described in the previous sections of this chapter. The input filter and drivers are, however, not represented in the circuit model as described in Sects. 2.2.2 and 2.2.3. Instead, some of the measured signals are employed in the simulation as excitation voltage sources, which represent the behavior at the terminals of the drivers and input filter. Thus, the simulation setup and validation process of the MOSFET model is considerably simplified.

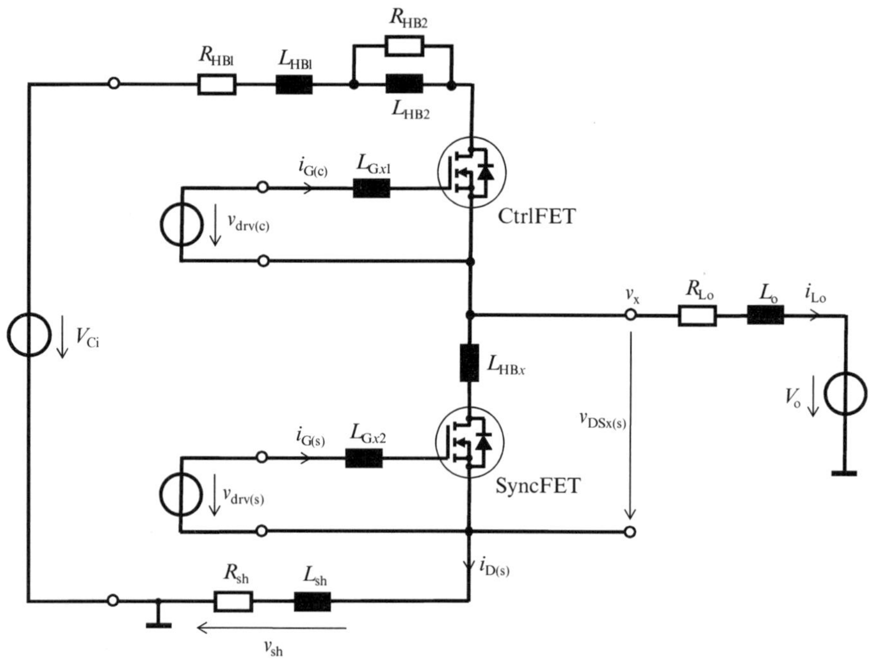

Fig. 2.32 Circuit model of a synchronous buck converter prototype for the validation of the proposed power MOSFET model. Extracted values: $R_{Lo} = 4.3\ m\Omega$ (total inductor + PCB), $L_o = 1$ μH. See Fig. 2.28 for the rest of the component values

In this arrangement, simulation results of the switch node voltage, gate currents, SyncFET drain current, and output inductor current waveforms are compared to their corresponding experimental waveforms.

Most of the measurements are performed with a digital oscilloscope and a number of high bandwidth differential probes. Steady-state switching signals are sensed corresponding to gate drive voltages $v_{drv(c)}$ and $v_{drv(s)}$, input and output voltages V_{Ci} and V_o, and switched node voltage v_x. Additionally, gate currents $i_{G(c)}$ and $i_{G(s)}$ as well as drain current $i_{D(s)}$ are monitored by voltage measurements across shunt resistors in the respective gates and half-bridge paths. Particularly, the SyncFET drain current shunt is modeled by means of resistor R_{sh} and parasitic inductor L_{sh}. The latter represents the ESL of the shunt device. Since the induced voltage across such inductance may be substantial during the switching times, its contribution must be subtracted from the measured voltage across the shunt in order to determine the correct magnitude of the current.

On the other hand, measuring output inductor current i_{Lo} may be alternatively realized with the use of transformer current probes since the relatively high inductive path of the output inductor allows the inclusion of clip-and-probe type of sensors. In these experiments, a 500 MHz DC current probe is conveniently clamped around the leads of the output coil to measure i_{Lo}.

Part of the validation process consists in evaluating the ability of the model to predict the switching performance of two commercial low-voltage power MOSFETs (SyncFET PH3330L and CtrlFET PH8030L from NXP semiconductors). The sample parts used in the experiments were previously characterized and modeled according to the methods presented in Sect. 2.1.

The prototype circuit is operated at 1 MHz switching frequency, 12 V input voltage and 1.5 V output voltage, unless otherwise specified. The ambient temperature operation is ~30°C and uniform throughout the PCB and components thereon. This temperature level corresponds to that of the MOSFETs characterization conditions, which is maintained throughout the experiments by operating the converter only for a limited number of switching periods necessary to perform noise-reduced-by-averaging readings in steady-state operation.

Figures 2.33 and 2.34 attest the accuracy of the model to precisely reproduce the switched node voltage and SyncFET drain current for a 10 A load current and 7 V input voltage. The model predicts the high ringing caused by the influence of the large half-bridge loop inductance, which is over 3.5 nH. The 100 mΩ shunt resistor for the SyncFET drain current sense contributes significantly to the rapid damping of such oscillation. The simulated output inductor current also matches the experimental data, which presents low ripple, thereby simplifying the loss calculations of the inductor losses later on.

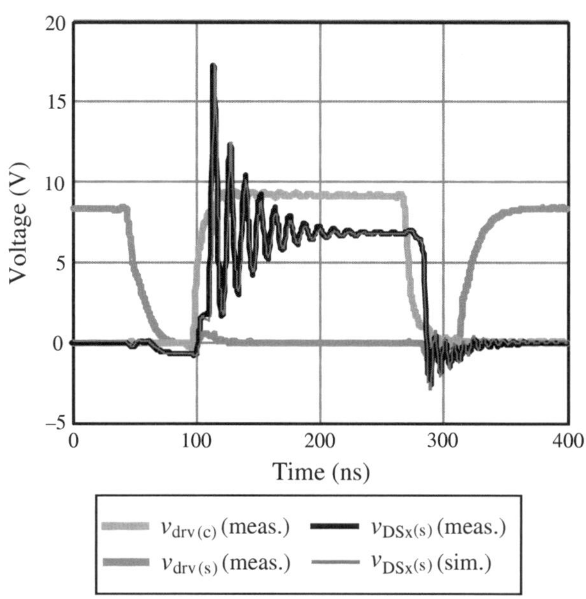

Fig. 2.33 Measurements and simulations of converter voltage waveforms. Switch node leading and falling edge (FE) transitions

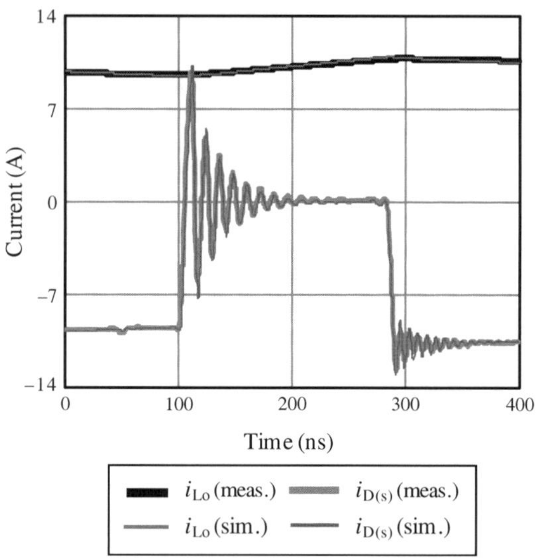

Fig. 2.34 Measurements and simulations of converter current waveforms. Switch node leading and FE transitions

A more detailed assessment of the model performance in terms of switching transient predictions is provided in Figs. 2.35–2.39. In Fig. 2.35, switch node LE transients are evaluated for various dead times. The strong impact of this driver parameter observed on the measured waveforms is precisely predicted by the model, even in cases of fast switching speeds leading to current commutations around 3.6 A/ns and RR peaks of 8 A. This gives strong credit to the basis of the model approach established by the superposition of responses from individually represented device characteristics.

Further proofs of the correct applicability of this principle is provided in Fig. 2.36, which shows matching waveforms of unsought switching effects like shoot-through and ABD.

Additional comparisons of the switched node LE transition are given in Fig. 2.37, from where it can be acknowledged that measured responses to variations of the load current as well as CtrlFET gate voltage are consistently reproduced in the simulator.

Similar performance predictions are deduced from evaluations of switch node FE transients. Figures 2.38 and 2.39 show how simulation and experimental waveforms match up well for a number of test conditions, which include variations of the CtrlFET gate voltage as well as the input voltage.

A detail description of the observed phenomena will be presented in the next sections based on simulation results of the converter operation.

The proposed model can also be harnessed for the purpose of quantifying and analyzing power losses. Loss predictions are assessed by means of carrying out power measurements of the test circuit. In this study, and in order to avoid excessive

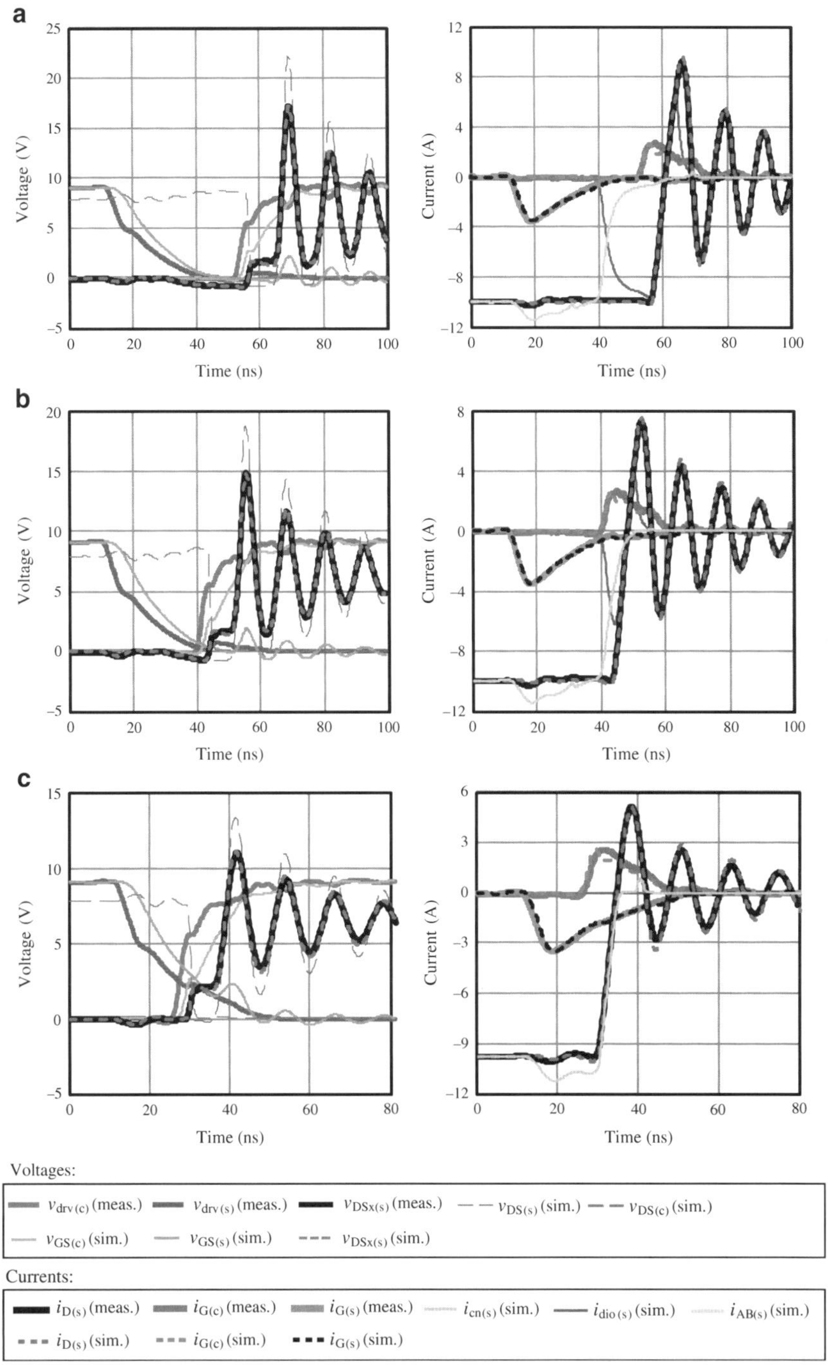

Fig. 2.35 Measurement and simulation waveforms of a SRBC test board. Switch node leading edge (LE) transient. Variation of dead time: (**a**) long dead time with RR, (**b**) moderate dead time with reduced RR, (**c**) short dead time with no RR

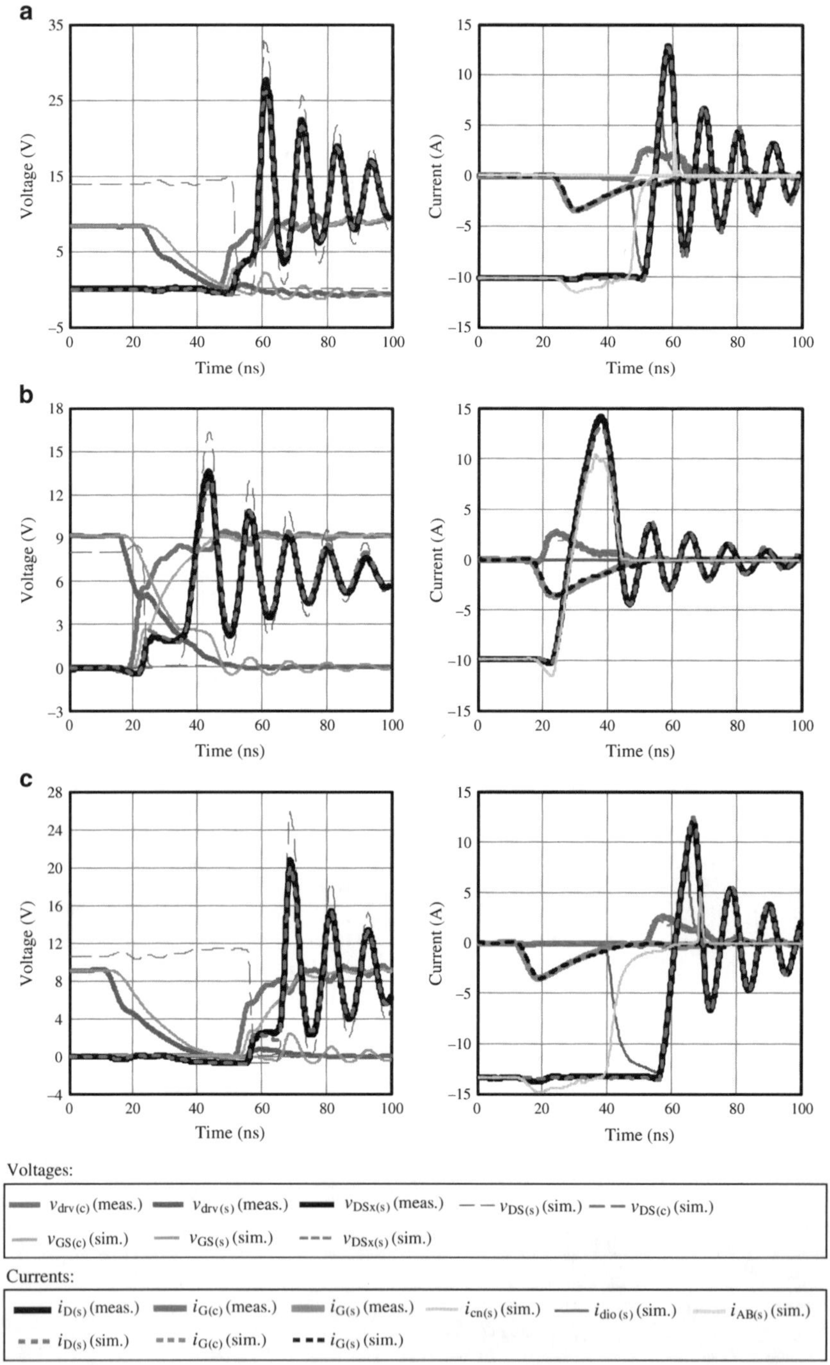

Fig. 2.36 Measurement and simulation waveforms of a SRBC test board. Switch node LE transient. (**a**) Avalanche breakdown (ABD); (**b**) Shoot-through due to short dead time; (**c**) SyncFET gate bounce (GB) shoot-through

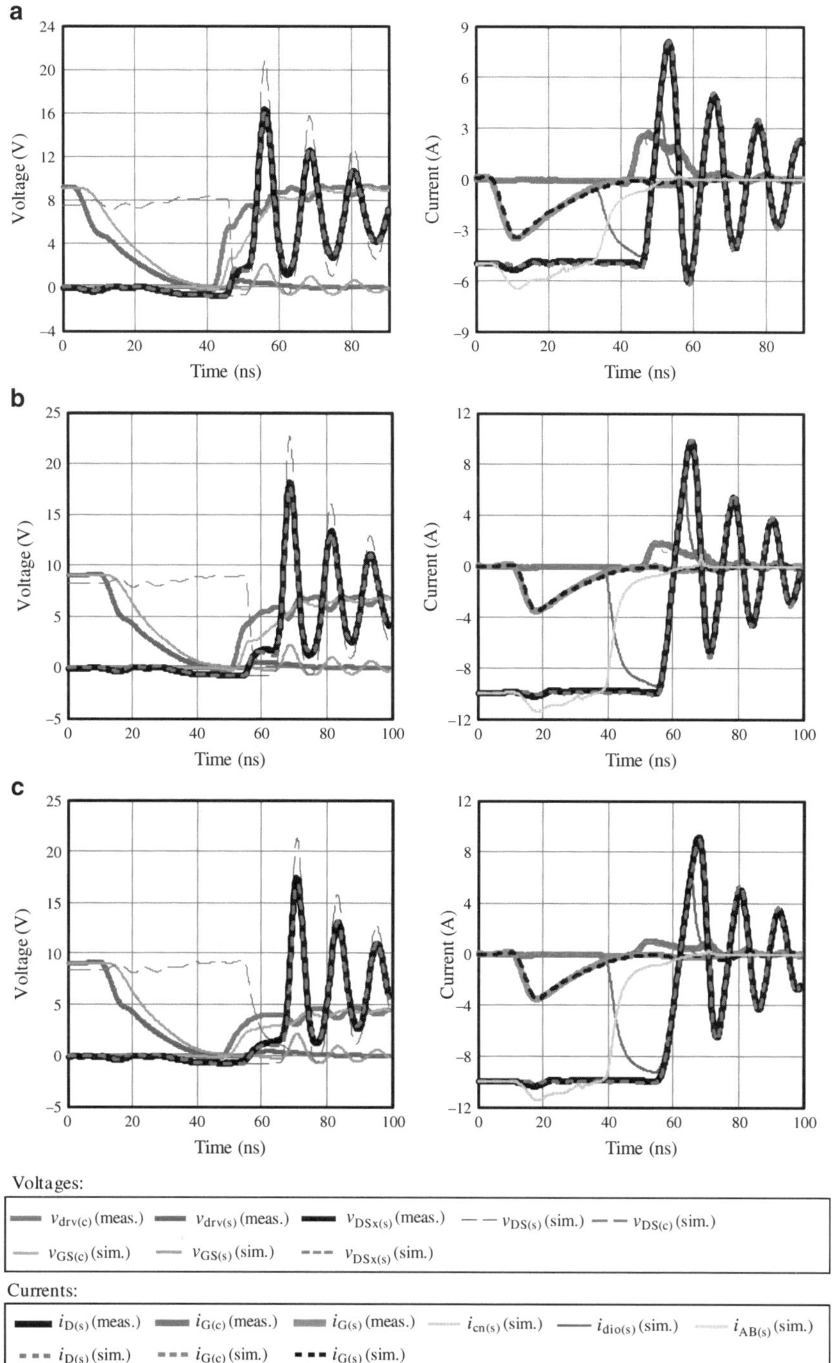

Fig. 2.37 Measurement and simulation waveforms of a SRBC test board. Switch node LE transient. (**a**) Moderate load current; (**b**, **c**) variation of CtrlFET gate voltage

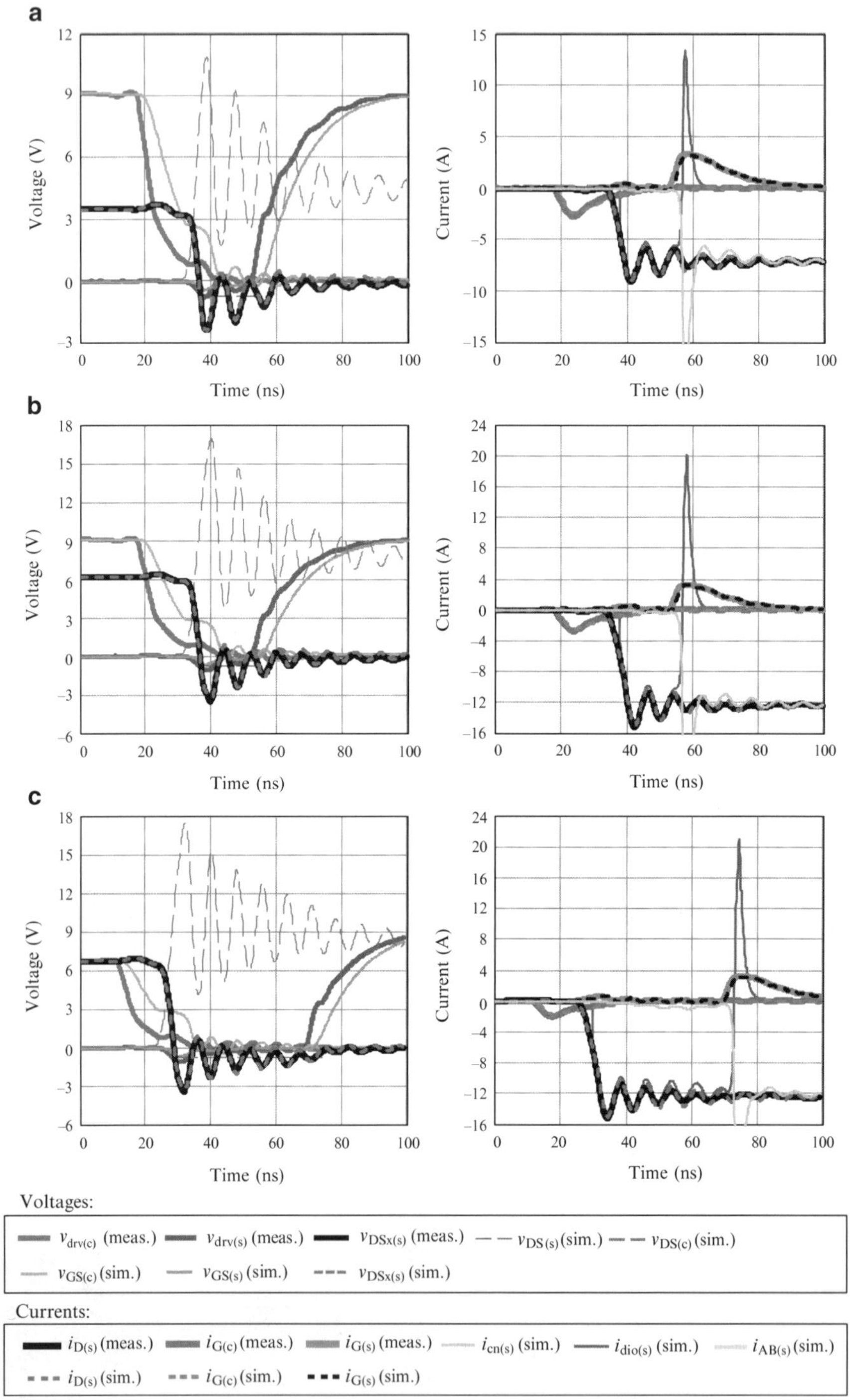

Fig. 2.38 Measurement and simulation waveforms of a SRBC test board. Switch node FE transient. (**a**, **b**) Variation of input voltage; (**b**, **c**) variation of gate voltage

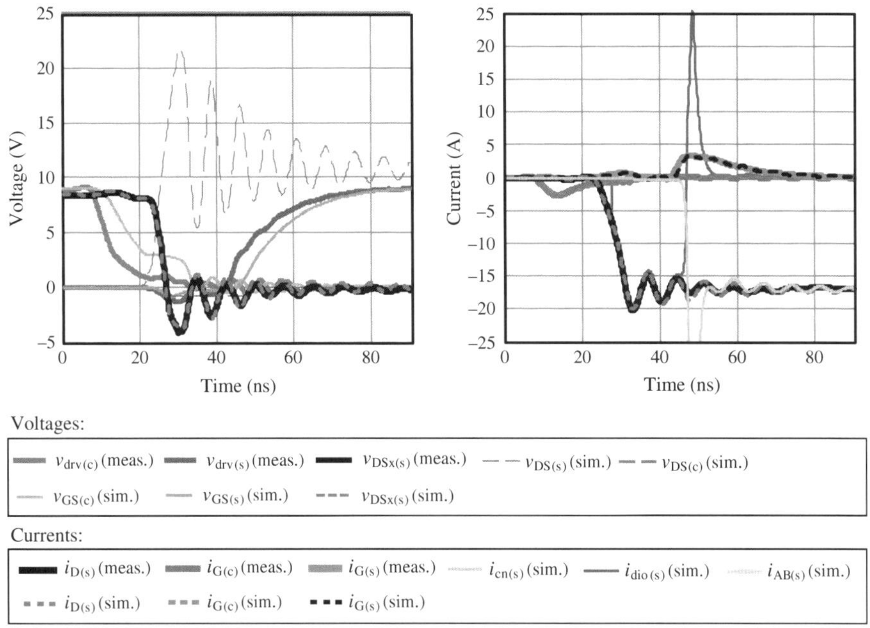

Fig. 2.39 Measurement and simulation waveforms of a SRBC test board. Switch node FE transient. Example of fast switching turn-off

losses, the shunt resistors (i.e., R_{sh} from Fig. 2.28) are removed from the circuit. The remaining resistance in the shunt path at 1 MHz is estimated in ~1.5 mΩ, whereas ESL L_{sh} reduces from 170 to 70 pH. This in turn translates in faster switching times and a lower damping of the switching node ringing compared to the waveforms of Figs. 2.33 and 2.34, as it can be observed from Figs. 2.40 to 2.42 corresponding to the waveforms at 21 A load current and 12 V input voltage. A close look at Fig. 2.41 reveals the good agreement between experimental and simulated waveforms of voltage $v_{DSx(s)}$, thus providing evidence of the model's capability to predict switching behavior under realistic high current, high frequency operating conditions.

The total power consumption is measured at the voltage supply of the gate drivers and converter input. The losses of the input inductor and interconnection cables are also accounted for. The resulting magnitude is combined with the additionally measured output power in order to determine the efficiency of the converter circuit.

Simulated power losses are readily obtained from the steady-state switching waveforms. Additional losses in the driver ICs are estimated from datasheet information. Losses in the input filter capacitors also need to be taken into account. These are basically calculated from the equivalent circuit of Fig. 2.26 and the simulated CtrlFET drain current.

Figure 2.43 compares simulated and measured results of the converter performance for a conversion rate of 12/1.5 V at 1 MHz. The good agreement between the

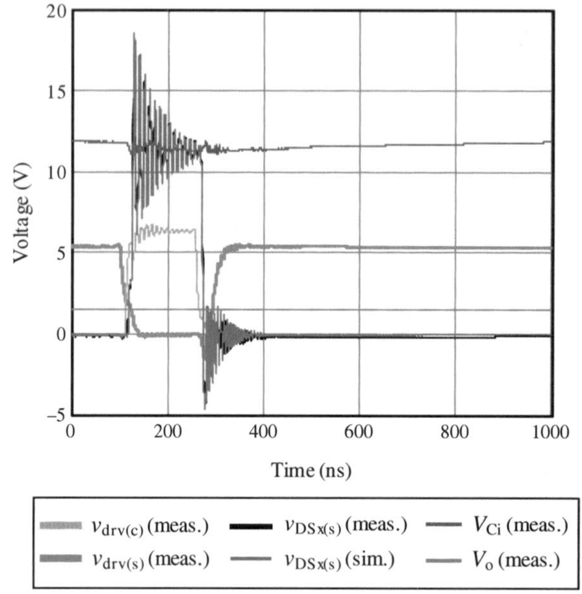

Fig. 2.40 Simulated and measured steady-state voltage waveforms of a SRBC test board

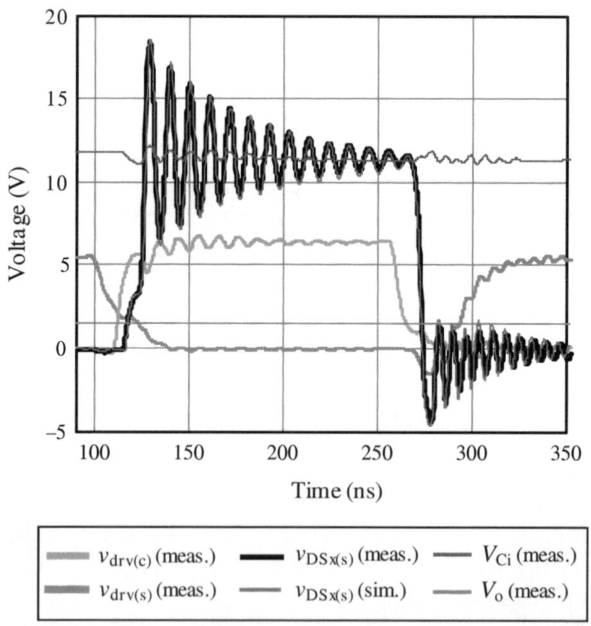

Fig. 2.41 Simulated and measured steady-state voltage waveforms of a SRBC test board. Details of transient response corresponding to waveforms of Fig. 2.40

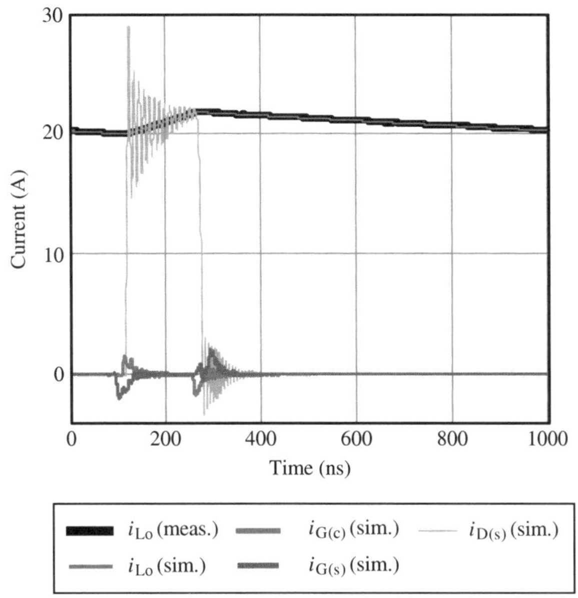

Fig. 2.42 Simulated and measured steady-state current waveforms of a SRBC test board

Fig. 2.43 Measured and simulated efficiency (12–1.5 V) of a SRBC test board. Junction temperature operation of the MOSFETs is forced to be 30°C (ambient temperature) in the entire load range

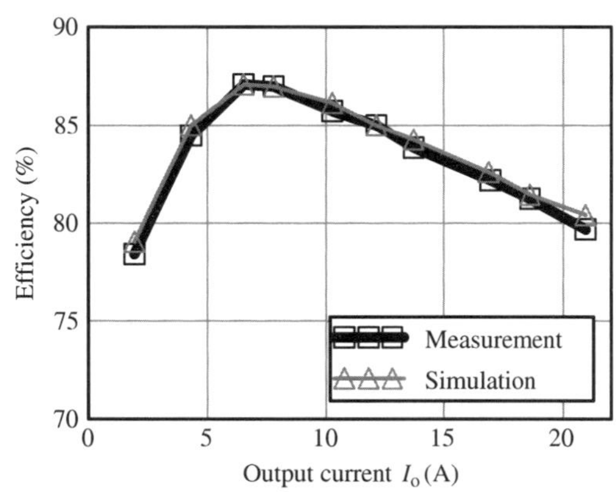

two curves in the entire load range pitches the accuracy of the MOSFET model to a level comparable to device physics simulations, with the additional advantage of the model to enable rather advanced studies on circuit analysis.

Total losses may readily be broken down into loss locations, as shown in the plot of Fig. 2.44. The curves show that major contributions are the heat generated in the power MOSFETs, with similar contributions of the CtrlFET and SyncFET at the highest load range. The significant loss contribution of the PCB is attributed to the rather long tracks of the layout design necessary to accommodate the shunt resistors and probing devices. The thickness of the copper tracks is only 40 μm

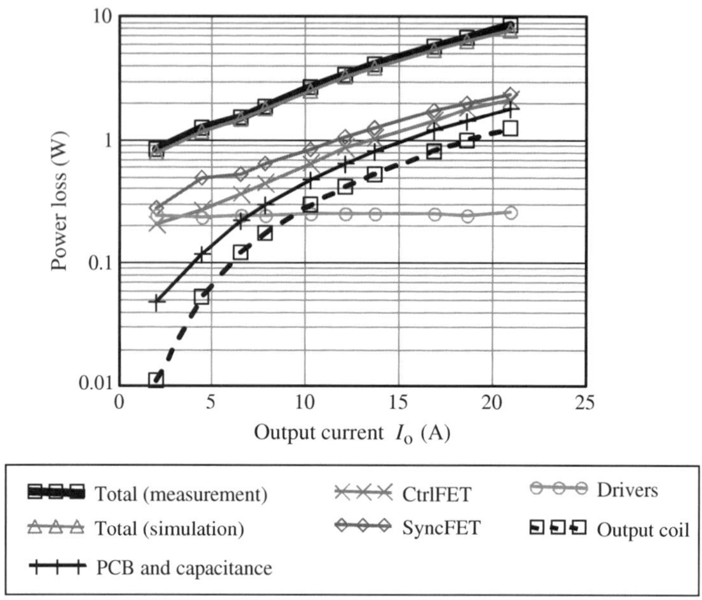

Fig. 2.44 Total and loss breakdown corresponding to efficiency curve of Fig. 2.43

so as to facilitate the soldering of parts, a task that needs to be performed rather frequently for the evaluations of multiple MOSFET technologies. Another significant loss contributor is the output inductor, which is attributed to a DC ESR of ~2.8 mΩ (i.e., the largest portion of the total R_{Lo} from Fig. 2.32). For the characteristics of the power choke, all other inductor losses are negligible.

A yet more rigorous power loss validation will be performed later on in the next chapters with the use of dedicated test boards for loss analysis. In these experiments, the contribution of the power MOSFETs will be predominant, thus allowing for a more accurate assessment of the model predictions.

Focusing the attention on the power MOSFET losses, Fig. 2.45 shows the simulated loss energy evolution of the two devices within a switching cycle under two different load currents. A couple of conclusions can be directly drawn from such pattern behavior. First, at high loads, CtrlFET turn-off loss predominates. And second, the CtrlFET turn-on induces some sort of losses in the SyncFET, likely attributed to either shoot-through or ABD. An examination of these and other relevant loss mechanisms is provided in the next section.

2.5 Analysis of Switching Behavior

This section exploits the developed models to study switching phenomena of power MOSFETs. The analysis aims at identifying the mechanisms of power losses under realistic operating conditions. Once determined and understood, the quantification

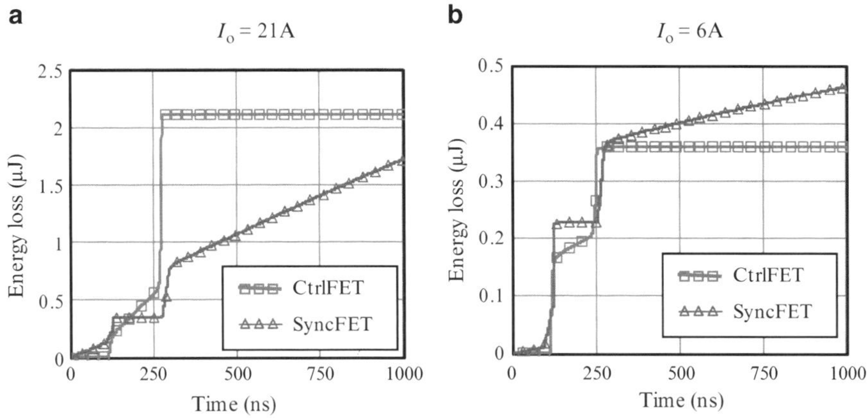

Fig. 2.45 Simulated energy loss profile at two different load currents corresponding to the estimated power losses of Fig. 2.44 (a) load current equals 21A, (b) load current equals 6A

of their effects will be consistently deduced from a method of loss separation, which is extensively described in Appendix D.

The methodology of analysis will be illustrated with three case examples that may be representative of VR applications. The first example will serve as basis for describing fundamental switching phenomena as well as identifying and quantifying loss mechanisms. Results of the loss quantification will lead to the identification of performance improvement options, which will be briefly examined.

The second case example will be applied to illustrate the strong interaction between complex loss mechanisms and the device characteristics. Special emphasis is given to the impact of RR and shoot-through-related losses with regard to the body-effect in power MOSFETs.

In the third illustration, losses are explored in a DrMOS compatible multichip module to demonstrate the importance of RR and its repercussion on other loss mechanisms such as ABD and gate bounce (GB) shoot-through.

2.5.1 Loss Breakdown

Table 2.4 lists specific parameter values of VRs for a laptop application. Corresponding simulated steady-state switching waveforms of one of the VR phases are shown in Fig. 2.46. As already observed in Fig. 2.45, the energy loss curves give again clear signs of significant losses occurring during the switching times. Unlike the losses in the ON conduction times, a thorough examination of the switching behavior is required in order to understand the mechanisms causing such sudden loss increase. This is the main purpose of the following subsections, which provide a comprehensive description of nanosecond-scale switching behavior as well as the identification and quantification of related loss mechanisms.

Table 2.4 Relevant parameter values of a case example to illustrate switching loss mechanisms. Other parameter values not shown are typical of VRs for laptops

General specifications	$V_{in} = 19$ V, $V_o = 1.5$ V, $I_o = 20$ A (max. per phase)	
Power MOSFETs	SyncFET	Trench MOSFET technology, 8.6 mm^2 active die area, LFPak
	CtrlFET	Trench MOSFET technology, 3.6 mm^2 active die area, LFPak
Gate drivers	SyncFET	Driving voltage: 0–5 V, 1.2 Ω turn-on resistance, 1.2 Ω turn-off resistance, 3.5 nH gate inductance turn-on, 2.5 nH gate inductance turn-off, 500 pH source inductance
	CtrlFET	Driving voltage: 0–5 V, 1.5 Ω turn-on resistance, 2 Ω turn-off resistance, 4 nH gate inductance turn-on, 2.5 nH gate inductance turn-off, 500 pH source inductance
Programmed dead times	Switch node LE: 30 ns, switch node FE: 30 ns	
Input/output filters	Total half-bridge inductance: 2.8 nH, $F_s = 300$ kHz, output inductance ripple current: $0.4I_o$	
Temperature operation	75°C uniform throughout the entire switched circuit	

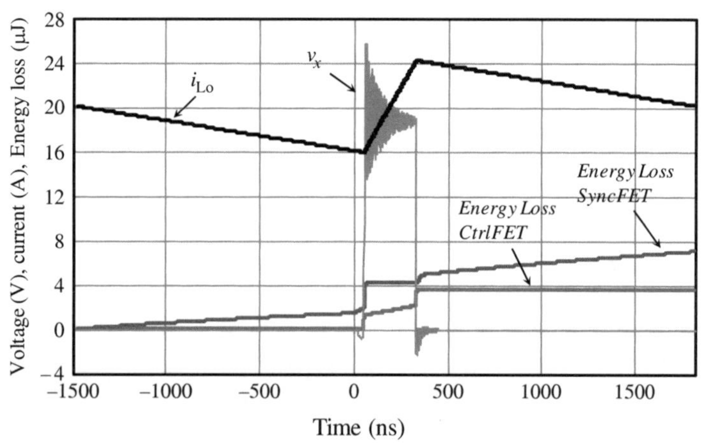

Fig. 2.46 Simulated switching period of a SRBC for laptop VRs

2.5.1.1 Switching Time Subintervals

Figures 2.47 and 2.48 show detailed switching waveforms of the switch node leading and FE transients corresponding to the simulation of Fig. 2.46. From the nature of the converter switching operation in continuous conduction mode, the switching period may be divided into the following time intervals:

– Dead time intervals t_0–t_1 and t_7–t_8 (DT): During the switch node LE transition, interval DT is defined between the FE of driving voltage $v_{drv(s)}$ and the beginning of CtrlFET conduction. Reversely, DT at the switch node FE transition starts as soon as the CtrlFET gate voltage falls below V_{TH}, and ends when $v_{GS(s)}$

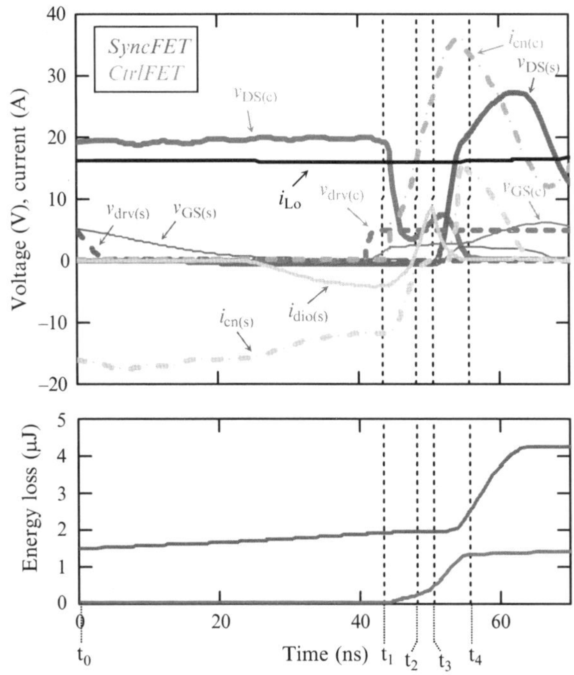

Fig. 2.47 Switch node LE transition of simulation of Fig. 2.46

reaches 90% of the ON state driving voltage. Note the difference between the definition of DT and the programmed dead times of the controller.

- Current hard-switching intervals t_1–t_2 and t_6–t_7 (CHS): It defines the di/dt switching transient. At the LE phase, CHS starts as soon as CtrlFET turns on, and ends when the SyncFET drain current reaches zero. CHS at the FE transition starts when the drain–source voltage across the SyncFET reverses and ends as soon as CtrlFET blocks. By this definition, CHS at the FE phase may not be present in a fast switching condition where the CtrlFET blocks before the SyncFET body diode is forward biased.
- Reverse recovery intervals t_2–t_3 (RR): It may occur at the LE transition due to reverse conduction of the SyncFET body diode. By definition, it starts at the end of the CHS interval and ends when the body diode reverse current reaches a maximum. The RR interval at the FE phase is not considered.
- Voltage hard-switching intervals t_3–t_4 and t_5–t_6 (VHS): It corresponds to the Miller plateau interval of the CtrlFET. At the LE transition, it starts at the end of the RR interval, and ends when the CtrlFET gets into the ohmic region. At the FE transition, the VHS starts when CtrlFET enters the active region, i.e., when voltage v_x starts rising, whereas it ends when the SyncFET body diode is forward biased, i.e., CHS starts. By definition, the VHS in the LE phase may

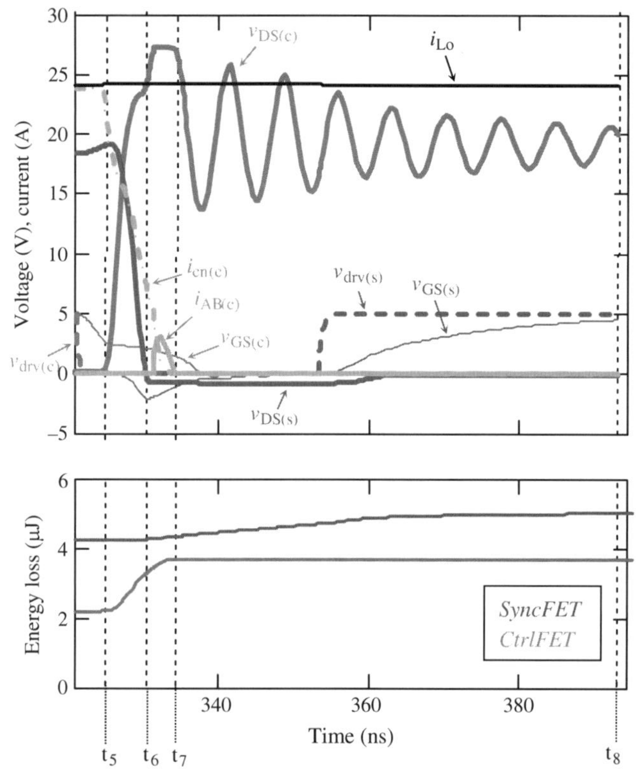

Fig. 2.48 Switch node FE transition of simulation of Fig. 2.46

not be present under conditions of fast switching where the CtrlFET enters ohmic region before v_x starts rising.

– Ohmic intervals t_4–t_5 and t_8–t_1: These are the intervals where the MOSFETs operate in the ohmic region.

The following subsections describe the switching operation of the MOSFETs during intervals CHS, RR, and VHS as these correspond to the switching periods of highest loss increase.

CHS and RR at LE Transition (t_1–t_3)

The CtrlFET takes over the output current at a speed (di/dt) that is determined by the characteristics of the gate drive and half-bridge circuits. Particularly, the source inductance plays a critical role as switching speed limiter. Namely, during the drain current rise the induced voltage across L_s feedbacks into the gate and lowers the driving charge current. A moderate di/dt may already produce significant losses in the CtrlFET since the SyncFET drain–source voltage remains clamped by the action of the body diode and thus the voltage stress across the CtrlFET may be high. The increase of di/dt reduces the CHS time and the voltage stress across the CtrlFET by the induced voltage across the parasitic half-bridge inductances, which thus behave as turn-on snubbers. Accordingly, the absolute maximum di/dt is approximately given by the ratio of the input voltage to the total half-bridge loop inductance. At this switching speed the voltage stress in the CtrlFET is minimized.

The CtrlFET drain current may continue rising beyond i_{Lo} due to the RR and output capacitance charge of the SyncFET, which may cause significant overshoot currents.

The RR charge effectively extends the hard-switching time by the time defined by interval RR. RR results from body diode conduction during DT and CHS. The example of Fig. 2.47 corresponds to a moderate RR effect due to a significant SyncFET channel conduction in the third quadrant prior to the switching event.

VHS at LE Transition (t_3–t_4)

The voltage across the SyncFET starts rising rapidly. The speed depends not only on the CtrlFET driver conditions but also on the half-bridge inductance, input voltage V_{in} and the RR current waveform, particularly the maximum reverse peak current and the steepness of the recovery tail. The higher the latter two values are, the larger the overshoot voltage will be across the SyncFET.

A high dv/dt may cause the input capacitance of the SyncFET to be charged above the threshold voltage, thereby turning on the device sporadically and producing shoot-through, as illustrated in Fig. 2.47. This loss mechanism is referred to as GB. When GB occurs, the dv/dt dramatically reduces as the drain current finds an alternative path through the channel. If, on the other hand, the gate voltage does not reach the threshold level, the output capacitance of the SyncFET may rise further until avalanching the device. Due to the nonlinearity of C_{oss} the maximum peak voltage may reach values well above $2V_{in}$ even at moderate switching speeds. The optimum turn-on switching speed is therefore conditioned by RR, avalanche, and GB.

The drain current through the SyncFET starts decreasing when the voltage across the output capacitance is sufficiently high. The negative di/dt induces a voltage across the source inductance of the SyncFET that tends to turn the device on. Thus, the L_s feedback worsens GB during the current decay.

The voltage across the CtrlFET decreases as the device is driven into the ohmic region. The stored energy in the output capacitance during this transient is mostly dissipated into the channel of the device. This current adds to i_{Lo}, C_{oss}, and RR currents from the SyncFET. Thus, although the voltage stress across the device may be lower than V_{in} at the beginning of the interval, the channel conducts a current higher than i_{Lo}.

VHS at FE Transition (t_5–t_6)

The output capacitances of both MOSFETs perform as turn-off snubbers since current i_{Lo} is shared between the CtrlFET channel and these capacitances. The stored energy in the SyncFET is recovered as it is delivered to the load. The capacitance current increases with the switching speed, thereby reducing the current through the channel of the CtrlFET. The SyncFET C_{GD} discharge current may flow through C_{GS} causing the gate voltage to become negative, as shown in Fig. 2.48. This stored energy in C_{GS} is dissipated in the gate circuit.

CHS at FE Transition (t_6–t_7)

The stored energy in the inductances of the half-bridge path causes an over voltage stress in the CtrlFET during the current commutation, which may result in ABD.

2.5.1.2 Identification of Switching Loss Mechanisms

From the previous analysis, several switching loss mechanisms can be identified and classified as: Snubbed hard-switching (SHS), half-bridge charging (HBC), gate charging (GC), RR, GB, and ABD. The following provides a definition and description of each one of them.

Load Current Hard-Switching (or Snubbed Hard-Switching SHS)

Hard-switching appears as a forced commutation of the output inductor current loop produced by the action of the switches. Particularly, for continuous unidirectional energy transfer from source to load, ZVS cannot be inherently established in the CtrlFET by active rectification. In contrast, the CtrlFET controls the state of the switched node voltage as in a diode-clamped inductive load configuration, where the SyncFET performs the rectification function. Fundamentally, during the current commutation phase (di/dt phase), the voltage across the switched node remains unchanged by way of the SyncFET freewheeling path. Only when the output current is fully taken by the CtrlFET, the switch node is unclamped and the voltage across the device starts changing, thus leading to the dv/dt phase. Such operating condition implies high stress and losses in the CtrlFET for high voltage across the device is present when the full load current flows through it.

The switching behavior of the CtrlFET differs from the aforementioned principle when the half-bridge parasitic elements are taken into account. As described in the previous section, the half-bridge inductances reduce the voltage stress across the device at turn-on, whereas it increases it at turn-off. Reciprocally, the output capacitances of the MOSFETs produce over currents through the CtrlFET at turn-on, yet they reduce them at turn-off. The MOSFET turn-on is therefore characterized by low voltage and high current, whereas the turn-off by low current and high voltage.

SHS-related loss accounts for a portion of the channel CtrlFET loss during the intervals CHS and VHS. This portion corresponds to the loss contribution of the input voltage and output (load) current components, and thus it disregards the contributions corresponding to both the overshoot current at turn-on (i.e., output capacitance and RR currents), and the overshoot voltage at turn-off (i.e., induced voltage across the half-bridge inductances), which are accounted for as different loss mechanisms.

Half-Bridge Charging (Both Capacitive and Inductive)

HBC is related to the losses produced by the output capacitances of the MOSFETs and the half-bridge loop inductances.

The SyncFET output capacitance is charged up from the input voltage supply at the LE transition, causing a resonant current flow in the half-bridge loop. The energy of the oscillation is dissipated in the parasitic resistances of the circuit.

The CtrlFET output capacitance further generates power losses in the Ctrl FET channel at the LE transient, when the device turns on. The energy dissipation corresponds to the stored charge in this capacitance prior to turn-on. The ringing produced in the FE phase is also related to the CtrlFET output capacitance, whose associated loss energy is accounted for.

Lastly, the stored energy in the half-bridge loop inductance is fully dissipated in the FE transition, mostly in the CtrlFET as it produces overshoot voltages capable of even avalanching the device.

Gate Charging

Power losses are produced in the resistances of the gate circuit at turn-on and turn-off. The induced voltage across L_s and the dv/dt across drain–source terminals further produce current in the gate, which contribute to the overall gate drive losses of each device.

Reverse Recovery

As described in the previous section, the RR enlarges the hard-switching interval by the time defined by times t_2 and t_3 from Fig. 2.47. Thus, the total power losses occurring in the half-bridge during this interval are considered RR related. Further losses result from the RR tail current during the subsequent intervals. The RR-related losses in the CtrlFET channel might be particularly significant during the hard-switching intervals.

Gate Bounce

GB-related losses may occur in the LE phase due to a spurious turn-on of the SyncFET during the VHS and subsequent intervals. Both MOSFETs are affected by the shoot-through. Note that GB-related losses might also be produced in the FE transition due to a spurious turn-on of the CtrlFET.

Avalanche Breakdown

Due to the voltage overshoot, ABD may occur in both SyncFET and CtrlFET during the LE and FE transitions, respectively.

A summary of all identified loss mechanisms and corresponding abbreviations is provided in Table 2.5.

2.5.1.3 Loss Quantification

The assessment of the above identified loss mechanisms involves assignment of quantities in a meaningful and systematic manner. This may be accomplished by regarding the losses in every dissipative element of the MOSFET model as the individual contribution of the different current loops of the switched circuit, each of which may be associated to one or more loss mechanisms at a given time instant. The way in which this is carried out is described in detail in Appendix D. The method allows breaking down the power losses by loss mechanisms, as illustrated in the example of Fig. 2.49.

The loss breakdown provides a powerful means to determine major loss contributions, thus aiding the identification of potential improvement options. In the given example, the major contributor is the ON conduction loss of the output

Table 2.5 Summary of loss mechanisms and abbreviations

Loss mechanism	CtrlFET	SyncFET
Load current ON conduction	OnCIoc	OnCIos
Load current hard-switching	SHS	–
Half-bridge capacitance charging	HBCc	HBCs
Half-bridge inductance charging	HBC(L)	
Gate charging	GCc	GCs
Reverse recovery	–	RRs
GB shoot-through	GBc	GBs
Avalanche breakdown	ABDc	ABDs

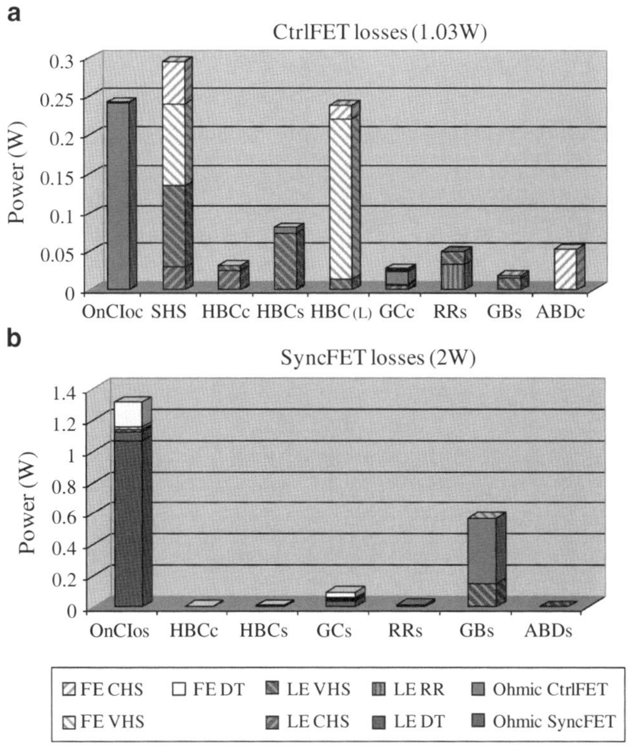

Fig. 2.49 Quantification of loss mechanisms corresponding to simulation of Fig. 2.46 (a) CtrlFET losses, (b) SyncFET losses

current in the SyncFET, which accounts for the 38.5% of the overall power dissipation in the circuit. Most of the losses occur during the ohmic time interval of the device. A significant portion though is caused due to body diode conduction during the FE dead time interval. The difference in power dissipation between LE and FE dead times is mainly attributed to the output current ripple, i.e., the output current during FE is about 8 A higher than during LE.

The second most relevant loss mechanism is GB, which accounts for 18.5% of the total losses. Note that most of these losses occur in the SyncFET.

SHS is on the other hand a predominant loss mechanism in the CtrlFET, followed by the ON conduction of the output current. Further significant losses are produced by the half-bridge loop inductance in the FE transient. The voltage stress in the CtrlFET is such that the device avalanches. The heat produced by this latter mechanism though is not large compared to the channel losses (i.e., HBC(L)). In the LE phase, the amount of output capacitance current and RR-related losses are moderated and primordially occur in the CtrlFET. GC is a minor contributor in both devices.

According to the above observations, measures may be taken to effectively improve the circuit performance. Clearly, a first major action is to reduce the

SyncFET ON conduction losses by, for instance, increasing the die size of the device. If the maximum chip size cannot be further increased in the package, parallel discrete devices may be used, instead.

In addition, the small contribution of gate charge loss suggests the increase of the SyncFET gate voltage to further reduce the ON channel resistance.

Another potential improvement target is to reduce SyncFET GB by either reducing the gate impedance or the C_{GD}/C_{GS} ratio or both. Achieving any of them without increasing C_{GD} allows faster switching, thus further enhancing switching performance. Note, however, that the reduction of GB may give rise to SyncFET avalanche due to an increase of the LE voltage overshoot across the device. Thus, the avalanche voltage level should be increased. As a rule of thumb, the ABD level should be about $2.5V_{in}$.

Another gate drive aspect observed in the loss chart is the insignificance of the gate charge loss compared to the channel losses, which suggest that the optimum gate voltage for the switches may be higher than 5 V in this particular case.

Finally, the reduction of FE CtrlFET losses may be achieved by increasing the output capacitance of the CtrlFET, thereby enhancing the turn-off snubber function.

An analysis of variations of these parameters yields a more concrete proposal of improvement options listed in Table 2.6. Note that the MOSFET dies are artificially scaled by means of parameter A_{die} from (2.1), (2.2), and (2.14). Likewise, each model parameter may be individually parameterized. The loss profiles of Fig. 2.50 show the effectiveness of the applied measures. The overall power loss is reduced by 0.85 W, which translates into a conversion efficiency increase from 90.6 to 93%.

2.5.2 *Influence of the Body-Effect on Switching Losses*

Gate bias refers to an offset voltage at gate-source during the OFF state of the MOSFET (voltage V_{drvn} from Fig. 2.22). Although such gate bias does not alter the ON state conduction losses, it may influence the transient behavior regarding RR, ABD, and GB by virtue of the MOSFETs body-effect.

Table 2.6 Proposed modifications of the SRBC parameters corresponding to case example of Table 2.4

Power MOSFETs	SyncFET	Three parallel discrete (LFPak) Trench MOSFETs of 8.6 mm^2 active die area each
		Reduction of C_{GD} by factor of 3
		Increase of avalanche voltage by 13 V
		Increase of ON resistance by factor of 2 (compromised in order to allow for points 2 and 3)
	CtrlFET	Increase of C_{DS} by factor of 3
Gate drivers	SyncFET	Increase of driving voltage: 0–8 V
	CtrlFET	Increase of driving voltage: 0–8 V

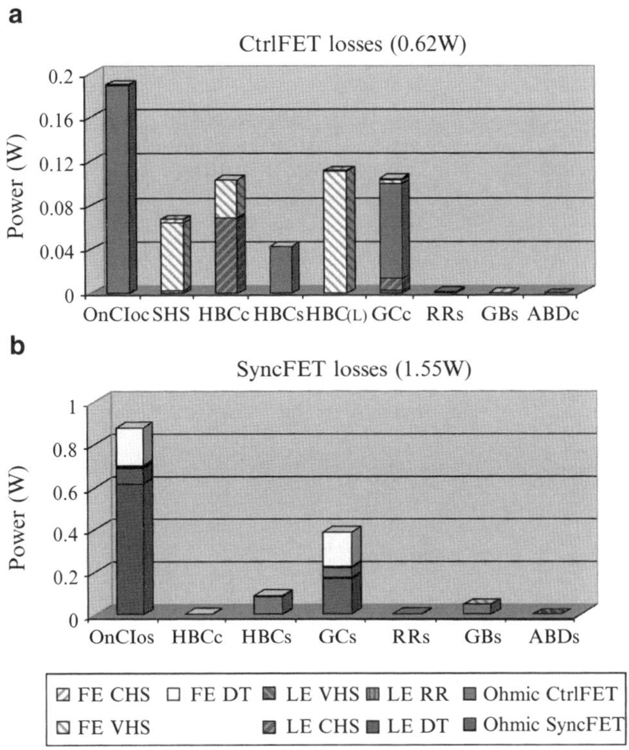

Fig. 2.50 Simulated loss breakdown of the laptop case example after applying proposed improvement options of Table 2.6 (a) CtrlFET losses, (b) SyncFET losses

To investigate this, the LE transient of the SRBC is analyzed by means of circuit simulations based on the proposed MOSFET model of Sect. 2.1. For the sake of highlighting the effects under consideration, a number of elements of the switched circuit are simplified. This is the case of the input filter, which is modeled with an ideal voltage source in series with an inductance. Since only the LE transition is simulated, the output filter and load is replaced by an ideal constant current source of value I_o. The gate driver circuits are represented by ideal trapezoidal voltage sources $v_{drv(c)}$ and $v_{drv(s)}$ in series with the gate parasitic elements and L_s.

Table 2.7 lists parameter values of the simulation setup, including the assignments for two cases discussed below (A, B). Parameter of variation in both cases is gate bias v_{drvn}. Note that the dead times are considerably long so as to ensure body diode conduction prior to CtrlFET turn-on.

Case A ($L_s = 0.5$ nH) deals with the impact of the gate bias on RR only. Increasing source inductance L_s in case B ($L_s = 2$ nH) raises the drain–source voltage $v_{DS(s)}$ in order to further introduce ABD and gate bouncing, which allows investigating their interaction with RR.

Table 2.7 Relevant parameter values of a case example for the analysis of the body-effect in switching transient behavior

General specifications	$V_{in} = 10$ V, $I_o = 10$ A	
Power MOSFETs	SyncFET	Trench MOSFET technology, 8.6 mm^2 active die area, LFPak (case A), and modified LFPak with $L_s = 2$ nH (case B)
	CtrlFET	Trench MOSFET technology, 3.6 mm^2 active die area, LFPak
Gate drivers	SyncFET	Driving voltage: v_{drvn} to $v_{drvn} + 5$ V, 1 Ω turn-on resistance, 0.4 Ω turn-off resistance, 2 nH series inductance
	CtrlFET	Driving voltage: 0–10 V, 1 Ω turn-on resistance, 0.4 Ω turn-off resistance, 2 nH series inductance
Programmed dead times	Switch node LE: 25–30 ns	
Input/output filters	Total half-bridge inductance: 1.2 nH (case A), 4.2 nH (case B) output inductance ripple current: 0 A	
Temperature operation	25°C uniform throughout the entire switched circuit	

Other parameter values not shown are typical of VRs for laptops

2.5.2.1 Impact on Reverse Recovery

Figure 2.51 shows simulations of a SRBC during the LE transition corresponding to the case A of Table 2.7 [16]. The six plots show voltages in the left column, whereas currents and cumulated loss power in the right column. Three gate bias values are considered as follows: $v_{drvn} = -1$, 0, and 1 V. Threshold voltage V_{TH} is 2 V.

The blue and red traces of the left plots show driver voltages $v_{drv(c)}$ and $v_{drv(s)}$, as well as resulting gate voltages $v_{GS(c)}$ and $v_{GS(s)}$ at the dies of the two transistors (bold traces). In all three simulations gate voltage $v_{GS(s)}$ reaches its OFF state value, i.e., bias voltage v_{drvn}, well before gate voltage $v_{GS(c)}$ starts rising. Usually, such sequence implies body diode conduction of the SyncFET. Gate voltage $v_{GS(c)}$ shows in all three cases the typical turn-on plateau, caused by the di/dt and dv/dt intervals. Only the amplitudes of switch node and die voltage v_x and $v_{DS(s)}$ show a strong response to the gate bias. The current waveforms of the right plots allow explaining this behavior. The bold traces show the total drain current $i_{D(s)}$, as well as the contributions of channel and diode $i_{cn(s)}$ and $i_{dio(s)}$. In case of the positive gate bias (still well below threshold), the current through the channel does not commutate to the body diode as a consequence of the body-effect. Accordingly, there is no sign of RR, and the current overshoot is substantially reduced, i.e., to as much as one-third compared to the case with negative gate bias $v_{drvn} = -1$ V. The reduction of the current overshoot in turn means less voltage ringing.

Although the body diode at zero gate bias still carries the full load current (the share of channel and body diode current strongly depends on the total current), RR is already much reduced compared to the transient at negative gate bias. Since the RR parameters are unchanged in both cases, it implies that a significant portion of RR current must freewheel internally through the channel already at $v_{GS(s)} = 0$ V.

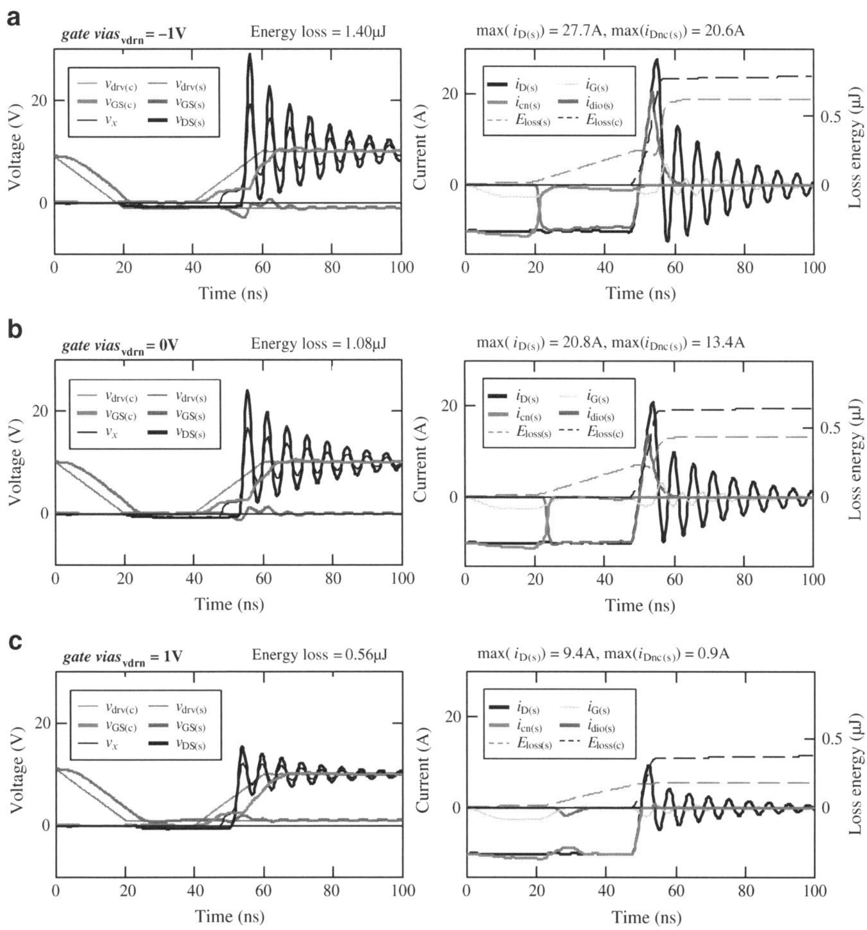

Fig. 2.51 Simulation of switch node LE switching transition. Parameters of Table 2.7 (case A, L_s = 0.5 nH) at various subthreshold gate bias values (a) Gate bias = −1 V, (b) gate bias = 0 V, (c) gate bias = 1 V

The dashed lines in the right plots represent the losses of both MOSFETs during the displayed switching transient ($E_{loss(c)}$, $E_{loss(s)}$). The decrease of the current overshoot leads to a significant decrease in losses (0.56 μJ at $v_{drvn} = 1$ V, compared to 1.4 μJ at $v_{drvn} = -1$ V). Because of the improvement in third quadrant conduction that the body-effect produces, the synchronous rectifier losses are also reduced at positive gate bias during the dead times.

From channel current $i_{cn(s)}$ and gate voltage $v_{GS(s)}$ waveforms at $v_{drvn} = 1$ V, it can be seen a slight shoot-through as consequence of GB, i.e., spurious turn-on of the channel as a result of the increased gate bias. For its importance, this is studied to a greater extent in the next section.

2.5.2.2 Impact on Gate Bounce

There are several sources of gate voltage fluctuations. Firstly, a dv/dt at drain–source terminals produces a gate current as a consequence of reverse transfer capacitance C_{rss}. Assuming a positive voltage ramp across v_{DS}, an initial gate voltage v_{drvn}, and an open gate, gate voltage v_{GS} varies according to the following expression:

$$v_{GS} = \int_{v_{DS0}}^{v_{DS1}} \frac{C_{rss}}{C_{iss}} \, dv_{DS} + V_{GS0}. \tag{2.38}$$

Figure 2.52 plots the results for three gate bias values and shows a strong dependence of GB on the bias, given by the crossings of the curves with the threshold voltage line V_{TH}.

A second cause for GB may be an inductive voltage drop at the source inductance due to a positive di/dt. This is illustrated in the simulation example of Fig. 2.53, where the source inductance of the SyncFET has been intentionally increased to highlight such effect. The curves clearly indicate that source inductance may potentially induce GB-related shoot-through. In contrast to the trends observed in Fig. 2.52, GB worsens at negative gate bias and virtually vanishes at positive v_{drvn}. This indicates the dominance of the di/dt-related influence on GB.

The increased inductive voltage drop across L_s may have an additional negative implication on switching losses due to ABD. Figure 2.54 plots the measured breakdown voltage vs. gate voltage, which shows, for the device under consideration, a decrease of the avalanche value as the gate voltage grows negative. Regarding both $v_{GS(s)}$ and $v_{DS(s)}$ at $v_{drvn} = -1$ V from Fig. 2.53, worst-case conditions are given because peak drain–source and minimum gate-source voltage occur at the same time.

The total losses during the LE transition are lower in case B compared to case A, resulting from decreased losses in the CtrlFET due to the snubber effect of the higher source inductance (which may in turn increase losses at the FE transient). The SyncFET, however, again benefits from the positive gate bias, since RR is reduced.

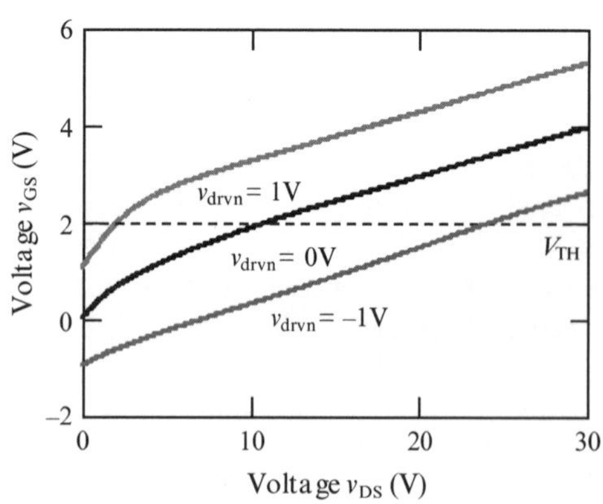

Fig. 2.52 Evaluation of (2.38). Comparison of various gate bias values with regard to GB caused by a dv/dt ramp at the drain–source terminals. The capacitance data refer to the characteristics shown in Fig. 2.18. Data refer to a 25 V logic level n-channel trench MOSFET at 25°C

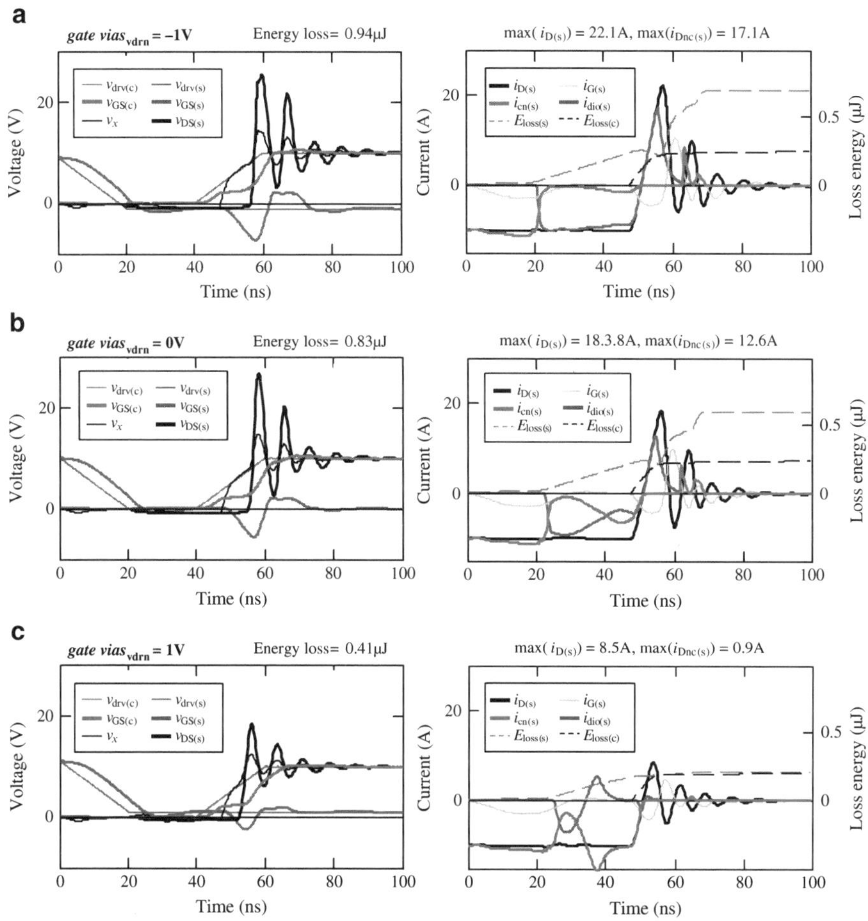

Fig. 2.53 Simulation of LE switching transition Parameters of Table 2.7 (case B, $L_s = 2$ nH) at various subthreshold gate bias values (a) Gate bias = −1 V, (b) gate bias = 0 V, (c) gate bias = 1 V

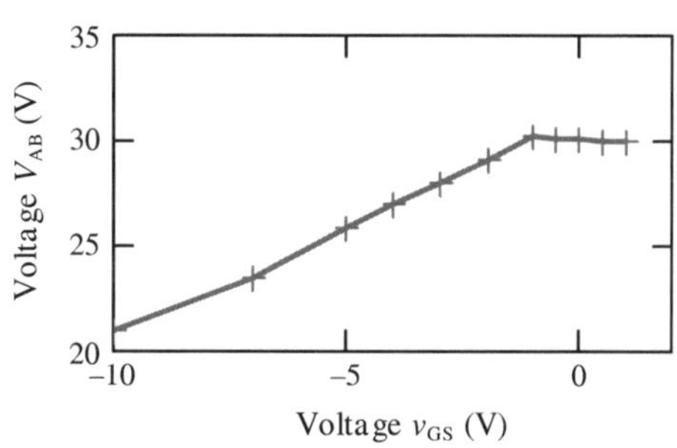

Fig. 2.54 ABD voltage as function of gate voltage. Measurement data of a 25 V logic level n-channel trench MOSFET at 25°C

The simulations at zero and positive gate bias further show that, during third quadrant operation, minor gate voltage variations may alter the drain current sharing between channel and body diode. In case of the positive gate bias of Fig. 2.53, RR occurs already during the dead time, which, however, cannot be observed if only total drain current $i_{D(s)}$ is monitored. This is not new as it was also manifested in the FE transitions of Figs. 2.38 and 2.39.

2.5.3 Loss Analysis of a Multichip Module

This section illustrates the use of the proposed MOSFET model to assess power loss and identify potential improvement options in a multichip powertrain module. The switched-circuit model is based on the IC PIP212-12M from NXP Semiconductors. Table 2.8 lists related parameter values relevant to the analysis. They refer to simplified representations of rather complex circuit models used for the gate drives, half-bridge layout and input filter. Note as well that some of the IC-related parameter values may not be accurate in correspondence with commercially available parts.

Parameter sweep simulations of the input filter inductance L_{HB}, CtrlFET gate resistance, and SyncFET RR peak have been carried out. The latter is possible by adequate adjustments of the coefficients from the RR model expression (see Sect. 2.1.2).

Figure 2.55 illustrates the impact that RR has on the SyncFET losses. The loss dependence is mainly attributed to channel losses at high $R_{G(s)}$ (Fig. 2.55a), and to ABD at low $R_{G(s)}$ (Fig. 2.52b). Further postprocessing of the simulated data

Table 2.8 Relevant parameter values of simulated MCM (PIP212-12M) unless otherwise specified

General specifications	$V_{in} = 12$ V, $V_o = 1$ V, $I_o = 30$ A (max. per phase)	
Power MOSFETs	SyncFET	Trench MOSFET technology, 25 mm^2 active die area
	CtrlFET	Trench MOSFET technology, 3.6 mm^2 active die area
Gate circuit	SyncFET	Driving voltage: 0–6 V, 1.2 Ω gate resistance, 2.5 nH gate inductance turn-on, 1.5 nH gate inductance turn-off, 100 pH source inductance
	CtrlFET	Driving voltage: 0–7 V, 1.7 Ω gate resistance, 3 nH gate inductance turn-on, 1.5 nH gate inductance turn-off, 100 pH source inductance
Package resistance (bond wires)	SyncFET	0.25 mΩ DC resistance (0.35 mΩ at 125°C)
	CtrlFET	1 mΩ at $F_s = 1$ MHz and $d = 0.08$
Programmed dead times	Switch node LE: 10 ns, switch node FE: 30 ns	
Input/output filters	Total half-bridge inductance (L_{HB}): 1.6 nH, $F_s = 1$ MHz, 150 nH output filter inductor	
Temperature operation	30°C uniform throughout the entire switched circuit	

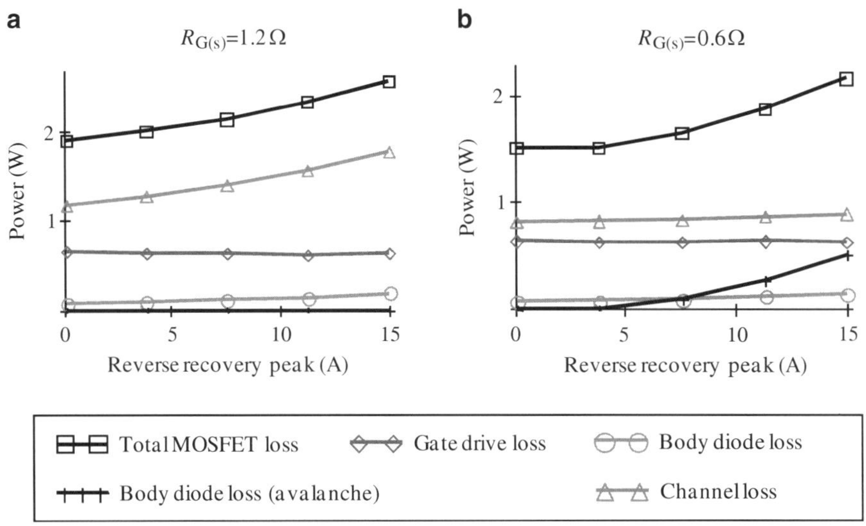

Fig. 2.55 SyncFET power loss as function of reverse recovery and various gate resistances with $L_{HB} = 3$ nH

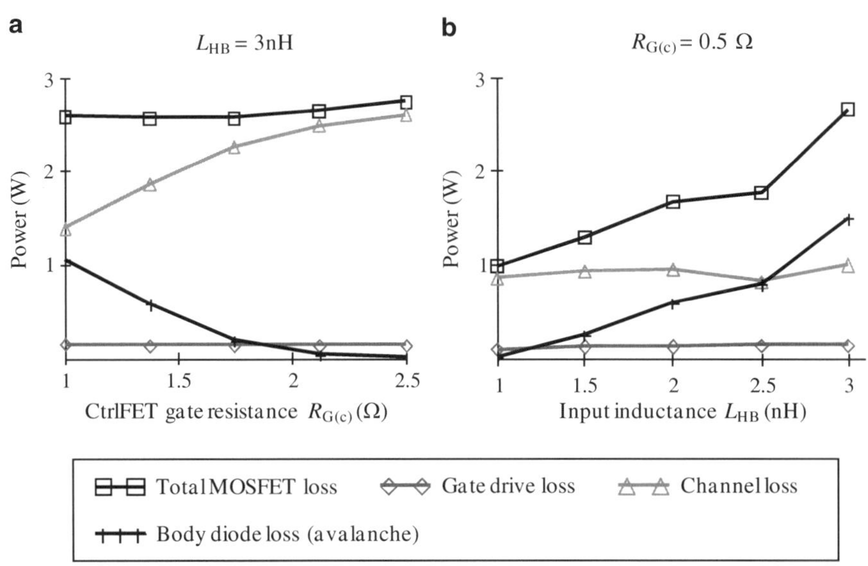

Fig. 2.56 CtrlFET power loss; (**a**) as function of CtrlFET gate resistance; (**b**) as function of total parasitic half-bridge loop inductance

yields the separation of power losses into loss mechanisms and time intervals, as shown in Fig. 2.56. Such a loss breakdown approach reveals that the increase of channel losses observed in Fig. 2.55a is mainly due to GB shoot-through (see Sect. 2.5.1.2).

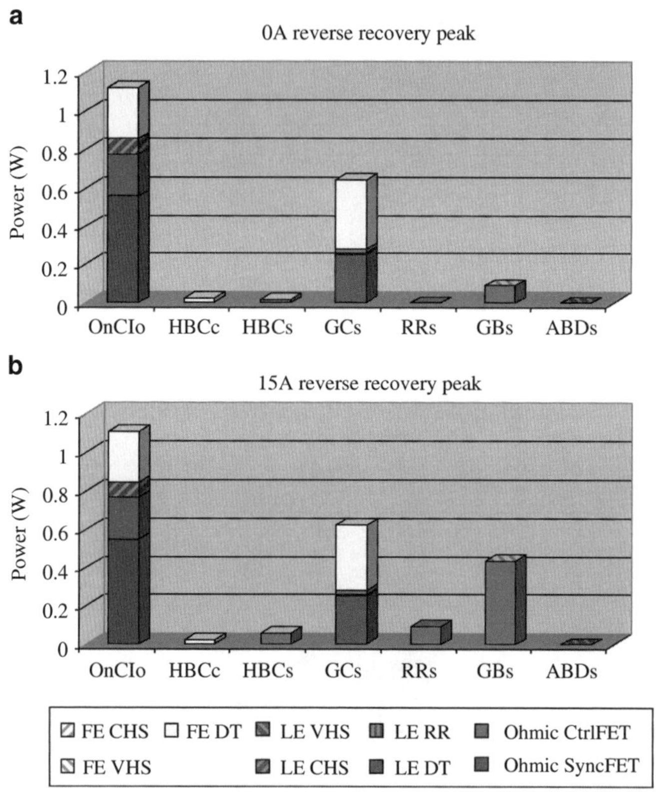

Fig. 2.57 SyncFET loss breakdown corresponding to simulations of Fig. 2.55a. See Table 2.5 for loss breakdown abbreviations (a) no RR, (b) 15A RR peak current

Figure 2.57a exemplifies that the increase of the CtrlFET switching speed may not always lead to an overall loss improvement, even though the gate drive loss may remain unchanged. That is, while the reduction of $R_{G(c)}$ increases the switching speed and thus minimizes hard-switching losses, it may also induce the ABD of the CtrlFET. This is observed in the loss bars of Fig. 2.58. Note that significant losses occur during the FE transition, which suggests the need for reducing L_{HB} to enable further significant improvements, as shown in Fig. 2.56b.

These results justify the implementation of an adaptive dead time control that effectively mitigates losses due to RR and related loss mechanisms. ABD, GB shoot-through, and RR represent major obstacles toward efficient fast switching. Their reduction is therefore essential to continue developing devices with lower Q_{Gt} and R_{DSon}. It suggests the need for refined figure of merits that take into consideration further insights of loss mechanisms in the application. Parameters involved in the optimization may be device related (e.g., Q_{rr}) and also circuit related (e.g., L_{HB}).

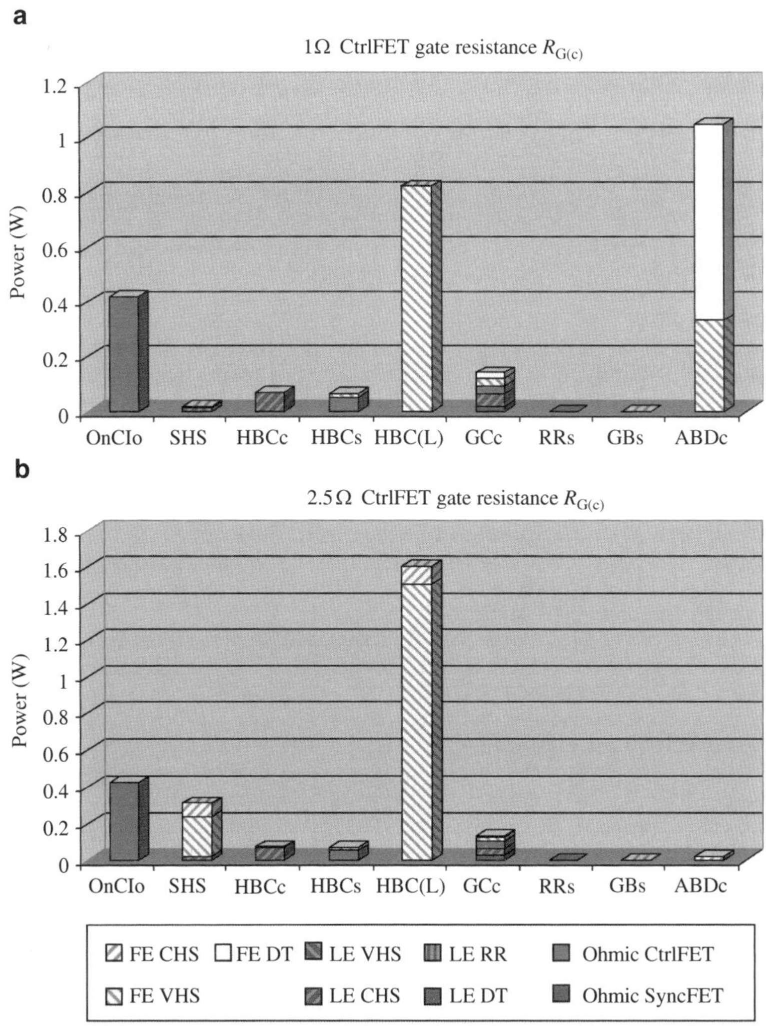

Fig. 2.58 CtrlFET loss breakdown corresponding to simulations of Fig. 2.56a. See Table 2.5 for loss breakdown abbreviations (a) 1 Ω gate resitance, (b) 2.5 Ω gate resistance

References

1. Dolny GM, Sapp S, Elbanhaway A, Wheatley CF (2004) The influence of body effect and threshold voltage reduction on trench MOSFET body diode characteristics. In: The 16th IEEE international symposium on power semiconductor devices and ICs, ISPSD 2004, pp 217–220
2. Lauritzen PO, Ma CL (1991) A simple diode model with reverse recovery. IEEE Trans Power Electron 6(2):188–189

3. Maxim A, Andreu D, Boucher J (1998) The analog behavioral SPICE macromodeling – a novel method of power semiconductor devices modeling. In: IEEE Proceedings of the 24th annual conference, IECON 1998, vol 1, pp 375–380
4. PSpice Application Note PSPA009. A nonlinear capacitor model for use in PSpice, 1998
5. Spence R (1970) Linear active networks. Wiley, London. ISBN 047181525X
6. Shizhong M, Ismail YI (2004) Modeling skin and proximity effects with reduced realizable RL circuits. IEEE Trans VLSI Syst 12(4):437–447
7. Kamgaing T, Myers T, Petras M, Miller M (2002) Modeling of frequency dependent losses in two-port and three-port inductors on silicon. In: IEEE symposium on RFIC, 2002, Seattle, WA, pp 307–310
8. Lefebvre S, Costa F, Miserey F (2002) Influence of the gate internal impedance on losses in a power MOS transistor switching at a high frequency in the ZVS model. IEEE Trans Power Electron 17(1):33–39
9. Jenkins KA, Sun JYC (1995) Measurement of I–V curves of silicon-on-insulator (SOI) MOSFET's without self-heating. IEEE Electron Dev Lett 16(4):145–147
10. López T, Elferich R (2007) Measurement technique for the static output characterization of high-current power MOSFETs. IEEE Trans Instrum Meas 56(4):1347–1354
11. Sze SM, Ng KK (2006) Physics of semiconductor devices, 3rd edn. Wiley, New York, 27 Oct 2006, ISBN: 0471143235
12. Budihardjo IK, Lauritzen PO, Mantooth HA (1997) Performance requirements for power MOSFET models. IEEE Trans Power Electron 12(1):36–45
13. López T, Sauerlaender G, Duerbaum T, Tolle T (2003) A detailed analysis of a resonant gate driver for PWM applications. In: The 18th Annual IEEE Applied power electronics conference and exposition, 2003, vol 2, pp 873–878
14. Sauerlaender G (2001) Loss analysis of integrated halfbridge drivers. In: European power electronics and drives conference, EPE 2001 http://www.epe_association.org/epe/index:php
15. López T, Duerbaum T, Tolle T, Elferich R (2004) PCB layout inductance modeling based on a time domain measurement approach. In: The 19th Applied power electronics conference and exposition, 2004, vol 3, pp 1870–1876
16. Elferich R, Lopez T (2006) Impact of gate voltage bias on reverse recovery losses of power MOSFETs. In: Applied Power Electronics Conference and Exposition, APEC '06. Twenty-First Annual IEEE, Philips Res. Labs., Aachen, Germany, 6 pp. DOI 10.1109/APEC.2006.162072

Chapter 3
Model Level 1: Piecewise Linear Analytical Switching Model

The accurate performance predictions of model level 0 are achieved by means of rather complex circuit representation techniques of both power MOSFETs and converter circuit. Consequently, the resulting nonlinear equivalent network can only be resolved by numerical methods usually involving high processing power and long simulations times.

Besides the complex model representation and solving algorithms, one important drawback of model level 0 lays on its unsuitability for analyzing switching performance at a fundamental level. Namely, the model's capability to thoroughly represent the switching converter behavior contrasts with its ability to readily establish relevant circuit parameter correlations. Variable dependencies and identification of extreme values can only be studied by rather tedious parameter sweep variations that allow exploring the design space. Without a proper sense of the improvement trends, such exploration might be grossly performed and typically restricted to a limited range of the entire design space. Therefore, conclusions drawn from such parametric studies may yield misleading prognostications and false generalizations.

These arguments justify the elaboration of an alternative simplified model devoted to examine the bases of switching phenomena. This may be effectively accomplished by analytical modeling aiming at simplifying complexity and concisely describing the essential and most relevant parameter correlations. The model thus neglects second-order effects and provides a compact mathematical description of the significant aspects of the circuit behavior while maintaining sufficient accuracy.

In order to allow for analyticity, the converter operation is divided in switching intervals and represented with coupled linearized circuit models from which waveforms can be described in closed-form. The ensuing equations elucidate the foundations of the converter switching without significant loss of generality and simulation accuracy. The resulting piecewise linear analytical (PLA) representation is extensively presented in the following sections.

T. López et al., *Voltage Regulators for Next Generation Microprocessors*,
DOI 10.1007/978-1-4419-7560-7_3, © Springer Science+Business Media, LLC 2011

3.1 Modeling Approach

The cornerstones of the proposed modeling approach are simplified equivalent circuits describing individual operation phases of the power MOSFETs in the converter circuit. Such phases coincide with those defined in Chap. 2, with the exception that the reverse recovery phase becomes part of the d*i*/d*t* phase. Figure 3.1 shows typical switching waveforms and highlights time intervals. Basically, four different times are distinguished: ON conduction (4) (SyncFET ON conduction is not shown), dead times (1) and (7), d*i*/d*t* (2) and (6), and CtrlFET d*v*/d*t* (3) and (5).

For each of the above time intervals, a number of basic equivalent circuits are derived to model the operation of the power switches as well as gate drivers.

The analysis of the converter operation carried out in Chap. 2 allows deducing the equivalent circuits of Table 3.1. Up to six different circuit models are deployed to represent the behavior of the MOSFETs in different time intervals of the switching cycle.

Circuit (a) models the CtrlFET hard-switching intervals, which include intervals (2), (3), (5), and (6) from Fig. 3.1. Relevant for the analysis of this operation phase are the parasitic capacitances and transconductance curves of the CtrlFET. The signal excitations are the gate and drain currents, both of which may be determined by circuital factors, such as the gate driver voltage and the half-bridge loop inductance. Within a sufficiently short time subinterval of the hard-switching phase, the CtrlFET gate current can be approximated by a constant value, whereas the drain current may be assumed to change at a constant rate. Combining these two approximations with the linearization of the capacitances and transconductance around the operating point of evaluation, analytical closed-form expressions may be deduced for the state variables.

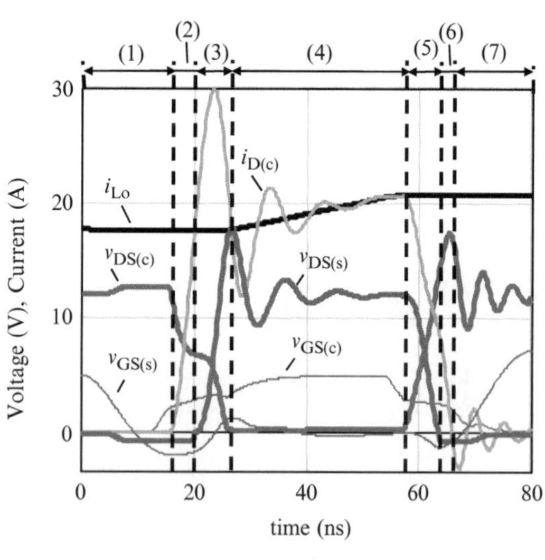

Fig. 3.1 Definition of time intervals (1)–(7) of applicability of the piecewise linear models. Simulation results of a SRBC

Table 3.1 Equivalent circuit models of MOSFET's switching behavior in the SRBC

Function	Equivalent circuit	Intervals[a]
(a) CtrlFET hard-switching	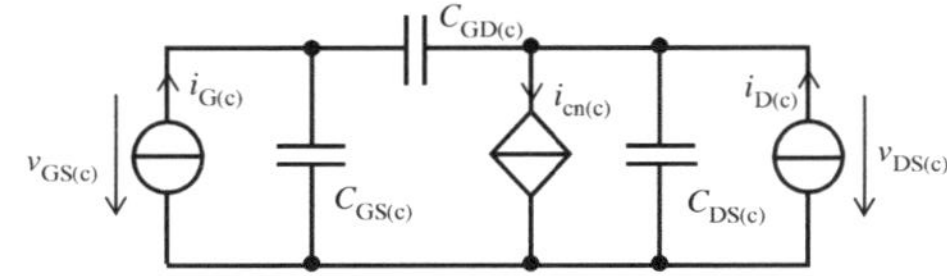	(2), (3), (5), and (6)
(b) Switched-node transient (including reverse recovery and gate bounce shoot-through	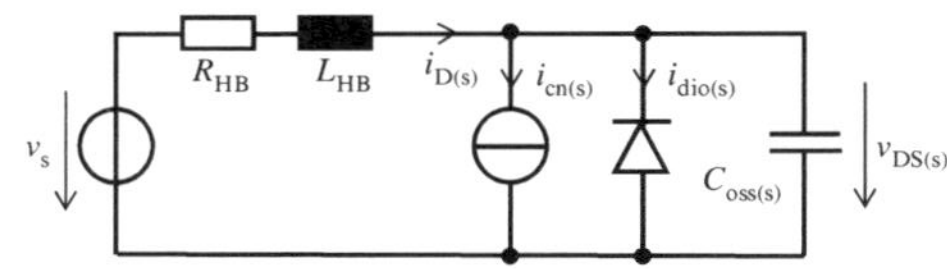	(3)–(6)
(c) Falling edge switched-node transient	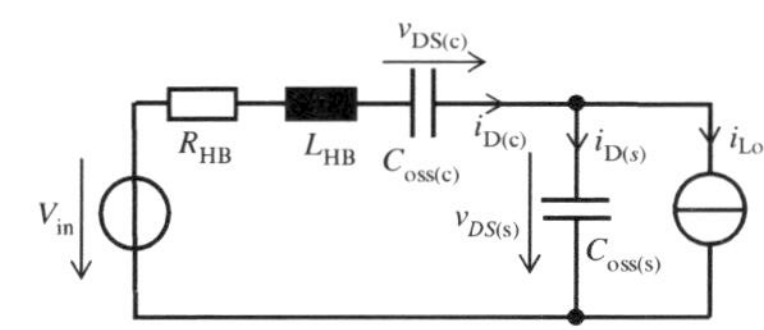	(7)
(d) CtrlFET OFF state and SyncFET reverse conduction	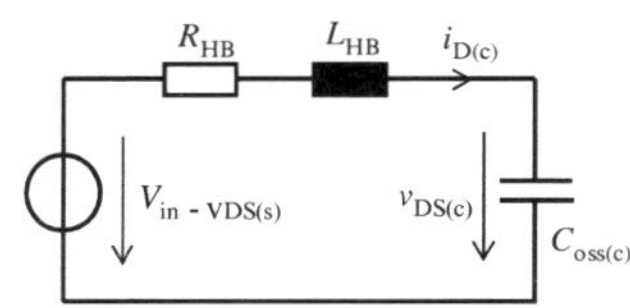	(7)

(*continued*)

Table 3.1 (continued)

Function	Equivalent circuit	Intervals[a]
(e) Avalanche breakdown	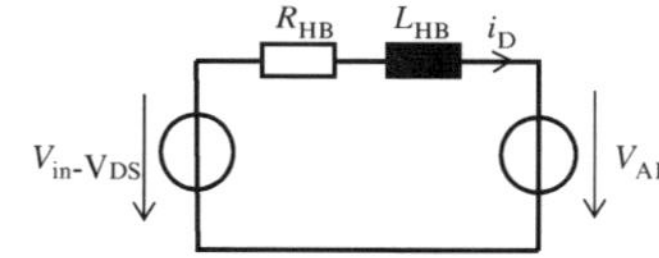	(3)–(7)
(f) Gate driver	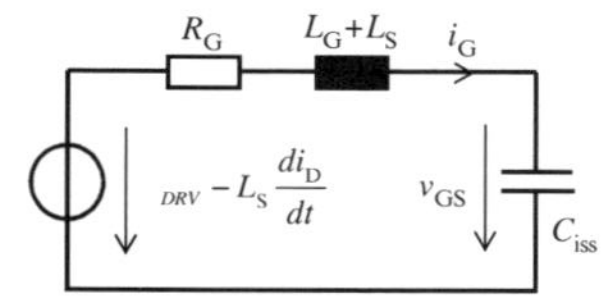	(1), (4), and (7)
(g) Gate driver during hard-switching	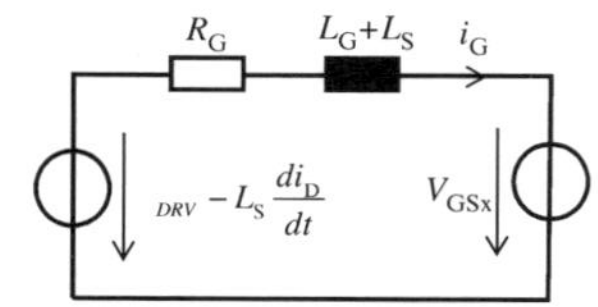	(2), (3), (5), and (6)
(h) Gate bounce	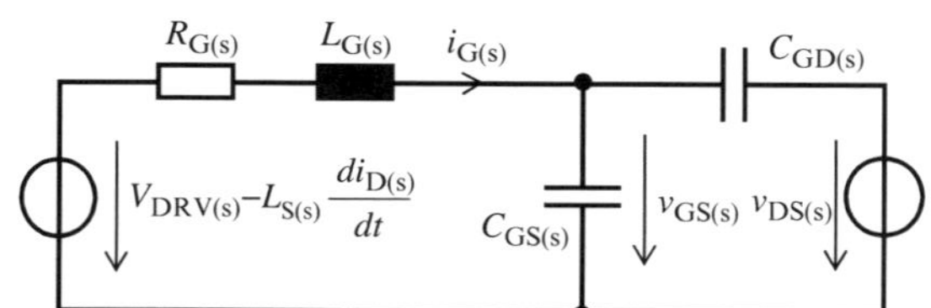	(3)–(7)

[a]See Fig. 3.1

The time-dependent solutions of a circuit model may be used as input variables to another so as to establish their interrelation in the converter operation as a recurrent dynamic map. For instance, circuit (b) requires the rate of change of the drain-source CtrlFET voltage as input excitation, which is part of the expression defining voltage source v_s and may be deduced from the results of model (a). In an ideal case, the CtrlFET switches instantly and so v_s may correspond to a step function of amplitude V_{in}. For a limited CtrlFET switching, though, v_s may be approximately represented with a ramp function.

Voltage v_s triggers an oscillation at the converter switched-node, which is established by the half-bridge loop inductance L_{HB} and the MOSFET's output capacitances (i.e., $C_{oss(s)}$ at the LE transition and $C_{oss(c)}$ at the FE transition). The ringing amplitude can be strongly dependent on the reverse recovery current, as already illustrated in the simulations of Chap. 2. In addition, the SyncFET channel current may be considered in case gate bounce shoot-through manifests. In such case, the channel current must be determined from the conditions of the gate drive circuit, whose corresponding model is represented as circuit (h).

Both reverse recovery and channel currents are modeled with analytical time-dependent current sources. The conditions for analyticity of the resulting equivalent circuit are similar to those of circuit (a) and are discussed later in this chapter.

It is important noticing that upon fast switching, the rise of the switched-node voltage at the LE phase may start well after the CtrlFET hard-switching has elapsed. Conversely, in the FE phase, the SyncFET body diode may still be reverse-biased (i.e., positive drain-source voltage) after the CtrlFET has fully switched off. In the latter case, the equivalent circuit representing the switched-node FE transient corresponds to circuit (c), where both MOSFETs are off and their output capacitances are positively charged. The equivalent circuit estimates the time at which the SyncFET is reverse-biased. When this occurs, the circuit representation may further be simplified to model (d), which may be readily solved by assuming that voltage $v_{DS(s)}$ is approximately constant.

Circuit models (a)–(d) can represent the voltage overshoot that the power devices may be exposed to under fast switching. With sufficient ringing energy, the MOSFETs may undergo avalanche breakdown, thereby producing additional power losses and high stress that may significantly degrade the lifetime of the semiconductor switches. During avalanche breakdown, the half-bridge circuit may be represented with circuit (e), where V_x describes the input voltage and the drain-source voltage of the non-avalanched device. Upon fast switching, the assessment of avalanche losses may be readily determined in cases where V_x is approximately constant.

Circuit models (f)–(h) represent the gate drive behavior under different operating conditions. Circuit (f) is employed during the times where the driven MOSFET is either in OFF state, reverse-biased or operating in the ohmic region. During the hard-switching intervals, the gate-source voltage is assumed constant and so the circuit simplifies to model (g). In both models (f) and (g), the effect of the induced voltage across feedback inductance L_s is considered as a constant offset level to gate drive voltage V_{DRV}. Finally, circuit model (h) is employed to study gate bounce as a consequence of the drain-source dv/dt and the di/dt induced voltage across L_s.

For the above equivalent circuits to be analytically describable, a linearization of associated MOSFET parameters is necessary. The strong nonlinearity of the capacitances, transconductance, and body diode characteristics require particular attention. One preferred linearization approach might be a piecewise discretization with the predefined linear intervals, the length of which is adapted according to the shape of the approximated function. Therefore, a high density of linear intervals may be extended through high nonlinear regions of operation, whereas a coarse coverage may be plausible elsewhere. In any case, an adjustable number of the total linear interval steps may be configured so as to compromise between accuracy and simulation speed. To illustrate this approach, Fig. 3.2 shows a piecewise linearization example of a typical C_{GD} capacitance function, where the linearization intervals vary in a logarithmic fashion, thereby effectively approximating the nonlinear capacitance function with a reduced number of linearization steps.

The linearization of the first quadrant static output characteristics involves two steps. Firstly, both ohmic and active regions are independently described by the ON resistance and transconductance curves, which are dependent on the gate voltage. The main simplification of this step lays on making these two parameters independent on the drain-source voltage. It results in curves like shown in the example of Fig. 3.3a. The gate-source dependence of the ON resistance and transconductance in this example are plotted in Fig. 3.4.

In a second step, the transconductance curve is linearized in the conducting region by calculating the derivative at a given operating point, which is defined as the local transconductance $g_m = g_m(v_{\mathrm{GS}})$. The zero current crossing of the line tangent to the transconductance curve at that given point is defined as threshold voltage V_{TH}. These two parameters are represented in Fig. 3.5 for the case example of Fig. 3.6. The piecewise linearization approach described above for the capacitance example can be similarly applied to represent the nonlinear g_m and V_{TH} curves. As one can see from Fig. 3.6 that the results of the linearization can accurately reproduce the transconductance curve with a sufficient number of discrete linear intervals and variable step size.

The third quadrant static output characteristics (see Fig. 3.3b) are represented with a simple diode model consisting of a voltage source and a series resistance. In contrast to a conventional diode though, both resistance and diode threshold voltage are gate-source voltage dependent. This is illustrated in Fig. 3.7 with parameters R_{DSon} and V_γ. From the curves, it can be observed that for gate-source voltages above the nominal V_{TH} of the device (e.g., 2.5V), V_γ falls to zero and R_{DSon} reaches the typical nominal values of the first quadrant. Such symmetry indicates that the channel is fully conducting while the body diode is off.

The linearization of the MOSFET characteristics enables the derivation of analytical expressions from the proposed linear circuit models within each linear interval. Other approximations are further applied that simplify the formulation of the equivalent circuits as well as their interrelation. The following ones are relevant to point out:

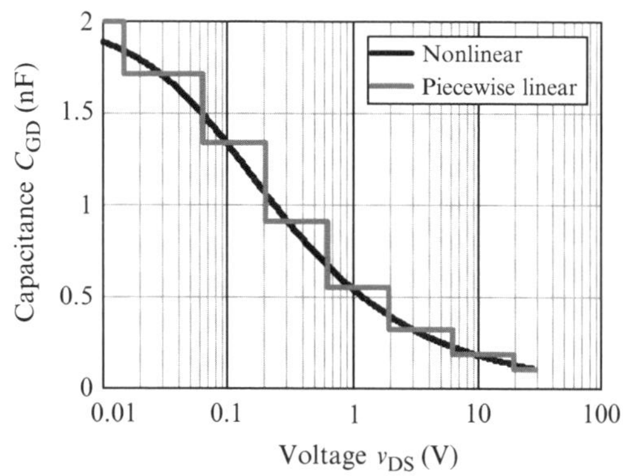

Fig. 3.2 Piecewise linearization into eight logarithmically spaced intervals of nonlinear capacitance C_{GD} of a power MOSFET

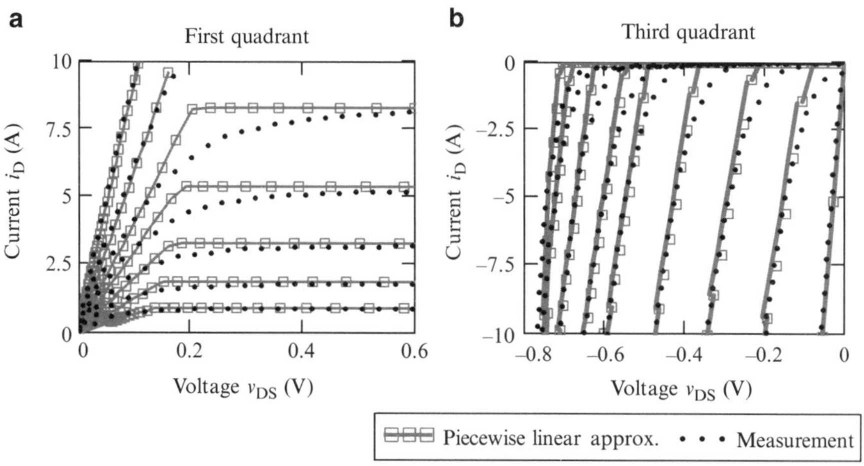

Fig. 3.3 Linearization of static output characteristics of power MOSFETs. (**a**) First quadrant characteristics and (**b**) third quadrant output characteristics. Measurements refer to a PH8030L (case (a)) and PH3330L (case (b)) from NXP semiconductors

- The nonlinear capacitances C_{DS} and C_{GD} are assumed to be uniquely dependent on voltage v_{DS}. Thus, the dependence on v_{GS} is neglected.
- Capacitance C_{GS} is assumed to be linear since its dependence on the interelectrode voltages is only marginal.
- The body diode reverse recovery peak current is provided as input parameter to the model, and as such, it is independent on the circuit conditions.
- The body diode reverse recovery tail is approximated by an exponential decay function, thereby allowing the analyticity of the equivalent circuit model.
- The gate drivers are assumed to provide ideal step functions.

Besides the linearization of the circuit elements, several additional aspects need to be addressed for solving the system of coupled equations that models the converter behavior. These are briefly summarized as follows,

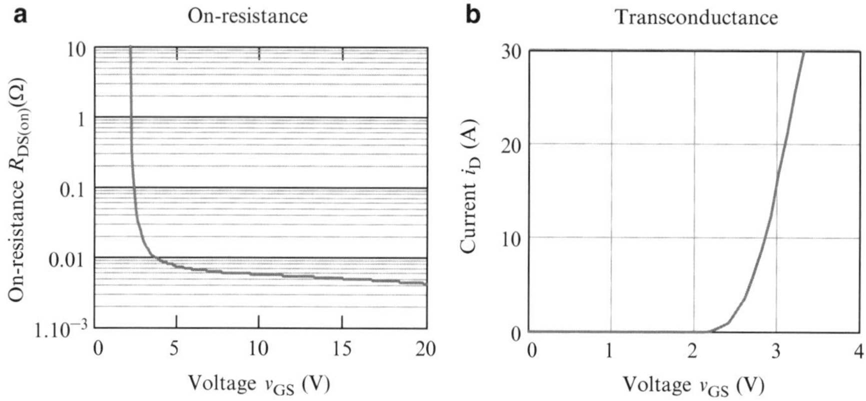

Fig. 3.4 Parameters of linearization of first quadrant output characteristics as a function of gate-source voltage. Data refer to Fig. 3.3a

Fig. 3.5 Piecewise linearization of parameters defining the transconductance curve of a power MOSFET. High density of linear intervals in the low drain current range. Data refer to Fig. 3.6

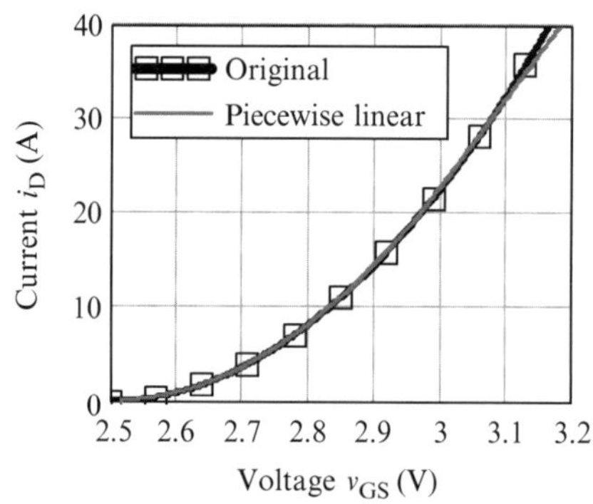

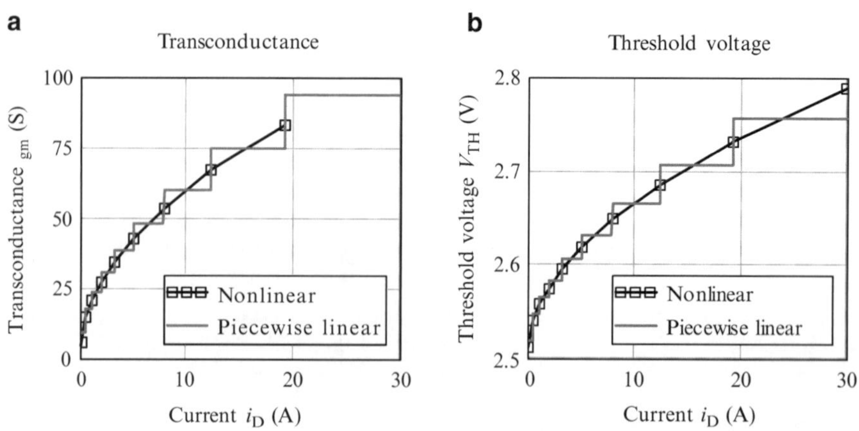

Fig. 3.6 Piecewise linearization of nonlinear transconductance curve of a power MOSFET

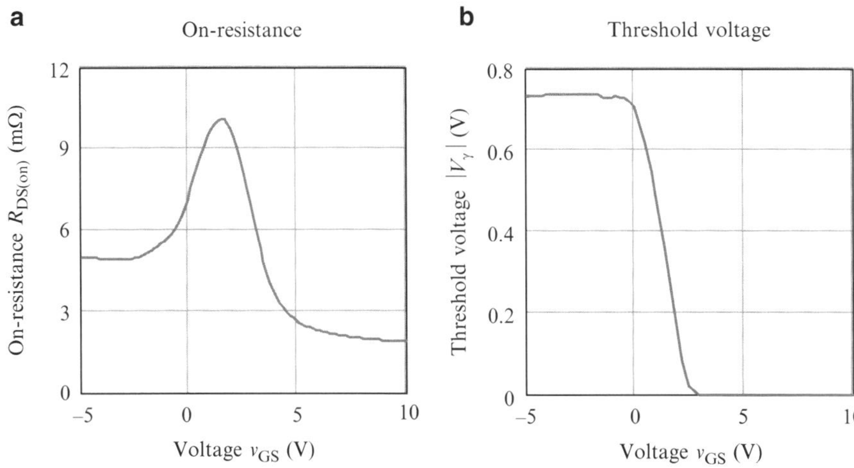

Fig. 3.7 Parameters of linearization of third quadrant output characteristics as a function of gate-source voltage. Data refer to Fig. 3.3b. Hard-switching model (CtrlFET switching)

- For every piecewise linear interval of switching operation, one must determine the limit values range of all state variables within which the above linear approximations apply. The length of these ranges depends on how fine the piecewise linear intervals of the MOSFET characteristics are defined.
- The boundary conditions that guarantee continuity between consecutive piecewise linear intervals need to be deduced according to the initial and final values of the state variables.
- Within each piecewise linear interval, the solutions to the state variables of the equivalent circuit models may be interrelated, where the excitation variable of one circuit may correspond to the output variable of the other and vice versa. In such case, the system of analytically coupled equations undergoes a recursive process by which solutions to the state variables are iteratively found.

The following sections describe the governing equations of each of the proposed equivalent circuits, illustrate the recursive algorithms for obtaining the switching waveforms, and investigate the circuit behavior based on the information contained in the mathematical expressions.

3.2 Hard-Switching Model

In Chap. 2, the causes for which the CtrlFET switching behavior may strictly differ from the conventional clamped inductive switching were mainly attributed to the effects of the half-bridge loop inductance and MOSFET's output capacitances. The following model takes into account such considerations upon the assumption of fast switching operation.

The hard-switching commutation of the CtrlFET is represented according to the equivalent circuit of Table 3.1(a). Within a linearized hard-switching interval, gate current $i_{G(c)}$ is assumed to be constant and equal to $I_{G(c)}$, whereas drain current $i_{D(c)}$ is approximated by a ramp function of initial value $I_{D(c)0}$ and slope $K_{id(c)}$. Thus, both drain-source voltage and drain current are allowed to commute simultaneously as opposed to the case of clamped inductive switching. Since this is partly given by virtue of the half-bridge loop inductance, such element will have to be considered for the estimation of input parameter $K_{id(c)}$, and it is established later in this section.

The following set of time domain equations results from the analysis of the circuit model,

$$v_{GS(c)}(t) = \frac{K_{id(c)}}{C_{x(c)}}\tau_x t + V_{GS(c)0}e^{-(t/\tau_x)} + \left(1 - e^{-(t/\tau_x)}\right)\cdot\ldots$$
$$\left(\frac{C_{oss(c)}}{C_{GD(c)}C_{x(c)}}I_{G(c)}\tau_x - \frac{K_{id(c)}}{C_{x(c)}}\tau_x^2 - \frac{I_{D(c)0}}{C_{x(c)}}\tau_x + V_{TH(c)}\right), \quad (3.1)$$

$$v_{DS(c)}(t) = -\frac{I_{G(c)}}{C_{GD(c)}}t + \left(\frac{C_{GS(c)}}{C_{GD(c)}} + 1\right)\cdot\left(v_{GS(c)}(t) - V_{GS(c)0}\right) + V_{DS(c)0}. \quad (3.2)$$

The channel current is proportional to (3.1) as,

$$i_{cn(c)}(t) = g_{m(c)}\cdot\left(v_{GS(c)}(t) - V_{TH(c)}\right). \quad (3.3)$$

The following relations are used in (3.1),

$$\tau_{x(c)} = \frac{C_{x(c)}}{g_{m(c)}}, \quad (3.4)$$

$$C_{x(c)} = \frac{C_{oss(c)}\cdot C_{iss(c)}}{C_{GD(c)}} - C_{GD(c)}, \quad (3.5)$$

In (3.3), $g_{m(c)}$ and $V_{TH(c)}$ are the linearized CtrlFET transconductance and threshold voltage, respectively. Likewise, $C_{iss(c)}$ and $C_{oss(c)}$ from (3.1) and (3.5) are the linearized input and output capacitances of the CtrlFET in the interval of operation. Note that the parameters of the current sources may be derived from the switched-node ringing and gate drive models of Table 3.1.

Figure 3.8 illustrates simulated results based on piecewise linear transitions of the proposed hard-switching model. As it can be observed, the initial assumptions of constant gate current and di_D/dt for each linear interval prevail for linear intervals

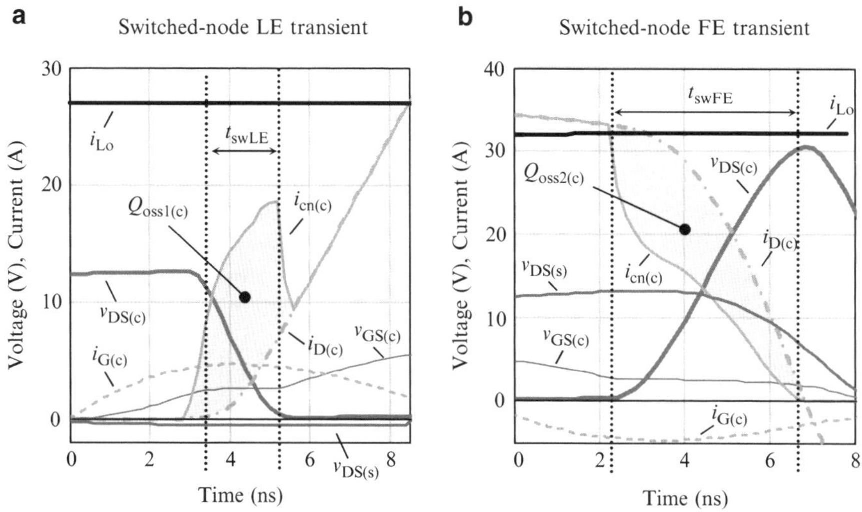

Fig. 3.8 Simulated CtrlFET hard-switching turn-on and turn-off waveforms based on the proposed PLA model

in the order of 500ps or less. The simulated curves outside the hard-switching intervals have been generated with other models from Table 3.1 and are addressed later on. The colored area in both plots defines the CtrlFET output charge, which increases with the switching speed. This and other dependencies affecting the switching performance of the device are addressed in the following subsections with the aid of governing mathematical expressions of the proposed hard-switching model.

3.2.1 *Leading Edge Transition*

This subsection describes the CtrlFET hard-switching turn-on transient based on the following simplifications and assumptions:

- All elements of the equivalent circuit are linear.
- The CtrlFET drain current rises at an average constant rate during the entire hard-switching time. The d*i*/d*t* is approximately determined by the following average expression in the interval of interest,

$$K_{id(c)}|_{LE} \cong \frac{V_{in}}{2L_{HB}}. \tag{3.6}$$

- For the above condition to be effective, the snubber effect of L_{HB} must be significant. Namely, the CtrlFET must commute to the ohmic region (i.e., the drain-source voltage is approximately zero) before the drain current reaches the level of the output inductor current, or in other words, before the SyncFET stops conducting. For this to happen, a sufficiently high d*i*/d*t* is necessary such that the voltage drop induced across L_{HB} counteracts supply voltage V_{in}. This can be achieved with adequate fast switching driving conditions.

- The equivalent gate drive circuit corresponds to that of Table 3.1(g). The effect of the source inductance on the driving voltage is though taken into account.

The first simplification does not represent an obstacle to qualitatively describe the fundamental behavior of hard-switching as it is demonstrated later on.

The second assumption is restrictive but typically given upon fast switching operation and, as it is argued in subsequent chapters, it turns out to be a necessary condition for optimum operation.

Figure 3.9a illustrates turn-on switching waveforms for a constant gate current and drain current slope. The effects of the exponential functions from (3.1) take place already in the first 500 ps since time constant $\tau_{x(c)}$is just 140 ps. Thus, voltage $v_{DS(c)}$ may be approximated by a constant slope of value $-(I_{G(c)}/C_{GD(c)})$. The channel current is the superposition of the drain current ramp and the constant output capacitance current of approximate value $(I_{G(c)}/C_{GD(c)})(C_{GD(c)}+C_{DS(c)})$.

Omitting for now the gate drive loss, the switching losses may be determined by computing the product of the channel current and the drain-source voltage, thus resulting in the waveform of Fig. 3.9b (HS loss). It is clear that the output stored charge, which is dissipated in this transition, contributes to the overall hard-switching loss. The product $0.5Q_{oss(c)}V_{in}$ may be used to estimate the charge loss contribution (marked as $Q_{oss(c)}$ loss in Fig. 3.9b), which accounts for 60% of the hard-switching loss, as Fig. 3.9b shows.

One may realize at this point that, as the drain current slope decreases relative to the absolute value of the $v_{DS(c)}$slope (e.g., by means of increasing the half-bridge

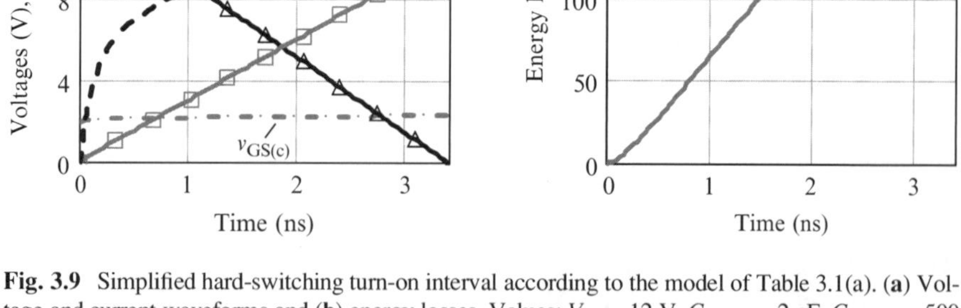

Fig. 3.9 Simplified hard-switching turn-on interval according to the model of Table 3.1(a). (**a**) Voltage and current waveforms and (**b**) energy losses. Values: $V_{in}=12$ V, $C_{GS(c)}=2$ nF, $C_{GD(c)}=500$ pF, $C_{DS(c)}=1$ nF, $g_{m(c)}=50$ S, $V_{TH(c)}=2$ V, $I_{G(c)}=2$ A, $K_{id(c)}=3$ A/ns ($L_{HB}=2$ nH)

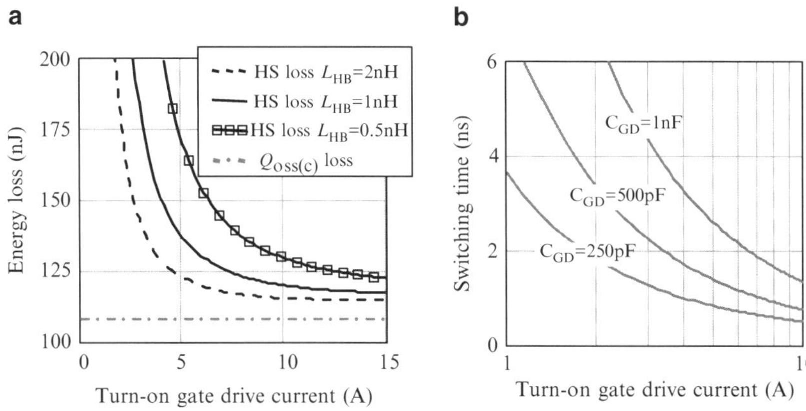

Fig. 3.10 Simplified hard-switching turn-on interval according to the model of Table 3.1(a). (**a**) Switching loss dependence on gate current for various half-bridge inductance values; (**b**) switching time dependence on gate current for various transfer capacitance values. Values (unless otherwise specified): $V_{in} = 12$ V, $C_{GS(c)} = 2$ nF, $C_{GD(c)} = 500$ pF, $C_{DS(c)} = 1$ nF, $g_{m(c)} = 50$ S, $V_{TH(c)} = 2$ V, $K_{id(c)} = 3$ A/ns (plot (**b**))

loop inductance), the hard-switching loss approaches the $Q_{oss(c)}$ loss, which is independent on the switching speed. Such trend can be observed in Fig. 3.10a, where the hard-switching loss is plotted as a function of the gate drive current and for various L_{HB}. Figure 3.10a further points out the benefits of the half-bridge loop inductance as snubber turn-on. The strong dependence of the gate-drain capacitance on the switching speed is highlighted in Fig. 3.10b.

The increase of the driving current so as to minimize the hard-switching loss may imply the need to increase the gate drive voltage in cases where the gate impedance cannot further be reduced.

Raising the gate voltage produces on the other hand a rapid increase of the gate drive loss. This fact must be therefore considered in the optimization process.

Figure 3.11 shows the total CtrlFET and associated gate drive loss as a function of the gate drive voltage. In the illustration, the trade-off between the hard-switching and gate drive loss yields an optimum gate drive voltage around 6V.

The optimum gate drive voltage may change considerably with the half-bridge loop inductance as depicted in Fig. 3.12. The reduction of L_{HB} leads to an increase of the hard-switching loss, which can be partially compensated with an increase of the driving voltage. A situation where driving losses start dominating as a result of this optimization (e.g., as L_{HB} lowers) may demand faster driving capabilities by the way of reducing the gate impedance. For instance, the mitigation of the gate resistance may effectively provide faster transitions without producing additional driving losses. Other lossless techniques such as resonant gate drivers may be an alternative solution, although such approach may end up exceedingly compromising the switching speed. This is addressed in more detail in Appendix F.

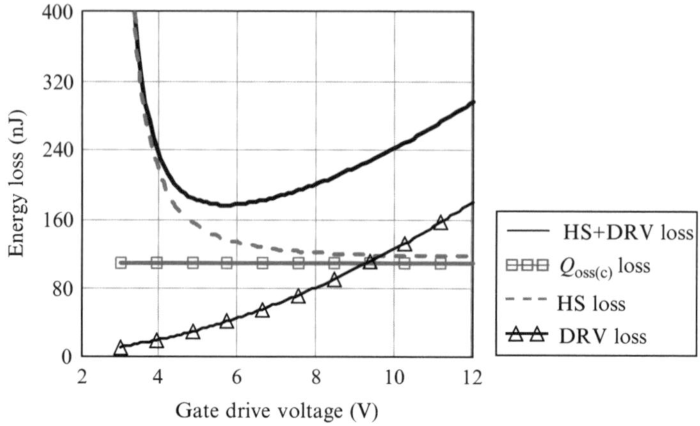

Fig. 3.11 CtrlFET turn-on losses as a function of gate drive voltage $V_{DRV(c)}$. Values: $V_{in} = 12$ V, $C_{GS(c)} = 2$ nF, $C_{GD(c)} = 500$ pF, $C_{DS(c)} = 1$ nF, $g_{m(c)} = 50$ S, $V_{TH(c)} = 2$ V, $K_{id(c)} = 3$ A/ns ($L_{HB} =$ 2 nH), $R_{G(c)} = 1\ \Omega$, $L_{S(c)} = 100$ pH, $L_{G(c)} = -L_{S(c)}$

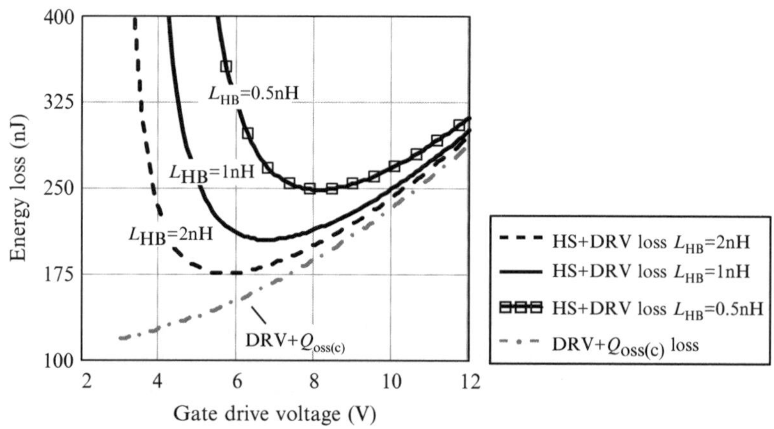

Fig. 3.12 CtrlFET turn-on losses as a function of gate drive voltage $V_{DRV(c)}$ and for various half-bridge inductance values. Values: $V_{in} = 12$ V, $C_{GS(c)} = 2$ nF, $C_{GD(c)} = 500$ pF, $C_{DS(c)} = 1$ nF, $g_{m(c)} = 50$ S, $V_{TH(c)} = 2$ V, $R_{G(c)} = 1\ \Omega$, $L_{S(c)} = 100$ pH, $L_{G(c)} = -L_{S(c)}$

As already pointed out, the conditions described thus far consider linear approximations of the MOSFET characteristics which may yield quantitative discrepancies compared with more accurate models. Nonetheless, the fundamental switching behavior must qualitatively prevail such that the conclusions drawn from the previous analysis remain valid.

Figure 3.13 shows simulated turn-on losses as a function of the gate drive voltage based on model level 0. Circuit conditions are similar to those from Fig. 3.11 with the considerations of the nonlinear capacitances and transconductance curves.

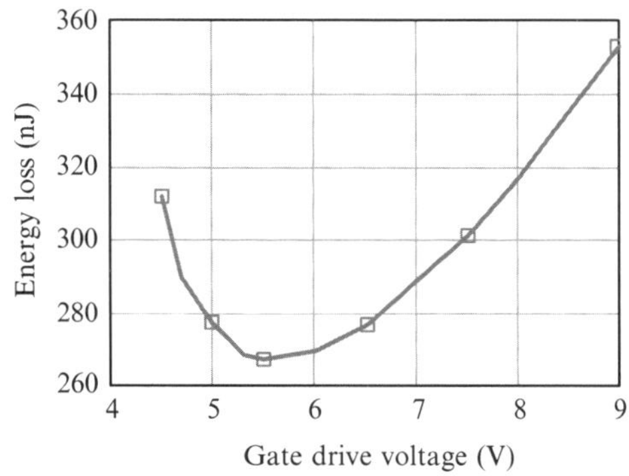

Fig. 3.13 CtrlFET turn-on loss including hard-switching, $Q_{oss(c)}$, and gate driving losses. PSpice simulations based on model level 0. Simulating conditions are similar to those described in Fig. 3.11

As in the simplified linear model, model level 0 predicts the tradeoff between the gate driver and hard-switching losses, which validates the suitability of the simplifications made to fundamentally describe switching phenomena.

3.2.2 Falling Edge Transition

The CtrlFET turn-off transient involves the inverse sequence of events described in the previous section. Thus, similarly as for the turn-on transition, the following simplifications and assumptions are taken.

- All elements of the equivalent circuit are linear.
- The hard-switching transient ends when the CtrlFET blocks, i.e., when the channel current drops to zero.
- The maximum drain-source voltage is always below the avalanche breakdown level. Thus, avalanche breakdown is neglected.
- The switching speed is such that the contributions of the half-bridge parasitic elements need to be taken into account. Therefore, the capacitance current can be high, e.g., $0.1 i_{o(max)} \ll C_{oss(s)} \left| dv_{DS(s)}/dt \right| + C_{oss(c)} \left| dv_{DS(c)}/dt \right|$. Similarly, the loop inductance may be significant such that $L_{HB} K_{id(c)}|_{FE} \gg 0.1 V_{in}$.
- The drain current slope is constant and approximately equal to

$$K_{id(c)}|_{FE} \cong \frac{V_{in} - V_{DS(c)end} - V_{DS(s)end}}{2L_{HB}}, \tag{3.7}$$

here $V_{DS(c)end}$ and $V_{DS(s)end}$ are the voltages across the respective MOSFETs at the end of the hard-switching phase. Since these magnitudes are dependent themselves on $K_{id(c)}$, expression (3.7) may be solved by numerical methods.

- The equivalent gate drive circuit corresponds to that of Table 3.1(g). The effect of the source inductance on the driving voltage is though taken into account.

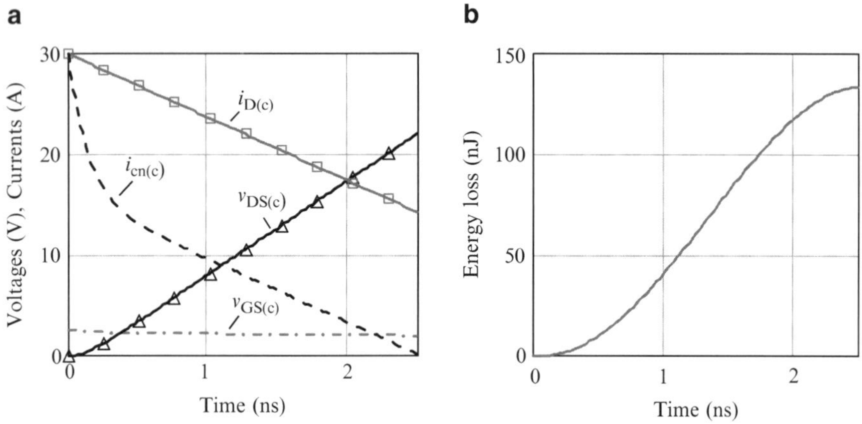

Fig. 3.14 Simplified hard-switching turn-off interval according to the model of Table 3.1(a). (**a**) Voltage and current waveforms and (**b**) hard-switching loss. Values: $V_{\text{in}} = 12$ V, $C_{\text{GS(c)}} = 2$ nF, $C_{\text{GD(c)}} = 500$ pF, $C_{\text{DS(c)}} = 1$ nF, $g_{m(\text{c})} = 50$ S, $V_{\text{TH(c)}} = 2$ V, $I_{\text{G(c)}} = -5$ A, $I_{\text{D(c)0}} = 30$ A, $L_{\text{HB}} = 1.5$ nH, $C_{\text{oss(s)}} = 6$ nF, $L_{\text{S(c)}} = 100$ pH, $L_{\text{G(c)}} = -L_{\text{S(c)}}$

Figure 3.14a illustrates turn-off switching waveforms for a constant gate current and a drain current slope. As before, the effects of the exponential functions from (3.1) take place only for the short period of 500 ps due to the low time constant $\tau_{x(\text{c})}$. Its effect may thus, in this case, be neglected to approximate $v_{\text{DS(c)}}$ by a linear ramp voltage of slope $I_{\text{G(c)}}/C_{\text{GD(c)}}$.

The CtrlFET drain current is split into the channel and output capacitance paths. Thus, the channel current may be effectively reduced with the increase of the output capacitance (e.g., additional $C_{\text{DS(c)}}$) and the switching speed as a consequence of the turn-off snubber effect. The product of the channel current and the drain-source voltage yields the hard-switching loss depicted in Fig. 3.14b.

At the end of the hard-switching interval, the drain current may differ from zero and as such the half-bridge loop inductance may contain stored energy. This energy may be expressed as follows,

$$w_{\text{LHB}} = \frac{1}{2} L_{\text{HB}} I^2_{\text{D(c)end}}, \tag{3.8}$$

where $I_{\text{D(c)end}}$ is the drain current at the end of the hard-switching interval. In the following analysis, it is assumed that such energy is fully dissipated in the form of a ringing oscillation measurable across the terminals of the CtrlFET after turn-off. It is shown though later on that this is just a first-order worst-case approximation since part of this energy may be transferred to the load (and therefore recovered) after the CtrlFET turn-off and during the time that the SyncFET output capacitance is discharged.

Another portion of the turn-off energy losses may be related to the stored charge in the CtrlFET output capacitance at the end of the hard-switching time. This stored energy may differ from the final energy state of the FE transition (i.e., $1/2C_{oss(c)}V_{in}^2$). The energy loss produced by this energy difference is, thus,

$$w_{Coss} = \frac{1}{2}C_{oss(c)}\left(V_{DS(c)end} - V_{in}\right)^2. \quad (3.9)$$

In (3.9), $V_{DS(c)end}$ is the CtrlFET drain-source voltage at the end of the hard-switching turn-off transition. Once again, the above expression is an approximation even for the case of linear capacitances, as it is argued later in Sect. 3.4.

Omitting for now the gate drive loss, Fig. 3.15 shows the FE CtrlFET losses as a function of the gate drive current. As it can be seen, the negative increase of gate current is effective to mitigate the hard-switching loss. In doing so, however, the inductance loss is increased until $I_{D(c)end}$ approaches $I_{D(c)0}$ (i.e., the load current).

While the half-bridge loop inductance is favorable at turn-on, it can be detrimental at turn-off in terms of switching losses and maximum voltage stress across the device (i.e., voltage $v_{DS(c)max}$). This can be deduced from the curves of Fig. 3.16, which show the switching loss and $v_{DS(c)max}$ as a function of the gate current and for various L_{HB}. Figure 3.17 further depicts the impact that both the gate current and L_{HB} have on the switching OFF time. The curves suggest that L_{HB}, though producing higher turn-off losses, may not significantly impact the switching time, which is mainly determined by both the gate current and the MOSFET's capacitances.

It is worth noticing that, as is shown later, the nonlinearity of MOSFET capacitances can worsen the maximum overshoot voltage. Therefore, adding an ideal linear capacitance in parallel to the MOSFET's intrinsic C_{DS} can effectively mitigate the voltage stress of the device in detriment of charging losses.

The above figures are given as a function of the turn-off gate current, which needs to be generated by the gate drive circuit. This is typically achieved by short-circuiting the gate and source terminals with a gate drive switch. Thus, the gate current discharge is mainly determined by the gate drive impedance and the Miller

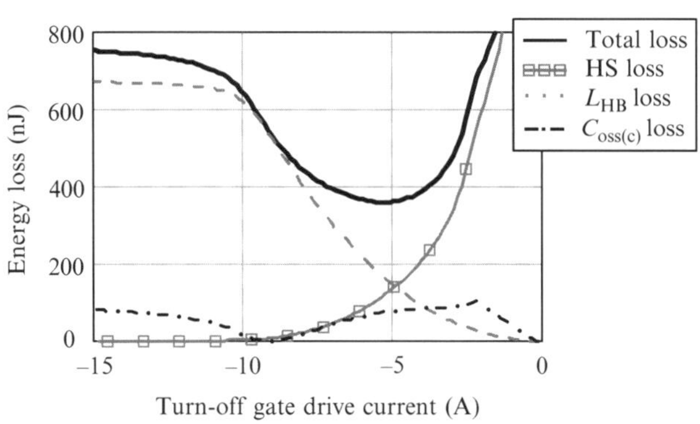

Fig. 3.15 Turn-off switching loss as a function of the CtrlFET gate current. Values: $V_{in} = 12$ V, $C_{GS(c)} = 2$ nF, $C_{GD(c)} = 500$ pF, $C_{DS(c)} = 1$ nF, $g_{m(c)} = 50$ S, $V_{TH(c)} = 2$ V, $I_{D(c)0} = 30$ A, $L_{HB} = 1.5$ nH, $C_{oss(s)} = 6$ nF

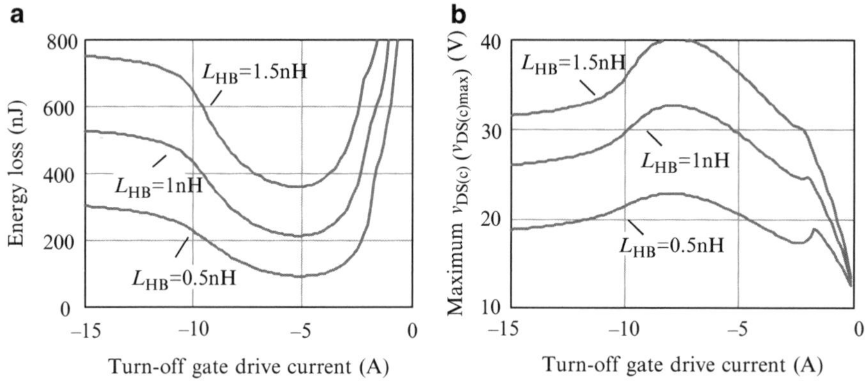

Fig. 3.16 Turn-off switching loss (**a**) and maximum drain-source voltage (**b**) as a function of the CtrlFET gate current (absolute values) and for various half-bridge loop inductance values. Values: $V_{in} = 12$ V, $C_{GS(c)} = 2$ nF, $C_{GD(c)} = 500$ pF, $C_{DS(c)} = 1$ nF, $g_{m(c)} = 50$ S, $V_{TH(c)} = 2$ V, $I_{D(c)0} = 30$ A, $C_{oss(s)} = 6$ nF

Fig. 3.17 Switching off time as a function of the CtrlFET gate current and for various half-bridge loop inductances. Values: $V_{in} = 12$ V, $C_{GS(c)} = 2$ nF, $C_{GD(c)} = 500$ pF, $C_{DS(c)} = 1$ nF, $g_{m(c)} = 50$ S, $V_{TH(c)} = 2$ V, $I_{D(c)0} = 30$ A, $C_{oss(s)} = 6$ nF

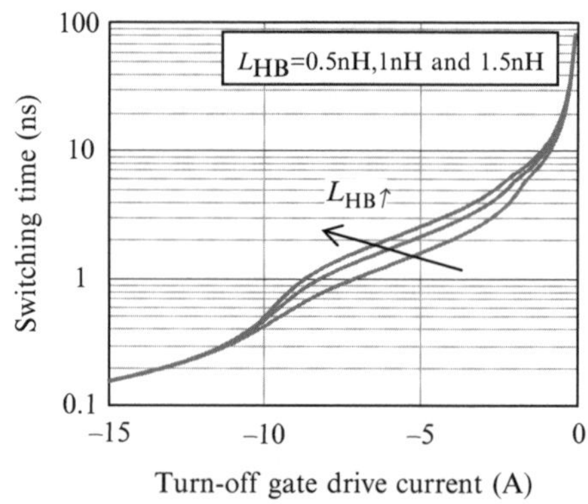

plateau voltage. For typical cases, the gate resistance required to generate gate currents of −10A during the hard-switching time must be well below 1 Ω.

These results can be validated with the model level 0 of Chap. 2. The data from Fig. 3.18 corresponds to PSpice simulations analogous to the conditions illustrated in this section with the consideration of the nonlinear MOSFET characteristics. As it can be observed, the resulting curve follows the same trends identified with the PLA model, thereby proving the validity of the simplifications adopted to study the underlying hard-switching phenomena.

3.2.3 Leading and Falling Edge Transitions Tradeoffs

Given the existing tradeoffs between the turn-on and turn-off hard-switching losses, the question arises as to what optimum combination of L_{HB}, additional snubber

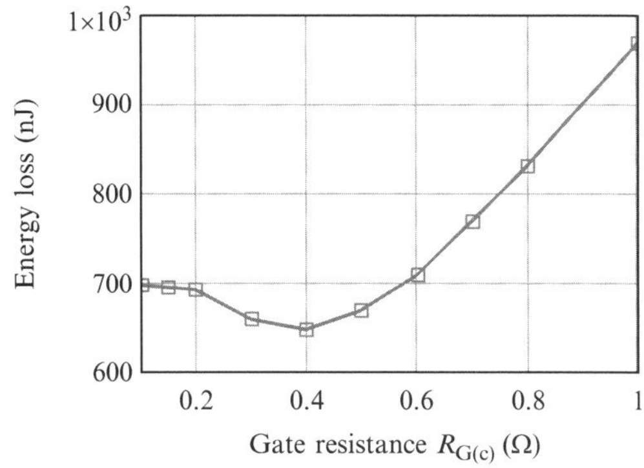

Fig. 3.18 CtrlFET turn-off loss including hard-switching, $Q_{oss(c)}$, and L_{HB} losses. PSpice simulations based on model level 0. Simulating conditions are similar to those described in Fig. 3.15

capacitance $\Delta C_{DS(c)}$ and driver parameters $V_{DRV(c)}$ and $R_{G(c)}$ can minimize the total switching losses.

In the following it is assumed that both $V_{DRV(c)}$ and $R_{G(c)}$ determine the turn-on gate current, whereas only $R_{G(c)}$ can adjust the turn-off gate current. Furthermore, the impedance of the gate path is common to both switching transitions.

Figure 3.19a illustrates the LE and FE CtrlFET switching losses as a function of L_{HB}. In order to minimize the sum of the two contributions, voltage $V_{DRV(c)}$ is optimally adjusted for every L_{HB}. The resulting total loss curve has a minimum at ~450pH. This optimum is expected because the increase of L_{HB} reduces the LE loss and worsens the FE loss. The latter has an exception at low L_{HB} (i.e., below 400pH) due to driving losses, which increases as L_{HB} lowers. This is a consequence of the increase of $V_{DRV(c)}$ necessary to shorten the turn-on switching time (and hence minimize the hard-switching losses) as L_{HB} reduces. The increase of $V_{DRV(c)}$, though not being the most efficient alternative, yields a net loss reduction due to the significance of hard-switching losses.

Figure 3.19b illustrates how the increase in excess of L_{HB} may additionally produce high overvoltage stress across the CtrlFET at turn-off. Thus, it is generally favorable to limit L_{HB} in order to minimize the switching losses at the same time that the use of lower voltage rating devices is enabled.

The converse effects of $C_{DS(c)}$ on the switching losses are illustrated in Fig. 3.20a. The minimum switching loss at ~1.7 nF results from the reduction of the hard-switching turn-off loss as $C_{DS(c)}$ increases. The achieved gain is only moderate because of the proportionality of $C_{DS(c)}$ on the LE loss.

The increase of $C_{DS(c)}$ has the additional advantage of reducing the voltage stress across the device, as shown in Fig. 3.20b.

A more general description of the identified effects is provided in Figs. 3.21 to 3.24. Figure 3.21 shows the total switching losses as a function of both $R_{G(c)}$ and L_{HB}. Similarly as in the previous example, both $\Delta C_{DS(c)}$ and $V_{DRV(c)}$ are optimized for every value in the contour plot in order to guarantee minimum switching losses. The results show that the existence of both LHB and $\Delta C_{DS(c)}$ can provide a net loss reduction despite of their detrimental effects at turn-off and turn-on, respectively.

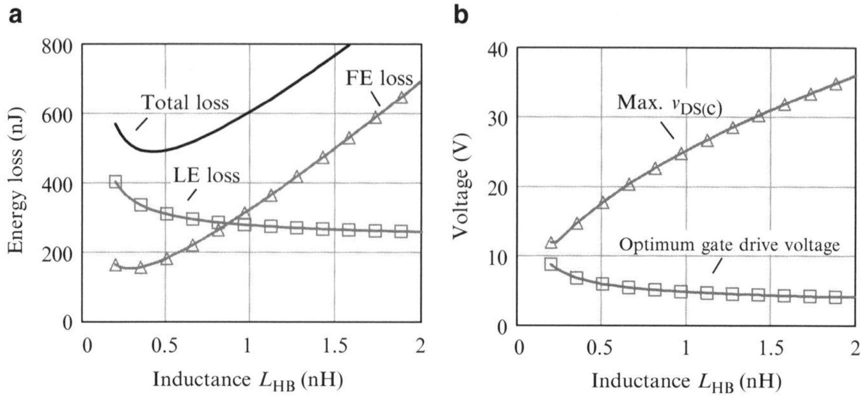

Fig. 3.19 (**a**) CtrlFET loss in both LE and FE transitions as a function of L_{HB}. (**b**) Maximum voltage stress across the CtrlFET and optimum gate drive voltage to achieve minimum losses. Values: $V_{in} = 12$ V, $C_{GS(c)} = 2$ nF, $C_{GD(c)} = 500$ pF, $C_{DS(c)} = 1$ nF, $g_{m(c)} = 50$ S, $V_{TH(c)} = 2$ V, $I_{D(c)0} = 30$ A, $L_{S(c)} = 100$ pH, $C_{oss(s)} = 6$ nF, $R_{G(c)} = 0.5\ \Omega$, $L_{G(c)} = -L_{S(c)}$

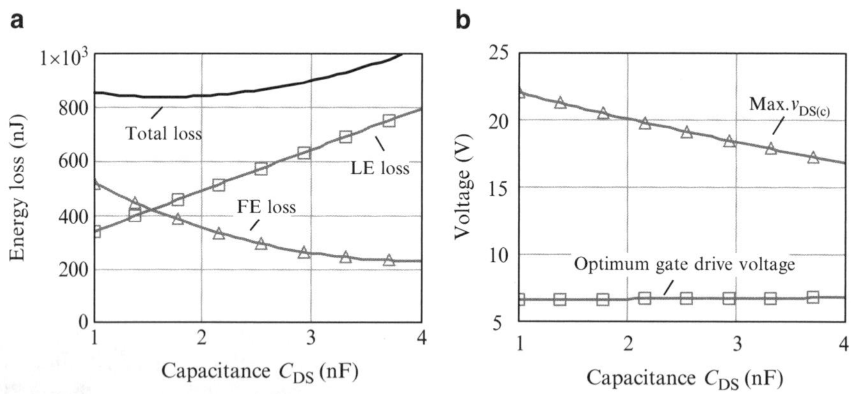

Fig. 3.20 (**a**) CtrlFET loss in both LE and FE transitions as a function of C_{DS}. (**b**) Maximum voltage stress across the CtrlFET and optimum gate drive voltage to achieve minimum losses. Values: $V_{in} = 12$ V, $C_{GS(c)} = 2$ nF, $C_{GD(c)} = 500$ pF, $g_{m(c)} = 50$ S, $V_{TH(c)} = 2$ V, $I_{D(c)0} = 30$ A, $L_{S(c)} = 100$ pH, $C_{oss(s)} = 6$ nF, $R_{G(c)} = 1\ \Omega$, $L_{G(c)} = -L_{S(c)}$, $L_{HB} = 1$ nH

This is due to the limitations of the gate driver regarding efficient fast switching capabilities. The existence of the source inductance is already sufficient to limit the maximum rate of change of the drain current. Thus, even with a zero gate resistance, the inductive turn-on snubber can provide overall loss savings.

Due to the fact that $L_{HB} > 0$, however, the optimum turn-off speed needs to be limited as illustrated in Fig. 3.15. This may force $R_{G(c)} > 0$ under optimum conditions, which in turn will also limit the turn-on speed. To compensate for the latter, the gate drive voltage, and hence, the gate drive losses, may need to be increased.

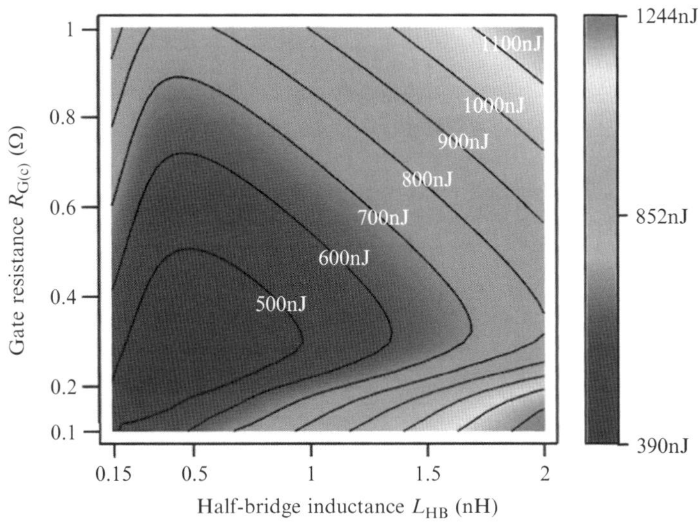

Fig. 3.21 Total CtrlFET switching loss of both LE and FE transitions as a function of gate resistance $R_{G(c)}$ and half-bridge loop inductance L_{HB}. Values: $V_{in} = 12$ V, $C_{GS(c)} = 2$ nF, $C_{GD(c)} = 500$ pF, $C_{DS(c)} = 1$ nF (minimum value; see added snubber capacitance from Fig. 3.22), $g_{m(c)} = 50$ S, $V_{TH(c)} = 2$ V, $I_{D(c)0} = 30$ A, $L_{G(c)} = -L_{S(c)}$, $L_{S(c)} = 100$ pH, $C_{oss(s)} = 6$ nF. The gate drive voltage is optimized for each value of $R_{G(c)}$ and L_{HB} (see Fig. 3.23)

This may justify the use of two independent gate current paths for the switching events. For the case illustrated in Fig. 3.21 (with a common gate current resistive path), the optimum $R_{G(c)}$ and L_{HB} result in ~0.3 Ω and ~0.5 nH, respectively. These values are significantly lower than those encountered in most VR, including system-in-package (SiP) based solutions (e.g., DrMOS).

For the same illustrated example, Fig. 3.22 shows the optimum $\Delta C_{DS(c)}$ that minimizes the hard-switching losses. The results indicate that the additional snubber capacitance is only beneficial for rather high $R_{G(c)}$ and L_{HB}. This is expected since a limited dv_{DS}/dt switching may be compensated with a larger output capacitance in order to reduce channel losses at turn-off. As the turn-off switching speed increases, any additional output capacitance can be unnecessary at turn-off and therefore detrimental for the overall net losses, as it is the case in the region around the minimum in the given example.

Given an optimum turn-on switching speed, the reduction of $R_{G(c)}$ enables enhancing the gate drive losses by means of lowering $V_{DRV(c)}$. On the contrary, the reduction of L_{HB} while keeping $R_{G(c)}$ constant may be compensated with the increase of $V_{DRV(c)}$ in order to maintain the effectiveness of the turn-on snubber. In some cases, a sharp augment of the driving voltage might be necessary as L_{HB} approaches very low values. This is reflected in the contour plot of Fig. 3.23, which represents the gate drive voltage that minimizes the total switching losses for each value in the graph of Fig. 3.21.

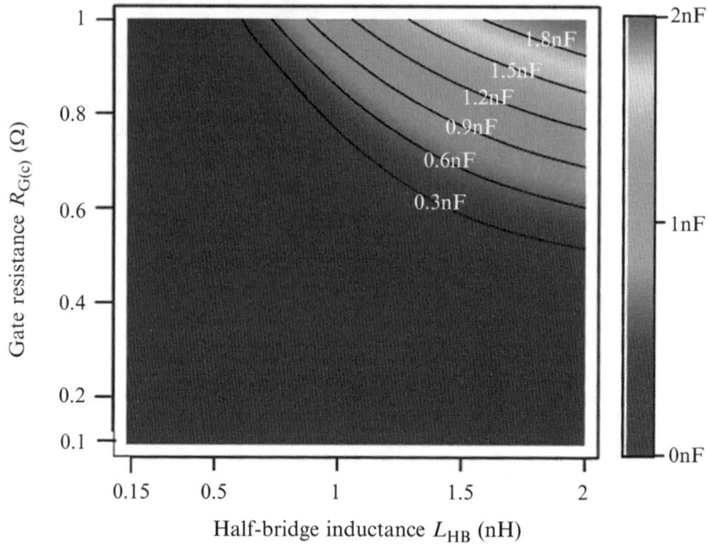

Fig. 3.22 Optimum additional CtrlFET drain-source capacitance $\Delta C_{DS(c)}$ corresponding to the simulations of Fig. 3.21

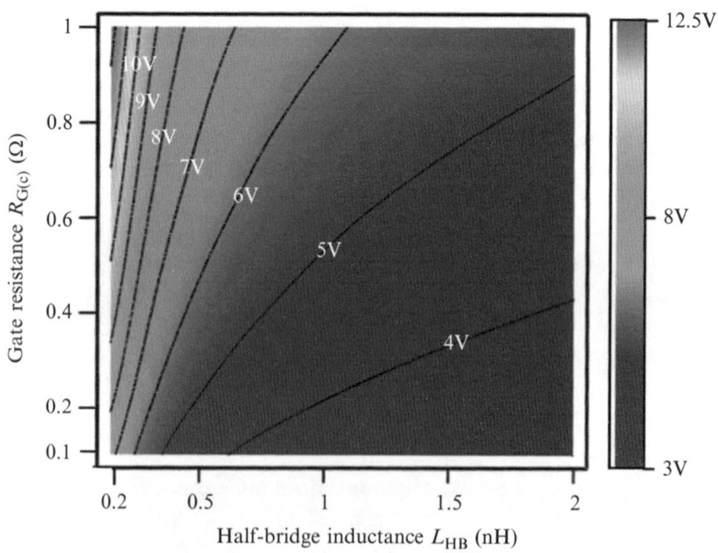

Fig. 3.23 Optimized CtrlFET gate drive voltage $V_{DRV(c)}$ corresponding to the simulations of Fig. 3.21

The negative influence that L_{HB} has on the voltage stress across the CtrlFET is depicted in Fig. 3.24. The maximum voltage stress is achieved with the highest L_{HB} and at rather low $R_{G(c)}$ (e.g., ~0.3 Ω). The maximum loop inductance to

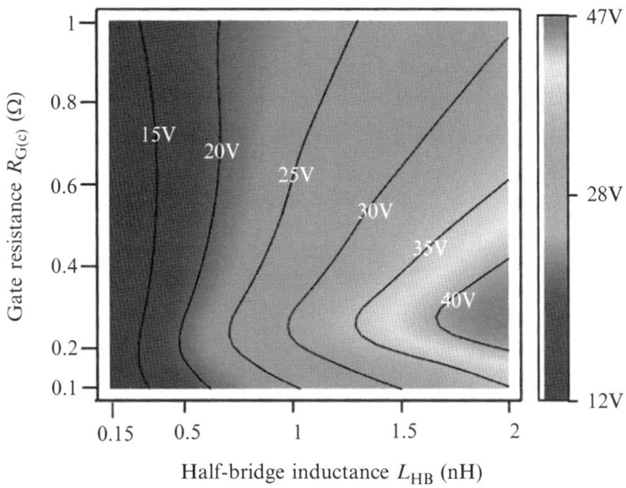

Fig. 3.24 Maximum CtrlFET drain-source voltage stress corresponding to the simulations of Fig. 3.21

avoid an overshoot higher than 30 V must be, under worst-case conditions, below 0.9 nH. From these resulting data, it is deduced that the most effective measure to mitigate the voltage stress at high current operation is to reduce L_{HB} rather than to increase $C_{DS(c)}$.

3.3 Leading Edge Switched-Node Ringing

The switched-node LE ringing transition is mainly represented with model (b) of Table 3.1. Other equivalent circuits, however, may need to be coupled such as gate drive model (h) and avalanche breakdown model (e), which is addressed later in this section. The following analysis thus focuses on model (b).

From the equivalent circuit, R_{HB} and L_{HB} represent the total half-bridge resistance and loop inductance, respectively. The model takes into account the SyncFET reverse recovery decay curve (assumed to be exponential), the gate bounce shoot-through channel current, and the $v_{DS(c)}$ switching transient. These excitation signals are expressed in terms of voltage and current sources to the resonant circuit as follows,

$$v_s(t) = V_{in} - (K_{v(c)}t + V_{DS(c)0}), \tag{3.10}$$

$$i_{cn(s)}(t) = K_{cn(s)}t + I_{cn(s)0}, \tag{3.11}$$

$$i_{dio(s)}(t) = I_{rr(s)} \cdot e^{-(t/\tau_{rr(s)})}. \tag{3.12}$$

In (3.10), voltage V_{in} is the converter input voltage, $K_{v(\text{c})}$ the voltage slope across the CtrlFET during commutation, and $V_{\text{DS(c)0}}$ the initial value. In each linear subinterval where shoot-through may occur, SyncFET channel current $i_{\text{cn(s)}}$ is approximated by a ramp function, the parameters of which derive from the model of Table 3.1(h). Solving the linear circuit with the above excitation functions yields the closed-form expressions of SyncFET drain current $i_{\text{cn(s)}}$ and drain-source voltage $v_{\text{DS(s)}}$. The earlier may be composed of five current contributions as,

$$i_{\text{D(s)}}(t) = i_{\text{D1(s)}}(t) + i_{\text{D2(s)}}(t) + i_{\text{D3(s)}}(t) + i_{\text{D4(s)}}(t) + i_{\text{D5(s)}}(t). \tag{3.13}$$

Each of the above current components corresponds to the three different current paths as well as initial conditions of the state variables. The transient analysis of the circuit (b) from Table 3.1 yields the following expressions for these current contributions. The first term corresponds to,

$$i_{\text{D1(s)}}(t) = \left(I_{\text{cn(s)0}} - C_{\text{oss(s)}}K_{v(\text{c})}\right)\left[1 - \frac{\mathrm{e}^{-\zeta\omega_n t}}{\sqrt{1-\zeta^2}}\sin\left(\omega_n\sqrt{1-\zeta^2}t + \phi\right)\right]. \tag{3.14}$$

The second term includes the initial conduction across the capacitances,

$$i_{\text{D2(s)}}(t) = C_{\text{oss(s)}}\left(V_{\text{in}} - V_{\text{DS(c)0}} - V_{\text{DS(s)0}}\right)\left[1 - \frac{\omega_n}{\sqrt{1-\zeta^2}}\mathrm{e}^{-\zeta\omega_n t}\sin\left(\omega_n\sqrt{1-\zeta^2}t\right)\right]. \tag{3.15}$$

The third term refers to the reverse recovery component as,

$$\begin{aligned} i_{\text{D3(s)}}(t) &= I_{\text{rr(s)}}\frac{\omega_n^2}{\tau_{\text{rr(s)}}^2 - 2\omega_n^2 + \omega_n^2\tau_{\text{rr(s)}}}\cdot\ldots \\ &\left(\mathrm{e}^{-(t/\tau_{\text{rr(s)}})} - \mathrm{e}^{-\zeta\omega_n t}\operatorname{sign}(A)\sqrt{1+A^{-2}}\sin\left(\omega_n\sqrt{1-\zeta^2}t + \psi\right)\right). \end{aligned} \tag{3.16}$$

The initial condition in the resonant inductor contributes as,

$$i_{\text{D4(s)}}(t) = -\frac{I_{\text{D(s)0}}}{\sqrt{1-\zeta^2}}\mathrm{e}^{-\zeta\omega_n t}\cdot\sin\left(\omega_n\sqrt{1-\zeta^2}t - \phi\right). \tag{3.17}$$

The last term in (3.13) accounts for the gate bounce shoot-through,

$$i_{\text{D5(s)}}(t) = K_{\text{cn(s)}}\left(t - \frac{2\zeta}{\omega_n} + \frac{\mathrm{e}^{-\zeta\omega_n t}}{\omega_n\sqrt{1-\zeta^2}}\cdot\sin\left(\omega_n\sqrt{1-\zeta^2}t + \varphi\right)\right). \tag{3.18}$$

The following relations are used above,

$$\varphi' = a\tan\left(\frac{2\zeta\sqrt{1-\zeta^2}}{2\zeta^2-1}\right), \tag{3.19}$$

$$\varphi = \varphi' + \pi \text{ if}\left(\frac{2\zeta\sqrt{1-\zeta^2}}{2\zeta^2-1} < 0\right) \text{else } \varphi = \varphi', \tag{3.20}$$

$$A = \frac{\omega_n\sqrt{1-\zeta^2}}{\zeta\omega_n - \tau_{\mathrm{rr(s)}}^{-1}}. \tag{3.21}$$

The resonant frequency and damping factors are, in this case,

$$\omega_n = \frac{1}{\sqrt{L_{\mathrm{HB}}C_{\mathrm{oss(s)}}}}, \tag{3.22}$$

$$\zeta = \frac{R_{\mathrm{HB}}}{2}\sqrt{\frac{C_{\mathrm{oss(s)}}}{L_{\mathrm{HB}}}}. \tag{3.23}$$

Furthermore, it follows that,

$$\phi = a\tan\left(\frac{\sqrt{1-\zeta^2}}{\zeta}\right), \tag{3.24}$$

$$\psi = a\tan(A). \tag{3.25}$$

As to the SyncFET drain-source voltage, it follows,

$$v_{\mathrm{DS(s)}}(t) = V_{\mathrm{in}} - V_{\mathrm{DS(c)0}} - K_{v\mathrm{(c)}}t - R_{\mathrm{HB}}i_{\mathrm{D(s)}}(t) - L_{\mathrm{HB}}\frac{\mathrm{d}i_{\mathrm{D(s)}}(t)}{\mathrm{d}t}. \tag{3.26}$$

The expression for the time derivative of the SyncFET drain current may also be obtained analytically as,

$$\frac{\mathrm{d}i_{\mathrm{D(s)}}(t)}{\mathrm{d}t} = K_{\mathrm{id(s)}}(t) = K_{\mathrm{id(s)1}}(t) + K_{\mathrm{id(s)2}}(t) + K_{\mathrm{id(s)3}}(t) + K_{\mathrm{id(s)4}}(t) + K_{\mathrm{id(s)5}}(t). \tag{3.27}$$

The five summands of the previous equation correspond to the time derivatives of the respective currents $i_{1(s)}(t)$ to $i_{5(s)}(t)$, defined in (3.13). The expressions for such derivatives are,

$$K_{\mathrm{id(s)1}}(t) = (I_{\mathrm{cn0}} - C_{\mathrm{oss(s)}}K_{v(\mathrm{c})})\frac{\omega_n}{\sqrt{1-\zeta^2}}\mathrm{e}^{-\zeta\omega_n t}\sin\left(\omega_n\sqrt{1-\zeta^2}t\right), \tag{3.28}$$

$$K_{\mathrm{id(s)2}}(t) = C_{\mathrm{oss(s)}}(V_{\mathrm{DS(s)0}} - V_{\mathrm{in}} + V_{\mathrm{DS(c)0}})\frac{\omega_n^2}{\sqrt{1-\zeta^2}}\mathrm{e}^{-\zeta\omega_n t}\sin\left(\omega_n\sqrt{1-\zeta^2}t - \phi\right), \tag{3.29}$$

$$K_{\mathrm{id(s)3}}(t) = -I_{\mathrm{rr(s)}}\frac{\tau_{\mathrm{rr(s)}}\omega_n^2}{\tau_{\mathrm{rr(s)}}^{-1} - 2\zeta\omega_n + \tau_{\mathrm{rr(s)}}\omega_n^2}\cdots \left(\begin{array}{c}\frac{\mathrm{e}^{-(t/\tau_{\mathrm{rr(s)}})}}{\tau_{\mathrm{rr(s)}}} - \mathrm{e}^{-\zeta\omega_n t}\mathrm{sign}(\psi)\omega_n\sqrt{1+A^{-2}}\ldots \\ \sin\left(\omega_n\sqrt{1-\zeta^2}t - \phi + \psi\right)\end{array}\right), \tag{3.30}$$

$$K_{\mathrm{id(s)4}}(t) = I_{\mathrm{D(s)0}}\frac{\omega_n}{\sqrt{1-\zeta^2}}\mathrm{e}^{-\zeta\omega_n t}\cdot\sin\left(\omega_n\sqrt{1-\zeta^2}t - 2\phi\right), \tag{3.31}$$

$$K_{\mathrm{id(s)5}}(t) = K_{\mathrm{cn(s)}}\left(\begin{array}{c} 2 + \frac{\mathrm{e}^{-\zeta\omega_n t}}{\sqrt{1-\zeta^2}}\cdot\ldots \\ \left[\begin{array}{c}\sqrt{1-\zeta^2}\sin\left(\omega_n\sqrt{1-\zeta^2}t + \varphi + \frac{\pi}{2}\right) - \cdots \\ \zeta\sin\left(\omega_n\sqrt{1-\zeta^2}t + \varphi\right) - \cdots \\ \sin\left(\omega_n\sqrt{1-\zeta^2}t + \varphi - \phi\right)\end{array}\right]\end{array}\right). \tag{3.32}$$

With the use of the above equations, Fig. 3.25 shows typical switched-node LE waveforms of the SyncFET transition that the proposed PLA model can

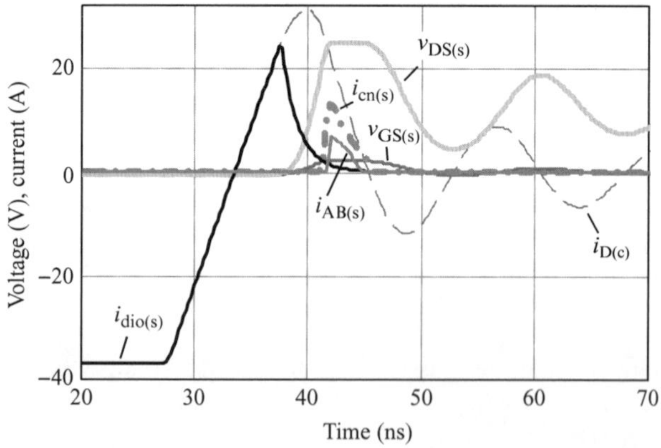

Fig. 3.25 Simulated switched-node LE transient based on the proposed PLA model (b), (e), and (h)

generate. In the illustration, the loss mechanisms of reverse recovery, gate bounce, and avalanche breakdown are simultaneously manifested. Note that current $i_{AB(s)}$ is the avalanche breakdown current contribution as defined in Chap. 2.

The simulated transient corresponds to a sequence of concatenated piecewise linear intervals where additional models (e) and (h) are coupled to reproduce the behavior during the periods of avalanche breakdown and gate bounce. Model (h) is required in order to determine the channel conduction as a consequence of the feedback effects of the transfer capacitance and/or source inductance. The following sections discuss these and other switching loss mechanisms with the use of the equivalent PLA circuit models.

3.3.1 Charging Loss

Power losses may be numerically computed from the current and voltage waveforms obtained in the previous section. This approach can offer accurate estimations even in complex scenarios as shown in Fig. 3.25, where multiple loss mechanisms may simultaneously manifest. Such cases may however be undesired in practice since, as it shall be argued in the following analysis, the mechanisms of reverse recovery, avalanche breakdown, and gate bounce drastically limit the converter efficiency at high switching frequency. One thus seeks for solutions where the power devices switch in the absence of the aforementioned loss mechanisms. Once achieved, the LE ringing loss may reduce to the charging loss of the MOSFET's parasitic capacitance, which may be estimated by means of a general closed-form expression. This loss calculation is through a direct analysis of the energy balance from the charging process as follows,

Whenever a constant voltage supply is connected to a capacitor via a switch that turns on and off in no time and where such electrical connection may be interfered by parasitic resistance and inductance elements, a charge transfer may be produced until a new permanent energy state is naturally reached. In such case, the work done to move a charge Q from a point of potential V_b to a point of potential V_e is,

$$w = \int_{V_b}^{V_e} Q \,\mathrm{d}V. \tag{3.33}$$

In a nonlinear voltage-dependent capacitor, the charge transfer is,

$$Q = \int_{V_b}^{V_e} C(V)\mathrm{d}V. \tag{3.34}$$

Combining expressions yields,

$$w = \int_{V_b}^{V_e} \int_{V_b}^{V_e} C(V')\mathrm{d}V'\mathrm{d}V. \tag{3.35}$$

Now, by changing the order of integration, it is straightforward to observe that,

$$w = \int_{V_b}^{V_e} C(v)(V_e - v)\mathrm{d}v. \tag{3.36}$$

For linear capacitors, the following familiar expression for the charge loss can be deduced from (3.36),

$$w_{\mathrm{lin}} = \frac{1}{2}C \cdot (V_e - v)^2. \tag{3.37}$$

Note that for (3.36) to be valid in the converter circuit, the ringing created by the switching action must be completely vanished before the next one begins. For resonant oscillations lasting longer than the ON time periods, the total charging loss may be higher or lower than the estimated by (3.36) depending on the energy state of the storage elements at the end of the transition phase.

Equation (3.36) reveals that the charging loss is independent on the impedance path. Also, it can be shown that the energy loss required to charge up a nonlinear capacitance may be lossier than charging a linear one to the same charge and voltage levels, as illustrated in the example of Fig. 3.26a. Right plot (b) shows the nonlinear capacitance under consideration and its mean value is in the interval 0–12V. Plot (a) further illustrates that the charging up phase is lossier than the discharge phase. This is a common case in nonlinear capacitances that vary inversely proportional to the applied voltage, such as the output capacitance of MOSFETs. The loss analysis in Chap. 4 further reveals that the average value of the LE and FE transient loss can be determined from the average capacitance of the charging interval. Thus, it will be demonstrated that ΔE1=ΔE2 in Fig. 3.26a.

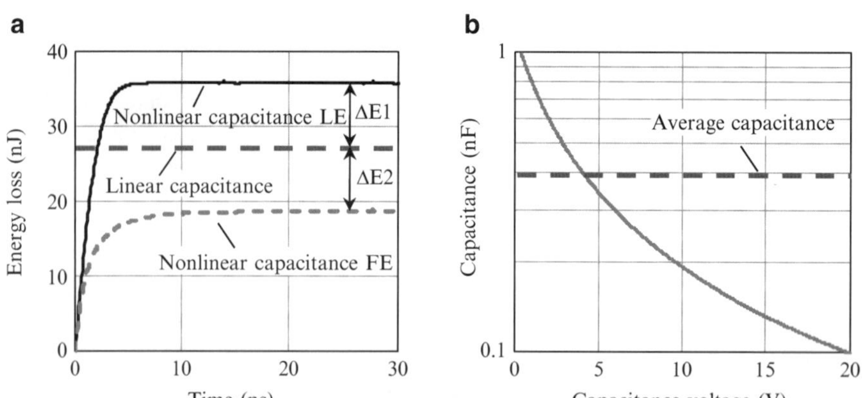

Fig. 3.26 Charging loss comparison between a linear and a nonlinear capacitance in the charging (LE) and discharging (FE) phases. Voltage charge from 0 to 12 V (LE transition) and from 12 to 0 V (FE transition). (**a**) Charging loss transient of a nonlinear *R–C* series circuit ($R = 8\ \Omega$, $C = C(V)$, as shown in plot (**b**)). (**b**) Nonlinear capacitance as a function of the terminal voltage and corresponding average capacitance in the interval 0–12 V

3.3.2 Overvoltage Stress

One key aspect of the converter design is the voltage rating of the power MOSFETs, which in the converter under consideration is usually well above the input voltage level. This has to do with the voltage overshoots across the blocking device during the switching ringing. In order to avoid avalanche breakdown, the MOSFETs must be rated to at least the maximum expected resonant peak voltage. Because lower voltage rated devices generally offer better performance and/or lower cost, it is desirable to minimize the maximum voltage stress.

For the case of the switched-node LE transition, the peak voltage across the SyncFET may be estimated from the equations derived from model (b) of Table 3.1. In this approach, a sufficient number of discretized linear intervals are necessary to cope with the nonlinearity of the MOSFETs capacitances. Neglecting for now the contributions of the body diode and channel currents, the worst-case voltage overshoot is given for an ideal CtrlFET turn-on. In the model, this translates into a voltage excitation corresponding to a step function. Figure 3.27 shows switching waveforms of a typical case example. The capacitance nonlinearity distorts the resonant oscillation, making wide valleys and sharp maximums in the voltage waveform, and slow ramp-ups and steep ramp-downs in the current waveform. This in turn produces additional charging losses, as shown in the previous section, as well as voltage spikes across the device that may well exceed twice the input voltage, which is the maximum limit in a linear resonator with zero initial conditions. This can be contrasted in Fig. 3.28. In Fig. 3.28a, the maximum peak voltage is given as a function of the total charge transfer. Note that in order to adjust the charge transfer, the capacitance is modified accordingly while keeping the input

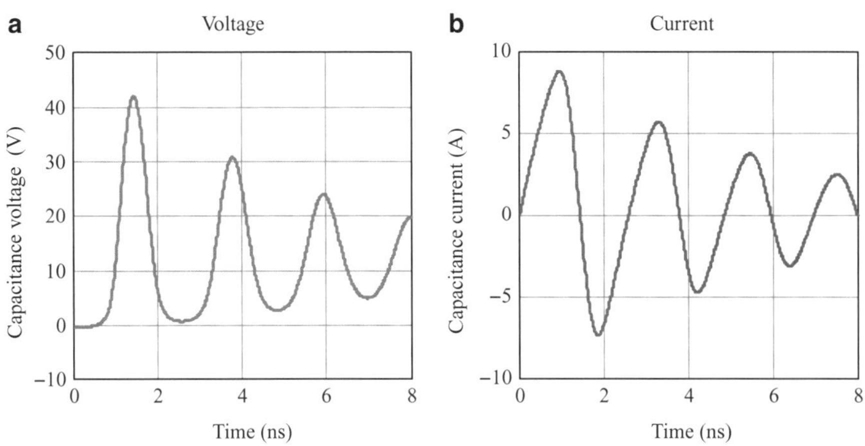

Fig. 3.27 SyncFET output capacitance charge-up ringing during the LE switched-node transition. Simulation conditions referred to model (b) of Table 3.1 with an ideal step function of −0.5 V to V_{in} at source voltage source v_s and at time equal to zero: $V_{in} = 12$ V, $i_{cn(s)} = 0$ A, $i_{dio(s)} = 0$ A, $R_{HB} = 0.4\ \Omega$, $L_{HB} = 1$ nH. See Fig. 3.28a ($Q_t = 7$ nC) for capacitance data

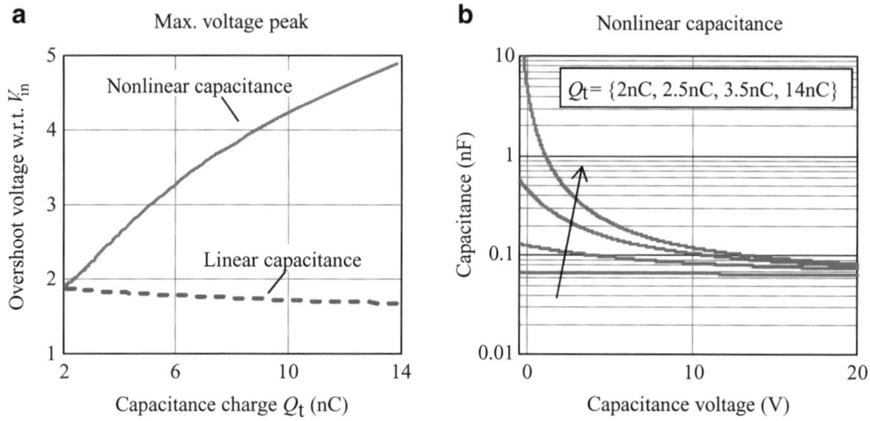

Fig. 3.28 (**a**) Maximum overshoot voltage relative to V_{in} for different capacitance values. Comparison between linear and nonlinear capacitances of equal stored charge Q_t, which refer to the charge transfer of a step voltage transition from -0.5 to 30 V. Nonlinear capacitances are shown in part (**b**) for various Q_t. Parameter values correspond to those of Fig. 3.27

voltage step equal for all cases (i.e., −0.5 to 12V). While this is trivial for the linear case, in the nonlinear case the capacitance may be increased in different ways, one of which is as shown in Fig. 3.28b. Such approach effectively produces an increase of the nonlinearity as the capacitance charge increases, which in turn yields an increase of the voltage spikes, as depicted in Fig. 3.28a. This goes against the behavior of a linear damped resonant oscillation, where the increase of the capacitance generally produces a decrease of the peak voltage. This later statement is also illustrated in Fig. 3.28a (dashed curve).

The effect of reverse recovery and gate bounce on the voltage stress is discussed in Sects. 3.3.4 and 3.3.5.

3.3.3 Avalanche Breakdown

The power converter must generally be designed such that the MOSFETs are prevented from avalanche breakdown for reasons concerning reliability, lifetime, and power loss savings. The latter is addressed in the following discussion to argue that avalanche can be a severe loss mechanism and thus must be avoided. With the aid of models (b) and (e) from Table 3.1, avalanche breakdown operation is analyzed and simple expressions deduced to estimate the associated losses.

Some modern power MOSFETs can feature such high levels of ruggedness that avalanche breakdown may be allowed as regular operation without compromising reliability. This has enabled designers to exploit this feature mostly to dump oscillations and absorb the energy of inductive loads even without the need of employing freewheeling diodes.

Damping the switched-node ringing in the SRBC could be one potential target application as a way to reduce EMI. That is, during avalanche breakdown, the energy stored in the half-bridge loop inductance is discharged into the avalanched device, thereby reducing the oscillation energy. This is illustrated in the simulations of Fig. 3.29, which compare the ringing amplitude and switching losses of two transitions with and without avalanche breakdown. As it can be observed, one characteristic of avalanche operation is that the ringing energy, which is fully lost in case of a natural damped oscillation, can be quickly dissipated without the need of producing excessive overshoots. On the downside, however, during avalanche breakdown, the direction of current flow produces energy consumption from the power supply during the entire period of avalanche. The following analysis demonstrates that these additional losses are proportional to the inductance energy and the avalanche time.

As in the previous section, model (b) from Table 3.1 is considered without the contributions of the channel and body diode currents. Also, voltage source v_s behaves as an ideal step-up function, which corresponds to the worst-case overshoot stress.

Figure 3.30 illustrates the case of a step response of a series resonant circuit interrupted by an avalanche breakdown transition that clamps the maximum peak voltage. The example is generalized for the case of initial conditions in both state variables. The switching transient is divided in the three intervals: A, B, and C. The total switching loss is the sum of the losses in each interval, i.e.,

$$w_{\mathrm{tLEr}} = w_{\mathrm{A\_LEr}} + w_{\mathrm{B\_LEr}} + w_{\mathrm{C\_LEr}}. \tag{3.38}$$

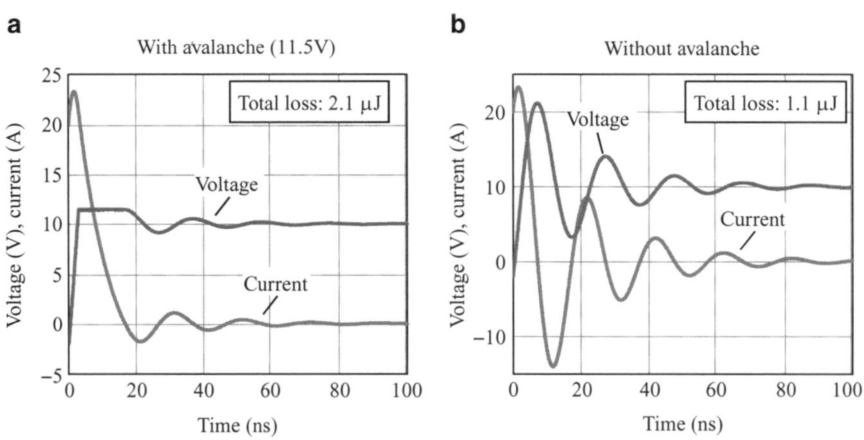

Fig. 3.29 Simulated step response of a series resonant circuit (see model (b) of Table 3.1) with and without avalanche breakdown. Parameter values: $V_{\mathrm{in}} = 10\mathrm{V}$, $i_{\mathrm{cn}(s)} = 0$ A, $i_{\mathrm{dio(s)}} = 0$ A, $I_{D(\mathrm{s})0} = 20$ A, $V_{\mathrm{DS(s)0}} = -2$ V, $R_{\mathrm{HB}} = 0.2\ \Omega$, $L_{\mathrm{HB}} = 2$ nH, $C_{\mathrm{oss(s)}} = 5$ nF (linear)

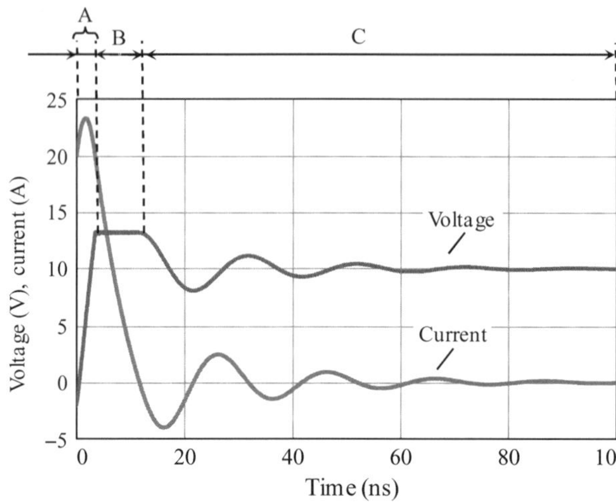

Fig. 3.30 Step response of a series resonant circuit with avalanche breakdown. Intervals A and C refer to model (b) from Table 3.1, whereas interval B corresponds to model (e). Parameter values: $V_{in} = 10$ V, $V_{AB(s)} = 13$ V, $i_{cn(s)} = 0$ A, $i_{dio(s)} = 0$ A, $I_{D(s)0} = 20$ A, $V_{DS(s)0} = -2$ V, $R_{HB} = 0.2\Omega$, $L_{HB} = 2$ nH, $C_{oss(s)} = 5$ nF (linear)

Considering that the resonant capacitance may be nonlinear, the energy loss in the first interval may be calculated as,

$$w_{A\_LEr} = \int_{V_{DS(s)0}}^{V_{AB(s)}} C_{oss(s)}(v) \cdot (V_{in} - v)\mathrm{d}v + \frac{1}{2} L_{HB}(I_{D(s)0}^2 - I_{AB(s)0}^2), \tag{3.39}$$

where $V_{DS(s)0}$ and $I_{D(S)0}$ are the initial conditions of interval A, $I_{AB(S)0}$ the initial condition of interval B, and $V_{AB(S)}$ the avalanche breakdown voltage level of the power device. The first term in (3.39) derives from (3.36). In interval (B), the SyncFET drain-source voltage is clamped to the avalanche level, and model (e) from Table 3.1 is applied with the assumption that the voltage drop across the CtrlFET drain-source terminals is negligible. Avalanche interval B lasts as soon as the current reaches zero. Losses $w_B$can thus be expressed as,

$$w_{B\_LEr} = V_{in}\tau_{AB}\left[I_{AB(s)0} - I_f\left(1 + \frac{I_{AB(s)0}}{I_f}\right)\right] + \frac{1}{2} L_{HB} I_{AB(s)0}^2. \tag{3.40}$$

The first term in (3.40) is delivered by the power supply, whereas the second one comes from the loop inductance. The two unknown variables of the first summand are,

$$\tau_{AB} = \frac{L_{HB}}{R_{HB}}, \tag{3.41}$$

$$I_{\mathrm{f}} = \frac{V_{\mathrm{AB(s)}} - V_{\mathrm{in}}}{R_{\mathrm{HB}}}. \quad (3.42)$$

Switching interval C begins with zero current condition. Therefore, the energy loss expression can be directly derived from (3.36), which yields,

$$w_{\mathrm{C\_LEr}} = \int_{V_{\mathrm{AB(s)}}}^{V_{\mathrm{in}}} C_{\mathrm{oss(s)}}(v) \cdot (V_{\mathrm{in}} - v)\mathrm{d}v. \quad (3.43)$$

Combining the above equations, it can be identified that the total switching loss may be broken down in two basic terms,

$$w_{\mathrm{tLEr}} = w_{\mathrm{noAB\_LEr}} + w_{\mathrm{AB(s)}}. \quad (3.44)$$

The first term refers to the charging loss, which equals to the total loss of a switching transient without avalanche. The following expression accounts for the charging loss and includes the initial conditions in both state variables.

$$w_{\mathrm{noAB\_LEr}} = \int_{V_{\mathrm{DS(s)0}}}^{V_{\mathrm{in}}} C_{\mathrm{oss(s)}}(v) \cdot (V_{\mathrm{in}} - v)\mathrm{d}v + \frac{1}{2} L_{\mathrm{HB}} I_{\mathrm{D(s)0}}^2. \quad (3.45)$$

The second term in (3.44) is therefore the actual additional loss contribution due to avalanche breakdown, that is,

$$w_{\mathrm{AB(s)}} = V_{\mathrm{in}} \cdot \tau_{\mathrm{AB}} \cdot \left[I_{\mathrm{AB(s)0}} - I_f \cdot \left(1 + \frac{I_{\mathrm{AB(s)0}}}{I_f}\right)\right]. \quad (3.46)$$

According to (3.46) and (3.42), the avalanche loss increases with V_{in} and $I_{\mathrm{AB(s)0}}$, whereas it decreases as the avalanche voltage level goes up. This is illustrated in the simulated example of Fig. 3.31, where the avalanche breakdown vanishes for $V_{\mathrm{AB(s)}} > 22\mathrm{V}$.

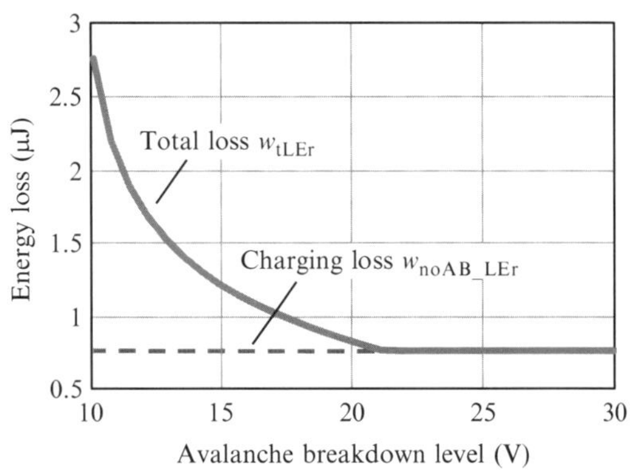

Fig. 3.31 Avalanche and charging losses as a function of the avalanche breakdown voltage level of the power MOSFET. Simulation conditions correspond to those of Fig. 3.30

An approximated expression for the CtrlFET avalanche breakdown in the FE transition can be analogously deduced as it is argued in Sect. 3.4.3.

3.3.4 Reverse Recovery

The linear switched-node ringing model (b) from Table 3.1 considers the contribution of the SyncFET body diode reverse recovery, whose detrimental effects on the converter switching shall be measured in terms of energy loss and switched-node ringing contributions. The later is not just an EMI concern alone, but also turns out to be a trigger to other loss mechanisms. This will become apparent from the following simulation results, which suggest the mitigation of reverse recovery as a major enabling condition for high switching frequency operation.

Figure 3.32 illustrates the simulation of a switched-node LE transient under the influence of the reverse recovery current from the SyncFET body diode. The curves show that, upon fast switching, the CtrlFET may be already fully on (i.e., operating in the ohmic region) before reverse recovery takes place, thereby avoiding the SyncFET current to flow through the CtrlFET during the hard-switching time.

During the time interval delimited by the zero crossing of the SyncFET drain current and the reverse recovery peak (i.e., time t_{rr0} from Fig. 3.32), the SyncFET drain-source voltage remains clamped to the body diode forward voltage as the stored charge is being removed from the *p-n* junction. Though this effect extends the LE time, stored charge energy can be partially recovered during t_{rr0}.

A far lossier transition may develop after interval t_{rr0} has elapsed. Then the reverse recovery decay current starts and the SyncFET drain-source voltage gets unclamped. These conditions give rise to the actual leading edge transition (LE) of

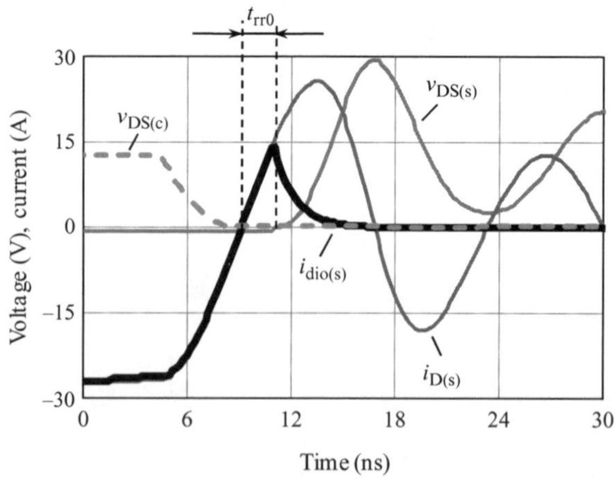

Fig. 3.32 Simulated switched-node LE transitions including reverse recovery and fast CtrlFET turn-on. Results based on the PLA model with multiple linear intervals

the switched-node voltage. As the SyncFET enters into the first quadrant operation, losses are generated in the body diode until the reverse recovery current vanishes.

As soon as this unclamped phase begins, the half-bridge loop inductance resonates with the SyncFET output capacitance, which can produce a significant overvoltage stress in the device. The resonant amplitude is strongly dependent on the stored energy in the inductance at the beginning of the unclamped phase. This energy may be dissipated in the resistance of the half-bridge loop and gate circuits during the ringing period. It can be estimated by the following expression,

$$w_{\mathrm{rr\_peak}} = \frac{1}{2} L_{\mathrm{HB}} \cdot I_{\mathrm{rr(s)}}^2. \tag{3.47}$$

Figure 3.33 shows how the reverse recovery current can critically impact switching losses. The example illustrates the contributions of the SyncFET charging and reverse recovery losses. Left plot (a) depicts the loss dependence on the reverse recovery peak current. The total losses are separated in the loss energy portion delivered from voltage supply V_{in}, and that of the half-bridge loop inductance at the time of the reverse recovery peak (i.e., $w_{\mathrm{rr\_peak}}$). For the particular case of $I_{\mathrm{rr(s)}}$=0 A (i.e., no reverse recovery), the total switching loss equals the charging loss given from (3.36). Right plot (b) shows the switching loss as a function of $I_{\mathrm{rr(s)}}$ and for various reverse recovery time constants. The dashed curve at $\tau_{\mathrm{rr(s)}}$=0 corresponds to the charging loss with initial conditions in L_{HB}, which derives from (3.45). The curves reveal the strong loss dependence on both reverse recovery parameters. Even for moderate L_{HB} and $\tau_{\mathrm{rr(s)}}$, a reverse recovery peak of 12 A can double the switching loss produced in case of no reverse recovery.

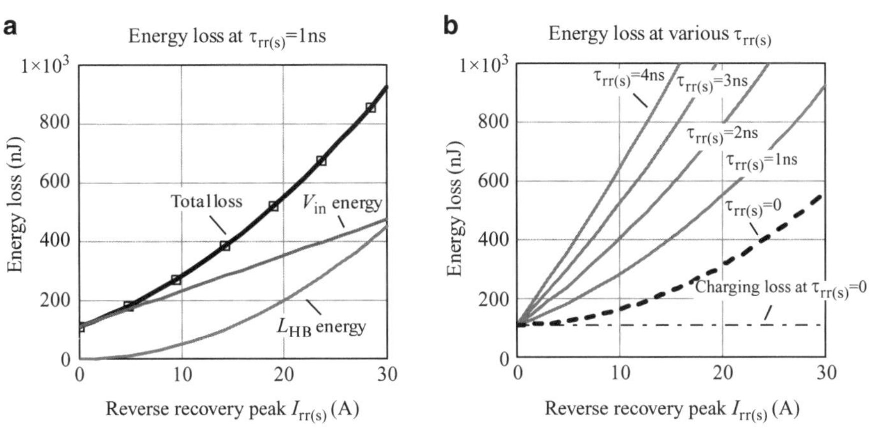

Fig. 3.33 Simulated step response of circuit model (b) from Table 3.1. Plot (**a**): energy loss as a function of the reverse recovery peak; plot (**b**): same as plot (**a**) but for various reverse recovery time constants. Parameter values: $V_{\mathrm{in}} = 12$ V, $i_{\mathrm{cn(s)}} = 0$ A, $V_{\mathrm{DS(s)0}} = 0$ V, $R_{\mathrm{HB}} = 0.3\ \Omega$, $L_{\mathrm{HB}} = 1$ nH, for $C_{\mathrm{oss(s)}}$ see Fig. 3.40

Relevant effects of the reverse recovery on the switched-node ringing voltage are further illustrated in Fig. 3.34. The two main parameters considered are the maximum peak voltage and the time derivative of the SyncFET drain-source voltage (dv/dt). These are relevant to the mechanisms of avalanche breakdown and gate bounce. According to the resulting curves, the peak voltage appears to be proportional to the dv/dt. Both parameters increase with $I_{rr(s)}$ only for sufficient low $\tau_{rr(s)}$. For moderate and high $I_{rr(s)}$, the reduction of $\tau_{rr(s)}$tends to worsen both ringing characteristics.

The same trends are observed when displaying the dependencies on $\tau_{rr(s)}$, as shown in Figs. 3.35 and 3.36.

It is common practice for MOSFET manufacturers to specify the reverse recovery in terms of a recovery charge (Q_r) and a reverse recovery time (t_{rr}), which, according to the proposed model, may be defined as follows,

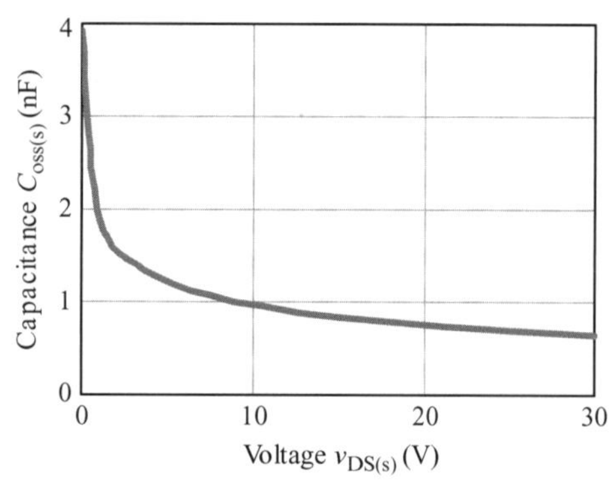

Fig. 3.34 Impact of reverse recovery on switched-node ringing. Plot (**a**): relative voltage overshoot; plot (**b**) maximum time derivative of switch-node voltage. Same parameter values as in example of Fig. 3.33

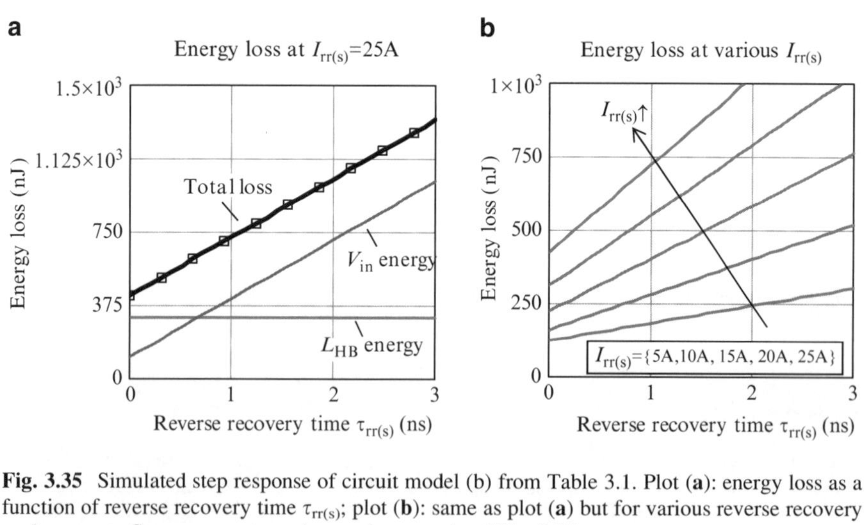

Fig. 3.35 Simulated step response of circuit model (b) from Table 3.1. Plot (**a**): energy loss as a function of reverse recovery time $\tau_{rr(s)}$; plot (**b**): same as plot (**a**) but for various reverse recovery peak currents. Same parameter values as in example of Fig. 3.33

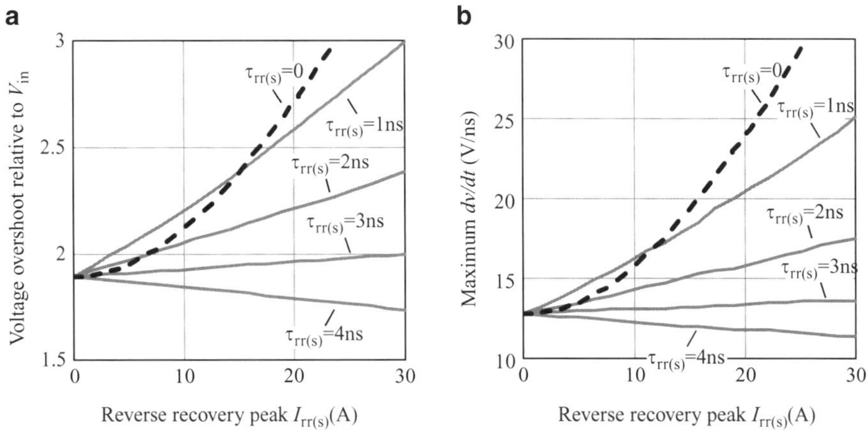

Fig. 3.36 Impact of reverse recovery on switched-node ringing. Plot (**a**): relative voltage overshoot; plot (**b**) maximum time derivative of switch-node voltage. Same parameter values as in example of Fig. 3.33

$$Q_{r(s)} = I_{rr(s)}\left(\frac{t_{rr0}}{2} + \tau_{rr(s)}\right), \tag{3.48}$$

$$t_{rr(s)} = t_{rr0} + \frac{3}{2}\tau_{rr(s)}. \tag{3.49}$$

The reverse recovery parameters are dependent on the switching speed (di/dt) and the load current. The relations may be established empirically either by accurate device physics simulations or, alternatively, by experiments. Results on this extraction procedure are provided in Appendix B.

3.3.5 Gate Bounce

Gate bounce shoot-through describes a loss mechanism by which the MOSFET's channel is spuriously turned on by the way of an induced gate voltage swing from the half-bridge circuit. Although the gate-source potential may oscillate in both MOSFETs due to coupled noise, gate bounce shoot-through is usually severely manifested only in the SyncFET device, as already identified in Chap. 2. The following study therefore focuses on this device.

The coupling between the gate and power circuits can be capacitive due to the transfer capacitance, but also inductive by the way of the source inductance. The two mechanisms are described as follows.

The worst-case capacitive coupling is produced during the switched-node voltage rise. That is, the fast dv/dt causes a charge transfer from the half-bridge to the SyncFET gate circuit such that this may charge up the gate above the threshold level, thereby inducing a spurious turn-on.

In the di/dt switching phase, the drain current decrease induces a voltage across $L_{S(s)}$ that negatively charges the gate. Although this effect may alone worsen the reverse recovery transient (see Appendix B), it does not directly lever gate bounce shoot-through losses. On the other hand, a negatively induced voltage across $L_{S(s)}$ may however be developed in the ringing transient during the drain current decay phases, particularly the first one as it is the highest in amplitude. Like in the capacitive gate bounce, such effect may trigger shoot-through, thus increasing the susceptibility of the device to gate bounce losses.

The analysis of both capacitive and inductive gate bounce can be partially carried out by means of circuit model (h) from Table 3.1, which describes the coupling effects of the feedback transfer capacitance and source inductance. For a linearized transient interval, both drain-source voltage and drain current excitations are assumed to vary at a constant rate. Accordingly, the time-dependent function for the gate current may be broken down into three components,

$$i_{G(s)}(t) = i_{G1(s)}(t) + i_{G2(s)}(t) + i_{G3(s)}(t). \tag{3.50}$$

The first term accounts for the time derivative of drain-source voltage $K_{v(s)}$ as,

$$i_{G1(s)}(t) = -C_{GD(s)} \cdot K_{v(s)} \cdot \left[1 - \frac{e^{-\zeta\omega_n t}}{\sqrt{1-\zeta^2}} \cdot \sin\left(\omega_n\sqrt{1-\zeta^2}t + \phi\right)\right]. \tag{3.51}$$

The second term in (3.50) includes the effect of the feedback source inductance as a consequence of drain current time derivative $K_{id(s)}$,

$$i_{G2(s)}(t) = -\left(L_{S(s)}K_{id(s)} + V_{GS(s)0}\right) \cdot C_{iss(s)} \cdot \frac{e^{-\zeta\omega_n t}}{\sqrt{1-\zeta^2}} \cdot \sin\left(\omega_n\sqrt{1-\zeta^2}t\right). \tag{3.52}$$

Finally, the last term from (3.50) considers the initial condition of the gate inductance,

$$i_{G3(s)}(t) = -\frac{I_{G(s)0}}{\sqrt{1-\zeta^2}} \cdot e^{-\zeta\omega_n t} \cdot \sin\left(\omega_n\sqrt{1-\zeta^2}t - \phi\right). \tag{3.53}$$

The natural resonant frequency (in radians per second) and damping factor of the second-order circuit are, for this particular case,

$$\omega_n = \frac{1}{\sqrt{(L_{S(s)} + L_{G(s)})C_{iss(s)}}}, \tag{3.54}$$

$$\zeta = \frac{R_{G(s)}}{2}\sqrt{\frac{C_{iss(s)}}{L_{S(s)} + L_{G(s)}}}. \quad (3.55)$$

where damping factor ζ is assumed to be lower than 1 in fast gate drivers. In OFF state (zero driving voltage), the gate-source voltage can be expressed as,

$$v_{GS(s)}(t) = -L_{S(s)}K_{id(s)} - R_{G(s)}i_{G(s)} - (L_{G(s)} + L_{S(s)})\frac{di_{G(s)}}{dt}. \quad (3.56)$$

The time derivative of the current can be derived from (3.50) to (3.53), which yield an expression of three summands as in (3.50),

$$\frac{di_{G(s)}}{dt} = \frac{di_{G1(s)}}{dt} + \frac{di_{G2(s)}}{dt} + \frac{di_{G3(s)}}{dt}. \quad (3.57)$$

where it follows that,

$$\frac{di_{G1(s)}}{dt} = -C_{GD(s)}\omega_n K_{v(s)}\frac{e^{-\zeta\omega_n t}}{\sqrt{1-\zeta^2}} \cdot \sin\left(\omega_n\sqrt{1-\zeta^2}t\right), \quad (3.58)$$

$$\frac{di_{G2(s)}}{dt} = (V_{GS(s)0} + L_{S(s)}K_{id(s)})C_{iss(s)}\omega_n^2\frac{e^{-\zeta\omega_n t}}{\sqrt{1-\zeta^2}} \cdot \sin\left(\omega_n\sqrt{1-\zeta^2}t - \phi\right), \quad (3.59)$$

$$\frac{di_{G3(s)}}{dt} = \frac{I_{G(s)0}}{\sqrt{1-\zeta^2}}\omega_n e^{-\zeta\omega_n t} \cdot \sin\left(\omega_n\sqrt{1-\zeta^2}t - 2\phi\right). \quad (3.60)$$

Neglecting for now inductive gate bounce and assuming zero initial conditions in all state variables, the maximum gate-source voltage due to capacitive gate bounce can be estimated by means of the following expression,

$$v_{GS(s)max} = K_{v(s)}R_{G(s)}C_{GD(s)} \cdot \left(1 + \frac{R_{G(s)}}{2\zeta}e^{-(\zeta/\sqrt{1-\zeta^2})\left(\pi - a\tan\left(\sqrt{1-\zeta^2}/\zeta\right)\right)}\right). \quad (3.61)$$

The second term within brackets accounts for the contribution of the voltage overshoot upon underdamped condition (i.e., $\zeta<1$). This term equals zero for $\zeta \geq 1$. Note that by virtue of the driver impedance, both $v_{GS(s)max}$ and steady state$v_{GS(s)}$ are independent on the drain-source voltage peak. This contrasts with the case of open gate, as it is analyzed later on in this section.

Despite the linear approximation (nonlinear effect shall be introduced later), (3.61) provides good insights of the parameter dependencies influencing gate bounce strictly caused by feedback capacitive coupling. Based on this relation, a simple condition can be readily deduced to prevent gate bounce shoot-through regardless of the maximum drain-source voltage, that is,

$$\tau_{\mathrm{DRVgb(s)}} = R_{\mathrm{G(s)}} \cdot C_{\mathrm{GD(s)}} \leq \frac{V_{\mathrm{TH(s)}}}{K_{v(\mathrm{s})}}, \tag{3.62}$$

$$\zeta \geq 1. \tag{3.63}$$

The first condition highlights the importance to keep time constant $\tau_{\mathrm{DRVgb(s)}}$ low. The values of $L_{\mathrm{G(s)}}$ and $C_{\mathrm{GS(s)}}$ are irrelevant as long as the underdamped condition is prevented. Unfortunately, in practice, the source inductance is usually large, which results in resonant voltage overshoots above $K_{v(\mathrm{s})}R_{\mathrm{G(s)}}C_{\mathrm{GD(s)}}$.

Figures 3.37 and 3.38 show the expected maximum gate voltage for the case of underdamped transients (governed by (3.61)) as a function of the different circuit parameters involved. As the curves show, also in this operating regime, the reduction of $\tau_{\mathrm{DRVgb(s)}}$ seems to be the most effective measure.

The values of $K_{v(\mathrm{s})}$ and $K_{\mathrm{id(s)}}$ are determined from the half-bridge circuit conditions. Both parameters increase (in absolute values) in proportion to the speed of the CtrlFET turn-on and the reverse recovery current. Estimating the excitations of model (h) thus involves the use of model (b). Furthermore, the nonlinearity of the capacitances as well as the initial conditions of the state variables may need to be considered for an accurate assessment of gate bounce.

Figure 3.39 illustrates a more accurate switched-node LE transient simulation including gate bounce shoot-through. The spurious channel turn-on is triggered by

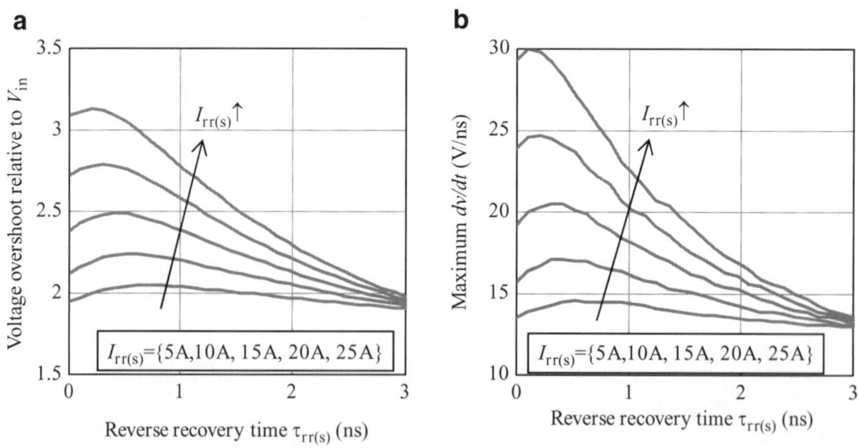

Fig. 3.37 Maximum gate voltage variation due to capacitive coupling according to (3.64). Parameter values: $C_{\mathrm{GS(s)}} = 4$ nF, $C_{\mathrm{GD(s)}} = 0.5$ nF, $L_{\mathrm{G(s)}} = 4$ nH (plot (**a**)), $R_{\mathrm{G(s)}} = 1\ \Omega$ (plot (**b**))

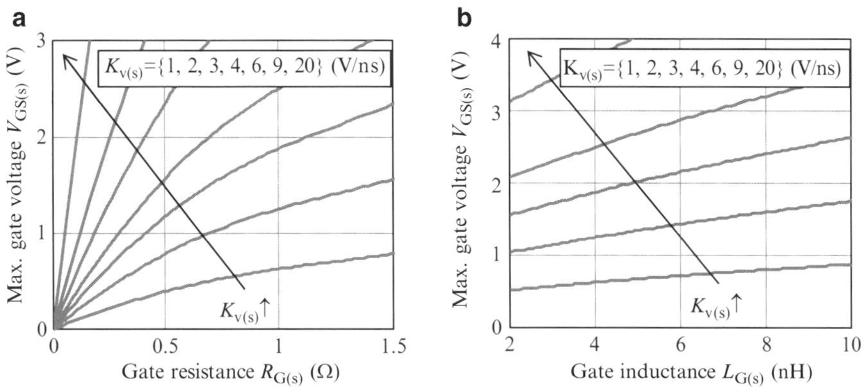

Fig. 3.38 Maximum gate voltage variation due to capacitive coupling according to (3.64). Parameter values: $L_{G(s)} = 4$ nH, $R_{G(s)} = 1\ \Omega$, $C_{GS(s)} = 4$ nF (plot (**a**)), $C_{GD(s)} = 0.5$ nF (plot (**b**))

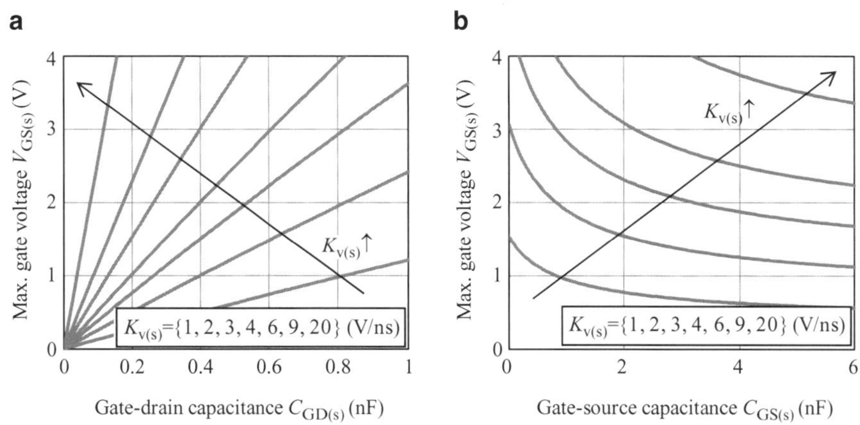

Fig. 3.39 Switched-node LE simulation including gate bounce shoot-through based on coupled models (b) and (h) from Table 3.1. Parameter values: $C_{GS(s)} = 3.8$ nF, $R_{G(s)} = 1.5\ \Omega$, $L_{G(s)} = 4$ nH, $L_{S(s)} = 200$ pH, $I_{rr(s)} = 20$ A, $\tau_{rr(s)} = 0.5$ ns, $L_{HB} = 1$ nH, $R_{HB} = 0.3\ \Omega$, $V_{in} = 12$ V. For the transconductance and rest of MOSFET capacitances, see Figs. 3.40 and 3.41

capacitive and inductive gate bounce, both of which are indirectly intensified by reverse recovery. Namely, the initial stored energy in L_{HB} causes a soaring $dv_{DS(s)}/dt$ as high as 20V/ns. At the same time, the drain current falls quickly toward zero as a consequence of the voltage spike, which induces a high voltage across $L_{S(s)}$. Both effects contribute strongly to gate bounce. The loss mechanism also leads to a damped oscillation and a somewhat distorted voltage spike, which are characteristic in switched-node transients under the influence of gate bounce shoot-through or avalanche breakdown mechanisms.

The illustrated simulation takes into account the nonlinearity of the transfer and output capacitance as well as the transconductance, both of which are represented in Figs 3.40 and 3.41. The voltage dependence of the capacitances is crucial to accurately estimate the maximum voltage coupled at the gate, as already learnt from the analysis on overvoltage stress in Sect. 3.3.2. Under the worst-case scenario of capacitive gate bounce where the gate is open and thus the driving current is zero, $v_{GS(s)max}$ is given by the drain-source voltage level and the capacitance divider formed by $C_{GD(s)}$ and $C_{GS(s)}$. When nonlinear capacitances are considered, then the gate voltage can be calculated as shown in Sect. 2.5.2.2 (see (2.38)). Based on this relation, the following definition is proposed to measure the capacitive gate bounce susceptibility,

$$\mathrm{GBS}_{GSO(s)} = \frac{V_{TH(s)} - V_{GS(s)0}}{V_{DSgb(s)} - V_{DS(s)0}}, \tag{3.64}$$

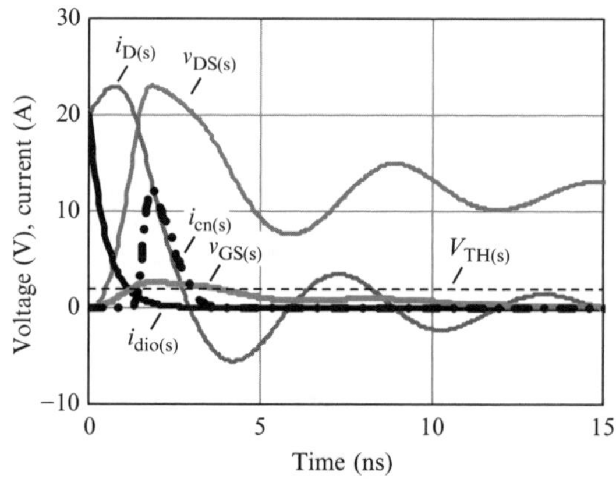

Fig. 3.40 Reference SyncFET output capacitance used in the analysis of reverse recovery

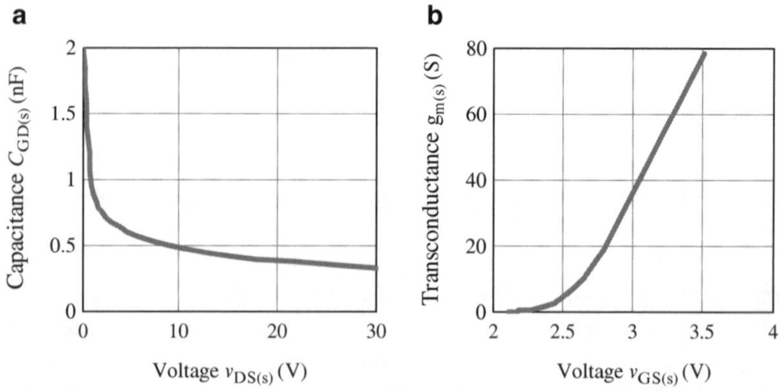

Fig. 3.41 Reference SyncFET transfer capacitance and transconductance values used in the analysis of gate bounce

where $V_{DSgb(s)}$ is the drain-source voltage above which the channel turns on. That is,

$$V_{TH(s)} = \int_{V_{DS(s)0}}^{V_{DSgb(s)}} \frac{C_{rss(s)}}{C_{iss(s)}} dv + V_{GS(s)0}. \tag{3.65}$$

Thus, $GBS_{GSO(s)}$ is a measure of the voltage divider ratio that the two capacitances form when the gate terminal is open. High values of $GBS_{GSO(s)}$ denote high susceptibility to capacitive gate bounce. For linear capacitances, (3.64) reduces to,

$$GBS_{GSO(s)} = \frac{C_{rss(s)}}{C_{iss(s)}}. \tag{3.66}$$

Ideally, the capacitive gate bounce susceptibility ratio should satisfy the following condition,

$$GBS_{GSO(s)} \leq \frac{V_{TH(s)} - V_{GS(s)0}}{V_{AB(s)} - V_{DS(s)0}}. \tag{3.67}$$

Therefore, by enforcing $V_{DSgb(s)}=V_{AB(s)}$ capacitive gate bounce can be prevented under all conditions. However, meeting condition (3.67) with existing silicon technologies appears to be difficult. For the case of 30V MOSFETs, typical values of $GBS_{GSO(s)}$ in existing commercial devices may typically vary around 0.17, whereas the ideal target value according to (3.67) might be 0.07. Figure 3.42 illustrates the variation of gate voltage as a consequence of the capacitance divider with open gate and under different initial conditions of a typical state of the art power MOSFET. The curves reveal potential susceptibility to gate bounce for drain-source voltages above 15V. The use of a low impedance path between gate and source can effectively mitigate the effect. Such impedance path should aim at

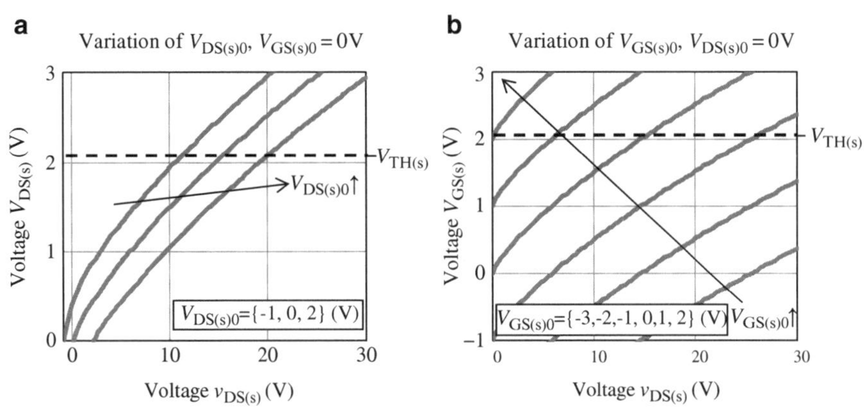

Fig. 3.42 Gate bounce susceptibility of SyncFET with $C_{GS(s)} = 3.8$ nF and $C_{GD(s)}$ from Fig. 3.41a. The line crossings with $V_{TH(s)}$provide $V_{DSgb(s)}$. Data refer to power MOSFET PH3330 from NXP semiconductors

the conditions established by (3.62) and (3.63), provided that inductive gate bounce is negligible.

The worst-case scenario of capacitive gate bounce turns out to be a favorable situation to avoid inductive gate bounce since the induced voltage across $L_{S(s)}$ cannot alter the input capacitance charge. As the gate impedance path reduces, the susceptibility to inductive gate bounce increases, and as such, conditions (3.65) and (3.63) might become, after all, detrimental for the overall performance. A gate impedance trade-off must thus be attained and shall be addressed with the aid of models (b) and (h) from Table 3.1 as follows.

Figure 3.43 shows the dependence of switching losses on the transfer capacitance at the LE transient. The total switching loss includes reverse recovery, $C_{OSS(s)}$ charging, and gate bounce shoot-through. The later is included in the graphs to point out its significant contribution as the ratio $C_{rss(s)}/C_{iss(s)}$ increases. Note that the mitigation of gate driving loss as a consequence of the reduction in $C_{GS(s)}$ may not compensate the increase in gate bounce. Figure 3.43b highlights the strong influence of $C_{GD(s)}$ on gate bounce shoot-through loss, which may dominate the total switching loss even with a moderate increase of $C_{rss(s)}/C_{iss(s)}$. Comparing plots (a) and (b) one observes that similar values of $GBS_{GSO(s)}$may yield to different gate bounce loss figures depending on $C_{GD(s)}$ and $C_{GS(s)}$.

The reduction of $C_{GD(s)}$ as a means to mitigate capacitive gate bounce seems to be more effective than the increase of $C_{GS(s)}$ since the earlier does not compromise gate charge losses. Figure 3.44 shows how the value of $C_{GD(s)}$ can influence gate bounce loss depending on the reverse recovery current. The maximum $Q_{GD(s)}$ allowed to avoid shoot-through lowers as the body diode reverse current increases.

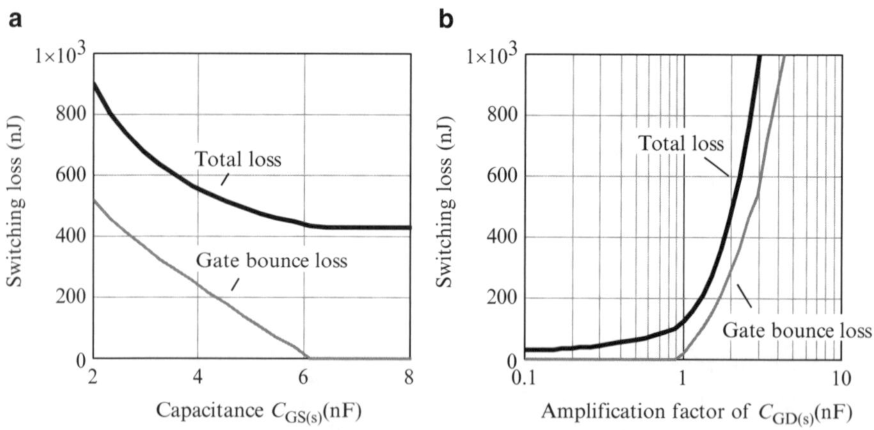

Fig. 3.43 Switched-node LE losses (excluding gate driving and CtrlFET switching) and gate bounce loss contribution as a function of, (**a**), $C_{GS(s)}$, and, (**b**), $C_{GD(s)}$. Simulations based on models (b) and (h) from Table 3.1. Parameter values: $V_{in} = 12$ V, $V_{GS(s)0} = 0$ V, $V_{DS(s)0} = 0$ V, $L_{HB} = 1$ nH, $R_{HB} = 0.3\ \Omega$, $I_{rr(s)} = 0$ A, $L_{G(s)} = 4$ nH, $L_{S(s)} = 200$ pH, $R_{G(s)} = 1.5\ \Omega$, $C_{GS(s)} = 3.8$ nF (only for plot(**b**)), see Figs. 3.40 and 3.41 for nonlinear MOSFET parameters. Model (b) response corresponds to an ideal step function

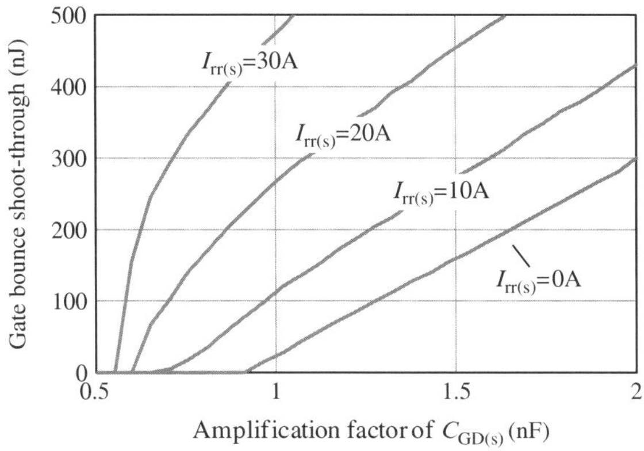

Fig. 3.44 Switched-node LE gate bounce loss as a function of $C_{GD(s)}$ and for various $I_{rr(max)}$. Simulations based on models (b) and (h) from Table 3.1. Parameter values: $V_{in} = 12$ V, $V_{GS(s)0} = 0$ V, $V_{DS(s)0} = 0$ V, $L_{HB} = 1$ nH, $R_{HB} = 0.3\ \Omega$, $\tau_{rr(s)} = 0.5$ ns, $L_{G(s)} = 4$ nH, $L_{S(s)} = 200$ pH, $R_{G(s)} = 1.5\ \Omega$, $C_{GS(s)} = 3.8$ nF (only for plot(**b**)), see Figs. 3.40 and 3.41 for nonlinear MOSFET parameters. Model (b) response corresponds to an ideal step function

Such relation is established through the time derivative of $v_{DS(s)}$(i.e., $K_{v(s)}$), which increases with reverse recovery. The proportionality of gate bounce with $K_{v(s)}$ expressed in (3.61) can be identified from Fig. 3.45 in terms of energy loss, where $K_{v(s)}$ is controlled by means of the reverse recovery current. Both $I_{rr(s)}$ and $\tau_{rr(s)}$ alter the speed of charge of $C_{OSS(s)}$. That is, while $I_{rr(s)}$defines the initial stored energy in L_{HB}, time constant $\tau_{rr(s)}$ influences the amount of charge flowing into $C_{OSS(s)}$ during the ramp up. Thus, increasing $\tau_{rr(s)}$ produces a $C_{OSS(s)}$ charge slow down as a consequence of the reverse charge increase.

As deduced from the equations of model (h), the expected detrimental effects on gate bounce with the increase of the gate impedance and negligible source inductance can be asserted from Fig. 3.46. In the high gate impedance range, the curves converge to a loss level corresponding to open gate, where the existence of shoot-through can be deduced from (3.38) (see Sect. 2.5.2.2). In the illustration, the half-bridge circuit and MOSFET susceptibility to gate bounce are such that the gate impedance required to avoid shoot-through is demandingly low (i.e., $L_{G(s)}$=0 and $R_{G(s)} < 0.5\ \Omega$).

A more interesting gate bounce behavior occurs as soon as the source inductance feedback becomes relevant. This is the case of Fig. 3.47, where gate bounce losses are displayed as a function of $L_{S(s)}$ and the gate impedance. The plot can be divided into two regions where one or the other gate bounce mechanisms dominates. At high $L_{S(s)}$, the increase of gate impedance yields a loss reduction, thereby suggesting that inductive gate bounce dominates. The opposite happens in the low $L_{S(s)}$ region as the increase of the gate impedance worsens the performance, which denotes capacitive gate bounce behavior. In a narrow interval at the boundary between the two regions, the gate impedance variation does not significantly impact the overall losses as the effects of both gate bounce mechanisms cancel out.

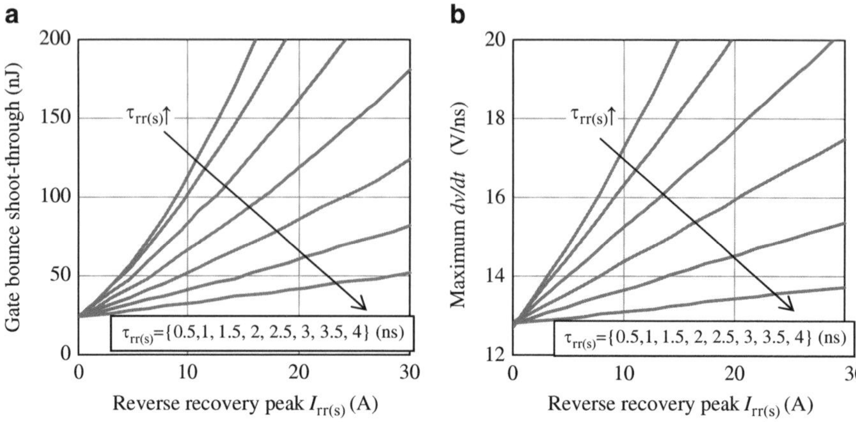

Fig. 3.45 Switched-node LE gate bounce loss and maximum $dv_{DS(s)}/dt$ as a function of $I_{rr(s)}$ and for various $\tau_{rr(s)}$. Simulations based on models (b) and (h) from Table 3.1. Parameter values: V_{in} = 12 V, $V_{GS(s)0}$ = 0 V, $V_{DS(s)0}$ = 0 V, L_{HB} = 1 nH, R_{HB} = 0.3Ω, $L_{G(s)}$ = 4 nH, $L_{S(s)}$ = 200 pH, $R_{G(s)}$ = 1.5 Ω, $C_{GS(s)}$ = 3.8 nF, see Figs. 3.40 and 3.41 for nonlinear MOSFET parameters. Model (b) response corresponds to an ideal step function

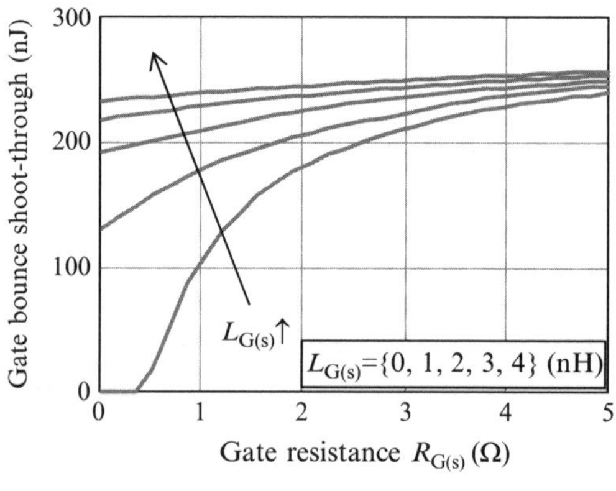

Fig. 3.46 Switched-node LE gate bounce loss as a function of $R_{G(s)}$ and for various $L_{G(s)}$. Simulations based on models (b) and (h) from Table 3.1. Parameter values: V_{in} = 12 V, $V_{GS(s)0}$ = 0 V, $V_{DS(s)0}$ = 0 V, L_{HB} = 1 nH, R_{HB} = 0.3 Ω, $L_{S(s)}$ = 0 H, $I_{rr(s)}$ = 20 A, $\tau_{rr(s)}$ = 0.5 ns, $C_{GS(s)}$ = 3.8 nF, see Figs. 3.40 and 3.41 for nonlinear MOSFET parameters. Model (b) response corresponds to an ideal step function

In order to address such trade-off, one can look at the variation of gate bounce losses as a function of the gate impedance, as illustrated in Fig. 3.48. The optimum gate resistance and inductance are determined for a variety of switching conditions affecting both the $dv_{DS(s)}/dt$ and the induced voltage across $L_{S(s)}$. The curves reveal that for typical conditions, the dominated gate bounce mechanism is primarily capacitive. Loss savings can therefore be achieved by the way of reducing the gate impedance. However, even for the favorable case of reduced reverse recovery, gate bounce can only be avoided with a drastic reduction of the gate impedance, which may result in a total gate resistance lower than 0.5 Ω and a gate inductance around 1 nH or less.

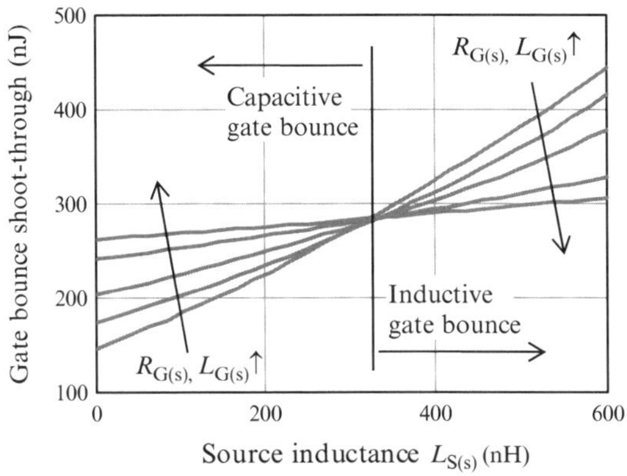

$R_{G(s)}$ (Ω)	$L_{G(s)}$ (nH)
3	8
1.5	4
0.75	2
0.5	1.334
0.375	1
0.15	0.4

Fig. 3.47 Switched-node LE gate bounce loss as a function of $L_{S(s)}$ and for various $R_{G(s)}$ and $L_{G(s)}$ while keeping constant ratio $L_{G(s)}/R_{G(s)}$. Simulations based on models (b) and (h) from Table 3.1. Parameter values: $V_{in} = 12$ V, $V_{GS(s)0} = 0$ V, $V_{DS(s)0} = 0$ V, $L_{HB} = 1$ nH, $R_{HB} = 0.3$ Ω, $I_{rr(s)} =$ 20 A, $\tau_{rr(s)} = 0.5$ ns, $C_{GS(s)} = 3.8$ nF, see Figs. 3.40 and 3.41 for nonlinear MOSFET parameters. Model (b) response corresponds to an ideal step function

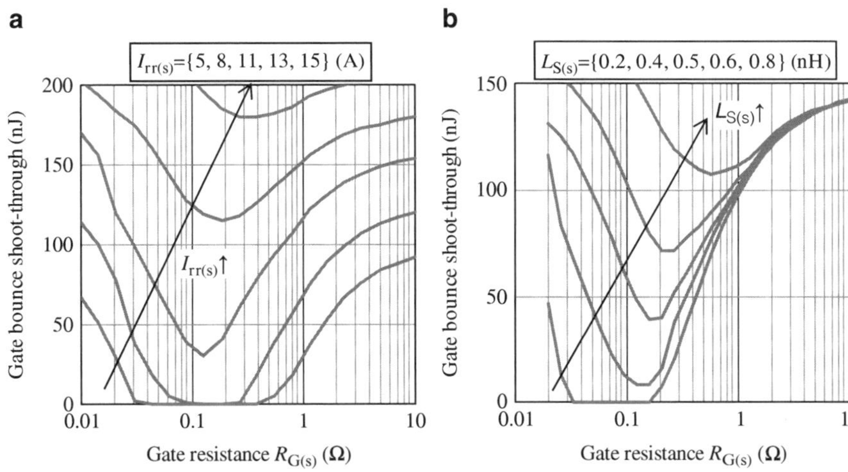

Fig. 3.48 Switched-node LE gate bounce loss as a function of $R_{G(s)}$ for various $I_{rr(s)}$ (plot (**a**)) and $L_{S(s)}$ (plot (**b**)). Ratio $L_{G(s)}/R_{G(s)}$ is kept constant and equal to 2.7 ns. Simulations based on models (b) and (h) from Table 3.1. Parameter values: $V_{in} = 12$ V, $V_{GS(s)0} = 0$ V, $V_{DS(s)0} = 0$ V, $L_{HB} = 1$ nH, $R_{HB} = 0.3$ Ω, $L_{S(s)} = 0.4$ nH (plot (**a**)), $I_{rr(s)} = 10$ A (plot (**b**)), $\tau_{rr(s)} = 0.5$ ns, $C_{GS(s)} = 3.8$ nF, see Figs. 3.40 and 3.41 for nonlinear MOSFET parameters. Model (b) response corresponds to an ideal step function

In addition, the required gate bounce susceptibility may be assessed as a function of the reverse recovery. According to the results of Fig. 3.49, fast switching operation may demand $GBS_{GSO(s)}$ as low as 0.07 for reverse recovery peaks approaching 30A.

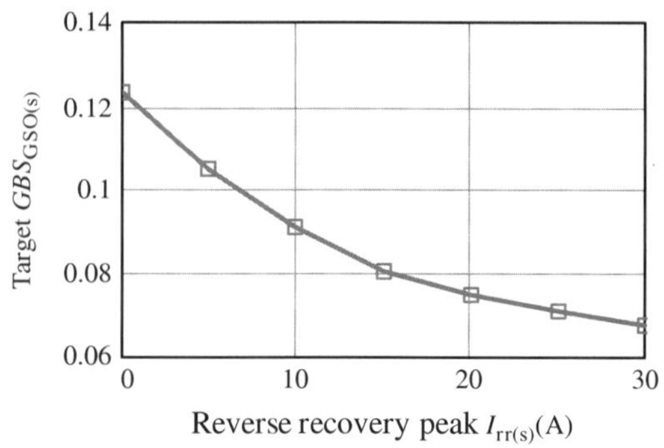

Fig. 3.49 Target gate bounce susceptibility as a function of the reverse recovery peak. Simulations based on models (b) and (h) from Table 3.1. Parameter values: $V_{in} = 12$ V, $V_{GS(s)0} = 0$ V, $V_{DS(s)0} = 0$ V, $L_{HB} = 1$ nH, $R_{HB} = 0.3\ \Omega$, $R_{G(s)} = 1.5\ \Omega$, $L_{G(s)} = 4$ nH, $L_{S(s)} = 0.2$ nH, $\tau_{rr(s)} = 0.5$ ns, $C_{GS(s)} = 3.8$ nF, see Figs. 3.40 and 3.41 for nonlinear MOSFET parameters. Model (b) response corresponds to an ideal step function. Note that gate bounce susceptibility is tuned by scaling $C_{GD(s)}$

In conditions where both capacitive and inductive gate bounce mechanisms are important, a more conservative expression for gate bounce susceptibility may be defined as follows,

$$\mathrm{GBS}_{\mathrm{GSOx(s)}} \geq \frac{V_{\mathrm{TH(s)}} + L_{\mathrm{S(s)}} \cdot K_{\mathrm{id(s)}}}{V_{\mathrm{AB(s)}} - V_{\mathrm{DS(s)0}}}, \tag{3.68}$$

$$V_{\mathrm{TH(s)}} > -L_{\mathrm{S(s)}} \cdot K_{\mathrm{id(s)}}. \tag{3.69}$$

Note that expression $\mathrm{GBS}_{\mathrm{GSOx(s)}}$differs from $\mathrm{GBS}_{\mathrm{GSO(s)}}$in that the initial condition of the gate-source voltage equals the induced voltage across $L_{S(s)}$. Satisfying (3.68) and (3.69) fulfills the requirements to avoid both capacitive and inductive gate bounce regardless of the gate impedance. The worst-case scenario for inductive gate bounce corresponds to the case where the induced voltage across $L_{S(s)}$is directly impressed across $C_{GS(s)}$. Thus, satisfying (3.69) is a sufficient condition only when capacitive gate bounce is negligible.

On the other hand, open gate conditions must meet (3.68) to avoid capacitive gate bounce considering that inductive coupling may have taken place prior to the dv/dt phase. This worst-case scenario corresponds to a precharge of $V_{GS(s)0} = -L_{S(s)}K_{id(s)}$ when the underdamped condition is avoided. In fast switching, $|K_{id(s)}|$ can reach levels of 20A/ns (such as in Fig. 3.40). This reduces $L_{S(s)}$to target values of 100pH or lower so as to fulfill (3.69), provided that the threshold voltage of typical power MOSFETs is maintained. The requirements may further increase at high temperature operation due to the reduction of V_{TH}.

3.4 Falling Edge Ringing Transition

In Sect. 3.2.2, the switched-node FE ringing losses are calculated as a function of the initial values of the state variables at the beginning of the resonant phase, which coincide with the end of the hard-switching interval t_{swFE}. The equations applied are accurate only when the SyncFET body diode is forward-biased prior to the resonant transition. Upon fast switching, however, such condition may not be met since the CtrlFET turn-off can be faster than the discharge of $C_{oss(s)}$. When this occurs, there exists a transition during which the SyncFET stored charge is transferred to the output (and therefore partially recovered) while both power MOSFETs are off. This particular transition phase is illustrated in the simulation example of Fig. 3.50. The switched-node FE time is divided into four intervals: $t_{on(c)}$, t_{swFE}, t_{R1} and t_{R2}. The last two times are part of the resonant transient. Time t_{R1} corresponds to the period where both devices are off while the SyncFET drain-source voltage remains positive. The particularity of this interval is that, unlike transition t_{R2}, expression (3.36) for the charge loss does not apply. This is completely ignored in the approximations of the ringing loss from Sect. 3.2.2, thereby resulting in certain loss calculation errors. During time, voltage $v_{DS(s)}$ is assumed to be clamped to a constant value (e.g., ~0V).

In this section, a more accurate model of the falling edge ringing transition is proposed and analyzed based on two simplified linear circuits of the half-bridge. These correspond to models (c) and (d) from Table 3.1. The first one represents the resonant behavior during t_{R1}, which allows more accurate loss predictions in this interval. As soon as the SyncFET enters third quadrant conduction, voltage $v_{DS(s)}$ clamps and a new operating phase starts, which can be represented with model (d).

Alternatively, and since CtrlFET avalanche breakdown may occur, equivalent circuit (e) shall be employed similarly as in the LE transient.

Combining these models should enable a proper description of the entire ringing transient in order to assess overvoltage stress and quantify losses associated to the

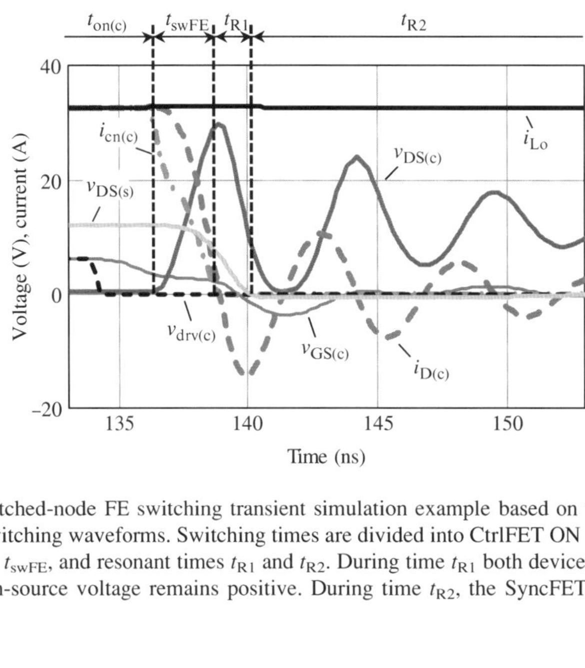

Fig. 3.50 Switched-node FE switching transient simulation example based on the PLA model. Typical fast switching waveforms. Switching times are divided into CtrlFET ON time $t_{on(c)}$, hard-switching time t_{swFE}, and resonant times t_{R1} and t_{R2}. During time t_{R1} both devices are off and the SyncFET drain-source voltage remains positive. During time t_{R2}, the SyncFET operates in the third quadrant

half-bridge stored energy and avalanche breakdown mechanisms. CtrlFET gate bounce is neglected in this transition phase.

As one may anticipate, many of the results from the LE switching transient are applicable to the FE transient. For instance, model (d) is equivalent to model (b) by imposing $i_{\mathrm{dio(s)}}=i_{\mathrm{cn(s)}}=0$, $v_{\mathrm{s}}=V_{\mathrm{in}}-V_{\mathrm{DS(s)}}$ and making $C_{\mathrm{oss(s)}}=C_{\mathrm{oss(c)}}$. Furthermore, in model (d), the third quadrant operation of the SyncFET may allow the simplification $v_{\mathrm{s}}\approx V_{\mathrm{in}}$. Equations governing model (d) can therefore be deduced from those of model (b).

As to model (c), time domain closed-form expressions result from the methods of circuit theory. Therefore, the CtrlFET drain current can be expressed as,

$$i_{\mathrm{D(c)}}(t) = i_{\mathrm{D1(c)}}(t) + i_{\mathrm{D2(c)}}(t) + i_{\mathrm{D3(c)}}(t). \tag{3.70}$$

The first summand in (3.70) accounts for the initial conditions of the voltages across both capacitances as,

$$i_{\mathrm{D1(c)}}(t) = \frac{V_{\mathrm{in}} - V_{\mathrm{DS(c)0}} - V_{\mathrm{DS(s)0}}}{L_{\mathrm{HB}}} \frac{\mathrm{e}^{-\zeta\omega_n t}}{\omega_n\sqrt{1-\zeta^2}} \sin\left(\omega_n\sqrt{1-\zeta^2}t\right). \tag{3.71}$$

The second term corresponds to the contribution of the load current,

$$i_{\mathrm{D2(c)}}(t) = \frac{i_{\mathrm{Lo}}}{(C_{\mathrm{oss(s)}}/C_{\mathrm{oss(c)}})+1}\left(1 - \frac{\mathrm{e}^{-\zeta\omega_n t}}{\sqrt{1-\zeta^2}} \sin\left(\omega_n\sqrt{1-\zeta^2}t + \phi\right)\right). \tag{3.72}$$

Whereas the third term considers the initial condition in L_{HB},

$$i_{\mathrm{D3(c)}}(t) = -I_{\mathrm{D(c)0}} \frac{\mathrm{e}^{-\zeta\omega_n t}}{\sqrt{1-\zeta^2}} \sin\left(\omega_n\sqrt{1-\zeta^2}t - \phi\right). \tag{3.73}$$

The natural resonant frequency (in radians per second) of this second-order circuit is, for this particular case,

$$\omega_n = \sqrt{\frac{1}{L_{\mathrm{HB}}}\left(\frac{1}{C_{\mathrm{oss(s)}}} + \frac{1}{C_{\mathrm{oss(c)}}}\right)}. \tag{3.74}$$

Whereas damping factor ζ equals to,

$$\zeta = \frac{R_{\mathrm{HB}}}{2}\sqrt{\frac{1}{L_{\mathrm{HB}}}\left(\frac{1}{C_{\mathrm{oss(s)}}} + \frac{1}{C_{\mathrm{oss(c)}}}\right)^{-1}}. \tag{3.75}$$

For the ringing to occur, condition $\zeta<1$ must be satisfied, which corresponds to an underdamped switching transient. Finally, it follows that,

$$\phi = a\tan\left(\frac{\sqrt{1-\zeta^2}}{\zeta}\right). \tag{3.76}$$

The SyncFET drain current readily results from subtracting i_{Lo} from $i_{D(\mathrm{c})}$. Integrating the current expressions yields the voltage across the MOSFETs. As for the case of the drain current, the CtrlFET voltage is broken down into four terms as follows,

$$v_{\mathrm{DS(c)}}(t) = v_{\mathrm{DS1(c)}}(t) + v_{\mathrm{DS2(c)}}(t) + v_{\mathrm{DS3(c)}}(t) + V_{\mathrm{DS(c)0}}. \tag{3.77}$$

The first term includes the initial conditions in the half-bridge output capacitances,

$$v_{\mathrm{DS1(c)}}(t) = \frac{V_{\mathrm{in}} - V_{\mathrm{DS(c)0}} - V_{\mathrm{DS(s)0}}}{1 + (C_{\mathrm{oss(c)}}/C_{\mathrm{oss(s)}})}\left(1 - \frac{\mathrm{e}^{-\zeta\omega_n t}}{\sqrt{1-\zeta^2}}\sin\left(\omega_n\sqrt{1-\zeta^2}t + \phi\right)\right). \tag{3.78}$$

The second summand from (3.77) is,

$$v_{\mathrm{DS2(c)}}(t) = \frac{i_{\mathrm{Lo}}}{C_{\mathrm{oss(c)}} + C_{\mathrm{oss(s)}}}\left(t - \frac{2\zeta}{\omega_n} + \frac{\mathrm{e}^{-\zeta\omega_n t}}{\omega_n\sqrt{1-\zeta^2}}\sin\left(\omega_n\sqrt{1-\zeta^2}t + \varphi\right)\right). \tag{3.79}$$

Phase angle φ is defined as,

$$\begin{aligned} &\varphi = a\tan(\Gamma) \text{ if } \Gamma \geq 0 \\ &\varphi = \pi + a\tan(\Gamma) \text{ Otherwise} \end{aligned} \tag{3.80}$$

where,

$$\Gamma = \frac{2\zeta\sqrt{1-\zeta^2}}{2\zeta^2 - 1}. \tag{3.81}$$

The third summand in (3.77) is,

$$v_{\mathrm{DS3(c)}}(t) = \frac{I_{\mathrm{D(c)0}}}{C_{\mathrm{oss(c)}}}\frac{\mathrm{e}^{-\zeta\omega_n t}}{\omega_n\sqrt{1-\zeta^2}}\sin\left(\omega_n\sqrt{1-\zeta^2}t\right). \tag{3.82}$$

Finally, voltage $v_{DS(s)}$can be expressed from $v_{DS(c)}$as follows,

$$v_{DS(s)}(t) = \frac{C_{oss(c)}}{C_{oss(s)}}(v_{DS1(c)}(t) + v_{DS2(c)}(t) + v_{DS3(c)}(t)) - \frac{i_{Lo}}{C_{oss(s)}}t + V_{DS(s)0}. \tag{3.83}$$

The following sections make use of these models to estimate relevant ringing characteristics such as overvoltage stress and energy losses.

3.4.1 Charging Loss

According to the switching behavior, the estimation of the charging losses may be divided into intervals t_{R1} and t_{R2}.While valid for t_{R2}, expression (3.36) in t_{R1} is not generally applicable because the charging process involves a current source. Thus, the RMS losses in t_{R1} may alternatively be determined by numerical methods.

The remaining resonant transient loss after t_{R1} has elapsed may be conveniently broken down into capacitive and inductive losses according to (3.36) and (3.8), respectively. Note that $I_{D(end)}$ from (3.8) must equal the CtrlFET drain current at the end of t_{R1}, which is defined as $I_{D(c)r}$. Similarly, V_b from (3.36) equals the final CtrlFET drain-source voltage at the end of t_{R1}, which is defined as $V_{DS(c)r}$. Figure 3.51 shows how these initial conditions vary as a function of the values of the state variables at the end of the hard-switching interval. As it can be observed

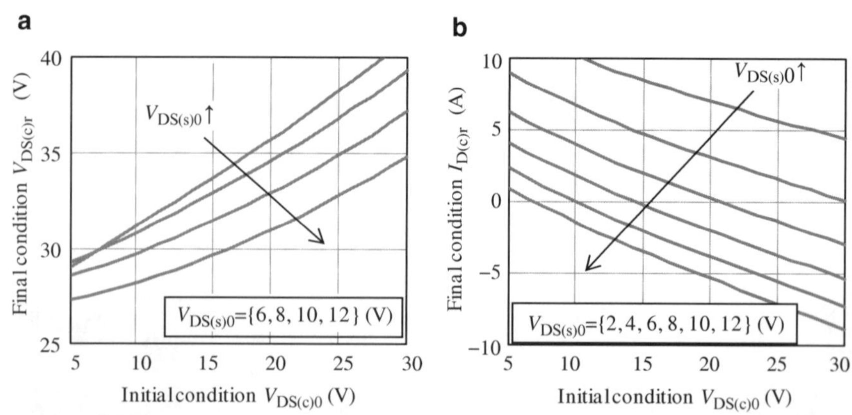

Fig. 3.51 Switched-node falling edge ringing. Calculation of state variables $V_{DS(c)}$ and $I_{D(c)}$ at the time the SyncFET body diode is forward-biased (beginning of interval t_{R2} as defined in Fig. 3.50). Simulation results based on model (c) from Table 3.1. Parameter values: $V_{in} = 12$ V, $I_{D(c)0} = 15$ A, $L_{HB} = 1$ nH, $R_{HB} = 0.3$ Ω. Output capacitances refer to Fig. 3.40 with applied scaling factors (SyncFET die area is three times larger than that of the CtrlFET)

from plot (a), minimizing $V_{DS(c)r}$ so as to reduce ringing loss involves a reduction of $V_{DS(c)0}$ while maintaining $V_{DS(s)0}$ high. Current $I_{D(c)r}$ may, however, lower with the increase of $V_{DS(c)0}$ to such an extent that it may grow negative. The optimum initial conditions that minimize ringing loss exist and may significantly vary depending on a number of parameters. Figure 3.52 shows the total FE ringing loss as a function of the initial voltage across the capacitances. The curves show that a favorable initial value for both $V_{DS(s)0}$ and $V_{DS(c)0}$ is V_{in}. Generally, condition $V_{DS(s)0} = V_{in}$ may be advantageous for discharge times of the SyncFET output capacitance lasting much longer than the resonant period of the FE ringing. To see this, model (c) may be simplified to a series RLC circuit (e.g., such as model (d)) where the excitation may be approximated to a ramp function representing the discharge of the SyncFET output capacitance. It is then straightforward to demonstrate that the energy loss for charging $C_{oss(c)}$ from $V_{DS(c)0}$ to V_{in} by means of a slow ramp function is lower than with a step function. Thus, $V_{DS(s)0}<0$ may correspond to one worst-case condition for the capacitance loss. Such is the case considered in the ringing loss approximation of Sect. 3.2.2 since, as already indicated earlier, the employed loss equation (3.9) (or more generally (3.36)) models the behavior of a step function response.

The loss curves of Fig. 3.52 may be de-embedded into the power losses within t_{R1}, and the capacitive and inductive charging loss of time t_{R2}. These contributions are displayed in Figs. 3.53–3.55. With regard to t_{R1}, Fig. 3.53 shows that losses may lower with the increase of both $V_{DS(s)0}$ and $V_{DS(c)0}$ by virtue of a reduction in RMS current through R_{HB}, which is the consequence of a faster current decay. This trend may only be maintained as long as $i_{D(c)}$ remains positive. In relative terms, and due to the high current and low quality factor of the resonance, t_{R1} losses may be significantly high despite of the typically short duration of this interval.

The inductive energy losses follow opposite trends depending on the direction of the drain current at the beginning of the t_{R2} phase. For the typical case of $I_{D(c)r}> 0$, the increase of $V_{DS(s)0}$ and $V_{DS(c)0}$ is proportional to the energy discharge speed of L_{HB}, which helps minimizing the inductive loss contributor.

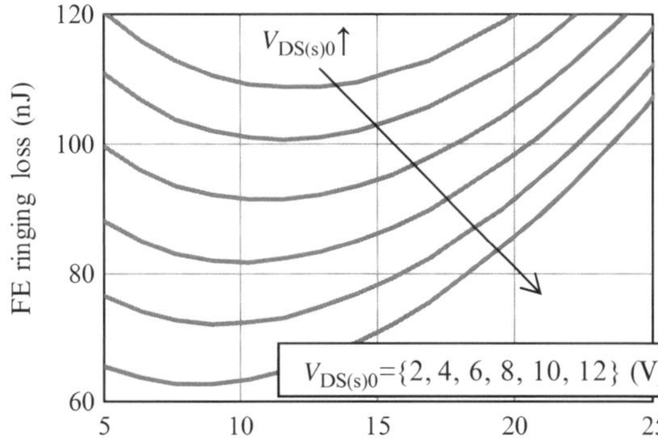

Fig. 3.52 Switched-node falling edge ringing loss (intervals t_{R1} and t_{R2} defined in Fig. 3.50) as a function of initial values of steady-state variables $V_{DS(s)}$ and $V_{DS(c)}$. Simulation results refer to those of Fig. 3.51

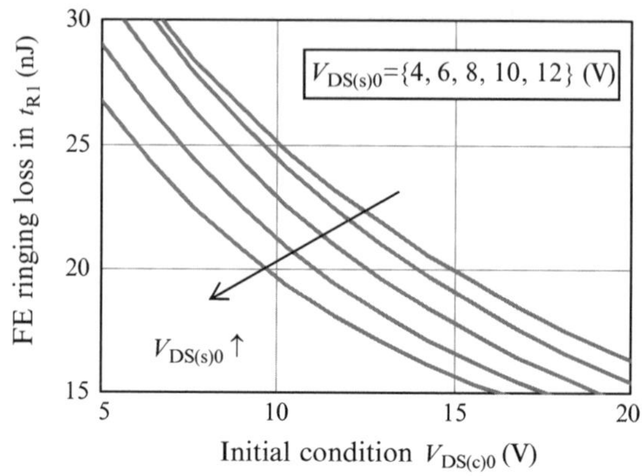

Fig. 3.53 Switched-node falling edge ringing loss in interval t_{R1} defined in Fig. 3.50. Simulation results refer to those of Fig. 3.51

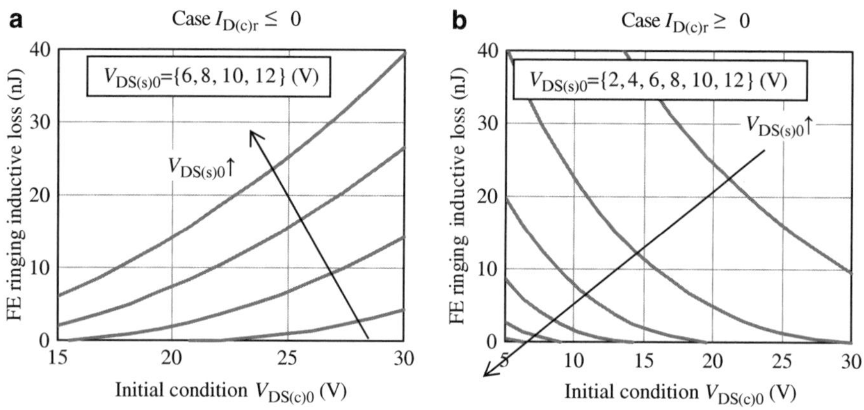

Fig. 3.54 Inductive loss contribution of the switched-node falling edge ringing loss as a function of various steady-state variable initial values (losses within interval t_{R2} defined in Fig. 3.50). (**a**) Case $I_{D(c)r} \leq 0$; (**b**) Case $I_{D(c)r} > 0$. Simulation results refer to those of Fig. 3.51

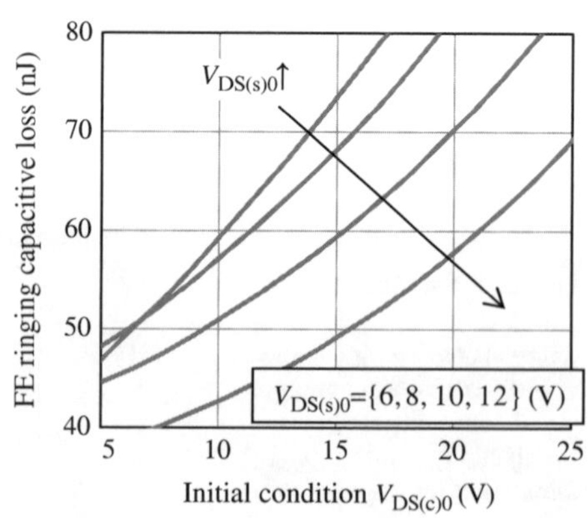

Fig. 3.55 Capacitive loss contribution of the switched-node falling edge ringing loss as a function of various steady-state variable initial values (losses within interval t_{R2} defined in Fig. 3.50). Simulation results refer to those of Fig. 3.51

One negative consequence of augmenting $V_{DS(c)0}$ is the increase of capacitive losses, which can be the most significant portion of the total ringing loss, as depicted in Fig. 3.55. Most generally, this loss contributor worsens with the increase of $V_{DS(c)0}$ above V_{in}. The reduction of $V_{DS(s)0}$ causes an indirect loss increase as it yields higher $V_{DS(c)0}$.

As Fig. 3.56 illustrates, initial current $I_{D(c)0}$ negatively influences the ringing loss. According to what it may be deduced from Fig. 3.57, the latter is due to a dramatic increase of the capacitive loss as well as inductive loss in case of moderate and high current levels.

Recalling the exercise of Sect. 3.2.2 where the ringing loss is approximately determined from (3.8) to (3.9) while ignoring time t_{R1}, the task now is to find out

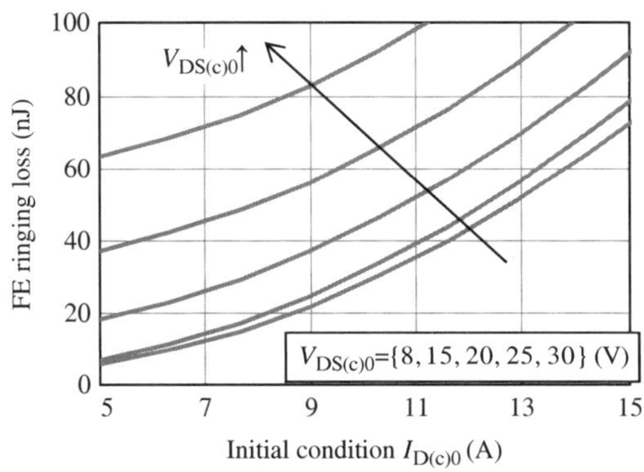

Fig. 3.56 Switched-node falling edge ringing loss (intervals t_{R1} and t_{R2} defined in Fig. 3.50) as a function of initial values of steady-state variables $I_{D(c)}$ and $V_{DS(c)}$. Simulation results based on models (c) and (d) from Table 3.1. Parameter values: $V_{in} = 12$ V, $V_{DS(s)0} = 10$ V, $L_{HB} = 1$ nH, $R_{HB} = 0.3\ \Omega$. Output capacitances refer to Fig. 3.40 with applied scaling factors (SyncFET die area is three times larger than that of the CtrlFET)

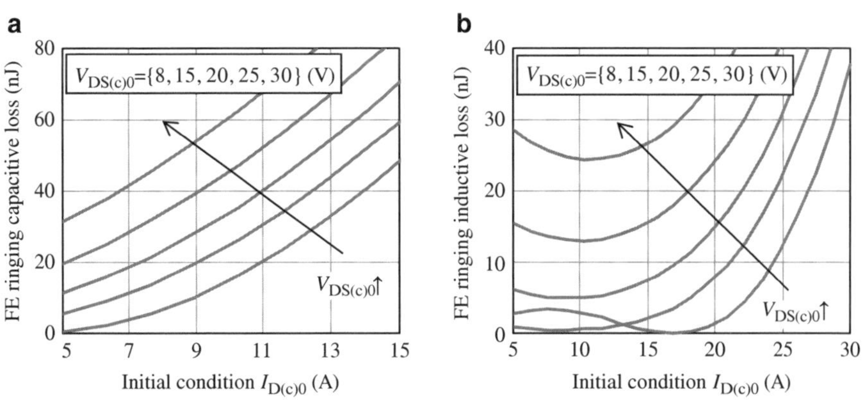

Fig. 3.57 Switched-node falling edge ringing loss breakdown as a function of initial values of steady-state variables $I_{D(c)}$ and $V_{DS(c)}$ (losses within interval t_{R2} defined in Fig. 3.50). Simulation results refer to those of Fig. 3.56

how such simplification impacts the accuracy of the estimations. For the sake of generalization, (3.9) shall be replaced by (3.36) in this analysis. Taking the proposed model of this section as a reference, Fig. 3.58a shows, as pointed out earlier, that the simple model approaches worst-case condition $V_{DS(s)0}<0$. The simplifications make the earlier model to further fail to predict losses at high initial current values. The trends, however, appear to be in line with the reference model.

3.4.2 *Overvoltage Stress*

While the overvoltage stressed device in the LE transition corresponds to the SyncFET, in the FE transition the overshoot voltage falls across the CtrlFET. There are two reasons that may make the CtrlFET voltage overshoot higher than the SyncFET's. Firstly, the energy in L_{HB} may be transferred to $C_{OSS(c)}$, thereby charging it up to a level depending on the load current and the size of the resonant elements. In a first-order approximation, it follows that,

$$V_{DS(c)} \geq \sqrt{\frac{L_{HB}}{C_{oss(c)}}} \cdot i_{Lo(max)}. \quad (3.84)$$

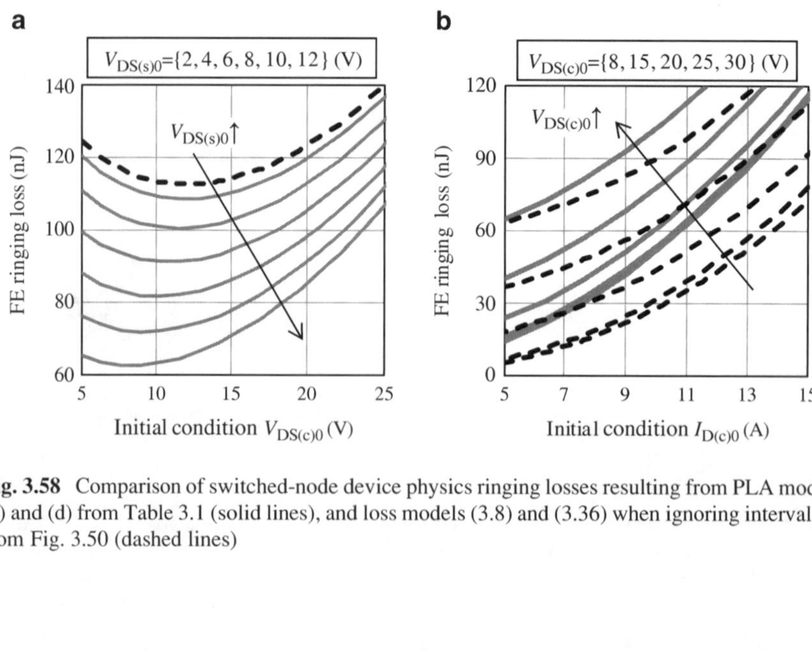

Fig. 3.58 Comparison of switched-node device physics ringing losses resulting from PLA models (c) and (d) from Table 3.1 (solid lines), and loss models (3.8) and (3.36) when ignoring interval t_{R1} from Fig. 3.50 (dashed lines)

The above equation assumes linearity and fast switching, where $i_{\mathrm{Lo(max)}}$ represents the maximum peak current through the output inductor.

The second reason suggesting a high overvoltage stress in the CtrlFET is the high resonant impedance of the half-bridge that at FE results from the usually low $C_{\mathrm{OSS(c)}}$. As it can be deduced from (3.84), the reduction of the resonant capacitance yields higher voltage stress. This is somewhat more relaxed at LE since the $C_{\mathrm{OSS(s)}}$ is usually significantly higher.

Figure 3.59 shows the maximum voltage stress across the CtrlFET as a function of initial condition $V_{\mathrm{DS(c)0}}$ and for various $V_{\mathrm{DS(s)0}}$. Simulations take into account the nonlinearity of the MOSFET capacitances. The results indicate that at high switching speeds, voltage overshoots can well exceed 30V, even at moderate $I_{\mathrm{D(c)0}}$ and L_{HB}. The increase of $V_{\mathrm{DS(s)0}}$ helps mitigating the voltage stress. Additional simulations shown in Fig. 3.60 reaffirm the detrimental effects that the increase of $V_{\mathrm{DS(c)0}}$ and $I_{\mathrm{D(c)0}}$ can have on the stress voltage across the device. The data suggest that drastic measures are required to lower the voltage stress below 30V at high-current switching

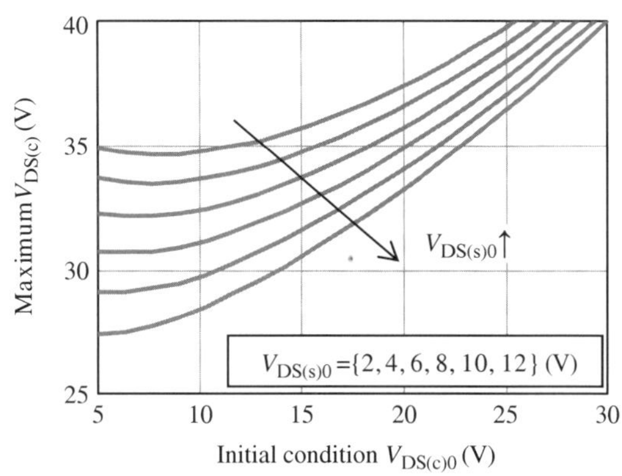

Fig. 3.59 Switched-node falling edge maximum peak voltage as a function of various steady-state variable initial values. Simulation results refer to those of Fig. 3.51

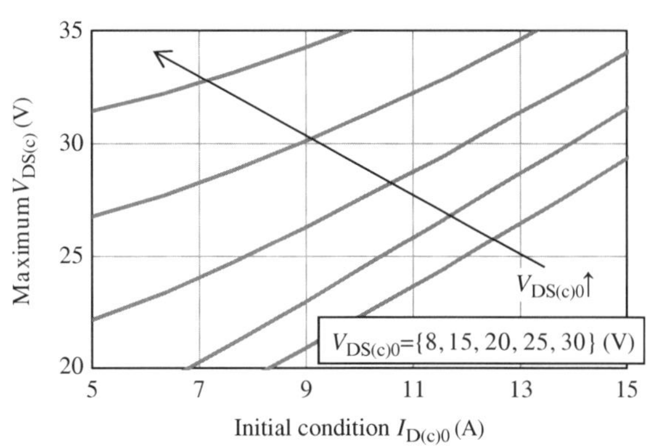

Fig. 3.60 Switched-node falling edge maximum peak voltage as a function of various steady-state variable initial values. Simulation results refer to those of Fig. 3.56

frequency, one of which can potentially be the reduction of the resonant impedance by the way of reducing L_{HB} and/or increasing $C_{OSS(c)}$.

3.4.3 Avalanche Breakdown

CtrlFET avalanche breakdown undergoes the same switching loss behavior analyzed in Sect. 3.3.3. Upon fast switching, the circuit model representing CtrlFET avalanche may be equivalent to that of the SyncFET case by approximating voltage $v_{DS(s)}$ by a constant value, either to V_{in} or around 0V. In such case, the equations derived for the SyncFET avalanche in Sect. 3.3.3 can be applied for the CtrlFET with the right corresponding variable readjustments.

On the other hand, when avalanche breakdown occurs during the transition phase of the SyncFET voltage discharge, then circuit model (c) may be used by making $C_{oss(c)} \rightarrow \infty$ and determining the time for the CtrlFET drain current to reach zero. Thus, equations of model (c) presented in Sect. 3.4 can be employed to calculate the avalanche loss.

3.5 Gate Driving

Following the approach of López [1] the switching action of the gate drive circuit is modeled by means of simplified models (f) and (g) from Table 3.1. The equivalent circuits fundamentally represent the commutation of the transistors. Model (f) reproduces the ZVS transient behavior of the SyncFET as well as the ohmic and cut-off transitions of the CtrlFET. On the other hand, model (g) considers the hard-switching intervals of the CtrlFET during which the gate-source voltage may be approximately constant, thus reducing the complexity of circuit description from a second- to a first-order differential equation. The influence of the source inductance is considered as an offset feedback voltage level proportional to the drain current commutation speed, which is added (or subtracted) to the driver voltage V_{DRV}.

Because of the feedback effects of both transfer capacitance and source inductance, the gate drive models must be coupled with the half-bridge equivalent circuits in every linear interval. A unique solution for the commutation speed of the drain current and drain-source voltage is therefore computed by means of iterative methods.

Perhaps the most relevant of the applied simplifications is the approximation of the excitation signal by an ideal step function commuting from a low to a high level or vice versa depending on the transition phase. Such assumption differs considerably from the proposed driver model of Chap. 2, where the limited commutation speed and saturation levels of the gate driver MOSFETs are taken into account.

In the simplified case, thus, the driving current limit only depends on the size of the gate resistance, loop inductance, as well as voltage levels.

The governing equations of the gate drive model (f) can be readily deduced from the already presented equivalent model (d) by means of a proper change of variables. As to model (g), the waveform expression for the gate current is,

$$i_{\mathrm{G}}(t) = I_{\mathrm{G1}} + (I_{\mathrm{G0}} - I_{\mathrm{G1}}) \cdot \mathrm{e}^{-\frac{t}{\tau_{\mathrm{GHS}}}} \tag{3.85}$$

where I_{G0} is the initial condition for the gate current. Also, it follows that,

$$I_{\mathrm{G1}} = \frac{1}{R_{\mathrm{G}}}\left(V_{\mathrm{DRV}} - L_{\mathrm{S}}\frac{\mathrm{d}i_{\mathrm{D}}(t)}{\mathrm{d}t} - V_{\mathrm{GSx}}\right), \tag{3.86}$$

$$\tau_{\mathrm{GHS}} = \frac{L_{\mathrm{G}} + L_{\mathrm{S}}}{R_{\mathrm{G}}}. \tag{3.87}$$

Voltage V_{GSx} is the gate voltage during the commutation period, which is approximately constant during the Miller plateau in clamped inductive switching. Since upon fast switching, the channel current may vary during hard-switching, an average value may be approximated as follows,

$$V_{\mathrm{GSx}} = g_m \cdot \left(\frac{I_{\mathrm{cn1}} + I_{\mathrm{cn0}}}{2}\right) + V_{\mathrm{TH}}. \tag{3.88}$$

In (3.88), I_{cn0} and I_{cn1} refer to the initial and final values of the channel current within the linear interval of interest.

Figure 3.61 illustrates the model performance as CtrlFET driver. Regarding hard-switching, an accurate representation of the transition through the transconductance curve requires a fine piecewise discretization of the region around V_{TH}, as depicted in Fig. 3.6. This in turn produces the largest number of linear intervals within the hard-switching time interval.

The fast current commutation reduces the linear intervals to fractions of nanoseconds, thus resulting in nonlinear transients that resemble the SPICE simulations of model level 1. To assess this more rigorously, the following section provides a performance comparison of the proposed PLA model and the SPICE model level 0 from Chap. 2.

3.6 Model Validation

Switching waveforms of a complete steady-state switching period are generated by properly concatenating the piecewise linear intervals. Namely, initial values of an interval correspond to the final values of its immediate preceding one.

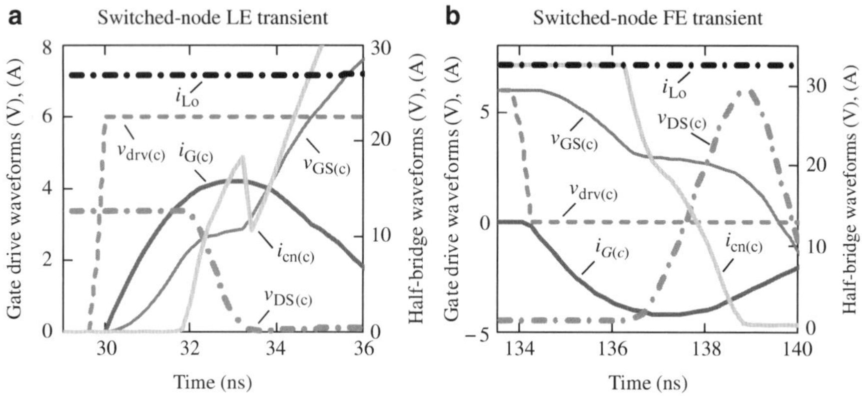

Fig. 3.61 Simulated CtrlFET turn-on and turn-off switching transients based on PLA models (a), (b), (d), (f), and (g) from Table 3.1. Parameter values: $L_{HB} = 1$ nH, $R_{HB} = 0.25\ \Omega$, $V_{in} = 12$ V, $V_{DRV(c)} = 6$ V, $R_{G(c)} = 0.5\ \Omega$, $L_{G(c)} = 2$ nH, $L_{S(c)} = 100$ pH

Figures 3.62 and 3.63 show matching simulated LE and FE transitions of the proposed model levels 0 and 1 in a variety of operating conditions, which include fast switching transients with and without loss mechanisms such as avalanche breakdown and reverse recovery. The good agreement between the two approaches validates the simplifications and assumptions the PLA model is based upon.

Figure 3.64 further shows that the loss predictions of the PLA model are consistent in a widespread range of load currents with the measured and simulated converter losses presented in Chap. 2, Sect. 2.4.

One additional advantage of the PLA model is that computations based on this approach implemented in engineering tools like Mathcad exhibit about one order of magnitude shorter times than model level 0 in PSpice. This has been tested for a large number of operating conditions and piecewise linear intervals.

To further illustrate the model capability to reproduce real behavior, the integrated power module PIP212-12M from NXP semiconductors is modeled according to the conditions given in [2]. Power measurements are obtained from an evaluation board of the IPM (see Appendix G, Sect. G3) and then reproduced with the use of the PLA model. Figure 3.65 shows matching results between experiments and simulations in a wide range of operating conditions, including load current and switching frequency variations. The displayed quantities refer to the overall losses of a single phase SRBC. Thus, the loss contribution of the output inductor (i.e., a total of ~0.5W at I_o=30A, F_s=1MHz) and other filter components are taken into account.

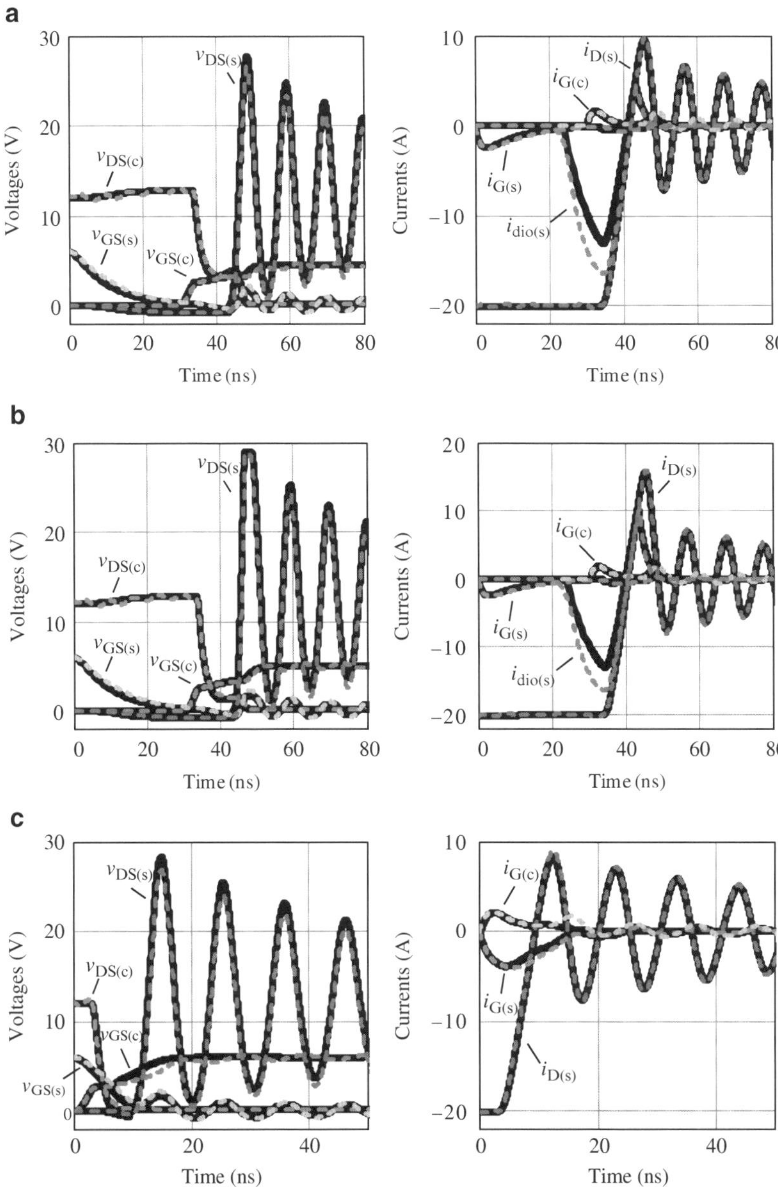

Fig. 3.62 Simulated waveforms comparison between PLA model (solid lines) and model level 0 from Chap. 2 (dashed lines). Switched-node leading edge transition. (**a**) Moderate switching speed ($V_{DRV(c)} = 4.5$ V); (**b**) fast switching with avalanche breakdown ($V_{DRV(c)} = 5$ V); (**c**) faster switching speed ($V_{DRV(c)} = 6$ V) with reduced dead time

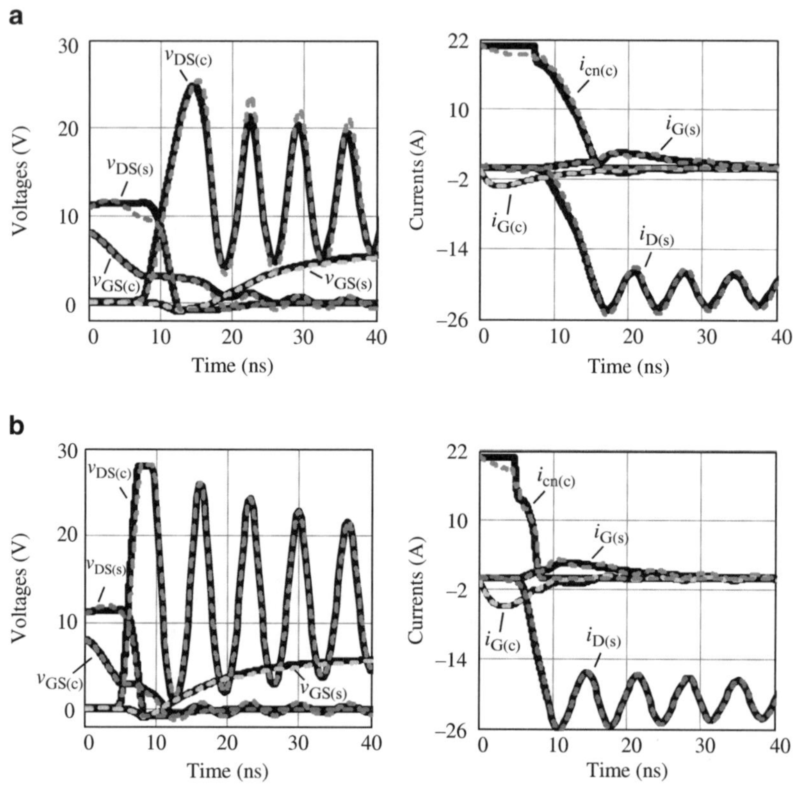

Fig. 3.63 Simulated waveforms comparison between PLA model (solid lines) and model level 0 from Chap 2 (dashed lines). Switched-node falling edge transition. (**a**) Moderate switching speed ($R_{G(c)} = 2\ \Omega$) and (**b**) fast switching with avalanche breakdown ($R_{G(c)} = 1\ \Omega$)

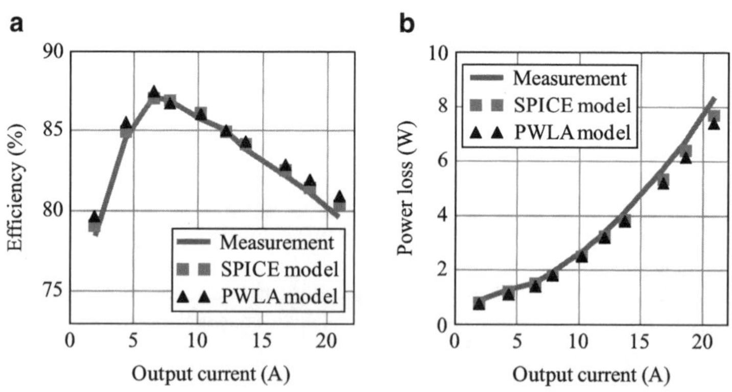

Fig. 3.64 Measured and simulated efficiency and power loss curves. Operating conditions described in Chap. 2, Sect. 2.4

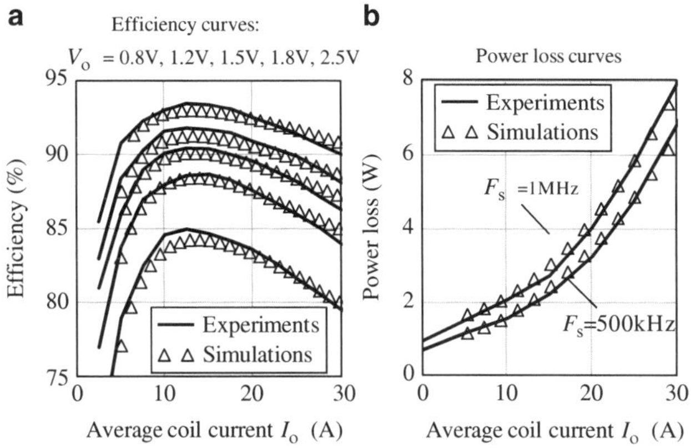

Fig. 3.65 Measurements [2] and simulations (PLA model) comparison of integrated power module PIP212–12M from NXP semiconductors. (**a**) Efficiency curves at $F_s = 500$ kHz for various output voltage levels. (**b**) Power loss at $V_o = 1.3$ V and various switching frequencies

References

1. López T (2002) High frequency resonant gate drivers forMOSFETs with gate charge recovery. Master of Science in Electrical Engineering, Polytechnical University of Catalunya, Barcelona, Spain
2. PIP212-12M DC-to-DC converter powertrain online documentation, http://www.nxp.com. Accessed Oct 2006

Chapter 4
Model Level 2: Power Loss Model

The proposed PLA model of Chap. 2 aids the derivation of power loss expressions that constitute model level 2 (see Sect. 1.4). While some of these equations have already been presented in the previous chapter, new ones will be derived in this chapter by way of simplifying and combining several of the analytical equations of the PLA model.

The aim of the new loss model is to compactly formulate fundamental circuit parameter dependencies of the identified individual loss quantities. Although the ensuing equations are less accurate than the numerical calculations of models 0 and 1, they provide a rationale for power losses, reveal trends toward efficient operation, and facilitate the derivation of figure of merits and design guidelines. As such, the proposed loss model underlies the converter design process, including the optimization of a number of key circuit and device parameters that will be addressed in Chap. 5.

One must bear in mind that the simplifications adopted herein restrict the applicability of the equations to particular operating conditions of interest, which need to be clearly defined. Also, the determination of extreme values (e.g., local minimum) in the optimization process may be subject to verification and/or refinements by means of more accurate models due to higher inaccuracies in the estimated quantities. Nonetheless, as it will be shown, the proposed loss model offers acceptable first-order approximations with drastically reduced computation power as opposed to model level 0.

While the converter design is left as a dedicated topic for Chap. 5, the focus of the present chapter is to deduce loss expressions that cover aspects of the switched converter besides those referring exclusively to the semiconductor switches. Namely, loss contributions from the power filter components, PCB tracks, and package resistance are taken into consideration as further limitations of the conversion efficiency that may condition the circuit design, particularly at high switching frequencies.

According to the device physics simulations of Appendix E, Eddy currents appear to largely dictate most of these additional loss contributions. This is observed in the output inductor coil, PCB track sections, and package leadframe of the semiconductor devices. Consequently, Eddy current-related losses will be accounted for in the loss model by means of approximate semiempirical equations, some of which retain relations to fundamental parameters such as the skin depth.

T. López et al., *Voltage Regulators for Next Generation Microprocessors*,
DOI 10.1007/978-1-4419-7560-7_4, © Springer Science+Business Media, LLC 2011

This chapter dedicates four sections to introduce loss expressions for the power MOSFETs, gate drive switches, filters, and PCB. Whenever indicated, package resistance may be individually addressed for each circuit element. The last section evaluates the performance of the model by direct comparisons to model level 1. Furthermore, an illustration is provided where the efficiency of an integrated power module is predicted and analyzed.

4.1 Power MOSFET Losses

Despite of the considerations on power losses in the passive elements of the converter, power MOSFET losses continue to be the main focus of attention. This section presents a set of equations that describe some of the MOSFETs loss mechanisms identified in Chap. 2. These are,

- Capacitive and inductive half-bridge charging
- Gate charging
- Load current hard-switching
- Load current ON conduction

The loss mechanisms of reverse recovery, gate bounce, and avalanche breakdown are assumed negligible. The conditions that allow for this are stated in Chap. 5, when establishing converter design guidelines.

4.1.1 Half-Bridge Charging Loss

As identified thus far, there is a number of switching events producing fundamental charging losses associated to the MOSFETs output capacitances and the half-bridge loop inductance. To quantify part of these loss mechanisms, it is convenient to recall (3.36) regarding the capacitance charging loss and declare the following related function,

$$w_{\mathrm{C}}(V_b, V_e) = \int_{V_b}^{V_e} C(v)(V_e - v)\,\mathrm{d}v. \tag{4.1}$$

Section 3.2.1 shows that a portion of the CtrlFET hard-switching turn-on loss is direct consequence of $Q_{\mathrm{oss(c)}}$. Regardless of the commutation speed, such contribution can be quantified as,

$$w_{\mathrm{cHBC(c)LE}} = w_{\mathrm{Coss(c)}}(V_{\mathrm{in}}, 0) = \int_{V_{\mathrm{in}}}^{0} C_{\mathrm{oss(c)}}(v)(0 - v)\,\mathrm{d}v. \tag{4.2}$$

Expression (4.2) is consistent with the general form of (4.1) and thus it may account for the nonlinearity of the MOSFETs capacitances. The total energy loss, which mainly dissipates in the CtrlFET channel, equals the energy stored in $C_{\mathrm{oss(c)}}$ prior to the switching action.

Similarly, the LE charging loss of the SyncFET output capacitance after the ringing vanishes is,

$$w_{\mathrm{cHBC(s)LE}} = w_{\mathrm{Coss(s)}}(0, V_{\mathrm{in}}). \tag{4.3}$$

Following the same procedure for the FE transient, one portion of the ringing loss after CtrlFET turn-off may be expressed as,

$$w_{\mathrm{cHBC(c)FE}} = w_{\mathrm{Coss(c)}}(V_{\mathrm{DS(c)end}}, V_{\mathrm{in}}), \tag{4.4}$$

where $V_{\mathrm{DS(c)end}}$ is the voltage level at the end of the hard-switching interval. The second portion of the FE ringing loss is proportional to the half-bridge loop inductance, that is,

$$w_{\mathrm{iHBC}} = \frac{1}{2} L_{\mathrm{HB}} I_{\mathrm{D(c)end}}^2. \tag{4.5}$$

Summing up (4.4) and (4.5) yields the overall ringing loss. According to the study of Sect. 3.4.1, this calculation is accurate as long as $v_{\mathrm{DS(s)}} \cong 0$ at the time the CtrlFET turns off.

In summary, the overall CtrlFET charging loss is hence,

$$w_{\mathrm{cHBC(c)}} = w_{\mathrm{cHBC(c)LE}} + w_{\mathrm{cHBC(c)FE}}. \tag{4.6}$$

The total SyncFET charging loss is simply,

$$w_{\mathrm{cHBC(s)}} = w_{\mathrm{cHBC(s)LE}}. \tag{4.7}$$

Finally, summing up all contributions yields the total half-bridge charging losses,

$$w_{\mathrm{HBC}} = w_{\mathrm{cHBC(s)}} + w_{\mathrm{cHBC(c)}} + w_{\mathrm{iHBC}}. \tag{4.8}$$

4.1.2 Gate Charging Loss

Integral (4.1) is general and can be applied to calculate the gate charging loss of each MOSFET. Nonetheless, a simpler expression can be deduced for the gate losses even when considering nonlinear capacitances and hard-switching operation.

Considering at first the switching transition of a capacitor voltage from V_b to V_e it follows, according to (4.1), that,

$$\begin{aligned} w_{\text{on}} = w_{\text{C}}(V_b, V_e) &= \int_{V_b}^{V_e} C(v)(V_e - v)\,\mathrm{d}v \\ &= \int_{V_b}^{V_e} C(v)V_e\,\mathrm{d}v - \int_{V_b}^{V_e} C(v)v\,\mathrm{d}v. \end{aligned} \tag{4.9}$$

Switching back the capacitance voltage from V_e to V_b involves the following energy loss,

$$\begin{aligned} w_{\text{off}} = w_{\text{C}}(V_e, V_b) &= \int_{V_e}^{V_b} C(v)(V_b - v)\,\mathrm{d}v \\ &= \int_{V_e}^{V_b} C(v)V_b\,\mathrm{d}v - \int_{V_e}^{V_b} C(v)v\,\mathrm{d}v. \end{aligned} \tag{4.10}$$

Now, summing up the losses of both switching transients yields,

$$w_{\text{t}} = w_{\text{on}} + w_{\text{off}} = (V_e - V_b)\int_{V_b}^{V_e} C(v)\,\mathrm{d}v. \tag{4.11}$$

Lastly, from the *fundamental theorem of calculus*, one deduces that,

$$w_{\text{t}} = (V_e - V_b)(Q_e - Q_b), \tag{4.12}$$

where Q_e and Q_b are the capacitance stored charges at voltages V_e and V_b, respectively. This means that,

$$w_{\text{on}} > \frac{1}{2}(V_e - V_b)(Q_e - Q_b) \text{ if } w_{\text{off}} < \frac{1}{2}(V_e - V_b)(Q_e - Q_b). \tag{4.13}$$

The above conditions are satisfied for the case of the MOSFETs output and reverse transient capacitances. Note that the condition has been encountered in Sect. 3.3.1.

It is easy to show that for linear (and time invariant) capacitances, it follows that,

$$w_{\text{on}} = w_{\text{off}} = \frac{1}{2}(V_e - V_b)(Q_e - Q_b). \tag{4.14}$$

It is now straightforward to formulate the total gate charging loss in the SyncFET in one switching cycle as,

$$w_{\text{DRV(s)}} = V_{\text{DRV(s)}} Q_{\text{iss(s)}}. \tag{4.15}$$

Expression (4.12) can be applied to the CtrlFET despite of hard-switching operation. That is, by defining Q_{Gt} as the total charge transfer per switching event, the total gate charging loss in the CtrlFET in one switching cycle is,

$$w_{\mathrm{DRV(c)}} = V_{\mathrm{DRV(c)}} Q_{\mathrm{Gt(c)}}. \tag{4.16}$$

Charge Q_{Gt} accounts for the input capacitance charge from 0 to $V_{\mathrm{DRV(c)}}$ as well as the charge transfer to $C_{\mathrm{GD(c)}}$ during the Miller plateau.

Equation (4.16) implies equal losses in both turn-on and turn-off under the assumption of symmetrical transients. In cases where such simplification is not applicable (e.g., large output current ripple), each gate charging transient may be individually represented by means of an equivalent nonlinear capacitance with a characteristic peak value at $V_{\mathrm{Miller}} \pm \Delta V$ that tends to infinite as ΔV approaches zero. Defining such equivalent capacitance as $C_{\mathrm{G(c)LE}}(v_{\mathrm{GS}})$ for the LE and $C_{\mathrm{G(c)FE}}(v_{\mathrm{GS}})$ for the FE, the following expression for the total gate driving loss is obtained,

$$w_{\mathrm{DRV(c)}} = w_{\mathrm{CG(c)LE}}(0, V_{\mathrm{DRV(c)}}) + w_{\mathrm{CG(c)FE}}(V_{\mathrm{DRV(c)}}, 0), \tag{4.17}$$

where now the loss contribution of every switching transient can be independently calculated as,

$$w_{\mathrm{CG(c)LE}}(V_b, V_e) = \int_{V_b}^{V_e} C_{\mathrm{G(c)LE}}(v)(V_e - v)\,\mathrm{d}v, \tag{4.18}$$

$$w_{\mathrm{CG(c)FE}}(V_b, V_e) = \int_{V_b}^{V_e} C_{\mathrm{G(c)FE}}(v)(V_e - v)\,\mathrm{d}v. \tag{4.19}$$

Note that $C_{\mathrm{G(c)LE}} = C_{\mathrm{G(c)FE}}$ for symmetric LE and FE switching transients, in which case (4.16) and (4.17) are equivalent. The value and shape of these capacitances depend on the switching current and the length of the Miller plateau. Both parameter dependencies can be estimated as described in the following section.

4.1.3 Load Current Hard-Switching

The equations of the PLA model describing hard-switching underlie the switching loss model presented in the following.

As a consequence of the impact of the parasitic half-bridge elements, the turn-on hard-switching transient differs from that of the turn-off. Thus, different models are developed for describing the losses of each transient individually. As such, the total load current hard-switching loss is defined as,

$$w_{\mathrm{ioHS(c)}} = w_{\mathrm{ioHS}} = w_{\mathrm{ioHS\_LE}} + w_{\mathrm{ioHS\_FE}}, \tag{4.20}$$

where $w_{\mathrm{ioHS\_LE}}$ and $w_{\mathrm{ioHS\_FE}}$ are the loss contributions in the LE and FE transients, respectively. These are described in detail in the following subsections.

4.1.3.1 LE Transition

The linear approximation of Sect. 3.2.1 is used here to express the drain current switching waveform as,

$$i_{\mathrm{D(c)}}(t) = \frac{V_{\mathrm{in}}}{2L_{\mathrm{HB}}}t. \tag{4.21}$$

Assuming that the exponential time constant $\tau_{x(c)}$ (see (3.4)) is much smaller than the switching time, expression (3.2) for the drain-source voltage might be approximated as,

$$v_{\mathrm{DS(c)}}(t) = V_{\mathrm{in}}\left(1 - \frac{I_{\mathrm{G(c)}}}{Q_{\mathrm{GD(c)}}}t\right). \tag{4.22}$$

Current $I_{\mathrm{G(c)}}$ is the averaged gate current during the hard-switching time, which defines the time required to bring the drain-source voltage from V_{in} down to zero. This time, according to Fig. 3.8a, is approximately,

$$t_{\mathrm{swLE}} \cong \frac{Q_{\mathrm{GD(c)}}}{I_{\mathrm{G(c)}}}. \tag{4.23}$$

In order to minimize the load current hard-switching loss, it is desirable to reduce the built-up current in L_{HB} during t_{swLE}. This implies to fulfill,

$$t_{\mathrm{swLE}} \ll \frac{2L_{\mathrm{HB}}}{V_{\mathrm{in}}} i_{\mathrm{Lo(min)}}. \tag{4.24}$$

Rearranging, one finds an expression for the gate current,

$$I_{\mathrm{G(c)}} = K_{\mathrm{ig}} \frac{Q_{\mathrm{GD(c)}} V_{\mathrm{in}}}{2L_{\mathrm{HB}} i_{\mathrm{Lo(min)}}}, \tag{4.25}$$

where $K_{\mathrm{ig}} \gg 1$. Integrating the product of (4.21) and (4.22) and replacing $I_{\mathrm{G(c)}}$ by (4.25) yield the equation for the switching loss contribution of the load current,

$$w_{\mathrm{ioHS\_LE}} = \frac{1}{3} L_{\mathrm{HB}} \left(\frac{i_{\mathrm{Lo(min)}}}{K_{\mathrm{ig}}}\right)^2. \tag{4.26}$$

The above expression disregards the contribution of the capacitive current into the CtrlFET channel, which is accounted for as the capacitive charging loss contribution of Sect. 4.1.1. It is important to note from (4.25) and (4.26) that the increase of L_{HB} yields a reduction of w_{ioHS_LE} as long as the gate current is kept constant. The proposed model ignores the effect of the MOSFET transconductance on the switching transition. This is a good approximation as long as $\tau_{x(c)}$ (see (3.4)) is much smaller than the switching time. This also holds for the FE transition, which is described in the following section.

4.1.3.2 FE Transition

As in the LE transition, the linearization of the hard-switching interval involves neglecting time constant (3.4). As seen in Sect. 3.2.2, this yields the following simplified expression for the CtrlFET drain-source voltage,

$$v_{DS(c)}(t) = -\frac{I_{G(c)}}{C_{GD(c)}} \cdot t. \tag{4.27}$$

Also, assuming that the drain current changes at a rate given by (3.7), the expression for the channel current may reduce to,

$$i_{cn(c)}(t) = \frac{V_{in} - V_{DS(c)end} - V_{DS(s)end}}{2L_{HB}} \cdot t + i_{Lo(max)} + \frac{C_{oss(c)}}{C_{GD(c)}} I_{G(c)}. \tag{4.28}$$

Expression (4.28) suggests that the hard-switching interval vanishes for $i_{Lo(max)} \leq -(C_{oss(c)}/C_{GD(c)})I_{G(c)}$. Thus, in order to limit the maximum gate current amplitude, the following condition for the gate current may be satisfied at turn-off,

$$|I_{G(c)}| \leq \frac{C_{GD(c)}}{C_{oss(c)}} \cdot i_{Lo(max)}. \tag{4.29}$$

The load current hard-switching loss is defined as the channel loss during the device commutation, that is,

$$w_{ioHS\_FE} = \int_0^{t_{swFE}} v_{DS(c)} i_{cn(c)} \, dt. \tag{4.30}$$

For very low $\tau_{x(c)}$, switching time t_{swFE} (see Fig. 3.8b) results from letting (4.28) equal to zero, which yields,

$$t_{swFE} \cong -\left(i_{Lo(max)} + \frac{C_{oss(c)}}{C_{GD(c)}} I_{G(c)} \right) \frac{2L_{HB}}{V_{in} - V_{DS(c)end} - V_{DS(s)end}}. \tag{4.31}$$

Solving (4.30) leads to the following analytical loss expression for the turn-off load current hard-switching,

$$w_{\text{ioHS.FE}} = \frac{C_{\text{GD(c)}}}{I_{\text{G(c)}}} \cdot V_{\text{DS(c)end}}^2 \cdot \cdots$$
$$\left[\frac{V_{\text{in}} - V_{\text{DS(c)end}} - V_{\text{DS(s)end}}}{6L_{\text{HB}}} \frac{C_{\text{GD(c)}}}{I_{\text{G(c)}}} V_{\text{DS(c)end}} - \frac{1}{2}\left(i_{\text{Lo(max)}} + \frac{C_{\text{oss(c)}}}{C_{\text{GD(c)}}} I_{\text{G(c)}}\right)\right]. \tag{4.32}$$

Using (4.31) into (4.27) allows determining $V_{\text{DS(c)end}}$ as,

$$V_{\text{DS(c)end}} = \frac{1}{2}(V_{\text{in}} - V_{\text{DS(s)end}}) + \cdots$$
$$+ \frac{1}{2}\sqrt{(V_{\text{in}} - V_{\text{DS(s)end}})^2 - 8I_{\text{G(c)}} \frac{L_{\text{HB}}}{C_{\text{GD(c)}}}\left(i_{\text{Lo(max)}} + \frac{C_{\text{oss(c)}}}{C_{\text{GD(c)}}} I_{\text{G(c)}}\right)}. \tag{4.33}$$

The current drawn from the SyncFET during the current commutation produces a discharge of $C_{\text{OSS(s)}}$. Voltage $v_{\text{DS(s)}}$ at the end of the hard-switching interval can be determined from,

$$V_{\text{DS(s)end}} = \frac{1}{C_{\text{oss(s)}}} \int_0^{t_{\text{swFE}}} (i_{\text{Lo(max)}} - i_{\text{D(c)}})\, \mathrm{d}t + V_{\text{in}}. \tag{4.34}$$

Solving the above integral yields,

$$V_{\text{DS(s)end}} = V_{\text{in}} + 2\frac{L_{\text{HB}}}{C_{\text{oss(s)}}}\left(i_{\text{Lo(max)}} + \frac{C_{\text{oss(c)}}}{C_{\text{GD(c)}}} I_{\text{G(c)}}\right)^2 \cdot \cdots$$
$$\left(\frac{V_{\text{in}} - V_{\text{DS(s)end}} - \cdots}{-\sqrt{(V_{\text{in}} - V_{\text{DS(s)end}})^2 - 8I_{\text{G(c)}} \frac{L_{\text{HB}}}{C_{\text{GD(c)}}}\left(i_{\text{Lo(max)}} + \frac{C_{\text{oss(c)}}}{C_{\text{GD(c)}}} I_{\text{G(c)}}\right)}}\right)^{-1} \tag{4.35}$$

Since $V_{\text{DS(s)end}}$ appears in both sides of the equal sign, expression (4.35) may be solved by numerical methods considering that $0 < V_{\text{DS(s)end}} < V_{\text{in}}$.

Section 3.2.2 shows the existence of an optimum gate current that minimizes turn-off losses provided that avalanche breakdown is avoided. For the latter condition to be fulfilled, one must estimate the maximum peak voltage across the CtrlFET. The following approximation establishes the energy transfer from L_{HB} to $C_{\text{OSS(c)}}$ after the hard-switching interval assuming $V_{\text{DS(s)}} = 0$, zero damping factor, and linear $C_{\mathrm{OSS(c)}}$. This yields a worst-case expression for the maximum voltage stress, which is always higher than or equal to $V_{\mathrm{DS(c)end}}$. That is,

$$V_{\mathrm{AB(c)}} > \sqrt{\frac{L_{\mathrm{HB}}}{C_{\mathrm{oss(c)}}}} \frac{I_{\mathrm{D(c)end}}}{\cos\left[a\tan\left(\frac{V_{\mathrm{DS(c)end}} - V_{\mathrm{in}}}{I_{\mathrm{D(c)end}}}\sqrt{\frac{C_{\mathrm{oss(c)}}}{L_{\mathrm{HB}}}}\right)\right]} + V_{\mathrm{in}}, \tag{4.36}$$

where $I_{\mathrm{D(c)end}}$ is the CtrlFET drain current at the end of the hard-switching interval, which corresponds to the output capacitive current,

$$I_{\mathrm{D(c)end}} = \frac{C_{\mathrm{oss(c)}}}{C_{\mathrm{GD(c)}}} \left|I_{\mathrm{G(c)}}\right|. \tag{4.37}$$

Clearly, the increase of the gate current yields a higher voltage stress, which may represent a limit to the maximum switching speed of the device.

4.1.4 Load Current ON Conduction

ON conduction loss involves heat dissipation in the ON resistance of the conducting device as a consequence of the output inductor current. Therefore, this loss contribution only accounts for the ON conduction times in the ohmic region of operation.

For the case of the CtrlFET, the ON conduction time extends between the end of the turn-on Miller plateau and the beginning of the turn-off one (see interval t_4 and t_5 from Figs. 2.47 and 2.48). Thus, two portions of the ON conduction losses may be distinguished, one corresponding to the ON resistance at full gate voltage, and the other to the remaining switching times between the Miller voltage and V_{DRV} in both LE and FE transients. Accordingly, the following expression is defined for the CtrlFET ON conduction loss,

$$w_{\mathrm{ioON(c)}} = w_{\mathrm{ioONs(c)}} + w_{\mathrm{ioONd(c)}}. \tag{4.38}$$

The first summand in (4.38) accounts for the ON conduction loss during the period of full gate drive voltage. It can be readily expressed as,

$$w_{\mathrm{ioONs(c)}} = R_{\mathrm{DSon(c)ac}}\left(f_{\mathrm{ON(c)}}\right) \cdot I^2_{\mathrm{RMS(c)s}} \cdot T_{\mathrm{s}}, \tag{4.39}$$

where $I_{\mathrm{RMS(c)s}}$ is the corresponding portion of the RMS drain current flow through the CtrlFET,

$$I_{\mathrm{RMS(c)s}} = \sqrt{\frac{1}{T_{\mathrm{s}}} \int_{t_4 + 2 \cdot R_{\mathrm{Gt(c)LE}} \cdot C_{\mathrm{iss(c)}}}^{t_5 - 2 \cdot R_{\mathrm{Gt(c)FE}} \cdot C_{\mathrm{iss(c)}}} i^2_{\mathrm{D(c)}} \, \mathrm{d}t}. \tag{4.40}$$

The term $R_{DSon(c)ac}$ is the total drain-source resistance of the CtrlFET, including the package contribution as illustrated in Fig. 2.4. Due to Eddy currents, the package resistance (leadframe, bond wires, etc.) depends on the switching dynamics. Although its contribution relative to the semiconductor resistance may be significant already in DC operation, in the megahertz range the package resistance may become a major limitation. This can be deduced from the device physics simulations of Appendix E.

The dynamic ON resistance may be obtained both experimentally or by harmonic calculations (see, e.g., Fig. 2.7). However, since the current flow differs in practice from a sinusoidal waveform, a Fourier expansion of the original periodic signal shall be deduced so as to estimate the total ON conduction loss. Thus, (4.39) represents an alternative approximation where only a single harmonic component is used, that is,

$$f_{ON(c)} = \frac{1}{2\pi t_{ON(c)}}, \tag{4.41}$$

where $t_{ON(c)}$ corresponds to the ON conduction time of the CtrlFET. According to the results of Appendix E, (4.39)–(4.41) are valid approximations for rather short conduction times relative to T_S, i.e., ($V_o \ll V_{in}$). Therefore, (4.39)–(4.41) are not applied to the SyncFET.

The second term in (4.38) refers to the ON conduction loss of the switching times, which are proportional to the time constant given by the input capacitance and the total gate resistance. Hence, $w_{ioONd(c)}$ is defined as,

$$\begin{aligned} w_{ioONd(c)} = {} & 2 \cdot R_{DSon(c)ac}(f_{ON(c)}) \cdot C_{iss(c)} \cdot \ldots \\ & \left(\left(1 - \frac{V_{TH(c)LE}}{V_{DRV(c)}}\right) \cdot R_{Gt(c)LE} \cdot \chi_{LE(c)} \cdot i^2_{Lo(min)} + \ldots \right. \\ & \left. + \left(1 - \frac{V_{TH(c)FE}}{V_{DRV(c)}}\right) \cdot R_{Gt(c)FE} \cdot \chi_{FE(c)} \cdot i^2_{Lo(max)} \right). \end{aligned} \tag{4.42}$$

Resistances $R_{Gt(c)LE}$ and $R_{Gt(c)FE}$ are the total gate resistance of the turn-on and turn-off transients, respectively. Parameters $\chi_{LE(c)}$ and $\chi_{FE(c)}$ are correction coefficients representing the effective ON resistance increase with respect to the $R_{DSon(c)}$ at full gate voltage. The values of these parameters are empirically determined and may vary between 1 and 4 depending on the MOSFET technology and driving voltage.

The ON conduction loss expression for the SyncFET is analogous to (4.38), that is,

$$w_{ioON(s)} = w_{ioONs(s)} + w_{ioONd(s)}. \tag{4.43}$$

Nonetheless, due to the long duration of the SyncFET ON time relative to T_S, the Eddy current effects are mainly attributed to the output ripple current (Appendix E). Thus, the first term in (4.43) is,

$$w_{\mathrm{ioONs(s)}} = \left[R_{\mathrm{DSon(s)ac}}(0) \cdot I^2_{\mathrm{RMS(s)s}} + R_{\mathrm{DSon(s)ac}}(F_{\mathrm{S}}) \cdot (I_{\mathrm{RMS(s)s}} - |I_{\mathrm{AVG(s)s}}|)^2\right] \cdot T_{\mathrm{s}}, \tag{4.44}$$

where $R_{\mathrm{DSon(s)ac}}(0)$ is the DC ON resistance, i.e., $R_{\mathrm{DSon(s)ac}}(0) = R_{\mathrm{DSon(s)dc}} = R_{\mathrm{DSon(s)}}$. The second summand accounts for the dynamic resistance, which approaches zero as the output ripple current vanishes, i.e., $I_{\mathrm{RMS(s)s}} = I_{\mathrm{AVG(s)s}}$. Thus, according to Figs. 2.47 and 2.48,

$$I_{\mathrm{RMS(s)s}} = \sqrt{\frac{1}{T_{\mathrm{s}}} \int_{t_8}^{T_{\mathrm{s}}-t_8} i^2_{\mathrm{D(s)}} \, \mathrm{d}t}. \tag{4.45}$$

Likewise, the average term $I_{\mathrm{AVG(s)s}}$ is,

$$I_{\mathrm{AVG(s)s}} = \frac{1}{T_{\mathrm{s}}} \int_{t_8}^{T_{\mathrm{s}}-t_8} i_{\mathrm{D(s)}} \, \mathrm{d}t. \tag{4.46}$$

This first harmonic approximation is only valid for ON conduction times close to T_{S} and has been tested up to a maximum frequency of 5 MHz (see Appendix E).
Similarly to (4.42), the second term from (4.43) can be expressed as,

$$w_{\mathrm{ioONd(s)}} = 2 \cdot R_{\mathrm{DSon(s)ac}}(0) \cdot C_{\mathrm{iss(s)}} \cdot \ldots$$
$$\left(\left(1 - \frac{V_{\mathrm{TH(s)LE}}}{V_{\mathrm{DRV(s)}}}\right) \cdot R_{\mathrm{Gt(s)LE}} \cdot \chi_{\mathrm{LE(s)}} \cdot i^2_{\mathrm{Lo(min)}} + \left(1 - \frac{V_{\mathrm{TH(s)FE}}}{V_{\mathrm{DRV(s)}}}\right) \cdot \right.$$
$$\left. R_{\mathrm{Gt(s)FE}} \cdot \chi_{\mathrm{FE(s)}} \cdot i^2_{\mathrm{Lo(max)}}\right), \tag{4.47}$$

where only the DC resistance is accounted for. The model further assumes ideal dead time operation so that body diode conduction can be neglected.

4.2 Losses of Gate Drive Switches

The gate drivers usually comprise small MOSFET switches that commute the driver voltage between the turn-on and turn-off levels (see Fig. 1.16). The total gate path resistance is partly determined by the ON resistance of these switches. Minimizing it implies an increase of the driver losses due to an increase of their intrinsic parasitic charge.

The typical ON resistance of the gate switches may be ~0.5 Ω, although in some cases it may even be lower as the switching frequency increases so as to minimize the commutation times. An optimum size of the gate drive switches needs to be determined with the use of an estimate of their related losses. For the case of the CtrlFET, the following expression may be used,

$$w_{\mathrm{DRVi(c)}} = V_{\mathrm{DRV(c)}}\left(Q_{\mathrm{Goss(c)hs}} + Q_{\mathrm{Goss(c)ls}}\right) + V_{\mathrm{DRVi(c)}}\left(Q_{\mathrm{Giss(c)hs}} + Q_{\mathrm{Giss(c)ls}}\right). \tag{4.48}$$

Parasitic charges $Q_{\mathrm{GOSS(c)hs}}$ and $Q_{\mathrm{GOSS(c)ls}}$ correspond to the output capacitances of the high and low side gate switches, respectively. Likewise, $Q_{\mathrm{GOSS(c)hs}}$ and $Q_{\mathrm{GOSS(c)ls}}$ refer to the input capacitances of the corresponding devices. Parameter $V_{\mathrm{DRVi(c)}}$ is the gate voltage of both gate switches.

The expression for the SyncFET loss contribution, i.e., $w_{\mathrm{DRVi(s)}}$, is analogous to (4.48).

4.3 Filter Loss

The loss contributions of both input and output filter elements are essentially attributed to their intrinsic parasitic ESR, which, as already stated, is frequency dependent. In the case of the magnetic devices, core losses may additionally be accounted for, particularly at high switching frequency, extreme duty cycle and high-current ripple operation. Thus, while the core losses of the input choke may be minor due to the low amplitude of the AC current, dynamic losses in the magnetic material of the output inductor may be significant and hence must be quantified and added to the portion of the copper (winding) loss.

Dedicated ferrite materials tailored for high flux, high-frequency operation feature reduced hysteresis curves and low electrical conductivities to mitigate hysteretic B–H loop areas and Eddy current losses, respectively. Thus, excess loss becomes, in most cases, the dominant loss contribution in such magnetic core materials.

Quantifying core losses may involve the use of parameters provided by the manufacturer in combination with Steinmetz's equation or other simplified loss expressions [1]. In addition, VR coil manufacturers may provide empirical loss models of both core and copper loss, which offer a first-order approximation of the overall inductor performance.

Accordingly, the loss analysis presented in this work employs manufacturer data to estimate inductor losses. In one particular case of interest, a detail loss analysis of a reference commercial power coil inductor is provided in Appendix E as an illustration. Such analysis is based on FE calculations of the coil structure. The results show that a great deal of the generated losses under stringent operating conditions (5 MHz, 50 A load current) are associated to the conductor. The loss mechanisms of hysteresis and induced Eddy currents in the magnetic material are minor contributions compared with the skin effects produced in the copper windings and the excess loss associated to the magnetic domains and domain walls.

Regarding filter capacitors, associated losses are AC related as the average DC current vanishes in steady state. Assuming a number N_p of interleaving phases and a total of N_{Ci} input capacitors of the same type, the following expression is applied to estimate the input capacitor losses,

$$w_{Cin} = \left[\frac{R_{ac(Cin)}(F_S N_p)}{N_{Ci}} \cdot I^2_{RMSi(ac)} \right] \cdot \frac{T_s}{N_p}, \tag{4.49}$$

where $R_{ac(Cin)}$ is the ESR of a single capacitor, which can be derived from manufacturer data, whereas $I_{RMSi(ac)}$ is the total AC RMS input current.

Analogously for the output capacitance,

$$w_{Co} = \left[\frac{R_{ac(Co)}(F_S N_p)}{N_{Co}} \cdot I^2_{RMSo(ac)} \right] \cdot \frac{T_s}{N_p}. \tag{4.50}$$

It is common to build the output filter by means of multiple capacitor technologies. However, it is assumed that only one capacitor type is employed to reduce steady-state ripple. Thus, Parameter N_{Co} is the total number of capacitors employed to mitigate steady-state ripple. These capacitors are assumed to be identical. All other possible capacitor types of the output filter are ignored. Furthermore, resistance $R_{ac(Co)}$ is the associated ESR resistance, whereas $I_{RMSo(ac)}$ is the current ripple.

4.4 PCB Loss

Multilayer PCBs are frequently employed to minimize the parasitic half-bridge loop inductance by means of a ground inner layer situated at a short distance from the top layer, where the MOSFETs are physically located. Simplified device physics simulations of such configuration are provided in Appendix E. The results clearly provide guidelines as to how wide and short the PCB tracks must be designed for to avoid excessive ESR. The track thickness becomes irrelevant at frequencies above cut-off, where the skin effect becomes dominant.

The PCB resistance adds up to the ESR of the different elements of the switched converter. The contributions of the various PCB sections are estimated in Appendix E. According to the simulation results, one resistive PCB section may potentially correspond to that of the CtrlFET path, since the associated AC ESR is inversely proportional to the short ON conduction time of the upper power switch (see Sect. 4.1.4). This yields stronger skin effects than in other PCB sections where the inverse proportionality may be with respect to T_S.

In some simplified geometrical cases, the frequency-dependent PCB resistance can be analytically formulated as a function of the skin depth. Combined with the empirical equations described in Sects. 4.1.4 and 4.3, the PCB loss contribution may then be expressed in closed-form. Further details of such procedure are provided in Appendix E.

4.5 Model Validation

This section provides a validation of the loss model for the power MOSFETs (i.e., equations of Sect. 4.1). The derivation of the filter and PCB loss equations is given in Appendix E. Further studies on the loss model from the gate drive switches are provided in [2, 3].

According to the loss expressions from Sect. 4.1, the gate driver and MOSFET losses of the SRBC can be calculated as,

$$\begin{aligned} w_{\mathrm{DRVs\&FETs}} = {} & w_{\mathrm{ioON(c)}} + w_{\mathrm{ioHS(c)}} + w_{\mathrm{cHBC(c)}} + w_{\mathrm{iHBC}} + w_{\mathrm{DRV(c)}} + \cdots \\ & + w_{\mathrm{DRVi(c)}} + w_{\mathrm{ioON(s)}} + w_{\mathrm{cHBC(s)}} + w_{\mathrm{DRV(s)}} + w_{\mathrm{DRVi(s)}}. \end{aligned} \tag{4.51}$$

The above equation applies for fast switching operation provided that gate bounce, avalanche breakdown, body diode conduction, and reverse recovery are negligible.

In order to assess the performance of the resulting loss model, expression (4.51) is used to estimate the power losses of the IPM from the illustration of Sect. 3.6. Given the satisfactory level of accuracy achieved with the PLA model, the quantities from (4.51) and expressions of each individual loss contribution are contrasted with the results of model level 1. Such comparison is carried out for a number of parameter sweep variations so as to cover a wide range of operating conditions. This is shown in the plots of Fig. 4.1. The loss curves related to charging losses, i.e., plots (a), (c), and (f) show good matching correlations between the two models, although the PLA model generally predicts slightly higher quanti ties. This is attributed to secondary effects such as capacitive and inductive gate bounce (case of plot (a)) and inaccurate initial charge conditions (case of plots (c) and (f)) that are not accounted for by the model level 2.

According to Fig. 4.1d, (4.38)–(4.42) also prove to be effective to estimate ON conduction losses, including those during the switching time intervals.[1]

Hard-switching losses of typical fast switching transients from IPM solutions can be predicted by model level 2 as shown in Fig. 4.1b, e. Nonetheless, the good matching of the LE transient loss contrasts with that of the FE transient loss. The discrepancies are partly due the linearization of the MOSFET capacitance in the loss model, which is based on the following expression,

[1]Other second order effects such as skin or proximity effects leading to an increase of the total MOSFET's ON resistance are not taken into account in neither of the two models in this particular case.

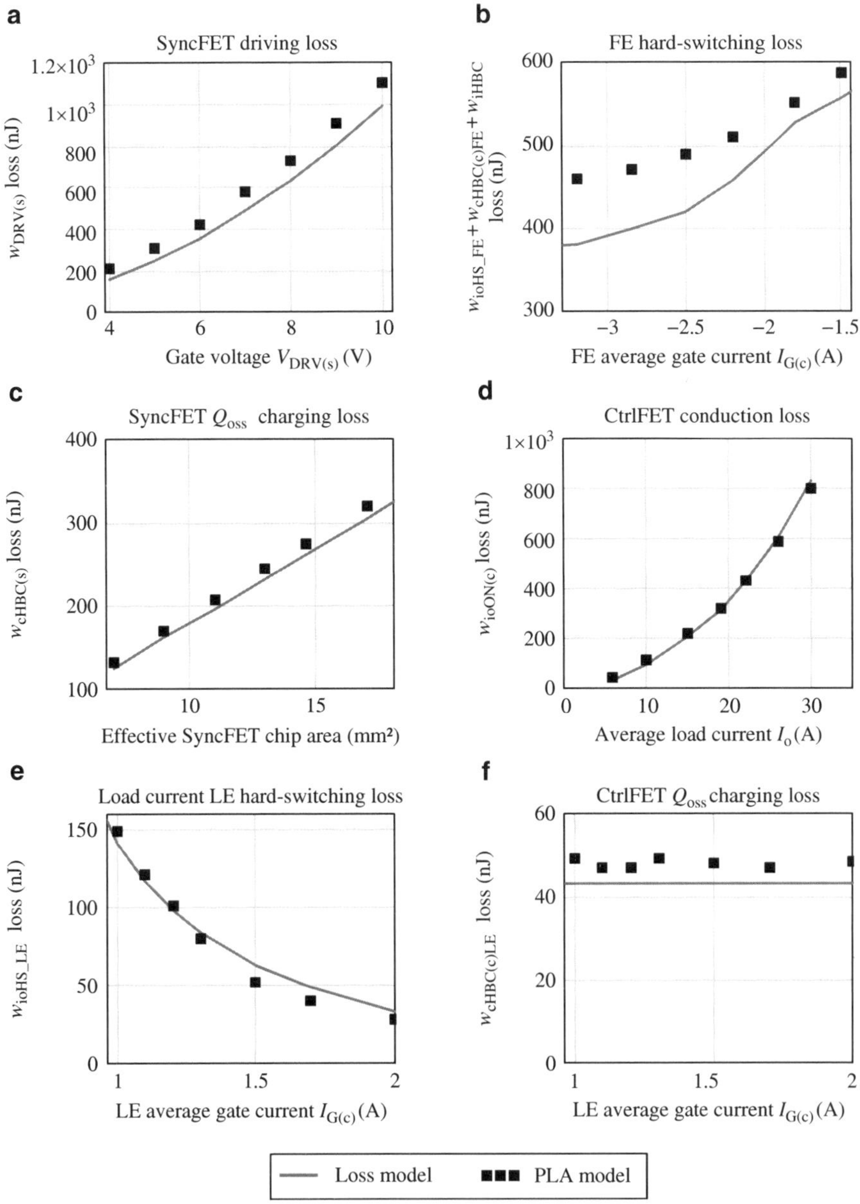

Fig. 4.1 Loss prediction comparison between models level 1 and level 2 of various loss mechanisms (a) SyncFET driving loss, (b) FE hard-switching loss, (c) SyncFET output charge loss, (d) CtrlFET conduction loss, (e) load current LED hard-switching loss, (f) CtrlFET output charge loss

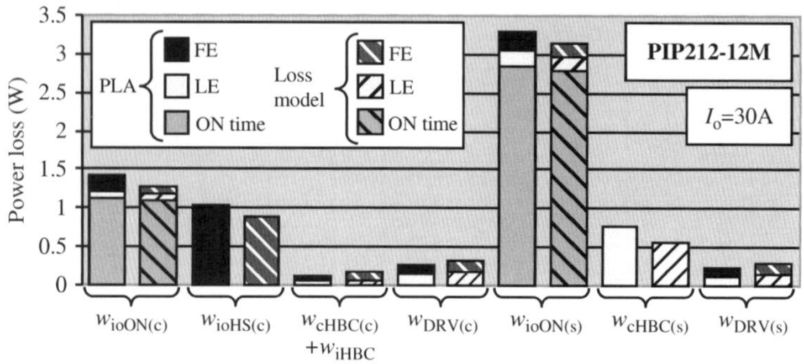

Fig. 4.2 Loss breakdown comparison of PLA model level 1 and loss model level 2 corresponding to powertrain module PIP212-12M from NXP semiconductors [4]. Total estimated converter losses are ~7.8 W (~7.3 W from IPM)

$$C_{x(\mathrm{avg})} = \frac{1}{|V_f - V_i|} \int_{V_i}^{V_f} C_x(v)\, \mathrm{d}v. \tag{4.52}$$

Another sensitive simplification that makes the loss model differ from the estimations of the PLA model is the averaging of the gate current during the hard-switching times, since the loss model assumes a constant gate current to determine the switching times and other relevant quantities according to the derived expressions.

Nonetheless, the performance of model level 2 seems to be acceptable for first-order design optimization and roadmap studies, as it is addressed in the next chapter.

Figure 4.2 further shows the loss breakdown of the PIP212-12M operating at 1 MHz switching frequency and 30 A load current as described in Sect. 3.6. Once again, the predictions from model levels 1 and 2 consistently compare for each individual loss quantity.

References

1. Li J, Abdallah T, Sullivan CR (2001) Improved calculation of core loss with nonsinusoidal waveforms. In: IEEE industry applications society annual meeting, Chicago, 30 Sept–4 Oct 2001, pp 2203–2210
2. López T, Sauerlaender G, Duerbaum T, Tolle T (2003) A detailed analysis of a resonant gate driver for frequency control applications. In European power electronics conference, 2003, http://www.epe-association.org/epe/index.php
3. López T, Sauerlaender G, Duerbaum T, Tolle T (2003) A detailed analysis of a resonant gate driver for PWM applications. In IEEE applied power electronics conference, APEC, Aachen, Germany, 9–13 Feb 2003, vol 2, pp 873–878
4. PIP212-12M DC-to-DC converter powertrain online documentation, www.nxp.com. Accessed Oct 2006

Chapter 5
Model Level 3: Optimization

This chapter presents a systematic approach for the design of main components of the VR, namely the input/output filters, gate drivers, and power switches. Regarding control schemes, the nonlinear hysteretic control approach is selected to fulfill the high bandwidth requirements imposed by the load transients. This selection will strongly condition the dynamic operation of the converter as well as the size of the output filter, both aspects of which will be addressed in the following sections.

In the design process, special emphasis is given to the selection of the switching frequency and the number of phases based on two correlated criteria: Converter size (i.e., power density and footprint) and efficiency. Other related aspects such as component count and maximum temperature operation shall be addressed along with the discussions to justify the suggested parameter settings of the converter.

The loss model presented in Chap. 4 underlies the derivation of various FoM for the optimization of the semiconductor devices. This is combined with loss estimations of other VR elements (such as the output coil) to set out a series of VR design guidelines that will allow later, in Chap. 6, an exploration of performance limitations of existing and forecasted technologies. Although the proposed design procedure may be generalized to a number of VR solutions, the main focus is put on IPM for their superior performance with respect to discrete alternatives.

The last section of the chapter investigates VR solutions for two different state-of-the-art computer systems: A desktop and a mobile laptop computer.

5.1 Output Filter

Of critical importance for the size as well as the transient and steady-state operation of the VR is the design of the output low-pass filter. Chapter 1 introduced the basic characteristics of such filter, which typically comprises a bank of decoupling capacitors in parallel to the load and as many inductors as interleaved phases. The implications of the number of phases and associated control algorithm on the output ripple current and dynamic transient behavior are crucial to the dimension of the filter components, which need to feature small volume and footprint so as to

T. López et al., *Voltage Regulators for Next Generation Microprocessors*,
DOI 10.1007/978-1-4419-7560-7_5, © Springer Science+Business Media, LLC 2011

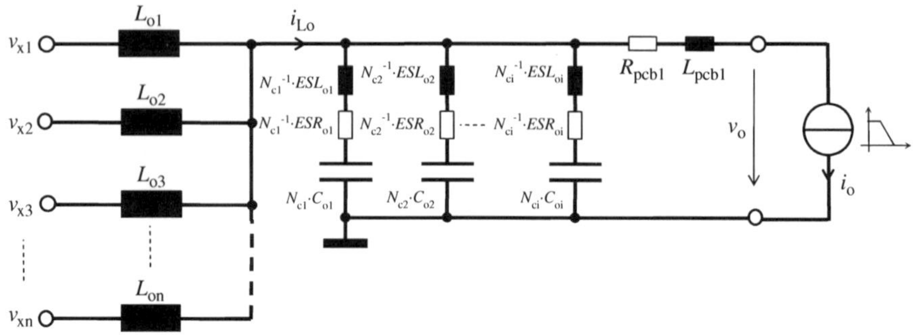

Fig. 5.1 Output filter model for steady-state and transient analysis

maximize the power density. A proper design procedure that compromises between the different filter demands must therefore be deduced.

This section proposes a simple model to study both the dynamic and the steady-state output voltage as a function of the output filter elements. Considering hysteretic control as the referent regulation technique for high-switching frequency, a design procedure is presented by which suitable filter component technologies can be identified. Analysis of worst-case scenarios further results in optimizations on component count and volume size. Minimizing them while ensuring compliance with load voltage tolerance specifications is one major goal of the filter design.

Figure 5.1 depicts the general form of the filter model, which includes the inductors of n interleaved phases, number i of capacitor types, and the parasitic elements interconnecting the VR with the load. The behavior of such load is represented with current source i_o.

Each individual capacitor model comprises an ideal capacitance in series to an ESR and an ESL. These elements together represent the equivalent circuit of a number N_{cx} of parallel capacitors with identical characteristics. Therefore, the filter capacitance increases in direct proportion to N_{cx}, whereas both ESR and ESL decrease with N_{cx} increase.

The proposed filter model, though being simple, intrinsically represents all fundamental effects of interest to a sufficient degree of accuracy and thus can be exploited for filter dimensioning. Such statement is based on results from [1–4]. When further accuracy is demanded to consider second-order effects, more complex lumped structure models such as those from [5–7] may be alternatively used.

Both steady-state and transient analysis of the proposed filter model are provided in the following subsections.

5.1.1 Steady-State Output Ripple

The steady-state output voltage may be expressed as a function of the output filter impedance and the harmonic composition of current i_{Lo} by means of Fourier expansions. Because of the interleaving phase shift operation, it is convenient to previously determine the shape of current i_{Lo} from an equivalent one-phase converter operating at $F'_s = N_p F_s$, $d' = N_p d$, $V'_{in} = V_{in}/N_p$, and a single equivalent output inductor $L'_o = L_o/N_p$. It is assumed that $L_o = L_{o1} = L_{o2} = \cdots = L_{on}$. This equivalent circuit is valid as long as CtrlFET ON conduction overlapping is avoided between the different phases (thus limiting the maximum N_p). Accordingly, the PWM signal at the switched-node voltage from the equivalent circuit may be expressed as,

$$v'_x(t) = d \cdot V_{in} + \cdots + \sum_{n=1}^{\infty} \left[\frac{2V_{in}}{n \cdot \pi \cdot N_p} \cdot \sin(\pi \cdot d \cdot N_p \cdot n) \cdot \cos\left(\frac{2\pi \cdot n \cdot N_p}{T_S} t - \pi \cdot d \cdot N_p \cdot n \right) \right]. \tag{5.1}$$

Now, given that the output voltage "seen" by the output inductors is virtually constant, current i_{Lo} from Fig. 5.1 can be approximated as,

$$i_{Lo}(t) \cong I_o + \cdots + \sum_{n=1}^{\infty} \left[\frac{2V_{in}}{n \cdot \pi \cdot N_p} \cdot \frac{\sin(\pi \cdot d \cdot N_p \cdot n)}{2\pi \cdot n \cdot L_o} \cdot T_S \cos\left(\frac{2\pi \cdot N_p \cdot n}{T_S} \cdot t - \pi\left(d \cdot N_p \cdot n + \frac{1}{2} \right) \right) \right], \tag{5.2}$$

where N_p is the number of phases and d the duty cycle. The later is output current dependent on a certain load line given by the following average output voltage equation,

$$V_{o(av)1} = V_{oi} + (R_{pcb1} - R_{LL}) \cdot I_{o(av)}. \tag{5.3}$$

Parameters R_{LL}, $I_{o(av)}$, and V_{oi} are the load line resistance, average load current, and output voltage at $i_o = 0$, respectively.

The product of the AC components from i_{Lo} with the total output capacitors' impedance gives the expression for the output ripple voltage. Adding to it the average output voltage yields,

$$v_o(t) = V_{o(av)1} + \cdots + \sum_{n=1}^{\infty} \left[\begin{array}{c} \dfrac{2V_{in}}{n \cdot \pi \cdot N_p} \cdot \dfrac{\left| Z_o\left(\dfrac{2\pi \cdot N_p \cdot n}{T_S} \right) \right|}{2\pi \cdot n \cdot L_o} \cdot T_S \cdot \sin(\pi \cdot d \cdot N_p \cdot n) \cdot \ldots \\ \cos\left(\dfrac{2\pi \cdot N_p \cdot n}{T_S} \cdot t - \pi\left(d \cdot N_p \cdot n + \dfrac{1}{2} \right) + \varphi_o\left(\dfrac{2\pi \cdot N_p \cdot n}{T_S} \right) \right) \end{array} \right]. \tag{5.4}$$

Output capacitor's impedance Z_o is expressed as,

$$Z_o(\omega) = \left[\sum_{q=1}^{N_{typ}} \left(\frac{\text{ESR}_{oq}}{N_{cq}} + j\omega\frac{\text{ESL}_{oq}}{N_{cq}} + \frac{1}{j\omega C_{oq}N_{cq}}\right)^{-1}\right]^{-1}, \tag{5.5}$$

where N_{typ} equals i in Fig. 5.1 and thus represents the number of different capacitor types used. The associated impedance phase is,

$$\varphi_o(\omega) = a\ \tan\left[\frac{\text{Re}(Z_o(\omega))}{\text{Im}(Z_o(\omega))}\right]. \tag{5.6}$$

Figure 5.2 illustrates steady-state switching waveforms corresponding to current i_{Lo} and voltage v_o. The case example considers four interleaved phases operating at 1 MHz, which down-convert 12 V to less than 1 V while driving a load current of 200 A. The output ripple voltage is maintained within a maximum voltage window, defined herein as $\Delta v_{os(max)}$, of 8 mV with the use of 40 nH coil inductors, 12 × 390 μF electrolytic capacitors and 55 × 4.7 μF ceramic capacitors. Both capacitance technologies are low ESR, SMD technology. Note that, according to the manufacturer, such ESR is frequency dependent, the magnitude of which may approximately correspond to that of frequency $N_p{\cdot}F_s$.

In steady state, the ripple voltage developed at the output capacitors is the sum of three voltage contributions produced by the AC components of current i_{Lo}. These are: the voltage drop across the ESR, the stored charge fluctuations in the capacitance, and the induced voltage across the ESL. While the first two are proportional to the amplitude of the ripple current, the ESL voltage only depends on the rate of change of i_{Lo}, which happens to be independent on the switching frequency. Thus, as the switching frequency increases and the ripple current diminishes, the remaining voltage ripple contribution is that of the ESL.

The close relation existing between the output inductor size and the required number of capacitors is illustrated in Fig. 5.3a for the cases of $N_{c1} = 0$ and 25. The

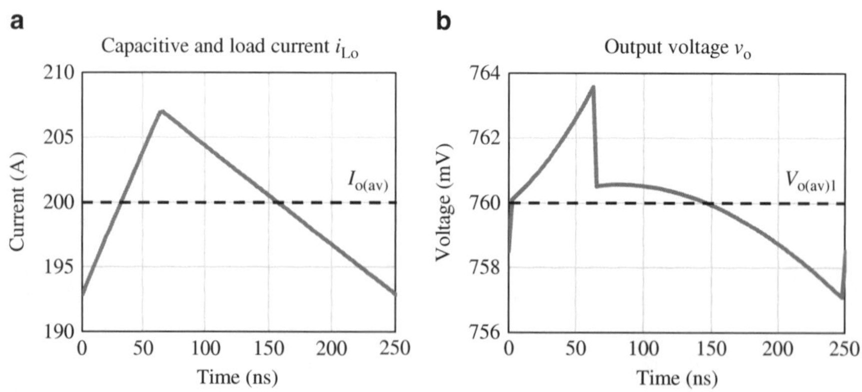

Fig. 5.2 Simulated steady-state output current i_{Lo} (a) and voltage v_o (b). Parameter values: $N_p = 4$, $F_s = 1$ MHz, $V_{in} = 12$ V, $L_o = 40$ nH, $N_{typ} = 2$, $C_{o1} = 390$ μF, $\text{ESL}_{o1} = 2$ nH, $\text{ESR}_{o1} = 8$ mΩ, $N_{c1} = 12$, $C_{o2} = 4.7$ μF, $\text{ESL}_{o2} = 600$ pH, $\text{ESR}_{o2} = 15$ mΩ, $N_{c2} = 55$, $R_{pcb1} = 0$ Ω

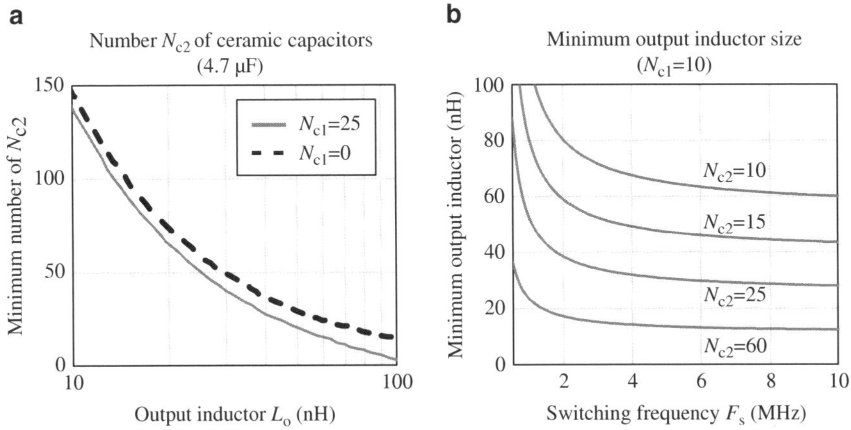

Fig. 5.3 Simulation results of (**a**), minimum number of 4.7 μF capacitors as a function of output inductor L_o, and (**b**), minimum required output inductor as a function of the switching frequency. Target output voltage tolerance, $\Delta v_{os(max)}$=10 mV. See legend of Fig. 5.2 for further parameter values

relatively small deviation observed between the two curves supports the aforementioned arguments that, at high-switching frequency and reduced current ripple, the capacitor types featuring the lowest ESL are the most effective despite of their relative low capacitance and high ESR.

Consequently, when the voltage across the ESL becomes the major ripple contribution, the minimum required output inductance L_o weakly depends on the switching frequency. This is shown in Fig. 5.3b for various N_{c2} and $N_{c1} = 10$, where the curves tend to quickly flatten out as the switching frequency enters into the multi-MHz range. At low F_s, the output inductance L_o, and hence the current ripple, become critical because the ripple output voltage is dominated by charge storage fluctuations, as illustrated in Fig. 5.4a. The predominant effect of the ESL at high F_s is exemplified in Fig. 5.4b. The sharp voltage edges at the switching instants are the consequence of the induced voltage across the ESL, which dominates the ripple voltage of the waveform at 8 MHz.

5.1.2 *Load Line Transient*

The load line transient response strongly depends on the adopted control scheme. When considering however an ideal nonlinear hysteretic control, loop delays may be neglected, and so the load line dynamics becomes mostly determined by the size of the output filter inductors, the load current magnitude, the output voltage, and the number of phases.

From the two possible load steps, the current step-down response is considered as worst-case scenario [1]. The stored energy of the output inductors has to be delivered to the output capacitors with the risk of producing an undue overvoltage. To mitigate this, a fast transient response may be achieved by deactivating the

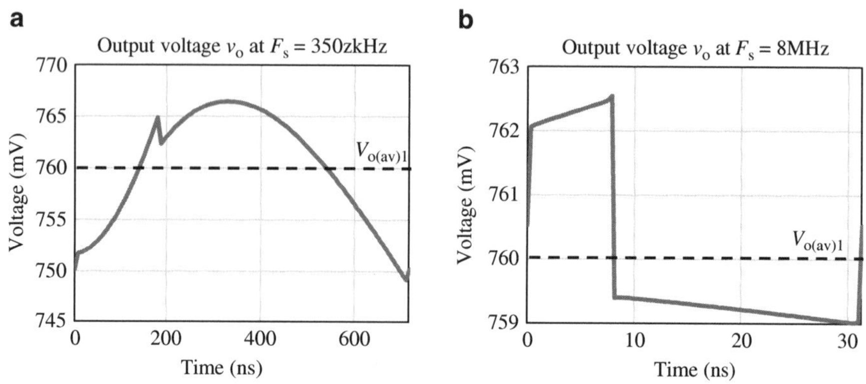

Fig. 5.4 Simulated output ripple voltage at opposite extreme switching frequencies. (**a**) Low F_s, where the output ripple becomes strongly dependent on capacitive AC charging voltage. (**b**) High F_s, where the output voltage ripple is dominated by the ESL contribution. See legend of Fig. 5.2 for further parameter values

CtrlFET of all phases simultaneously at the moment of the load change[1] [10]. Therefore, the inductors discharge at a maximum rate of $v_o N_p / L_o$. Accordingly, the reduction of v_o, N_p or the increase of L_o will slow down the transition time.

The worst-case overshoot scenario is found at the end of the CtrlFET conduction phase (i.e., a maximum in i_{Lo}) when the load switches from heavy to light load at a maximum slew-rate (SR) [1].

Figure 5.5 shows simulated load transient waveforms of a VR with hysteretic control and active droop compensation. From the output voltage waveform it can be seen that the benefit of the load line implementation lies on the reduction of the voltage difference between the maximum overshoot level and the target output voltage after the load transition. Ideally, a load transient without overshoot occurs when the active voltage positioning precisely cancels out the voltage swing across the output capacitance (e.g., $\Delta v_2 = 0$ in Fig. 5.5b).

Generally, two different kinds of voltage overshoots can be distinguished. In Fig. 5.5b, these are identified as Δv_1 and Δv_2. The first sharp spike has a parasitic inductive nature and hence depends on the load current SR. On the other hand, voltage overshoot Δv_2 develops at a much slower pace during the discharge of the output inductors. That is, the output voltage rises as the magnetic energy is transferred to the output capacitors, which is the only impedance path available for the inductance current at zero load.

In order to effectively reduce these voltage overshoots by capacitive decoupling, at least two different capacitor types may be employed. On the one hand, ceramic SMD capacitors of limited capacitance are typically connected in close proximity to

[1] Further deactivating the SyncFET switches (and hence the synchronous rectification functionality) during the load transient may notably enhance the transient response considering that the body diode reverse voltage can be even higher than the output voltage.

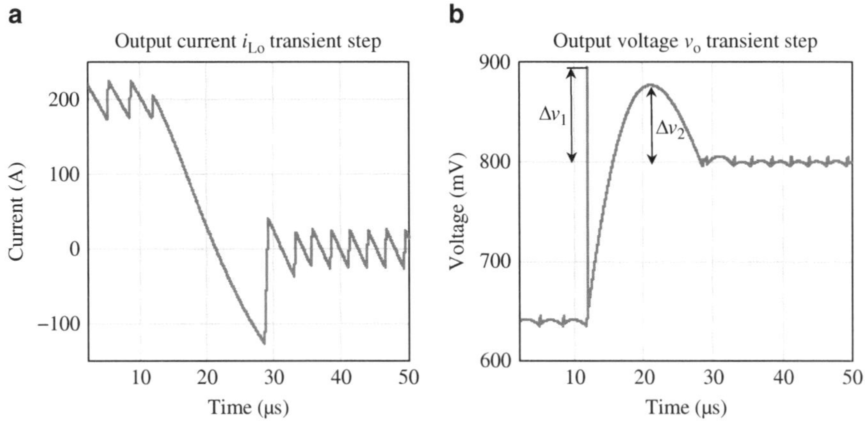

Fig. 5.5 Simulated example of a load current transient step, (a) current waveform, (b) voltage waveform. Parameter values: −10 A/ns load current slew rate, $V_{in} = 12$ V, $v_o = V_{oi} = 0.8$ V for $i_o = 0$ A, $R_{LL} = 0.8$ mΩ (load line), $R_{pcb1} = 0.3$ mΩ, $L_{pcb1} = 20$ pH, steady-state tolerance band $\Delta v_{os(max)}$=10 mV, 40 nH equivalent output inductor and 40 × 100 μF output capacitors with ESR=2 mΩ and ESL=1 nH per part

the load. The reason for this is that these capacitors feature small footprints and low ESL, which effectively allows restraining Δv_1. On the other hand, dealing with Δv_2 implies the use of high capacitive types so as to mitigate the voltage swing during the charge transfer. Since ESL is not as critical in this case, bulkier capacitor types than the ceramic ones are usually employed, such as electrolytic capacitors, by which much higher capacitance per unit volume can be achieved in detriment of ESL.

In view of these filtering demands, the following discussion assumes $N_{typ} = 2$, where N_{c1} will always refer to the required number of bulk capacitors dealing with Δv_2, whereas N_{c2} will denote the number of ceramic capacitors for high frequency signal decoupling (i.e., Δv_1 and steady-state output voltage ripple).

5.1.2.1 Output Current Slew Rate

Whenever the load current changes, induced voltages develop across the ESL of both capacitances and PCB that produce transient deviations in the output voltage (see Δv_{o1} from Fig. 5.5b). The maximum voltage swing must not exceed a specified level, defined herein as $\Delta v_{ot(max)}$. Considering the inductive effects of two different capacitor types and PCB, the model of Fig. 5.1 can be used to estimate the required number of ceramic capacitors. That is, provided that N_{c1} and ESL_{o1} from the bulk capacitors are given, the following equation can be derived,

$$N_{c2} = \frac{ESL_{o2}}{\Delta v_{ot(max)} + (R_{LL} - R_{pcb1}) \cdot i_{o(max)} + L_{pcb1} \cdot SR} \left| \frac{V_{in} - N_p \cdot V_{o(av)1}}{L_o} - SR \right| - \frac{ESL_{o2}}{ESL_{o1}} N_{c1}. \tag{5.7}$$

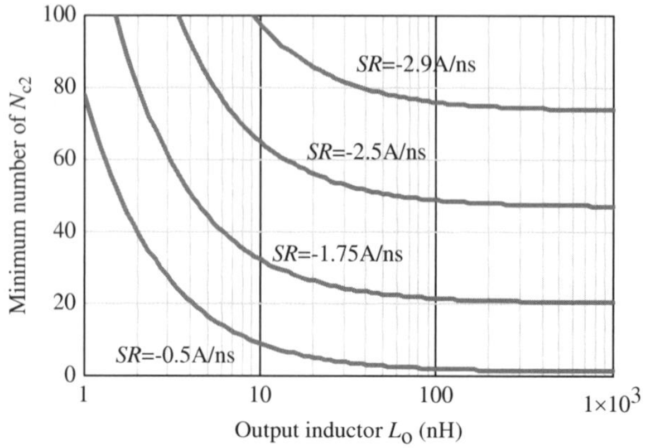

Fig. 5.6 Minimum number N_{c2} of ceramic capacitors as function of L_o and various slew-rates. Parameter values: $ESL_{o1} = 2.5$ nH, $ESL_{o2} = 0.6$ nH, $N_{c1} = 15$, $N_p = 4$, $\Delta v_{ot(max)} = 40$ mV, $V_{in} =$ 12 V, $i_{o(max)} = 200$ A, $v_o = V_{oi} = 0.8$ V for $i_o = 0$ A, $R_{LL} = 0.8$ mΩ (load line), $R_{pcb1} = 0.3$ mΩ, $L_{pcb1} = 20$ pH

The denominator of the first term accounts for the maximum induced voltage across v_o, which differs from $\Delta v_{ot(max)}$ due to the effects of the load line and PCB parasitic elements. Note the significance of the di/dt from i_{Lo} (first term within the absolute sign) relative to the load current SR. This is illustrated in Fig. 5.6, where the impact of the slopes of i_{Lo} on the capacitor count becomes significant as L_o lowers below 100 nH. The effect, as such, is independent on the switching frequency, and hence, on the ripple amplitude. The curves flatten out for values above 100 nH, where the SR portion dominates over the SR of i_{Lo}. The trends further indicate a dramatic increase on capacitor count with SR.

Note that (5.7) may become negative. If this occurs because of a large L_{pcb1}, it suggests that the adopted solution cannot fulfill the target requirements since the voltage overshoot contribution of the layout inductance alone is high enough to break the allowed limit.

This PCB inductance problem, combined with the proportionality of ESL_{o2} in (5.7) and the demands of steady-state operation, provides grounds to justify ESL free on-chip filter capacitance integration as one potential solution for present and future VR solutions.

In any case, and regardless of the switching frequency, a minimum L_o in combination with a maximum |SR| must be defined to limit N_{c2}.

5.1.2.2 Output Inductors' Discharge

Voltage overshoot Δv_2 (see Fig. 5.5) is assumed to be largely dependent on the bulk capacitors, and as such, the effect of the ceramic capacitors can be negligible. Thus, the following relation can be readily established from the model provided in Fig. 5.1 assuming that $i_o = 0$, $N_{typ} = 1$, $v_{x0}=v_{x1}=v_{x2}=\cdots=v_{xn}=0$, and with initial conditions in the inductors being $i_{o(max)} + \Delta i_{Lo}/2$,

$$N_{c1(h1)} = \frac{1}{\Delta v_{ot(max)} + (R_{LL} - R_{pcb1}) \cdot i_{o(max)}} \cdot \ldots$$
$$\left| \frac{L_o \left(i_{o(max)} + \frac{\Delta i_{Lo}}{2}\right)^2}{2C_{o1} \cdot N_p \cdot V_{o(av)2}} + \frac{N_p \cdot V_{o(av)2}}{L_o} \left(\frac{ESR_{o1}^2 \cdot C_{o1}}{2} - ESL_{o1}\right) \right|. \tag{5.8}$$

In (5.8), $N_{c1(h1)}$ is the required number of bulk capacitors to keep $\Delta v_2 \leq \Delta v_{ot(max)}$, $\Delta v_{ot(max)}$ being the maximum allowed transient voltage overshoot. Current $i_{o(max)}$ is the maximum load current and Δi_{Lo} the i_{Lo} ripple current, which is expressed as,

$$\Delta i_{Lo} = \frac{V_{in} - N_p \cdot V_{o(av)1}}{L_o} dT_S = \frac{V_{in} - V_{o(av)1}}{L_o} \frac{1 - d \cdot N_p}{1 - d} dT_S. \tag{5.9}$$

Furthermore, average voltage $V_{o(av)2}$ is defined as,

$$V_{o(av)2} = V_{o(av)1} + \frac{(R_{LL} - R_{pcb1}) \cdot I_{o(av)} + \Delta v_{o2(max)}}{2}. \tag{5.10}$$

Equation (5.8) considers the worst-case scenario, namely when i_{Lo} is maximum and the microprocessor goes from full to light load in no time. The inductors' discharge period starts right after the load current has dropped to zero, that is, when the load's $di/dt = 0$. Thus, current i_{Lo} is assumed to flow entirely through the output capacitances.

During the output voltage transient, the control system turns off all CtrlFET switches simultaneously and so the equivalent output inductance effectively reduces to $L_o N_p^{-1}$. This minimizes the energy transfer time, thereby improving the transient dynamic response.

Equation (5.8) is only valid as long as the following condition is met,

$$\frac{i_{o(max)} + \frac{\Delta i_{Lo}}{2}}{V_{o(av)2}} > \frac{ESR_{o1} \cdot C_{o1}}{L_o} \cdot N_p \tag{5.11}$$

Not satisfying (5.11) indicates that the output voltage maximum Δv_2 occurs at $t = 0$ (i.e., at the time the load current reaches zero). In such case, the expression for the number of capacitors becomes,

$$N_{c1(h2)} = \frac{1}{\Delta v_{ot(max)} + (R_{LL} - R_{pcb1}) \cdot i_{o(max)}} \cdot \left| ESR_{o1} \left(i_{o(max)} + \frac{\Delta i_{Lo}}{2} \right) - \frac{V_{o(av)1}}{L_o} \cdot ESL_{o1} \cdot N_p \right|. \tag{5.12}$$

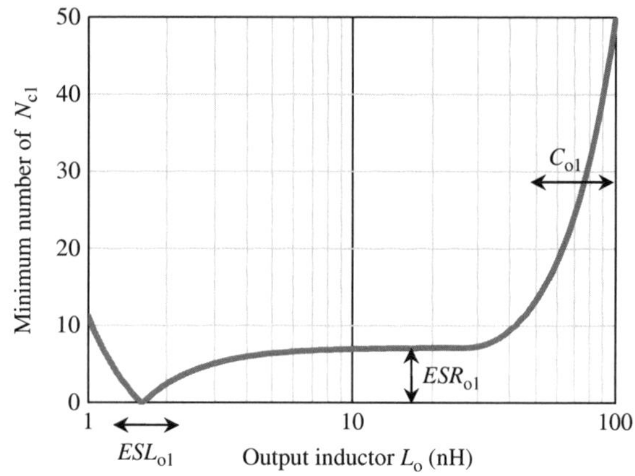

Fig. 5.7 Minimum number N_{c1} of bulk capacitors as function of output coil inductor L_o

Finally, combining expressions, the number of bulk capacitors N_{c1} can be defined as,

$$N_{c1} = \left| \begin{matrix} N_{c1(h1)} \text{if} \left(\dfrac{i_{o(max)} + \frac{\Delta i_{Lo}}{2}}{V_{o(av)2}} > \dfrac{\text{ESR}_{o1} \cdot C_{o1}}{L_o} \cdot N_p \right) \\ N_{c1(h2)} \text{ otherwise} \end{matrix} \right| . \tag{5.13}$$

Figure 5.7 illustrates the dependence of N_{c1} on L_o. For large L_o, condition (5.11) is satisfied and the first command of (5.8) dominates. It can then be clearly identified that the position of the slope to the right of the plateau can be defined by C_{o1}. Particularly, the plateau may be extended for larger L_o as capacitance C_{o1} increases. It is therefore convenient to select a sufficiently large C_{o1} such that the selected L_o falls in the plateau region, with a consequent minimization in N_{c1}. This then implies that C_{o1} must increase in proportion with the demanded L_o. When operating in the plateau, the value increase of the first command from (5.8) cancels out with that of the second. The established level becomes then largely dependent on ESR_o. Note that, as it will be shown later, such plateau may vanish with a sufficiently low C_{o1}.

For low L_o, condition (5.11) may no longer be fulfilled, in which case N_{c1} is governed by (5.12). In this regime, N_{c1} can theoretically approach zero as the voltage drop across ESR_{o1} compensates the induced voltage of ESL_{o1}. The location of this minimum on the curve can be adjusted with ESL_{o1}. Note, however, that small tolerances causing a shift to the right of the minimum can lead to a rapid increase of N_{c1}. Thus, operating in this regime appears to be reliably unsuitable.

Figure 5.8 shows an example of N_{c1} as function of L_o for different commercial capacitor types. All examined parts are polymer electrolytic capacitors [7] except for the 100 μF ones, which correspond to the X5R ceramic technology.

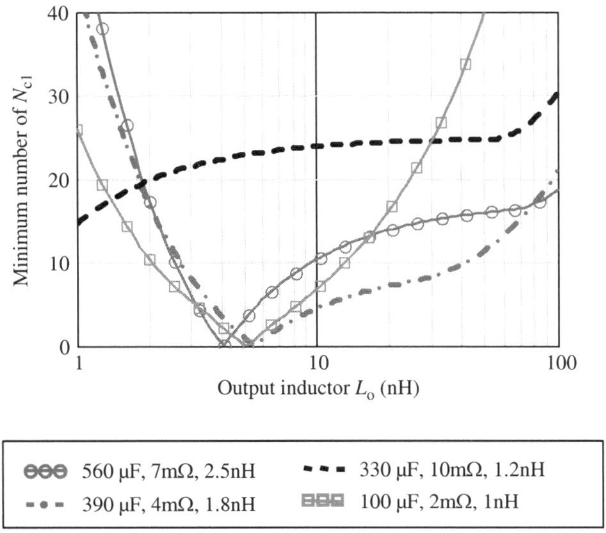

Fig. 5.8 Required number N_{c1} of bulk capacitors (high capacitive type) as a function of the output inductor. Parameter values: $V_{in} = 12$ V, $v_o = V_{oi} = 0.8$ V for $i_o = 0$ A, $R_{LL} = 0.5$ mΩ (load line), $R_{pcb1} = 0.3$ mΩ, $\Delta v_{ot(max)} = 40$ mV, $L_o = 40$ nH, $N_p = 4$, $i_{o(max)} = 210$ A, $F_s = 1$ MHz

As it can be seen, in the range 10–60 nH, the least number of capacitors required for $\Delta v_{ot(max)} = 40$ mV is obtained with the 390 μF capacitors. Though featuring lower ESR, the ceramic capacitors are less adequate due to an insufficient capacitance, that is, no presence of the plateau. The 560 μF capacitors feature the highest capacitance and thus become the best choice at high L_o. The relatively high ESR of the 330 μF ones makes them unsuitable for this particular case.

This comparison suggests that, as L_o shrinks, the ESR becomes a key parameter together with the capacitance value. Furthermore, a proper increase of ESL can help minimize N_{o1}, yet slight variations due to tolerances can lead to a rapid increase of N_{o1} when operating around the minimum. For large L_o the relevance of ESR fades and the capacitor featuring the largest capacity per unit volume becomes the most suitable.

5.1.3 Component Selection Procedure

In the discussions thus far the estimation of the output filter size have been studied individually for a number of particular cases involving either steady-state or transient behavior. This section considers these cases altogether to minimize the output filter size while ensuring compliance with the load requirements under all operating conditions.

For that purpose, it is convenient at this point to define the following variables: $N_{c1(d)}$, $N_{c2(ss)}$, and $N_{c2(sr)}$. Variable $N_{c1(d)}$ defines the required number of bulk capacitors to avoid undue voltage overshoots during the output inductance discharge transient, which is given by (5.13). On the other hand, $N_{c2(ss)}$ is the number of ceramic capacitors required for proper steady-state operation (see Fig. 5.3a). Regarding load current SR, a different number of required ceramic capacitors may be derived from (5.7), which is assigned to variable $N_{c2(sr)}$. Therefore, the final number of bulk (N_{c1}) and ceramic (N_{c2}) capacitors for all the operating conditions considered can be expressed as,

$$N_{c1} = N_{c1(d)}, \tag{5.14}$$

$$N_{c2} = \max(N_{c2(ss)}, N_{c2(sr)}). \tag{5.15}$$

It becomes clear from (5.14) that the absolute number of bulk capacitors is uniquely defined by (5.13).

Having fixed the type and number of bulk capacitors, a maximum L_o per phase can be expressed as function of N_{c1}, i.e.,

$$L_{o(max)} = f_1(N_{c1}). \tag{5.16}$$

Similarly, a minimum L_o per phase can be determined once all output capacitors have been defined, i.e.,

$$L_{o(min)} = f_2(N_{c1}, N_{c2}). \tag{5.17}$$

Expressions (5.16) and (5.17) are the actual inverse functions of (5.14) and (5.15), respectively, and can be readily resolved by numerical methods.

Based on (5.16), Fig. 5.9 illustrates how the upper limit of L_o relaxes with the increase of N_{c1} and number of interleaved phases N_p. It is important to further notice that $L_{o(max)}$ is virtually independent on F_s as long as Δi_{Lo} is much lower than $i_{o(max)}$, as it can be readily deduced from (5.8) and (5.12).

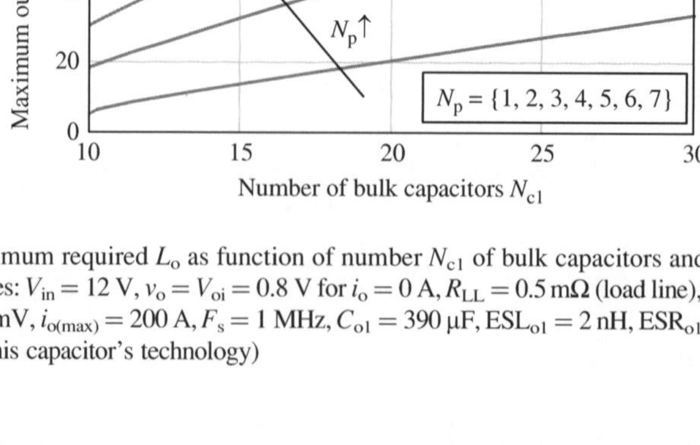

Fig. 5.9 Maximum required L_o as function of number N_{c1} of bulk capacitors and for various N_p. Parameter values: $V_{in} = 12$ V, $v_o = V_{oi} = 0.8$ V for $i_o = 0$ A, $R_{LL} = 0.5$ mΩ (load line), $R_{pcb1} = 0.3$ mΩ, $\Delta v_{ot(max)} = 40$ mV, $i_{o(max)} = 200$ A, $F_s = 1$ MHz, $C_{o1} = 390$ μF, $ESL_{o1} = 2$ nH, $ESR_{o1} = 4$ mΩ (see [7] for details on this capacitor's technology)

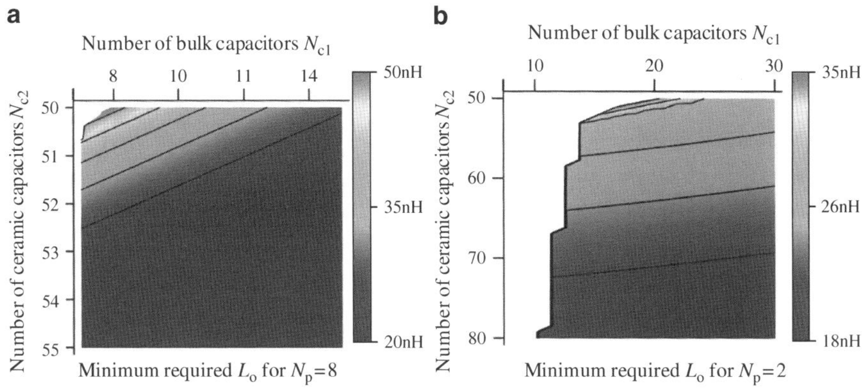

Fig. 5.10 Minimum required L_o as function of N_{c1} and N_{c2} for: (**a**) four phases VR, (**b**) two phases VR. Parameter values: $V_{in} = 12$ V, $v_o = V_{oi} = 0.8$ V at $i_o = 0$ A, $R_{LL} = 0.5$ mΩ (load line), $R_{pcb1} = 0.3$ mΩ, $L_{pcb1} = 20$ pH, $\Delta v_{ot(max)} = 40$ mV, $\Delta v_{o1(max)} = 5$ mV, SR = −2.5 A/ns, $i_{o(max)} = 200$ A, $F_s = 1$ MHz, $C_{o1} = 390$ μF, $ESL_{o1} = 2$ nH, $ESR_{o1} = 8$ mΩ (max.), $C_{o2} = 4.7$ μF, $ESL_{o2} = 0.6$ nH, $ESR_{o2} = 15$ mΩ. Note that the *white areas* are regions of operation not compliant with the given specifications

The lower boundary of L_o is shown in the example of Fig. 5.10 as a function of N_{c1} and N_{c2}. The two cases compared highlight the benefits of phase interleaving to reduce the output filter size. For the eight phase VR case-example (Fig. 5.10a), the output filter can be implemented with 8 × 390 μF SMD electrolytic capacitors, 50 × 4.7 μF ceramic SMD capacitors, and 8 × 48 nH output coils.

On the contrary, a higher number of bulk capacitors is demanded in the two phase design, where one possible solution involves 15 × 390 μF SMD electrolytic capacitors, 55 × 4.7 μF ceramic SMD capacitors, and 2 × 29 nH output coils.

Among the possible values of L_o (i.e., $L_{o(min)} \leq L_o \leq L_{o(max)}$), the design target from a volume size viewpoint may generally be $L_o = L_{o(min)}$. Note, however, that enabling the use of extremely low L_o (just a few tenths of nanohenries) may strongly depend on power losses due to either high-ripple current or high-switching frequency, or both. Thus, an evaluation of the converter efficiency needs to be carried out once the filter elements have been selected. This procedure, which will be addressed in Sect. 5.4, enables to systematically determine the optimum number of phases, the switching frequency, and, ultimately, the filter components.

The following guidelines offer a selection of the output filter elements provided that a number of circuit parameters are adopted by initial estimation. Note that worst-case scenarios given by component tolerances, lifetime degradation, control delays, and safety margins are not rigorously taken into account. Therefore, the resulting values may be correspondingly corrected so as to compensate for all potential variances. The ultimate filter size may thus end up being larger than the initially estimated one by the following procedure depending on the conservativeness of the design.

1. Selection of a capacitor technology offering the highest capacitance per unit volume with the lowest possible ESR (e.g., bulk electrolytic capacitors from [7]). The capacitance value per part is typically 100 μF or higher in typical state-of-the-art VR applications.
2. Selection of a ceramic capacitor type offering the lowest possible ESL and footprint. The capacitance value per part is typically 100 nF or higher in state-of-the-art VR applications.
3. Extraction of the corresponding worst-case parameter values of the selected capacitor types considering tolerances, temperature, and lifetime degradation.
4. Use of (5.14) and (5.15) to find the number of bulk and ceramic capacitors. To achieve this, not only the count but also the volume size and footprint of the selected components may be taken into account as main selection criteria.

In step 4, the number of phases, switching frequency, and output filter inductance must be provided for the calculations. The proposed loss model from Chap. 4 can be used to estimate the values of these parameters, which strongly depend on the minimum targeted converter efficiency and power density operation, as it shall be shown in Sect. 5.4.

5.2 Input Filter

The specifications of the low-pass input filter design according to VRD guidelines (see Chap. 1) are generally less stringent than the tight regulations demanded for the output filter. It is established however that the SR of the input current must currently be limited to a maximum of 0.5 A/μs. Without the input filter, the approximately rectangular current pulses drawn from each of the phases could potentially interfere to nearby devices and cause system malfunctions.

The most common input filter circuit comprises a decoupling input capacitance (resulting from a number of high-frequency parallel capacitors) and a power choke as illustrated in Fig. 1.18 of Chap. 1. The contribution to the total size and efficiency of the VR may depend to a great extend on the selection and location of the input filter elements.

First, the input capacitors must feature low ESL and ESR and physically be in close proximity to the half-bridge of each phase to minimize the associated half-bridge loop inductance. These requirements combined with the low volume size are of utmost importance, even to the extent that the total capacitance may be compromised, for instance, when selecting the capacitor package and size that offer the lowest ESR and ESL.

The implications of a too low input capacitance on the operation and performance of the VR may become relevant as the ripple amplitude increases significantly, as it shall be addressed later.

Once the input capacitors have been selected, a power choke is designed to comply with the current SR requirements. In order to calculate the value of the inductance, a similar model to that of the output filter is used to estimate the input

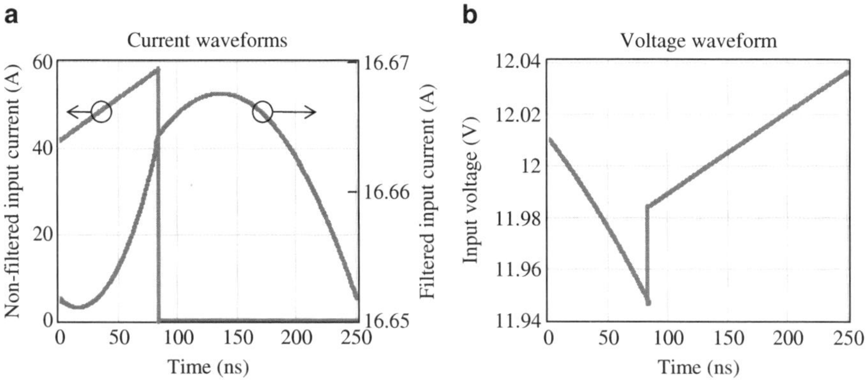

Fig. 5.11 Simulated input filter response to a pulsed current signal from a four phase VR. Parameter values: 200 A load current, $V_{in} = 12$ V, $V_o = 1$ V, 16×3.3 μF input capacitance with 10 mΩ ESR (ESL omitted), 130 nH input inductor, and 4×40 nH output inductors

waveforms. Figure 5.11 illustrates simulated waveforms resulting from Fourier series calculations. The capacitors are arranged such that the half-bridge of each phase is in close proximity to a group of four parallel capacitors. The high-switching frequency and interleaving operation allows minimizing the size of the capacitance while keeping a negligible input voltage ripple. Figure 5.11a shows the input current waveforms. The filtered input current has a maximum SR not greater than the specified 0.5 A/μs with the use of just a 130 nH inductor. The low ESR of such inductor produces 140 mW. It is important to design the input filter so that the resulting inductor features small size and low DC ESR. This is usually readily achieved at high-switching frequency and with a sufficiently large input capacitor bank.

The relation between the required inductor, input capacitance, and switching frequency is shown in Fig. 5.12. For each curve, a capacitance exists above which any inductance reduction becomes marginal as a consequence of a negligible voltage ripple. The increase of the switching frequency helps to reduce the input filter size mainly only at relatively low capacitance, where the input voltage ripple is large. In the given example, the value of the inductance is strongly compromised for a total input capacitance below 1 μF, particularly at low frequencies. This goes in detriment of the size and losses in the inductor and as such may need to be avoided.

Nonetheless, a moderate ripple of several volts may be beneficial when falling edge losses caused by hard-switching are the main loss contributions. In such case, lowering the input voltage at the time the CtrlFET turns off can mitigate both hard-switching and half-bridge charging related loss. While beneficial at the FE, the input voltage ringing is detrimental for the LE transition as the input voltage increases. Consequently, in cases where LE losses are high, the increase of the input ripple voltage may not lead to an overall efficiency gain and thus, higher input capacitance may be most adequate.

The amplitude of the input voltage ripple is shown in the example of Fig. 5.13 as function of the capacitance and for various F_s. An input oscillation of a few volts

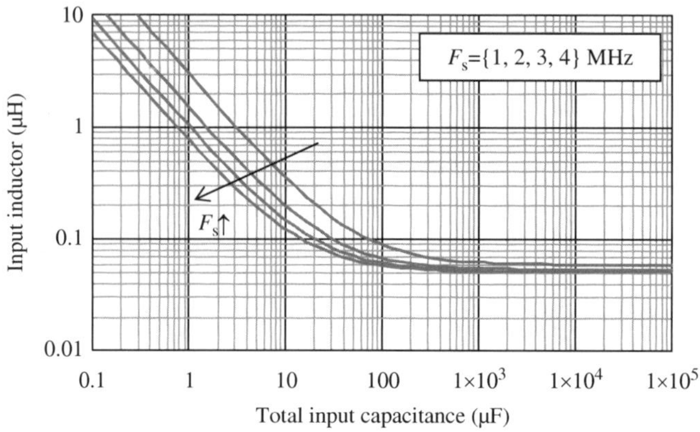

Fig. 5.12 Relation of input inductor and equivalent input capacitance to meet a maximum slew rate of 0.5 A/μs at various switching frequencies. Parameter values: see example of Fig. 5.11

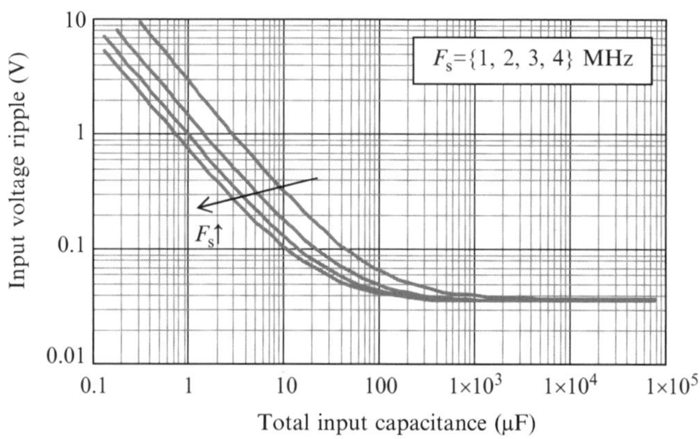

Fig. 5.13 Voltage ripple at the half-bridge input as a function of the input capacitance and for various switching frequencies. Parameter values: see example of Fig. 5.11

amplitude is possible with a total input capacitance below 1 μF. This compromises strongly the size and losses of the input inductor as seen before, which may motivate the use of a cascade arrangement of L–C filter elements, as shown in Fig. 5.14. The proposed filter structure solves the problem by setting the resonance frequency of the $L_{i1} - C_{i1}$ lower than that for $L_{i2} - C_{i2}$. Thus, $L_{i2} - C_{i1}$ aims at limiting the current SR, while $L_{i2} - C_{i2}$ allows for the desired ripple voltage.

5.3 Power MOSFETs and Gate Drivers

This section presents new power MOSFET FoM based on the loss model of Chap. 4. Unlike the standard FoM, the proposed figures include relevant parameter dependencies relative to the converter circuit, such as the parasitic inductances and

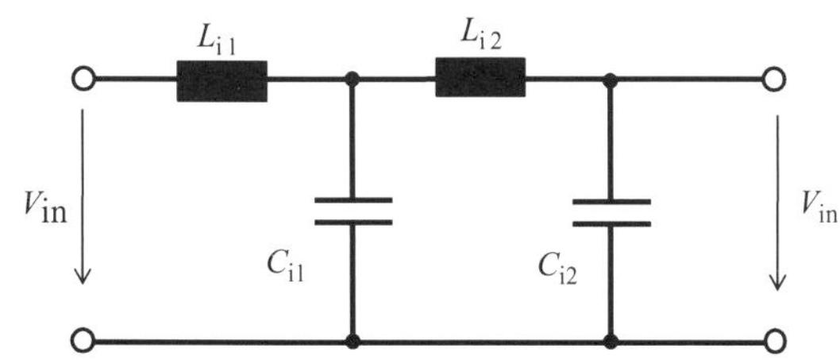

Fig. 5.14 Proposed input filter structure to allow for significant input voltage ringing

the switching frequency. Thus, these FoM allows the assessment of MOSFET technologies for a particular application under specific operating conditions.

To this end, some of the loss equations from model level 2 are slightly modified for expressing loss quantities relative to the MOSFETs' chip areas. This is convenient as the devices optimization heavily involves the minimization of the amount of semiconductor material required for the power conversion. Accordingly, equations representing switching losses associated to the MOSFETs capacitances are expressed as,

$$w_{\text{cHBC(c)}} = A_{\text{die(c)}} w'_{\text{cHBC(c)}}, \tag{5.18}$$

$$w_{\text{cHBC(s)}} = A_{\text{die(s)}} w'_{\text{cHBC(s)}}, \tag{5.19}$$

$$w_{\text{DRV(s)}} = A_{\text{die(s)}} w'_{\text{DRV(s)}}, \tag{5.20}$$

$$w_{\text{DRV(c)}} = A_{\text{die(c)}} w'_{\text{DRV(c)}}. \tag{5.21}$$

Note that $A_{\text{die(c)}}$ and $A_{\text{die(s)}}$ are the effective die areas of the CtrlFET and SyncFET, respectively. With regard to CtrlFET hard-switching, the LE transient is independent on the die area as long as (4.26) applies. It implies an increase of the CtrlFET gate current in proportion to the die area.

On the other hand, the FE hard-switching transient loss (4.32) is assumed to be approximately proportional to the die area as,

$$w_{\text{ioHS}_{\text{FE}}} \approx A_{\text{die(c)}} w'_{\text{ioHS}_{\text{FE}}}. \tag{5.22}$$

The above expression is imprecise because the area increase enlarges $Q_{\text{oss(c)}}$, thereby reducing the channel current, and with it, hard-switching losses. Nonetheless, the impact of $Q_{\text{GD(c)}}$ appears to be more dominant, hence making (5.22) a valid approximation in most operating conditions of interest.

With regard to ON conduction loss, the contributions of the semiconductor and package resistance may be expressed as (see Sect. 4.1.4),

$$w_{\text{ioONs(c)}} = w_{\text{ioONs(c)sem}} + w_{\text{ioONs(c)pak}}, \tag{5.23}$$

where

$$w_{\text{ioONs(c)sem}} = R_{\text{DSon(c)}} \cdot I^2_{\text{RMS(c)s}} \cdot T_{\text{s}}, \tag{5.24}$$

$$w_{\text{ioONs(c)pak}} = R_{\text{DSpak}}\left(f_{\text{ON(c)}}\right) \cdot I^2_{\text{RMS(c)s}} \cdot T_{\text{s}}. \tag{5.25}$$

Resistance $R_{\text{DSon(c)}}$ and $R_{\text{DSpak}}(f_{\text{ON(c)}})$ correspond to the semiconductor and package contributions, respectively. The contribution of semiconductor resistance $R_{\text{DSon(c)}}$ is given by,

$$w_{\text{ioONs(c)sem}} = \frac{w'_{\text{ioONs(c)sem}}}{A_{\text{die(c)}}}. \tag{5.26}$$

And similarly for the SyncFET,

$$w_{\text{ioONs(s)sem}} = \frac{w'_{\text{ioONs(s)sem}}}{A_{\text{die(s)}}}. \tag{5.27}$$

Conversely, the ON conduction loss during the switching times is assumed to be independent on the chip area provided that the semiconductor resistance is predominant (see (4.42) and (4.47)).

Computing the first derivative of (4.51) with respect to $A_{\text{die(s)}}$ and $A_{\text{die(c)}}$ when considering the above dependencies yields the following expressions for the optimum chip areas,

$$A_{\text{die(s)opt}} = \sqrt{\frac{w'_{\text{ioONs(s)sem}}}{w'_{\text{cHBC(s)}} + w'_{\text{DRV(s)}}}}, \tag{5.28}$$

$$A_{\text{die(c)opt}} = \sqrt{\frac{w'_{\text{ioONs(c)sem}}}{w'_{\text{cHBC(c)}} + w'_{\text{ioHS}_{\text{FE}}} + w'_{\text{DRV(c)}}}}. \tag{5.29}$$

Finally, substituting (5.28) and (5.29) into (4.51) with the use of (5.18)–(5.27) yields the following proposed FoM for the power MOSFETs,

$$\text{FoM}_{(\text{s})} = 2\sqrt{w'_{\text{ioONs(s)sem}}\left(w'_{\text{cHBC(s)}} + w'_{\text{DRV(s)}}\right)} + w_{\text{ioONs(s)pak}} + w_{\text{ioONd(s)}} + w_{\text{DRVi(s)}}, \tag{5.30}$$

$$\begin{aligned} \text{FoM}_{(\text{c})} = {} & 2\sqrt{w'_{\text{ioONs(c)sem}}\left(w'_{\text{cHBC(c)}} + w'_{\text{DRV(c)}} + w'_{\text{ioHS}_{\text{FE}}}\right)} + \cdots \\ & + w_{\text{ioONs(c)pak}} + w_{\text{ioONd(c)}} + w_{\text{DRVi(c)}} + w_{\text{ioHS}_{\text{LE}}} + w_{\text{iHBC}}. \end{aligned} \tag{5.31}$$

Table 5.1 FoM comparison of various commercial devices

Reference device	SyncFET		CtrlFET	
	$FoM_{(s)}$ (μJ)	FoM $Q_{iss}{\cdot}R_{DSon}$ (nC mΩ)	$FoM_{(c)}$ (μJ)	FoM $Q_{GD}{\cdot}R_{DSon}$ (nC mΩ)
PSMN1R7-30YL (NXP)	2.46	85	1.75	14.8
BSB012N03MX3 (Infineon)	2.45	156	1.88	21.6
FDMS8660AS (Fairchild)	2.7	92.4	1.8	10.9
IRF6727 (IR)	2.73	105.4	2.1	27.2

The above figures have units of Joules and express the total MOSFET and driver losses per switching cycle as a function of all identified loss mechanisms except the neglected reverse recovery, body diode conduction, and avalanche breakdown.

Table 5.1 compares various state-of–the-art MOSFETs in terms of the standard (see [136, 137], Chap. 1) and proposed FoM. Each referenced device part is simulated both as SyncFET and CtrlFET in an IPM application. Since only the semiconductor switches are subject to comparison, the IPM package, PCB layout, and driver characteristics (except R_G) are common among all analyzed cases. The chip areas are individually optimized for the particular operating conditions of $F_S = 1$ MHz, $V_i = 12$ V, $V_o = 1.25$ V and $I_o = 30$ A (one phase).

The resulting data illustrate how the standard FoM may be misleading. Namely, for case of the SyncFET, the worst performing device in terms of $Q_{iss}{\cdot}R_{DSon}$ (the one with the lowest R_{DSon} and highest Q_{iss}) exhibits the lowest $FoM_{(s)}$. Because of the very low R_{DSon}, this device is expected to produce the least loss at heavy load since the ON conduction loss prevails over gate charge loss under optimum $V_{DRV(s)}$. As to the CtrlFET, the presumably least lossy device part (i.e., the lowest in $FoM_{(c)}$) is 4 nC mΩ higher than the best part in terms of $Q_{GD}R_{DSon}$. Such discrepancy lays on the fact that the best $FoM_{(c)}$ can switch faster for its $R_{G(c)}$ is 0.5 Ω lower than that of the best $Q_{GD}R_{DSon}$. The results of Table 5.1 thus exemplify the need to consider all crucial key parameters of the converter besides those exclusively related to the semiconductor material, such as the total gate driver resistance and the loop inductance.

5.3.1 *Design Guidelines*

The following design procedure encompasses the optimization of all relevant design parameters of the power MOSFETs and drivers, including voltage ranges, chip areas, and gate drivers. The method employs the proposed FoM from the previous section as the basis for the optimization. Throughout the process steps, it is assumed that the controller manages to effectively optimize the dead times so as to avoid body diode conduction, reverse recovery, and shoot-through. Furthermore, avalanche breakdown and gate bounce shoot-through are neglected (i.e., $v_{GS(s)} < V_{TH(s)}$ during CtrlFET conduction and dead times).

1. Derivation of the output filter parameters according to Sect. 5.1. This includes setting up guess values for the output filter inductor, the switching frequency, and the number of phases.
2. Selection of the nominal load. The design is based upon a single current operation from the entire load range, which may typically correspond to a mid range value so as to save silicon die area and improve light load condition [8]. When optimizing for minimum temperature operation however, the design may be considered for the maximum load current. In any case, reliable operation must be guaranteed under the worst-case scenario of maximum load and board temperature.
3. Selection of the SyncFET voltage rating such that avalanche breakdown is avoided. This may typically imply that $V_{AB(s)} \geq 2.5V_{in}$. This rule of thumb may also be applied to the CtrlFET as initial value. Voltage $V_{AB(c)}$ may later be refined with (4.36).
4. Optimization of the SyncFET die area and gate voltage according to (5.28) and (5.30).
5. Optimization of the CtrlFET die area and gate voltage according to (5.29) and (5.31).
6. Validation of condition (4.36) and determination of minimum L_{HB} needed to avoid avalanche breakdown in both switches. In case L_{HB} cannot be further reduced, then either the increase of $V_{AB(c)}$ or the reduction of the switching speed may be enforced.
7. Checking gate bounce susceptibility with PLA model (Sect. 3.3.5) and improve gate impedance correspondingly whenever necessary.

The procedure may be repeated from point 2 on with refined values of the gate drive parameters and avalanche breakdown levels. Results may be later validated or further readjusted with predictions from model levels 0 and 1.

5.3.2 *Optimization Case Example*

This section illustrates the proposed design procedure to optimize the performance of the IPM PIP212-12M from NXP Semiconductors [9]. The estimated maximum efficiency derived from the proposed method is compared to typical performance data in order to identify improvement options.

The IPM is studied under the following conditions: $V_{in} = 12$ V, $V_o = 1.25$ V, $F_S = 1$ MHz, $I_o = 30$ A, $T_j = 140°$C, and $L_o = 320$ nH. Following the guidelines of the previous section, resulting loss calculations suggest modifications of actual PIP212-12M parameter values to significantly boost the converter efficiency. The proposed new design involves the increase of the MOSFET die areas and the reduction of half-bridge loop inductance L_{HB}, MOSFET drivers' voltage V_{DRV}, and gate resistance R_G.

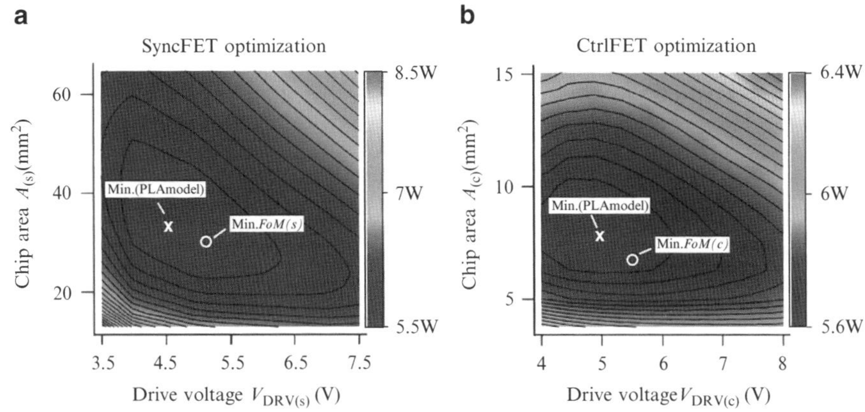

Fig. 5.15 Contour plots of simulated losses of an IPM as function of the drive voltage and chip area of the power MOSFETs at 30 A load current (a) SyncFET optimization, (b) CtrlFET optimization

Table 5.2 Parameter values of the compared design examples

		Normalized chip area	V_{DRV} (V)	R_G (Ω)	L_{HB} (nH)
PIP212-12M (existing design)	SyncFET	4.3	5	1.3	1.6
	CtrlFET	1	12	2.3	
Proposed IPM design	SyncFET	7	4.5	0.5	1
	CtrlFET	1.7	5	0.5	

The contour plots of Fig. 5.15 indicate optimum IPM operation at reduced gate drive voltages and increased chip areas. Regarding the SyncFET, the optimum chip size at $I_o = 30$ A ranges over 30 mm^2, which represents more than twice the area utilized in the actual PIP212-12M. As indicated earlier, for the sake of cost effectiveness and improved light load conditions, it is common practice to alternatively optimize at currents lower than the maximum specified. Nonetheless, at $I_o =$ 20 A, the optimum SyncFET die area is still larger than in the existing design, as shown in Table 5.2. This suggests that conduction loss may be the dominating factor even at moderate load currents.

Table 5.2 further compares parameter values of the present design to those resulting from the optimization procedure. Among the listed parameters, the most critical ones involving switching losses are the gate resistances, for they provide the strongest lever in reducing the switching times (both turn-on and turn-off) while compromising neither ON conduction nor gate drive losses. It is clear from (4.36) that the reduction of $R_{G(c)}$ is only appropriated provided that L_{HB} is sufficiently low such that CtrlFET avalanche is prevented. In this particular case, the gate resistance reduction becomes a potential benefit for $L_{HB} < 1$ nH and $I_o = 30$ A.

According to the simulations of Fig. 5.16, the proposed IPM design outperforms the existing PIP212-12M design in the entire operating regime except at light load, where the efficiency differences are marginal. At full load, the energy savings reach

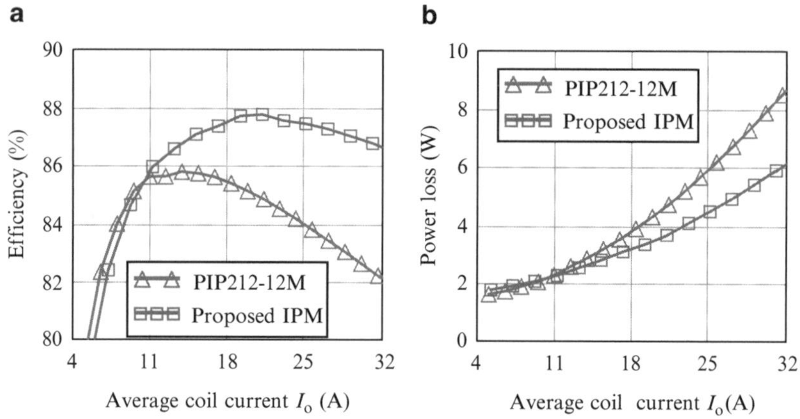

Fig. 5.16 Efficiency (**a**) and total converter power loss (**b**) comparisons between the PIP212-12M from NXP Semiconductors and the proposed IPM design

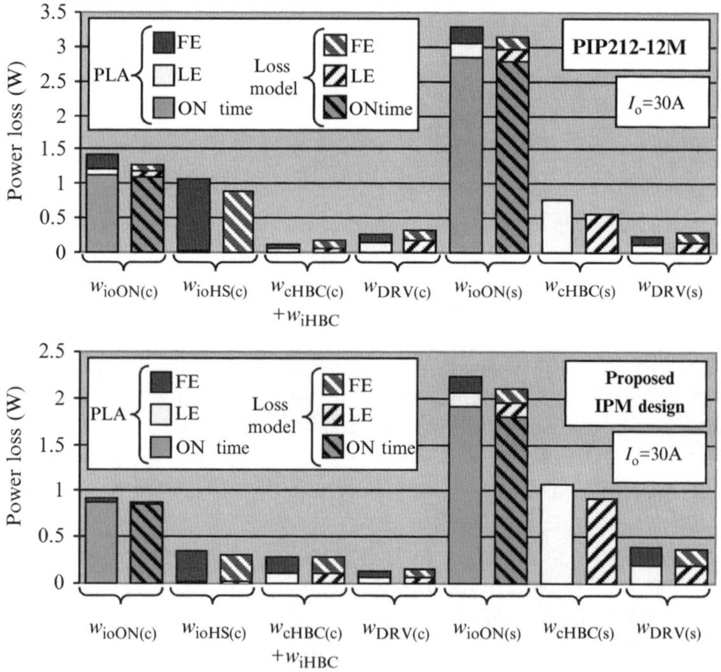

Fig. 5.17 Loss breakdown derived from both PLA and loss models of the PIP212-12M (7.75 W loss) and proposed IPM design (5.9 W loss)

25%. The loss breakdown from Fig. 5.17 provides the grounds for such a gain (~0.6 W from the filter and PCB loss contributions are not included). The double predictions shown derive from level 1 and level 2 models, which consistently compare each individual loss quantity. As expected for the existing design, ON conduction ($w_{ioON(s)}$) is the main contribution to SyncFET loss with 65% rate, followed by $w_{cHBC(s)}$ with 19%. The portion of $W_{DRV(s)}$ loss is less relevant than $w_{cHBC(s)}$ because, although $C_{iss(s)} > C_{oss(s)}$, it prevails that $V_{in} > V_{DRV(s)}$. The CtrlFET loss is dominated by ON conduction ($w_{ioON(c)}$) and hard-switching ($w_{ioHC(c)}$) with 39 and 37%, respectively.

As a consequence of the area increase in the proposed IPM design, the SyncFET ON conduction loss reduces by 1 W at full load, which in turn leaves an equal share of conduction and switching losses. In this optimum condition, $W_{DRV(s)}$ loss is 0.4 W, whereas $w_{cHBC(s)}$ becomes the main switching loss contribution with more than 1 W. The increase of $Q_{iss(s)}$ is compensated with the reduction of $R_{G(s)}$, which avoids the increase of the switching times and with that, the ON conduction loss in these intervals. On the other hand, improvements in L_{HB} and $R_{G(c)}$ translate into a significant reduction of the hard-switching loss to as much as 0.7 W.

5.4 Selection of F_s, L_o, and N_p

The design guidelines presented in the previous sections set on the basis of predefined values for the switching frequency, output inductance, and number of phases. To complete the VR design, a systematic procedure for the selection of these three key parameters of the power converter is proposed according to either of the following two design criteria:

- Criteria A: Minimization of power losses for a specified converter footprint.
- Criteria B: Minimization of the converter footprint for a specified converter efficiency.

Depending on the system demands, one may decide to base criteria A and B on the power density (or volume size) of the converter as opposed to its area.

Implicitly in the above statements are the thermal requirements, particularly the maximum temperature operation that, under worst-case conditions, any of the VR components and PCB must not exceed.

The selection procedure is illustratively described with the representative case-examples of the next subsections.

5.4.1 Case-Example 1: State-of-the-Art Desktop Application

VR specifications for state-of-the-art desktop computer microprocessors were introduced in Chap. 1 (see Table 1.1). To meet the requirements of this application, the DrMOS compatible IPM PIP212-12M from NXP Semiconductors is adopted [9] as main solution approach.

Regarding the passive filter elements, and according to the guidelines of Sects. 5.1 and 5.2, the following two commercial capacitor technologies are selected and modeled,

- 390 μF special polymer electrolytic capacitors from Panasonic [11], used in the output filter mainly as a means to filter out load transient effects.
- 4.7 μF X7R ceramic capacitors from Murata [12], used in the input and output filters for high-frequency signal decoupling.

Having selected the filter capacitors and semiconductor technologies, the next is to determine the values of the switching frequency, output inductance, and number of phases. This is carried out through the use of the models from the previous subsections, which reflect the close interrelation between these three parameters and the power losses. The use of contour plots aids the graphical representation of such dependencies and the final selection of the associated values. Regarding the proposed example, Fig. 5.18 shows the contour plots of the converter efficiency and power losses for the solution case of five phases. For every point in the explored design space, the guidelines of Sects. 5.1–5.3 are followed to estimate the optimum number of filter capacitors and adjust the parameter settings of the IPM.

The wide inductance range shown in the graphs is covered through the use of different inductor types that vary both in loss profile and volume size. Accordingly, various families of commercially available power coils are selected and characterized to estimate their performance in the operating conditions of interest. Since one main purpose of reducing the inductance is the size shrinkage, the selection of the coil types is such that the volume increases in proportion with the inductance while the total losses are kept constant. Of relevant importance is the thermal resistance, which is assumed to worsen as the inductor size shrinks, as it is usually the case in most commercial products. The later will have a significant impact on the permissible margins of operation within the design space.

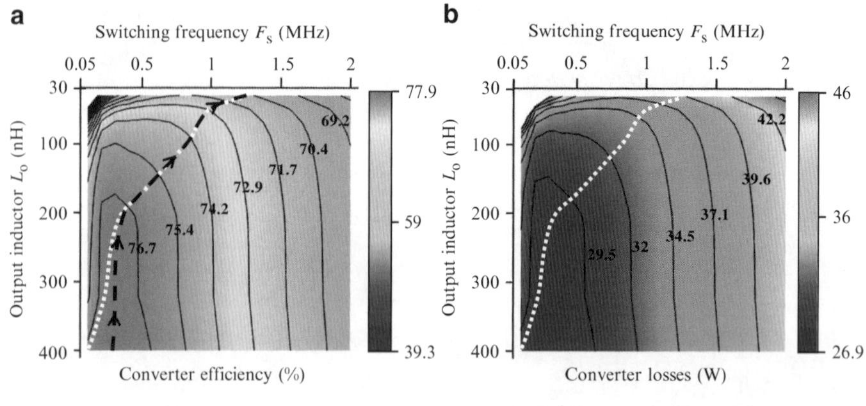

Fig. 5.18 VR efficiency at heavy load (a), and total converter losses (b) as function of the switching frequency and output inductor. Case-example of five phases equipped with commercial IPM PIP212-12M from NXP Semiconductors. Simulated conditions: see 2008 specifications from Table 1.1 of Chap. 1. Additional parameter values of IPM: $T_j = 140°C$, $V_{DRV(c)} = 12$ V

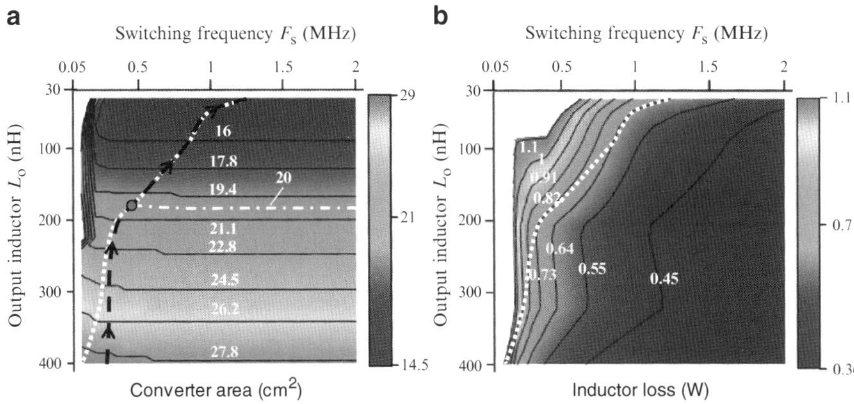

Fig. 5.19 VR footprint (a) and inductor losses (per individual coil) (b) as function of the switching frequency and output coil inductance. See Fig. 5.18 for further details on simulated conditions

In the range of 100–470 nH, various families of power cube and power bead coils from Pulse [13] are chosen as referent parts for the design. Below 100 nH, small form factor, high-frequency inductor types are selected from Coilcraft [14]. The characterization of all these inductors results in a database of interpolated data containing information of the volume, footprint, maximum power dissipation, and total estimated losses as function of the ripple current amplitude, duty cycle, switching frequency, and DC current (see Appendix E for details on loss estimations).

With this information the plot of Fig. 5.19a is generated. The graph takes into account the footprint of all filter elements and IPMs of the VR. Note that the heatsinks area (in case of being used) and PCB tracks are not included.

The dotted line of Figs. 5.18 and 5.19 represents the power dissipation boundary of the selected power coils. The line is originated in Fig. 5.19b and projected on the other plots for further analysis. According to these results, the inductors below 100 nH can barely dissipate more that 0.5 W due to severe limitations for removing the heat of a small volume of magnetic material. As such, the minimum switching-frequency operation for the inductors below 50 nH is in the MHz range since the large ripple currents otherwise produces excessive AC losses. For inductors in the range of 500 nH, the switching frequency can go below 100 kHz while a dissipation of 1 W and above can be handled by the larger volumes of these coils.

The dark dashed line in Fig. 5.18a shows the optimum trajectory that reduces footprint with a minimum possible efficiency penalty. For inductances above 200 nH, the lowest decay gradient in efficiency as the area shrinks occurs in the region around 300 kHz, where the converter efficiency is maximized. As the inductance drops below 200 nH, the trajectory gradually deviates from the minimum gradient and proceeds along the boundary limits of operation set by the inductor (dotted line). This suggests that the converter performance can be significantly bounded by the limited heat dissipation capabilities of the coils.

According to criteria A, the output inductor and switching frequency can be readily selected by projecting the trajectory of Fig. 5.18a into the map of Fig. 5.19a to find the intersection with the aimed isoline. In this case-example, the optimum trajectory meets the 20 cm^2 isoline with $L_o = 180$ nH and $F_s = 450$ kHz.

A similar procedure can be followed regarding design criteria B, except that the chosen trajectory that maximizes the gradient decay of the footprint must intersect with a specified isoline of Fig. 5.18a. In the given example, the minimum achievable footprint for an efficiency of 74% is 16.5 cm^2, which is achieved with $L_o = 100$ nH and $F_s = 850$ kHz at the limit boundary of the inductor.

Repeating this same process for various N_p yields the bar plots of Fig. 5.20, where, according to design criteria A, the optimum number of phases results in 7. The efficiency however is only slightly compromised when reducing the number of

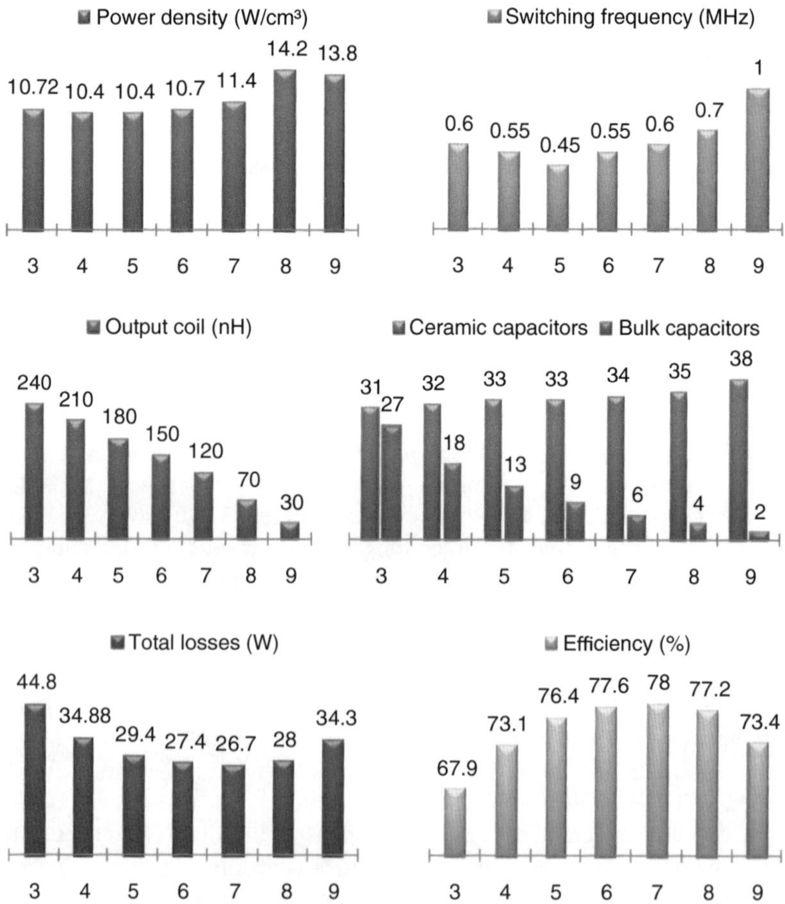

Fig. 5.20 VR optimization results as function of the number of phases. Design criteria A: Maximum efficiency out of 20 cm^2 footprint. See Fig. 5.18 for further details on simulated conditions

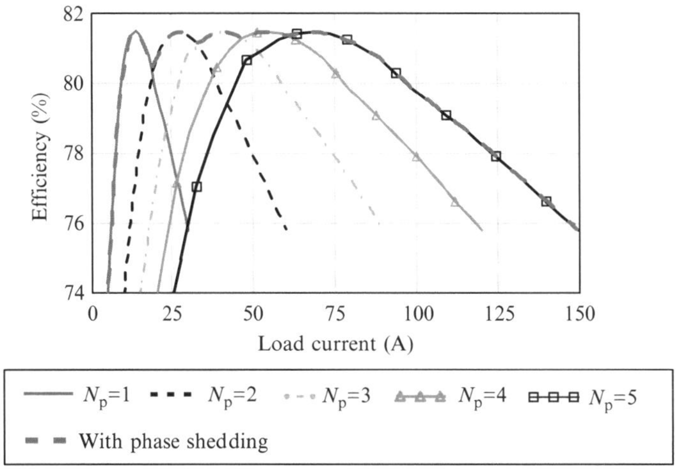

Fig. 5.21 Efficiency curves of the five phases VR design from Fig. 5.20 for various configurations of active phases

phases down to $N_p = 5$. Below this number, the heat dissipation increases above 9 W per phase, which may impose the use of bulky heatsinks and fans.

The increase of the number of phases leads to the reduction of bulk capacitors and the increase of the switching frequency. The later is required to enable the use of smaller footprint inductors that compensate for the area increase of the new phases. For $N_p > 7$ phases, the number of bulk capacitors reduces below 5 and the inductance drops below 100 nH. All this translates into a significant boost of the power density conversion. Further increasing the number of phases ends up with excessive switching losses that inevitably cause an efficiency drop.

Simulated efficiency curves of the five phase solution from Fig. 5.20 is shown in Fig. 5.21 for various active phase configurations. As it can be seen, the use of phase shedding techniques provides potential energy savings at moderate and light load conditions. In the later case, i.e., when only one phase is active, other techniques such as PFM may further boost the efficiency. In doing so, the converter design can be exclusively tailored for heavy load operation without exceedingly compromising light load efficiency, thereby minimizing power losses (and hence T_j) under worst-case conditions. Further analysis on the advantages of some of these operation modes are provided in Chap. 6.

5.4.2 Case Example 2: State-of-the-Art Laptop Application

As a second example, the 2008 specifications of Table 1.2 are taken as reference for the VR design of state-of-the-art laptop applications. A discrete solution must be adopted here as the specified input voltage is greater than the maximum rating level of typical IPM commercial products.

The discrete MOSFET and gate driver technology equal to those employed in the IPM module analyzed in the previous case-example. The difference between the two solutions therefore lies on the parasitic inductances (compare values of Table 2.4 with those of Table 2.8). MOSFET paralleling is allowed to effectively improve ON conduction when required. This is applied to the SyncFET, which is usually composed of two parallel devices. The following analysis employs a half-bridge configuration consisting of one discrete device for the CtrlFET function and two parallel ones for the SyncFET unless otherwise specified.

Furthermore, the PCB layout arrangement and packaging technology (i.e., LFPak unless otherwise specified) are assumed to provide low parasitic inductances in the half-bridge and gate driver paths. With a proper design, the resulting parasitic inductances may approach the values of the IPM solution from the previous example. The gate drivers are likewise assumed to feature automatic dead time reduction so that body diode conduction and reverse recovery can be neglected.

The design procedure described in the previous example is again applied here and generalized as follows,

1. Generation of contour plots showing the dependence of the VR efficiency and footprint (or alternatively power density) as a function of the switching frequency and the output inductance for a predefined number of phases. The generated loss data are based on the model of Chap. 4. Every point in the graphs derives from the guidelines of Sects. 5.1–5.3.
2. Estimation of the limit boundaries of the converter according to the maximum temperature operation of any element of the VR. Particularly critical are the output inductor and MOSFET temperatures.
3. According to design criteria A: Projection of the aimed isoline of the footprint (or power density) plot into the efficiency plot to find its intersection with a maximum within the permissible limits of operation.
4. According to design criteria B: Projection of the aimed isoline of the efficiency plot into the footprint (or power density) plot to find its intersection with a minimum (or a maximum in case of power density) within the permissible limits of operation.
5. Extraction of the output inductance and switching-frequency values from the intersected operating point.
6. Steps 1–5 may be repeated for various number of phases.
7. Selection of the number of phases that offer the highest efficiency (criteria A), the highest power density (criteria B), or the smallest footprint (criteria B).

Figure 5.22 summarizes the simulation results of the proposed VR optimization procedure according to design criteria A, where the targeted footprint is set to 10 cm^2. Upon such premises, the number of phases that maximizes the converter efficiency is $N_p = 3$. In this solution, the number of bulk capacitors is reduced to just 2, and the maximum power dissipation per phase is ~4 W. The switching frequency increases along with the number of phases, reaching 700 kHz at $N_p = 5$, which results in an optimum output inductance of just 80 nH.

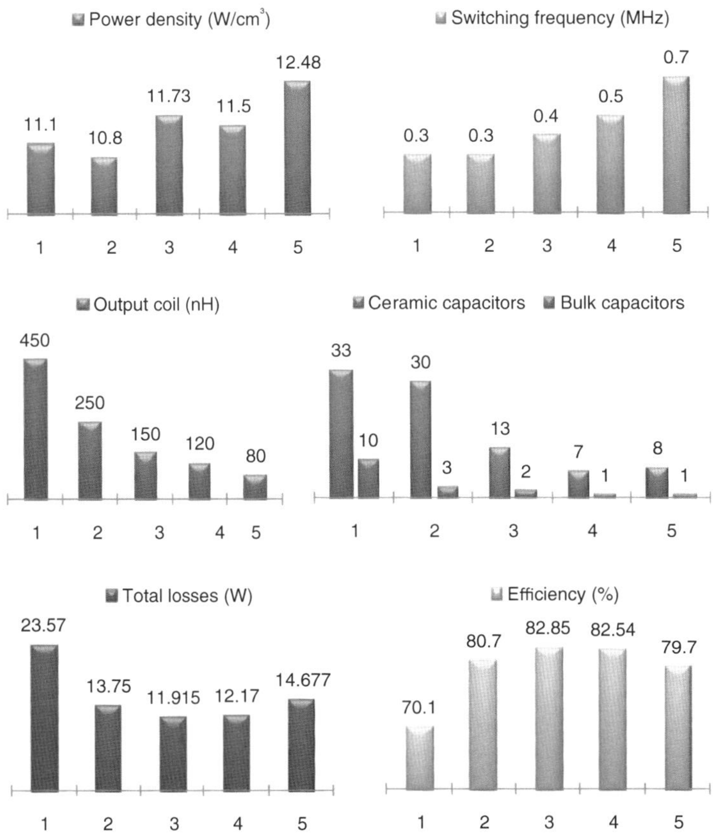

Fig. 5.22 VR optimization results as function of the number of phases. Design criteria A: Maximum efficiency out of 10 cm^2 footprint. See Fig. 5.18 for further details on simulated conditions

Under these operating conditions the current ripple per phase reaches over 15 A peak-to-peak, which is higher than the maximum average inductance current. This suggests that high-current ripple operation in combination with phase interleaving might potentially be an effective approach to boost power density conversion while relaxing the demands of high-switching frequency operation.

Figure 5.23 further shows efficiency curves for the design case of $N_p = 2$ from Fig. 5.22. Similarly as illustrated previously for the desktop application example, the performance of phase shedding is positively evaluated based on the efficiency gains at light and moderate load. It is worth noticing that at light load ZVS operation in both SyncFET and CtrlFET may be encountered as a consequence of a negative current flow through the output coil during the LE transition. The loss savings of this switching mode operation can be identified in the efficiency curve of $N_p = 2$ with the sudden step around 12 A. Such benefit already at that current level can also be attributed to the relatively high current ripple.

Fig. 5.23 Efficiency curves of the two phases VR design from Fig. 5.20 for various configurations of active phases

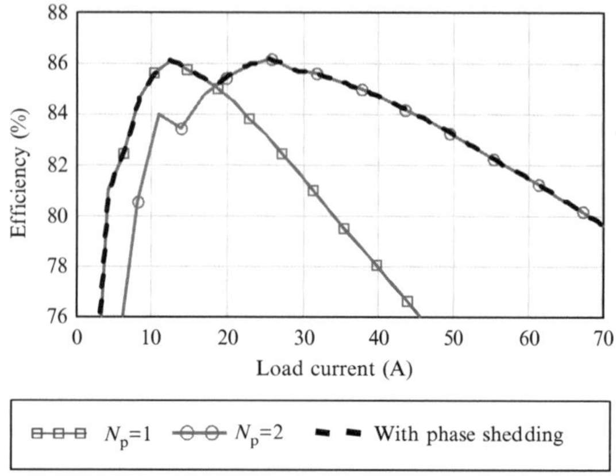

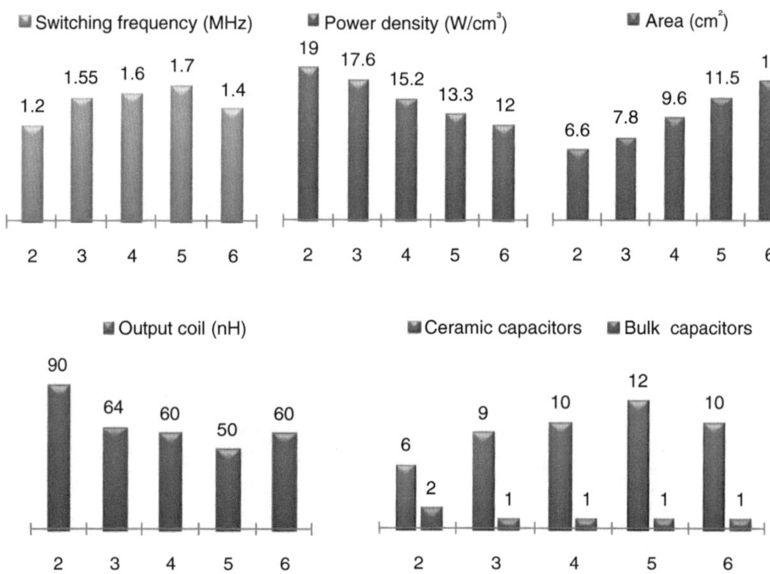

Fig. 5.24 VR optimization results as function of the number of phases. Design criteria B: Minimum achievable footprint with 75% efficiency (19 W losses). See 2008 specifications from Table 1.2, Chap. 1

Results of the VR design based on criteria B are summarized in Fig. 5.24 for the case of a target efficiency of 75%. The number of phases featuring the smallest footprint coincides with the highest power density at $N_p = 2$. In all cases, the demanded switching frequency is above 1 MHz, which enables the use of reduced form factor inductors. The required footprint dramatically increases with the number of phases, doubling the number of $N_p = 2$ at $N_p = 6$.

The minimum achieved footprint is strongly dependent on the predefined target efficiency. Demanding lower power dissipation generally implies the use of larger areas unless improvements in the technology are provided. With regard to the latter, the following chapter discusses potential improvement options to most effectively achieve significant performance enhancements.

References

1. Miftakhutdinov R (2000) Optimal output filter design for microprocessor or DSP power supply. In: Texas Instruments Analog Applications Journal, Aug 2000, pp 22–29. Available at http://focus.ti.com/lit/an/slyt162/slyt162.pdf
2. Miftakhutdinov R (1999) Analysis of synchronous buck converter with hysteretic controller at high slew-rate load current transients. In: Proceedings of high frequency power conversion conference, pp 55–69
3. Miftakhutdinov R (2000) Analysis and optimization of synchronous buck converter at high slew-rate load current transients. In: Proceedings of the IEEE power electronics specialists conference, PESC, Galway, Ireland, 18 June to 23 Jan 2000, pp 714–720
4. Miftakhutdinov R (2001) Optimal design of interleaved synchronous buck converter at high slew-rate load current transients. In: Proceedings of the IEEE power electronics specialists conference, PESC, Vancouver, 17–21 June 2001, pp 1714–1718
5. Yao K (2004) High-frequency and high-performance VRM design for the next generations of processors. PhD thesis at Virginia Polytechnic Institute and State University, Blacksburg, Virginia, 14 April 2004
6. Smith LD, Hockanson D (2001) Distributed SPICE circuit model for ceramic capacitors. In: 51st electronic components and technology conference, 2001, pp 523–528
7. Panasonic Industrial Company (2004) Specialty polymer aluminum electrolytic capacitors. Technical guide 2004, www.panasonic.com
8. Rutter P (2008) Design considerations for integrates powertrains. In: Presentation at international workshop on power supply on chip, Cork, Ireland, 22–24 Sept 2008
9. PIP212-12M DC-to-DC converter powertrain online documentations, www.nxp.com
10. Qiu W, Miller G (2007) Dual-edge PWM improves multiphase regulators. In: Power electronics technology magazine, July 2007, http://powerelectronics.com/issue-20100701
11. Capacitor Reference EEFSD0D391R from Panasonic Industrial Company (2004) Specialty polymer aluminum electrolytic capacitors. Technical guide 2004, www.panasonic.com
12. Capacitor reference GRM32DR71E475KA61 from Murata, www.murata.com
13. SMT power inductors, power beads, PA051XNL, PA121XNL, PA151XNL series from Pulse, www.pulseeng.com
14. SMT power inductors, SCL7649 series from Coilcraft, www.coilcraft.com

Chapter 6
Roadmap Targets

It is convenient at this point to recall the *virtual design loop* concept introduced in Chap. 1, which aids the development process of future power management technologies. Throughout the previous chapters, this idea has been realized with the use of dedicated device and circuit models that, in combination with a series of optimization guidelines, represent the cornerstones of the adopted analysis and development methodology.

By adopting the workflow from the scheme of Fig. 1.32, state-of-the-art switching converters can be characterized, modeled, and consistently validated with the support of extensive experimental measurements and device physics simulations. The performed validation analysis thus far demonstrates that the proposed models can accurately predict a number of relevant performance aspects related to already existing VR solutions.

One crucial step further consists of identifying whether these models can also correctly quantify the potential benefits that future technologies might bring into the application before their actual realization. This implies that the modeling characterization can only rely on simulated "virtual" data as opposed to a combination of both experimental and simulated results, as shown in all earlier cases.

The main purpose of this chapter is to establish meaningful roadmap targets for future technology developments. With the main focus set on power semiconductor devices, both state of the art and research technologies are evaluated as candidate solutions to fulfill the demands of VRs for next generation computer microprocessors. Among the main characteristics assessed, the limits on power density, footprint requirements, and conversion efficiency are fundamental in considerations of selection criteria of relevant roadmap targets and improvement options.

The methodology of analysis establishes a baseline of comparison between a state-of-the-art technology and alternative future technologies. This comparative analysis is based on parametric studies of an optimized VR solution that allows the identification of the two following key unknowns:

- Critical parameters of the converter influencing the VR performance.
- Quantified improvement factor of each critical parameter required to achieve a certain system performance.

T. López et al., *Voltage Regulators for Next Generation Microprocessors*,

DOI 10.1007/978-1-4419-7560-7_6, © Springer Science+Business Media, LLC 2011

As it will be shown, the latter point may yield a multiple number of improvement options from which roadmap targets can be based upon.

Following this conceptual circuit/device co-design approach, a new developed trench MOSFET technology generation is presented as the successor of the existing Trench4 generation from NXP Semiconductors. During this development process, the benefits of the new technology could be estimated from the FE device and circuit models. Validation results of the virtual loop analysis were available only after the devices were fabricated and experimental results matched the model predictions.

The explored improvement options of this chapter cover aspects not only related to the power semiconductor device but also to the gate drivers, packaging technology, filter components, and control.

Although most of the findings concern computer desktops and workstations, the last section of the chapter is fully devoted to mobile laptop applications, the particular requirements of which shall be emphasized.

Throughout the analysis, the following assumptions are taken unless otherwise stated,

- The roadmap refers to the 2014 target load specification of Table 1.1.
- The employed passive filter components correspond to those of the illustration of Sect. 5.4.1.
- The optimized designs are based upon worst-case conditions, i.e., heavy load and maximum junction temperature of 140°C.
- The VR parameter designs derive in each case from the guidelines of Chap. Chap. 5.
- The analysis assumes IPM solutions as reference designs. Only in Sect. 6.7 (i.e. laptop application) discrete solutions are considered.

6.1 Power Switches

The technology roadmap of power semiconductor devices is of vital importance to define power conversion solutions for next generation computer CPUs. This section starts by analyzing the performance limitations of VR solutions based on the state-of-the-art Trench4 technology from NXP Semiconductors. A sensitivity analysis will follow to identify critical MOSFET and circuit parameters as basis for the definition of roadmap targets. This will lead to the development of the next generation Trench6 MOSFET technology from NXP Semiconductors, the performance aspects of which will be studied and compared to the initially defined target requirements. Further performance studies on future technology developments will be carried out to complete the roadmap projections. These predictions will ultimately reveal the potential of the chosen converter topology as candidate for future VR solutions.

Throughout this section the following defined parameters shall be frequently used: Q_{GSsp}, Q_{GDsp} and Q_{DSsp} are, respectively, the specific gate-source, gate-drain, and drain-source parasitic FET charges in units C/m^2. Also, parameter FoM_t is defined as,

$$FoM_t = FoM_{(s)} + FoM_{(c)}. \tag{6.1}$$

6.1.1 State-of-the-Art Trench4 MOSFET Technology

The performance of the existing IPM PIP212-12M from NXP Semiconductors is evaluated in a VR system for next generation computer microprocessors (see 2014 specifications from Table 1.1). The target VR solution under consideration aims at minimizing footprint while maintaining a converter efficiency not lower than 75%, as expected from Intel® VR guidelines (see [1], Chap. 1). To determine the minimum area requirements upon these conditions, the data of Fig. 6.1 are generated from the optimization procedure of Chap. 5, where the existing IPM design of Table 5.2 is taken as reference. The bars refer to optimized parameters as a function of the number of N_p phases. It can be seen that, although the minimum footprint is found at $N_p = 12$, little penalty in area requirements is paid for the ten phases solution despite of the fact that the number of bulk capacitors increases significantly from 21 to 34.

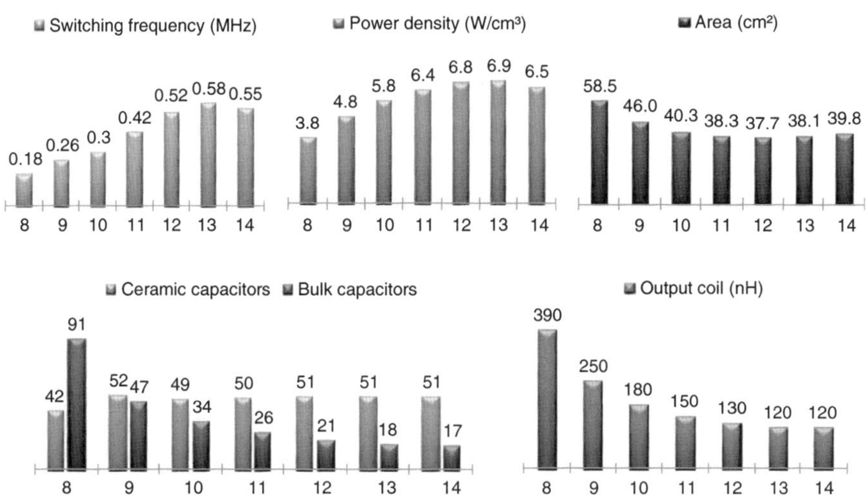

Fig. 6.1 Trench4-based VR solution for next generation Desktop computer microprocessors. Optimization results as function of the number of phases N_p. Design criteria: minimum achievable footprint with 75% converter efficiency (33 W losses). See 2014 specifications from Table 1.1 of Chap. 1. See Table 6.5 for details on passive filter components

A look at the loss breakdown of Fig. 6.2 corresponding to $N_p = 10$ reveals SyncFET ON conduction as the major loss contributor, even with the use of the largest possible semiconductor die area fitting in the package of the IPM.

As already seen in the previous chapter, the optimization results may yield operating conditions characterized by high ripple inductor currents. Considering, for instance, the ten phase solution with a selected output inductor of 180 nH and a switching frequency of 300 kHz, the peak-to-peak current ripple results in 10 A. As a main consequence of maximizing the overall power density, the inductor losses may well reach the power dissipation limits of the part (e.g., ~730 mW at full load).

According to these results, Trench4 may offer VR solutions of power densities not greater than 7 W/cm^3. This can only be obtained with a high number of phases ($N_p = 13$) and a limited efficiency (75%).

Further improvements in MOSFET technology can effectively boost system performance in the way reflected in Figs. 6.3 and 6.4. The total MOSFET loss reduction factor, which is relative to Trench4 under the same examined conditions of Fig. 6.2, shows that the power density can double with a 30% loss reduction in both power devices. This is achieved as the switching frequency approaches the MHz range and the output inductance lowers well below 100 nH.

Of relevant importance is the barrier of 50% loss reduction, as the power density hits the level of 15 W/cm^3, the footprint shrinks to 20 cm^2 and the number of phases reduces to 6. As it will be shown later on in Sect. 6.6, these conditions may enable the location of all VR phases in the PCB area right underneath the microprocessor's heatsink. This dramatically reduces extra space requirements on the main board while allowing a large exploitation of the forced convection heat transfer mechanism produced by the heatsink's integrated fan of the CPU. The latter appears to be of great attractiveness as direct airflow from the

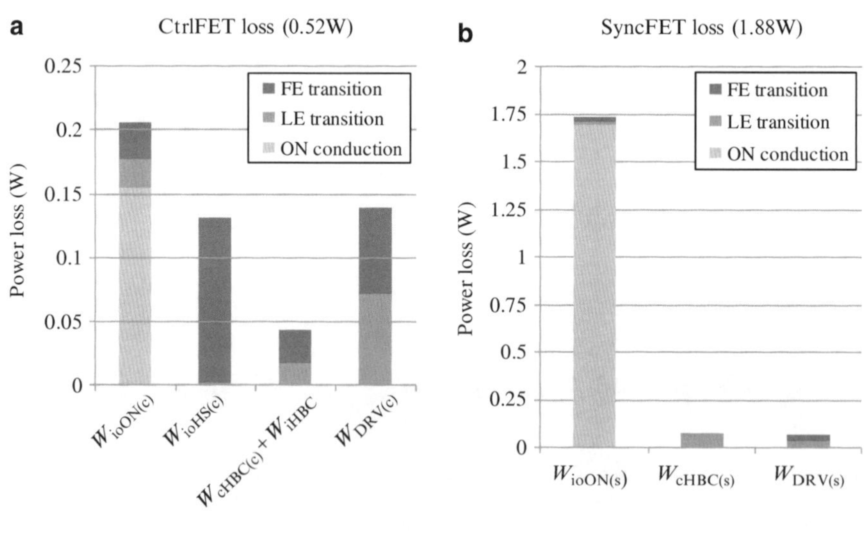

Fig. 6.2 Trench4 power MOSFET loss breakdown in the case of $N_p = 10$ from Fig. 6.1

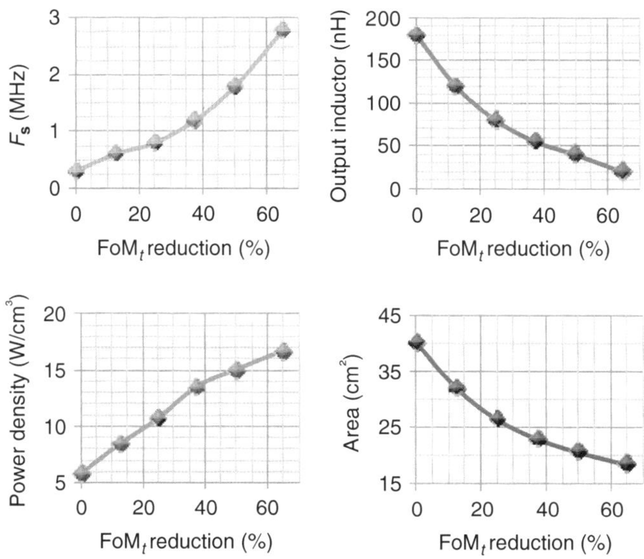

Fig. 6.3 VR performance benefits from improvements in MOSFET technology relative to Trench4. Power density increase and footprint reduction

Fig. 6.4 VR performance benefits from improvements in MOSFET technology relative to Trench4. Reduction in number of phases

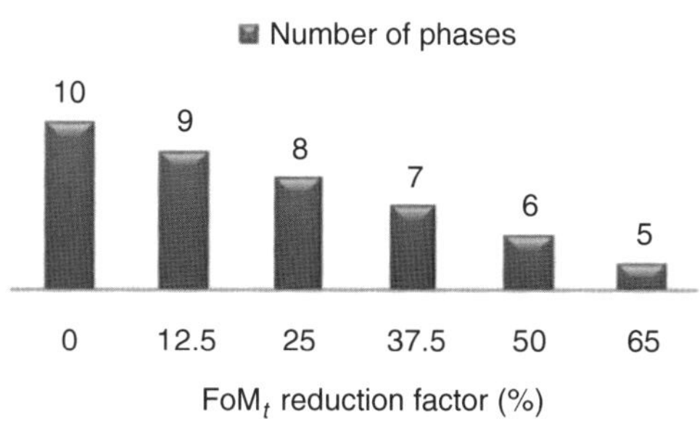

microprocessor's fan into the VR parts may avoid the use of heavy and bulky heatsinks and further enable the introduction of smaller form factor inductors and limited thermal conductivity.

In addition to the apparent benefits in power density, improvements in MOSFET technology further offer a progressive reduction in bulk capacitor count, which is primarily attributed to the reduction of the output filter inductance. As illustrated in Fig. 6.5, achieving a 40% loss reduction of MOSFET losses allows a significant cut down of 20 polymer electrolytic capacitors while virtually keeping the same number of ceramic capacitors.

These results justify the definition of a potential performance target for the next technology generation, which particularly aims at reducing by at least 50% the

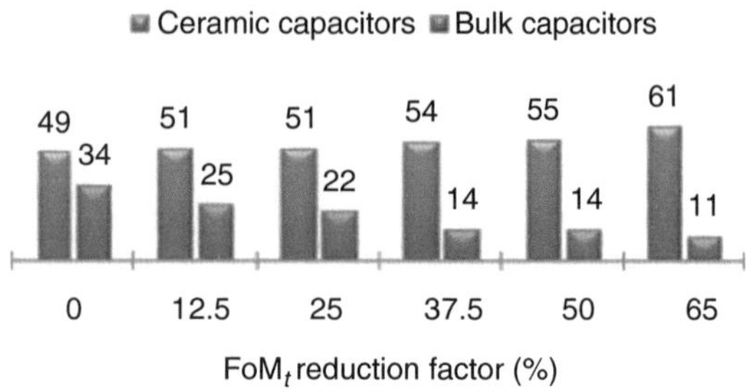

Fig. 6.5 VR performance benefits from improvements in MOSFET technology relative to Trench4. Reduction in output filter capacitor count

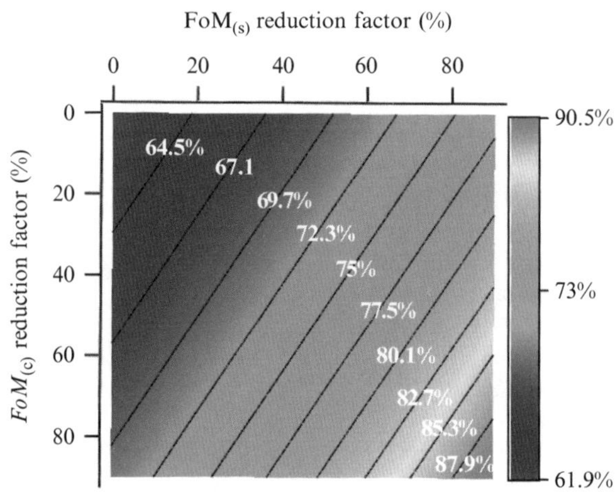

Fig. 6.6 VR efficiency boost from improvements in MOSFET technology relative to Trench4. Increase in converter efficiency under the conditions of Table 6.1

MOSFET losses produced with Trench4 under the examined conditions of Fig. 6.2. Meeting this goal will enable setting up the configuration of Table 6.1. According to the above predictions, the power density level achieved with such arrangement will allow the VR phases to fit underneath the microprocessor heatsink, with the consequent aforementioned advantages. Therefore, unless otherwise specified, the following analysis of present and future MOSFET technologies will be based upon the conditions of Table 6.1.

Paying close attention as to how loss savings can be most effectively attained, Fig. 6.6 shows the VR efficiency as a function of both FoM$_{(c)}$ and FoM$_{(s)}$ relative to Trench4 under the defined target conditions of Table 6.1. The contour plot reveals that performance improvements in the SyncFET yields larger efficiency boosts than those associated to the CtrlFET, which suggests a dominant contribution of ON conduction losses, also in the multi-MHz switching frequency range.

The target isoline of 75% can be met by a number of measure combinations, among which includes an improvement of ~80% in FoM$_{(s)}$ (w.r.t. Trench4) without

Table 6.1 VR target solution to achieve 15 W/cm^3 power density and 20 cm^2 footprint in next generation Desktop computer microprocessors

Switching frequency (MHz)	1.8
Number of phases	6
Output filter inductance (nH)	40
Output filter bulk capacitor count	14
Output filter ceramic capacitor count	55
VR efficiency (%)	≥75
IPM package	SOT684-4 compatible

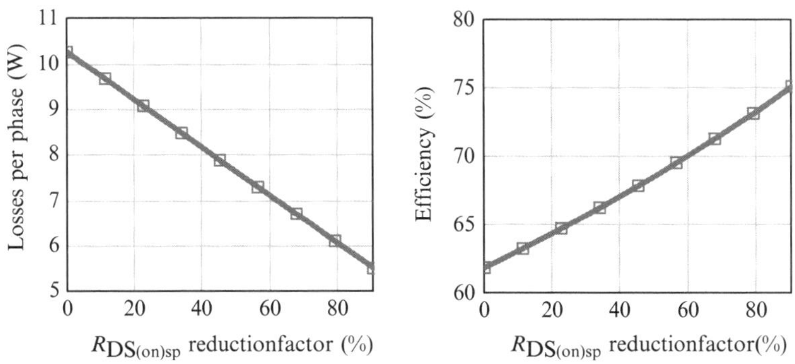

Fig. 6.7 Power losses and converter efficiency as a function of $R_{DS(on)sp}$ relative to Trench4 under the conditions of Table 6.1

boosting the CtrlFET performance. Limiting however the SyncFET performance gain to 50% implies an additional ~57% reduction in FoM$_{(c)}$. Over 90% efficiency may be achieved with a dramatic 90% reduction in both MOSFETs FoM.

The relevant contribution of the SyncFET losses on the overall VR efficiency suggests that reducing the MOSFETs specific ON resistance[1] ($R_{DS(on)sp}$) might be the most determining aspect of the technology roadmap. More concretely, Fig. 6.7 shows that the expected efficiency gain may linearly change at a rate of $-0.5\%/\mathrm{m}\Omega\ \mathrm{mm}^2$.

Further measures for enhancing system efficiency include the mitigation of the MOSFETs parasitic charges. Figure 6.8 (right) illustrates that reducing the input charge is notably more effective than reducing the output charge. Nevertheless, the performance gain of the input charge mitigation relative to the benefits in $R_{DS(on)sp}$ reduction is of minor significance. For instance, a 50% reduction in $R_{DS(on)sp}$ may result in a 6% efficiency increase, whereas the same reduction in gate charge Q_{GSsp} produces an efficiency boost of 1.4%. The latter however may involve detrimental effects on gate bounce susceptibility, which also needs to be assessed, as it shall be seen later.

[1]The specific ON resistance of Trench4 is 30 mΩ mm^2 at $T_j = 25°C$.

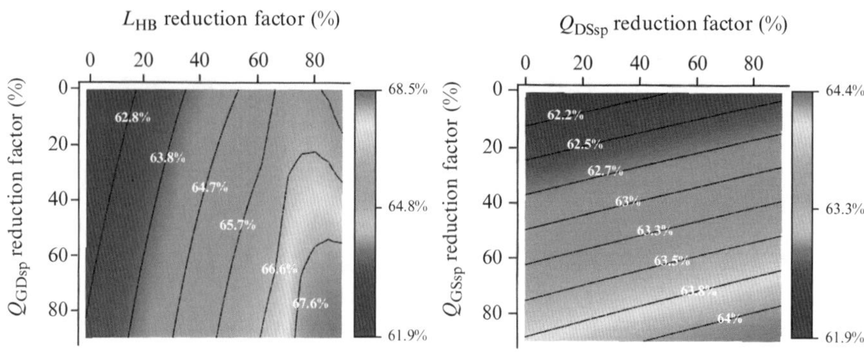

Fig. 6.8 Converter efficiency as a function of Q_{GDsp} and L_{HB} relative to Trench4 and the recommended PIP212-12M layout design

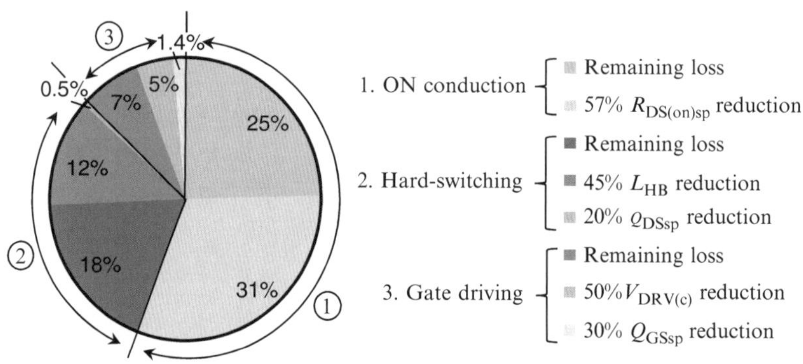

Fig. 6.9 Improvement options for a 50% loss reduction relative to the existing Trench4-based IPM solution under the conditions of Table 6.1

The transfer capacitance and half-bridge inductance strongly influence the hard-switching loss of the converter, which may generally represent the major loss contribution in the CtrlFET. From the latter two circuit elements, however, the parasitic inductance appears to be the most relevant in the given target scenario as it can be interpreted from the left contour plot of Fig. 6.8. A great profit of featuring reduced half-bridge loop inductance is obtained when the CtrlFET turn-off loss can be effectively mitigated without compromising CtrlFET turn-on loss. Therefore, it may turn out convenient to reduce both parameters simultaneously such that the resulting increase in commutation speed compensates the reduction of L_{HB} at turn-on. This is also reflected in the 2D plot, where an excessive L_{HB} reduction produces an overall efficiency loss due to a high increase in turn-on losses. The reduction of the relatively large transfer charge eliminates this problem as it enables faster turn-on switching.

The potential of these identified measures can be quantified as shown in Fig. 6.9. The chart displays the loss savings produced by five loss-dependent parameters related to the MOSFET technology (i.e. $R_{DS(on)sp}$, Q_{DSsp}, and Q_{GSsp}) and other two from the half-bridge circuit and gate driver (i.e., L_{HB} and $V_{DRV(c)}$). Among these parameters, improvements in $R_{DS(on)sp}$ offer the highest loss savings, as expected. Thus, it is intended that the specific ON conduction resistance undergoes the largest reduction, aiming at a 43% $R_{DS(on)sp}$ of Trench4 so as to save as much as 31% losses.

The second most sensitive parameter on power losses is L_{HB}. In the illustrated scenario, a reduction from 1.6 to 0.7 nH allows avoiding 12% loss associated to hard-switching turn-off. Of minor relevance with regard to hard-switching is the Q_{DSsp} charge, as a 20% reduction of this parasitic charge produces less than 1% loss savings.

Gate driving loss, which represents more than 13% of the overall VR consumption, can be effectively lowered down to 7% by halving the CtrlFET driving voltage and reducing the input charge by 30%.

The resulting solution halves the losses of the existing Trench4 solution (i.e., PIP212-12 M), thereby meeting the targeted 75% efficiency with an equally shared loss profile between switching and conduction losses.

6.1.2 Trench6 MOSFET Technology

Advances in silicon technology enables the accomplishment of roadmap targets established according to the proposed virtual loop approach, namely a combination of device physics and circuit simulations that underlies a co-design approach where the power MOSFET performance is assessed in the converter circuit to best determine the technology requirements. This approach replaces the usual experimental/empirical efforts of physicists to elucidate the pathways toward the most effective technological solutions.

In line with the conclusions drawn from the analysis of improvement options, Trench6 emerges as the successor of Trench4 to meet the specifications of next generation CPUs. It has been possible after several virtual loop iterations had been run and the most suitable design chosen. As opposed to numerous experimental wafer runs, simulation results have set the basis of the technology platform for the implementation of a selected device structure candidate that has been virtually analyzed.

As shown in Table 6.2, the result of the new technology platform has made possible the implementation of a trench MOSFET device featuring significant superior characteristics than its predecessor. Particularly, $R_{DS(on)sp}$ is reduced by 50%, thereby nearly reaching the required target loss reduction determined from the analysis of the previous section (57% reduction).

Significant further improvements have been achieved in gate bounce susceptibility, which will enable tighter dead time control schemes and faster switching speeds in the final application.

Table 6.2 30 V technology comparison: Trench4 vs. Trench6

	Trench4	Trench6
$Q_{GD}R_{DSon}$ (nC mΩ)	23.5	14.5
$Q_{Gt}R_{DSon}$ (nC mΩ)	104	60.5
$R_{DS(on)sp}$ (mΩ mm^2)	30	15
Gate resistance R_G (Ω)	0.7	0.77
Gate bounce susceptibility GBS_{GS0} [a]	0.88	0.064
$FoM_{(c)}$ (μJ) [b]	1.9	1.2
$FoM_{(s)}$ (μJ) [b]	3.1	1.8

Parameter values unless otherwise specified: $V_{GS} = 4.5$ V, $V_{DS} = 12$ V, $I_D = 10$ A and $T_j = 25$°C

[a]Zero initial conditions across v_{GS} and v_{DS}. T_j=140°C. See Chap. 3 for details on GBS_{GS0}

[b]VR design based on IPM PIP212-12M with $N_p = 6$, $L_o = 40$ nH, $F_s = 1.8$ MHz, V_i=12 V, T_j=140°C and full load. See 2014 specifications from Table 1.1 of Chap. 1

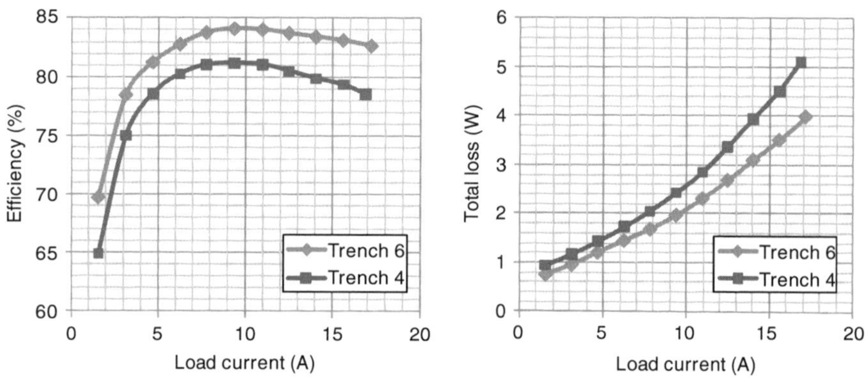

Fig. 6.10 Efficiency and total loss comparison of a PoL converter based on Trench4 and Trench6 technologies. Measurement results

In terms of parasitic charge, the resulting input capacitance is comparable to that of Trench4, which differs from the target by 30% reduction in Q_{GDsp} proposed in the previous subsection. This will consequently lead to a performance deficit in the application upon consideration as it shall be shown later. Nonetheless, the $FoM_{(c)}$ and $FoM_{(s)}$ related to such application already indicate over 40% loss savings.

The efficiency gains of the new trench technology in the converter circuit could be predicted before experimental realization, as it can be seen from the efficiency plots of Figs. 6.10–6.12. The one phase PoL prototype converts 12–1.1 V with F_s = 500 kHz and 2 m/s airflow. Further details of the experimental setup are available in Appendix G.

Similarly, as carried out in the analysis of the Trench4 from the previous subsection, Fig. 6.13 summarizes optimization results of a Trench6 based IPM solution for the 2,014 specifications of Table 1.1. The estimations indicate a maximum power density of ~14 W/cm^3 and minimum footprint of ~23 cm^2 with $N_p = 7$, $F_s = 1.7$ MHz, $L_o = 50$ nH, and 15 output bulk capacitors. Although this

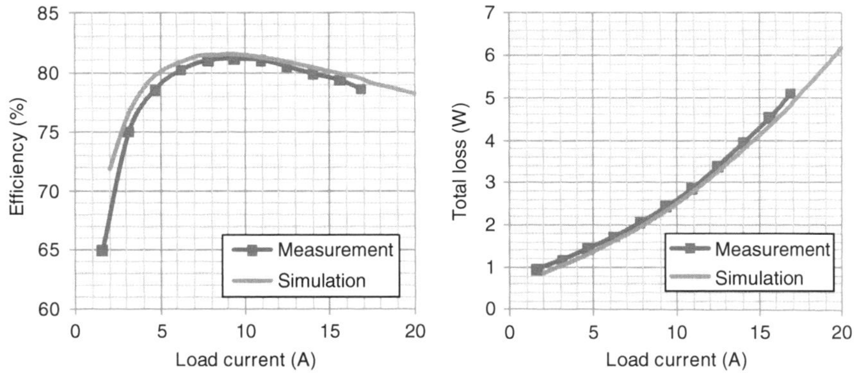

Fig. 6.11 Trench4 performance. Simulation and measurement results

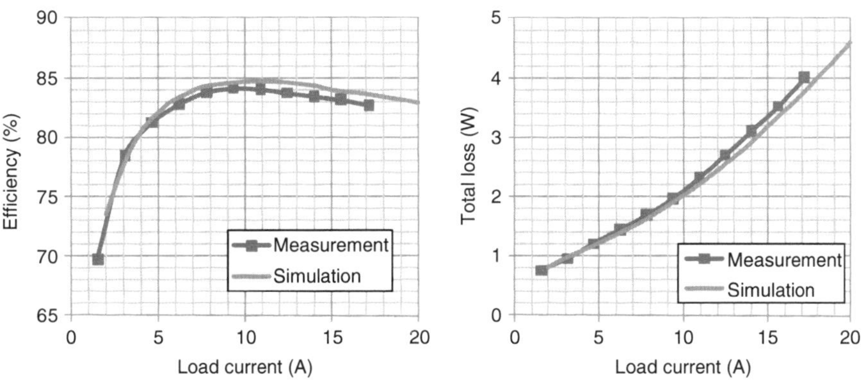

Fig. 6.12 Trench6 performance. Simulation and measurement results

solution does not meet the initially proposed VR targets (see Table 6.1), it practically doubles the performance of the existing Trench4 based solution in terms of power density and main board area requirements.

This achievement, however, is conditioned to additional changes in the IPM PIP212-12M that allow a better exploitation of the Trench6 technology as concluded from the results of Fig. 6.9. It is thus assumed that inductance L_{HB} reduces from 1.6 nH down to 1 nH, whereas CtrlFET driving voltage $V_{DRV(c)}$ lowers from 12 to 6 V. MOSFETs die areas remain unchanged with respect to the Trench4 solution. Once again, the optimum SyncFET die area turns out to be slightly larger than the limits established by the package dimensions of the IPM, which consequently determines, as in the Trench4 case, the size of the largest switch.

Despite the area limitations, a look at the loss breakdown of Fig. 6.14 reveals that ON conduction loss is no longer the main loss mechanism since it accounts for less than 34% of the overall losses during the ON conduction times, whereas switching

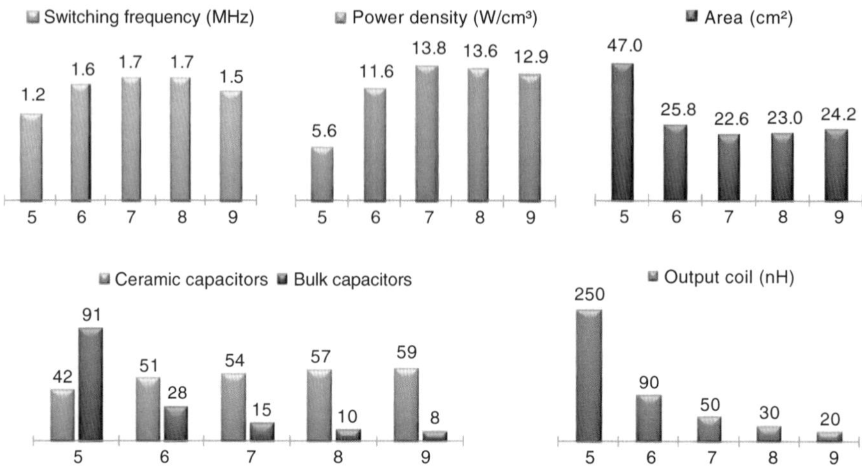

Fig. 6.13 Trench6-based VR solution. Optimization results as a function of the number of phases. Design criteria: minimum achievable footprint with 75% converter efficiency (33 W losses)

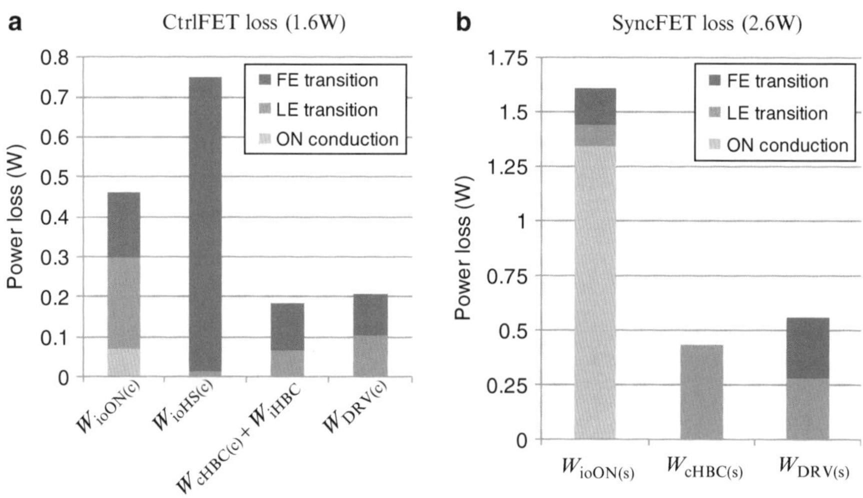

Fig. 6.14 Trench6 power MOSFET loss breakdown for the case of $N_p = 7$ from Fig. 6.13

losses contribute with 29% in the LE transition and 37% in the FE transition. The latter contribution is largely dominated by load current hard-switching losses ($w_{ioHS(c)}$).

The gate driver settings are optimized so that $w_{ioHS(c)}$ cannot be reduced any further by means of increasing the gate current, and hence the switching speed, as otherwise the overvoltage spike across the CtrlFET will reach the level of avalanche breakdown. The turn-off gate current is thus limited to less than 2 A during the hard-switching time.

This analysis suggests that the way to go about reducing switching losses is by further lowering L_{HB}. As additional measure, turn-off snubber enhancement by way of increasing the CtrlFET output capacitance appears effective as long as the turn-on loss is significantly lower than the turn-off loss.

Further conclusions may be drawn from the loss breakdown of Fig. 6.14. On the one hand, it can be observed that the CtrlFET ON conduction loss is dominated by the switching transient times as opposed to the ON conduction time, as it is typically the case at low F_s. The reason for that is attributed to the extreme conversion ratio combined with high F_s, thus leading to a very short ON conduction time, even shorter than the switching times. Therefore, the converter approaches the lower duty cycle limit, which increases as F_s raises.

As it shall be addressed later in Sect. 6.2, in order to further expand the conversion ratio while maintaining (or even increasing) F_s, it may be required to enable faster switching speeds by improvements in the gate driver technology and the MOSFETs gate impedance. The latter is particularly important as it may not only enable faster switching, but also a reduction of the gate driving loss, which becomes increasingly relevant at high F_s. In this case-study, the gate driving loss represents a considerable ~28% of the total switching losses.

Of equal importance appears to be the charging loss associated to the SyncFET output capacitance (i.e., $w_{cHBC(s)}$), even though the output parasitic charge may be significantly lower than the input parasitic charge. Two different aspects make this possible: One is the switched-node voltage swing, which may typically be slightly higher than twice the value of the gate driver voltage in Desktop computer applications, and even higher in laptop applications. The other aspect is related to the strong nonlinearity of the output capacitance, the shape of which worsens the charging-up loss as it was shown in Chap. 3. Thus, $w_{cHBC(s)}$ may become a main contributor at high F_s and as such the reduction of the SyncFET C_{oss} may be a potentially advantageous improvement option (as opposed to reducing the CtrlFET C_{oss}).

Furthermore, since the SyncFET losses are higher than the CtrlFET losses, measures to reduce $\mathrm{FoM}_{(s)}$ may be the most effective. This can be interpreted from the contour plot of Fig. 6.15, which shows the required loss savings in the switches to meet a certain efficiency target under the conditions of Table 6.1. For the minimum demanded 75% efficiency, $\mathrm{FoM}_{(s)}$ may need to be reduced by as much as 20% in case of keeping $\mathrm{FoM}_{(c)}$ unchanged. On the contrary, the $\mathrm{FoM}_{(c)}$ needs an improvement of 30% to meet the same performance in case the $\mathrm{FoM}_{(s)}$ remains constant.

According to these estimations, future technology developments may aim at a minimum 43% reduction in either $\mathrm{FoM}_{(s)}$ or $\mathrm{FoM}_{(c)}$ or both to achieve 80% VR efficiency, which represents more than a 7% efficiency boost with respect to the Trench6 based solution.

As in the Trench4 technology, and despite of high-switching frequency operation, the $R_{DS(on)sp}$ continues to be the most relevant technologic parameter influencing the VR performance even for values below 10 mΩ mm^2. As it can be deduced from Fig. 6.16, a target efficiency of 80% may be achieved by a 77% reduction of the $R_{DS(on)sp}$ of Trench6 without altering any other parameter of the VR.

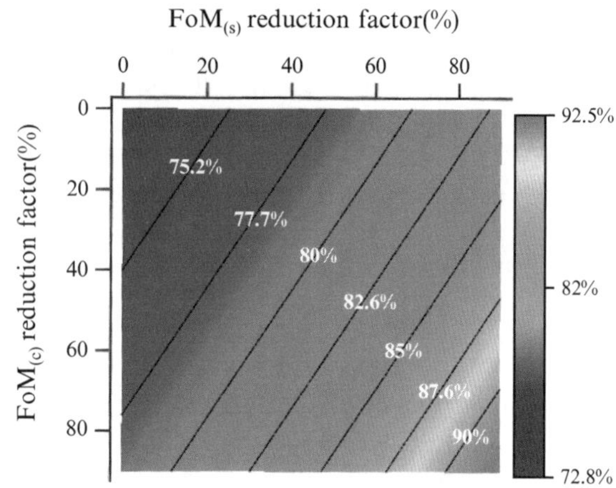

Fig. 6.15 VR performance benefits from improvements in MOSFET technology relative to Trench6. Increase in converter efficiency under the conditions of Table 6.1

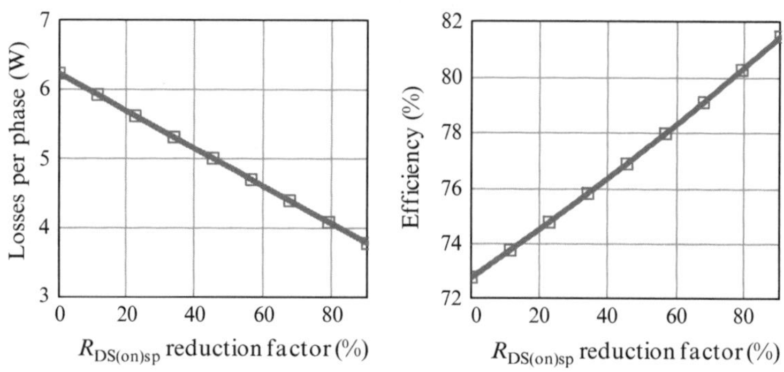

Fig. 6.16 Power losses and converter efficiency as function of $R_{DS(on)sp}$ relative to Trench6 (i.e., 15 mΩ mm^2)

Less dramatic are the efficiency improvements resulting from reducing the MOSFET capacitances in the same order as applied to $R_{DS(on)sp}$. For the particular case of the transfer capacitance (or Q_{GDsp}), the loss reduction is strongly limited by L_{HB} since, in order to avoid avalanche breakdown, the switching off time cannot lower in proportion to Q_{GDsp}. Therefore, inductance L_{HB} continues to be one main parameter influencing switching losses, as shown in Fig. 6.17. The left plot reveals however that for an effective measure, the L_{HB} reduction may need to be combined with improvements in Q_{GDsp} since the turn-on snubber may become ineffective unless such reduction is compensated with an increase of the switching speed.

The input capacitance associated to Q_{GSsp} can be identified from Fig. 6.17 (right plot) as yet another relevant MOSFET parameter primarily due to the significant loss contribution of the SyncFET gate driver. Compared to Q_{GSsp}, reducing Q_{DSsp} is not as effective because the loss reduction achieved on $w_{cHBC(s)}$ largely cancels out with the loss increase from $w_{ioHS(c)}$. Thus, maintaining a minimum snubber turn-off

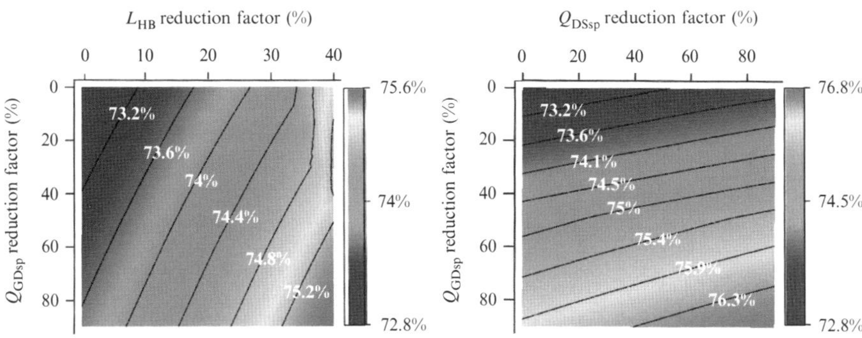

Fig. 6.17 Converter efficiency as function of Q_{GDsp} and L_{HB} relative to Trench6 and the recommended PIP212-12M layout design (by which it is assumed that $L_{HB} = 1$ nH)

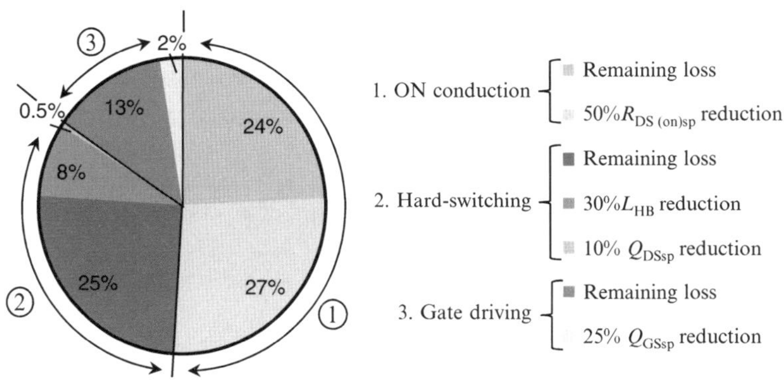

Fig. 6.18 Improvement options on Trench6 to achieve 80% efficiency in next generation IPM solutions. Reference PIP212-12M layout design recommendation for Trench6 with $L_{HB} = 1$ nH

capacitance across the CtrlFET while lowering Q_{DSsp} of the SyncFET may represent an alternative effective measure as long as $w_{ioHS(c)}$ is dominant over $w_{cHBC(c)}$.

Figure 6.18 displays the impact that a number of selected measures have on the converter losses. These are broken down into three components: Hard-switching, gate driver, and ON conduction losses. The parasitic elements previously analyzed are reduced according to their individual identified levels of relevance. Therefore, the proposed Trench6 based improvement scenario assumes a 50% $R_{DS(on)sp}$ reduction combined with 30% less L_{HB}, 25% lower Q_{GSsp}, and a 10% improvement in Q_{DSsp}. This yields a 37.5% loss savings, 27% of which are ON conduction, 8.5% hard-switching, and 2% gate driver loss. Overall, it represents an efficiency gain of 7.2%, most of which results from improvements in $R_{DS(on)sp}$ and L_{HB}.

Figure 6.19 shows an alternative approach that allows achieving 80% efficiency. This considered option may become appropriate when limitations in the MOSFET technology does not allow the simultaneous reduction of both parasitic capacitances and $R_{DS(on)sp}$. The proposed solution reduces gate drive losses from the SyncFET by

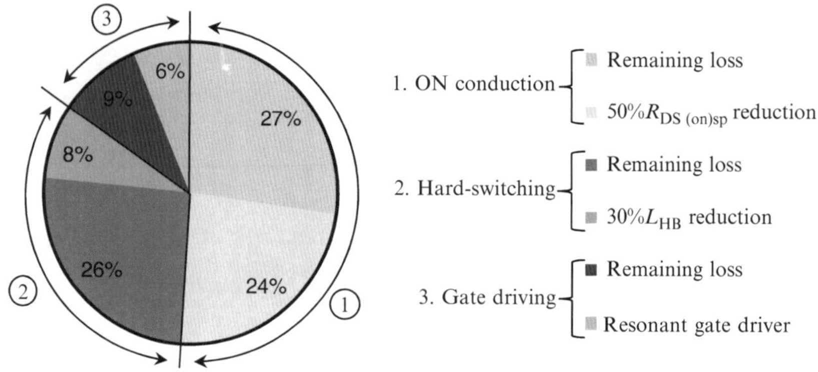

Fig. 6.19 Improvement options on Trench6 to achieve 80% efficiency in next generation IPM solutions with the use of a resonant gate driver topology for the SyncFET

means of a resonant gate driver (RGD). The advantages and drawbacks of a number of representative RGD topologies are analyzed and compared to the conventional gate driver (CGD) in Appendix F. According to the results of that study, an RGD topology (driver circuit (a) from Appendix F) is selected as the most suitable driving scheme for this application. Optimizing the parameters of both driver and MOSFET as described in Appendix F, the gate driver losses can be reduced by as much as 40%. Improvements in $R_{DS(on)sp}$ and L_{HB} are maintained as in case of Fig. 6.18. Nonetheless, savings in ON conduction losses are 3% less due to the longer switching times intrinsically produced by the chosen driver topology. Among the modifications required to make the RGD an effective alternative solution, it includes the reduction of the MOSFET gate resistance down to 0.2 Ω. Further details regarding the configuration of this driver are available in Appendix F.

6.1.3 Next Generation Technologies

Looking at the technology trends in terms of standardized figures, Table 6.3 shows the relative performance of a number of commercially existing and under development technologies suitable for VR applications.

With regard to conventional 30 V trench MOSFET technologies, investigations point out toward a progressive saturation of $R_{DS(on)sp}$ around 15 mΩ mm^2 (see Fig. 1.13). As discussed in Chap. 1, lower specific resistance can be further achieved by exploiting the RESURF principle with structures such as the split-gate RSO and superjunctions. These technologies can offer $R_{DS(on)sp}$ below 10 mΩ mm^2 and competitive switching performance characteristics compared to conventional trench MOSFET technologies.

Table 6.3 30 V power device technology trends

	$Q_{GD} \cdot R_{DSon}$ (nC mΩ)	$Q_{Gt} \cdot R_{DSon}$ (nC mΩ)	$R_{DS(on)sp}$ (mΩ mm^2)
Trench4 (NXP Semiconductors) [1, 2]	23.5	104	30
Trench6 (NXP Semiconductors) Field balanced TrenchMOS [3, 4]	14.5	60.5	15
Split-gate RESURF [5, 6]	4	85	8
Trench7 (NXP Semiconductor) Superjunction RESURF[a]	17	56	9
GaN(1) based platform [8–11]	3	15	9
Improved GaN(2) based platform [8–11]	2	10	7
TBD	1.3	8.5	3

See Table 6.2 for further details
[a]Approximated values. More accurate numbers are given in [7]

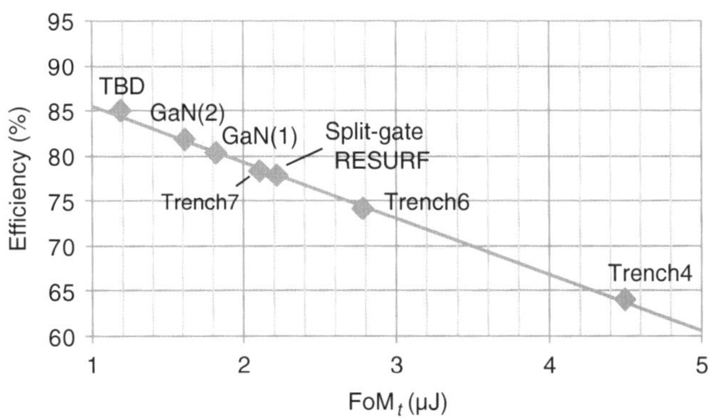

Fig. 6.20 Projected 30 V power device technology performance

A new paradigm opens up with the introduction of alternative semiconductor materials for power conversion applications. Recent studies from International Rectifier (IR) have demonstrated breakthroughs in $R_{DS(on)sp}$ and switching speed capabilities with GaN semiconductor materials. Based on HEMT structures, GaN technology may potentially offer $R_{DS(on)sp}$ well below 10 mΩ mm^2 and switching FoM outperforming those of the best Si trench MOSFETs.

Translating the above figures in terms of conversion efficiency under the conditions of Table 6.1 and Fig. 6.20 shows the projected performance as function of the FoM$_t$ from the chosen technologies. According to the model predictions, RESURF technologies enable a further exploitation of the Si material to boost the converter efficiency above 75%. Achieving efficiencies above 80% may though be only possible with future advanced GaN based devices. Extrapolated data indicate that in order to meet an efficiency of 85% at full load, the required technology (still to be determined) may need to feature 3 mΩ mm^2 of $R_{DS(on)sp}$ combined with 1.3 and 8.5 nC mΩ in $Q_{GD} \cdot R_{DSon}$ and $Q_{Gt} \cdot R_{DSon}$, respectively. In the application, this translates into a total energy loss in the power switches of less than 1.2 μJ per switching cycle in every VR phase. It is relevant pointing out that these estimations result from combining these emerging technologies with improved packaging technologies that enable the reduction of the parasitic ESL and ESR of both half-bridge and gate driver paths. These measures are discussed in Sect. 6.3.

Aiming at the same target load specification, Fig. 6.21 illustrates a series of improved circuit and device parameters of a Trench6-based solution that allow the increase of the switching frequency while maintaining 75% converter efficiency. Based on the conditions of Table 6.1, the resulting simulations assume a constant ripple current per phase, thereby implying a reduction of the output filter inductance as the frequency rises. Accordingly, for a programed ripple of ~7 A, L_o may lower from the selected 40 nH at 1.8 MHz to just 8 nH at 9 MHz. To achieve this latter operating frequency, major improvements in the device technology may include the reduction of $R_{DS(on)sp}$, Q_{GSsp}, and Q_{GDsp} to realize 4.5 mΩ mm^2, 1 and 0.5 nC/mm^2, respectively.[2]

With regard to the circuit parameters, the most important element from the loss savings stand-point is the half-bridge loop inductance, which may need to undergo a dramatic reduction from what it could be achieved with existing MCM (i.e., values around 1 nH) down to 300 pH.

The most obvious approach to go about reducing L_{HB} to such a level consists of achieving a higher level of system integration. This leads to the concepts of PSiP and PwrSoC introduced in Chap. 1 that include the integration of the passive components either in a package or on a chip. These concepts turn out to be potential candidate solutions to mitigate not just the ESL of the half-bridge and gate circuits, but also the ESL of the output capacitance, thereby allowing an effective

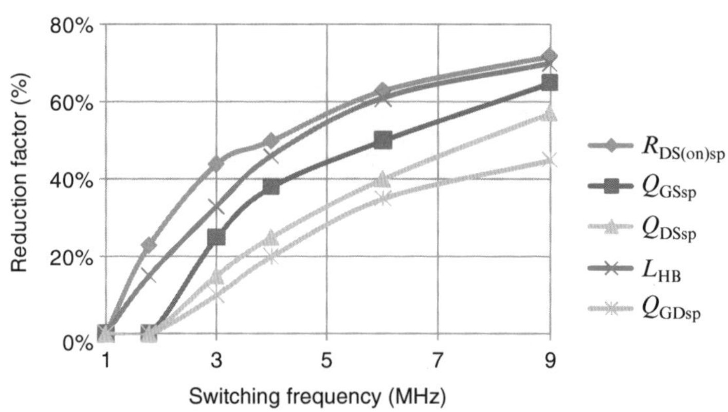

Fig. 6.21 Trench6 parameter improvements as function of the switching frequency so as to keep 75% converter efficiency at heavy load. See 2014 specifications from Table 1.1 regarding load conditions

[2]See Table 6.2 for comparison to existing technologies and test conditions for relative charge values.

reduction of the required output capacitance as the switching frequency increases (see Sect. 4.1).

6.2 Gate Drivers

Improvements in the MOSFET technology have direct implications on the gate drivers of both CtrlFET and SyncFET switches. As seen previously, changes may involve not just mere readjustments of the gate voltage, but also the optimization of the gate impedance to achieve faster switching speeds, the increase of the gate bounce susceptibility, and the implementation of extended designs that enable asymmetric LE and FE gate current transitions.

Of particular importance to achieve high-switching frequency and extremely low-duty cycle operation is to feature fast switching gate driving capabilities in the CtrlFET switch. Figure 6.22 shows the maximum switching frequency as a function of the CtrlFET gate resistance, which limits the charging times from the Miller plateau voltage to the ON state driving voltage and vice versa. The maximum switching frequency for a given duty cycle must be such that the sum of the charging times equals the ON time of the CtrlFET. According to the conditions defined in the legend of Fig. 6.22, the maximum F_s varies significantly with the selected MOSFET technology. In future laptop applications (i.e., voltage conversion from 19 to 0.6 V) for instance, the maximum F_s practically triples when changing from the Trench4 to the Trench7 MOSFET technology at $R_{G(c)} = 1\ \Omega$.

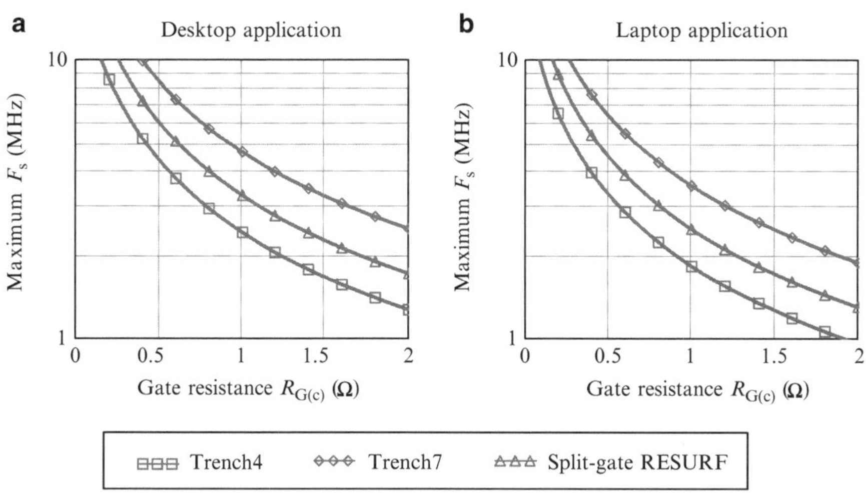

Fig. 6.22 Maximum switching frequency as function of the CtrlFET gate resistance (common to both turn-on and turn-off transients) and for various power MOSFET technologies. Both cases (**a**) and (**b**) assume parasitic inductances as in PIP212-12M (see Table 2.8) and linear gate resistance. Hard-switching corresponds to a 30 A load. See Table 6.3 for details regarding the MOSFET technologies

In future desktop applications (e.g., voltage conversion from 12 to 0.5 V) current state-of-the-art technologies may only allow switching frequencies below 2 MHz. The reduction of the total gate resistance down to 0.5 Ω may enable a maximum F_s higher than 8 MHz in case of the Trench7 technology. The even lower duty cycle operation and limited switching speed (i.e., to avoid avalanche breakdown) in laptop applications may restrict the frequency to less than 6.5 MHz if all other driving conditions prevail as in case (a).

It is relevant to point out the challenges of reducing the gate resistance well below 0.5 Ω, first because of the gate spreading resistance effects across the MOSFET area (see Chap. 2), and secondly, because the AC resistance of gate bond wires may become significantly large as the switching times reduce to just a few nanoseconds.

One assumption of these simulations is that the parasitic gate inductance remains unchanged regardless of the gate resistance value. In doing so, however, problems may occur at low $R_{G(c)}$ due to excessive underdamped ringing that may produce driving malfunctions. Therefore, in order to maintain an adequate gate voltage ringing, the total parasitic gate inductance should be designed according to the following rule of thumb,

$$\zeta' = \frac{R_{G(c)}}{2}\sqrt{\frac{C_{iss(c)}}{L_{S(c)} + L_{G(c)}}} > 0.4. \tag{6.2}$$

With a damping factor greater than 0.4, the gate voltage ringing will be kept within acceptable amplitude levels to avoid spurious turn-on and turn-off.

With regard to the CtrlFET source inductance, presently achieved values around 100 pH with advanced packaging techniques appear to be suitable for switching frequencies up to 5 MHz. The reason for this is the presence of L_{HB}, which is considerably higher than $L_{s(c)}$, and which primarily limits the rate of change of the drain current during the hard-switching times. The current source effect of the gate inductance also helps mitigating the feedback action of the source inductance and, as a consequence, a significant reduction of L_G may be detrimental to the overall efficiency of the converter in some cases.

On the other hand, optimization results clearly indicate different driving demands for the turn-on and turn-off transients of the CtrlFET. Namely, while the hard-switching turn-on transient can be extremely fast, e.g., less than 2 ns, the turn-off time must be moderated so as to avoid avalanche breakdown triggered by the action of L_{HB}. Consequently, the optimum gate driver current for turn-on and turn-off can be drastically different, as illustrated in Fig. 6.23 for the Trench4 design case of Sect. 6.1.1 and Fig. 6.24 for the Trench6 design case of Sect. 6.1.2. In either case, the demanded gate current increases as L_{HB} lowers or the transfer capacitance increases since it implies faster switching requirements in order to maintain equal switching losses. The turn-on and turn-off gate currents may differ by more than a factor of 2. As a consequence of the parasitic gate charge reduction, Trench6 technology offers the advantage of lower gate current requirements compared to Trench4.

A possible circuit implementation for asymmetric gate current operation may employ MOSFET switches of different ON resistance to form the gate driver

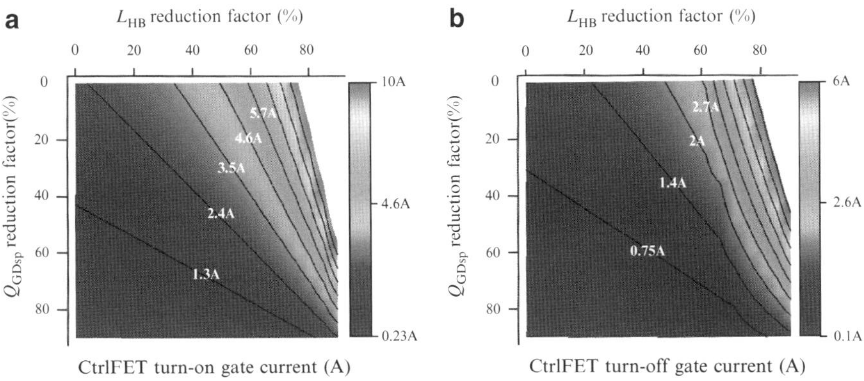

Fig. 6.23 Gate driver current requirements (in absolute values) as function of Q_{GDsp} and L_{HB} relative to Trench4 and the recommended PIP212-12M layout design (i.e., L_{HB} = 1 nH) (a) CtrlFET turn-on gate current, (b) CtrlFET turn-off gate current

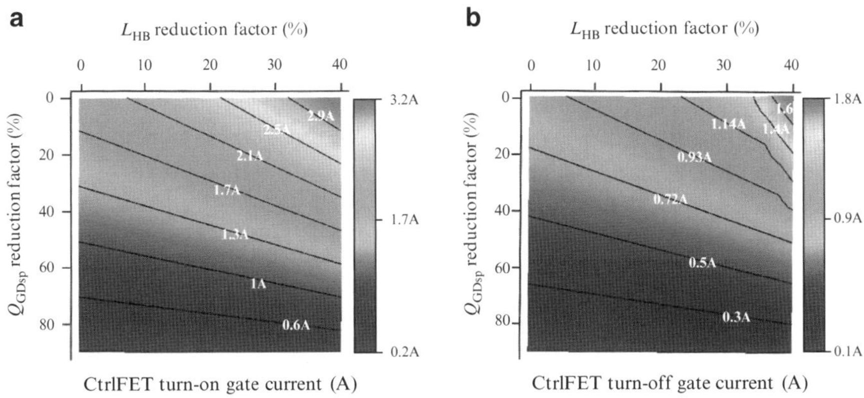

Fig. 6.24 Gate driver current requirements (in absolute values) as function of Q_{GD} and L_{HB} relative to Trench6 and the recommended PIP212-12M layout design (i.e., L_{HB} = 1 nH) (a) CtrlFET turn-on gate current, (b) CtrlFET turn-off gate current

half-bridge. In particular, the upper gate-driver switch may feature a lower resistance than the lower switch.

In line with the aforementioned requirements, Table 6.4 summarizes basic gate driver specifications for IPM solutions to address the demands of next generation computer microprocessors. In comparison to the existing solutions, next generation gate drivers may need to feature improved gate current capabilities, asymmetric turn-on/turn-off driving speeds, lower CtrlFET driver voltage and reduced parasitic gate impedance.

In addition to the CGD topology, RGDs have been identified as alternative solutions to reduce the gate driver losses of the SyncFET. The characteristics of these gate driver schemes, and more particularly, the proposed topology for desktop applications (see Sect. 6.1.2), are described in Appendix F.

Table 6.4 Existing and proposed gate driver specifications for IPM

		Existing solution	Next generation
Total gate resistance (Ω)	CtrlFET	2.3	0.75 (LE), 1.5 (FE)
	SyncFET	1.2	0.5
Driver voltage (V)	CtrlFET	12	6
	SyncFET	5	5
Source inductance (nH)	CtrlFET	0.1	0.1
	SyncFET	0.1	0.1
Gate inductance (nH)	CtrlFET	3	1
	SyncFET	3	1

6.3 Packaging

According to the identified improvement options, future IPM package developments should aim at reducing the parasitic inductances of both half-bridge and gate driver loops. As already identified, the predicted performance of Fig. 6.14 is partly achieved provided a reduction of L_{HB} from 1.6 to 1 nH. Further lowering this inductance down to 700 pH yields additional performance gains as Figs. 6.18 and 6.19 illustrate. Recent studies on packaging technologies have shown that L_{HB} can be as low as 500 pH by means of techniques such as embedded technologies [12, 13], which allow avoiding the use of wire bonds. These same techniques can be applied for the interconnection of the gate driver with the power MOSFETs, thereby reducing the gate inductance by at least a factor of three.

With the dramatic reduction in $R_{DS(on)sp}$ offered by the emerging MOSFET technologies, packaging parasitic resistance may additionally start playing a major role, particularly, when skin effects in the package leads and wire interconnects become significant at the operated switching frequency. Thus, the use of packages like the SOT684-4 from the PIP212-12M may be inadequate as the actual bond wire resistance of the SyncFET path (see Table 2.8) may contribute with more than 50% of the losses in that device.

Besides the benefits on loop inductance reduction, wireless packaging techniques like [14] and other copper straps based approaches can potentially minimize the ON conduction losses in cases where similar footprints are to be maintained and where the selected MOSFET technology may feature low $R_{DS(on)sp}$ in the range of 5–10 mΩ mm^2. Upon these conditions, and in order to achieve a good exploitation of these ultra-low ohmic MOSFETs, the AC resistance of future packaging solutions shall be below 100 μΩ.

6.4 Passive Filters

Two development activities on passive filter components demand special attention. The first one makes reference to the output inductor coil and the need to optimize miniaturized power cubes featuring (as deduced in Sect. 6.1.1) a target inductance of 40 nH and a nominal current of 35 A. Although some existing commercial solutions [15] already offer compact designs and reduced footprints, the expected power dissipation of these parts implies the need of enhanced thermal dissipation means such as forced convection.

To eliminate this likely inconvenience, the reduction of skin effect losses in the coil windings is presented as a main improvement measure given the significance of its contribution (e.g., up to 40% at heavy load). This is, according to the device physics simulations from Appendix E, partly attributed to a strong current crowding localized in the area nearby the airgap. Related negative implications of airgaped inductors on power losses are reported in [16, 17]. These results further suggest that the use of alternative magnetic cores with distributed airgaps may enhance current distribution and hence reduce the intrinsic dynamic resistance [18–20]. This is particularly of primary importance given the relatively high ripple content in the inductor current that is enforced by design in order to minimize the filter size.

Combined with a wider geometric design that guarantees proper magnetic shielding, conduction losses can be considerably mitigated, thus reducing the overall inductor loss by as much as 300 mW, according to the model estimations.

The second most relevant development activity related to passive filter components are the advances in bulk capacitor technologies. As previously stated, bulk aluminum electrolytic capacitors in can package form are progressively being replaced by polymer electrolytic capacitors featuring high and stable capacitance, low ESR and surface mount with reduced height and footprint. Existing products are already suitable to fulfill next generation VR requirements in terms of area occupancy and filtering capabilities against load transient responses. These capacitors can be effective for the suppression of the longest transient times of the output voltage. This is the case even for designs leading to relatively high ESL, thus making feasible their location outside the microprocessor socket area, and even on the bottom side of the mainboard. The number of bulk capacitors strongly depends on the number of phases as well as on the output filter inductance. As the latter lowers below 50 nH, the bulk capacitor count can be reduced to ten and below, as shown in Fig. 6.13.

SMD ceramic capacitors continue to be essential to filter out the high frequency output voltage ripple as well as the load current slew rate. Of vital importance is to develop extremely low ESL and ESR ceramic capacitors with minimized footprint so that they can fit inside the CPU socket cavity. In line with the previous estimations, the available X7R materials can already deliver the demanded performance provided that both top and bottom mainboard layers can be utilized. This, however, raises the concern as to whether the capacitors soldered on the bottom layer may show excessively high ESL, in which case lower footprint packages might be required in order to fit all capacitors on the top layer and within the socket cavity.

Table 6.5 summarizes the differences between an existing and the proposed solution to address the challenges of next generation computer microprocessors.

Table 6.5 Existing and proposed passive filter component characteristics for next generation computer microprocessors

		Based on existing technology[a]	Proposed solution[b]
Output coil	Inductance (nH)	150	40
	Footprint (cm^2)	1.1	0.57
	Volume (cm^3)	0.78	0.28
	Maximum power loss (W)	0.8	0.5
Bulk capacitors	Capacitance (μF)	560	390
	ESL (nH)	5	3
	Min. ESR (mΩ)	4	2
	Footprint (mm^2)	40	31.4
	Volume (mm^3)	0.15	0.09
	Required number	26	14
Ceramic capacitors	Capacitance (μF)	10	4.7
	ESL (nH)	0.8	0.6
	Min. ESR (mΩ)	3	4
	Footprint (mm^2)	9	8
	Volume (mm^3)	0.025	0.016
	Required number (output filter only)	50	55

[a]Trench4 MOSFET technology. See Fig. 6.1 for the case of $N_p = 11$
[b]According to Table 6.1

6.5 Control

The VR optimization at heavy load, though leading to the minimum possible temperature rise, can severely penalize the performance at moderate and light loads due to excessive switching losses. To mitigate this effect it is common practice to perform the optimization at medium load.

An alternative approach to keep the highest possible efficiency in a much wider load range without compromising the performance at heavy loads is phase shedding operation. Introduced in Chap. 1, phase shedding offers a control of the active phases as function of the load current. At light load, ON conduction losses become minor, and as such, the number of active phases can be reduced to effectively enhance conversion efficiency by reducing the switching losses, which dominate at low and moderate load.

One main drawback of this technique is that the advantage of ripple current cancellation of the multiphase interleaving operation may be jeopardized since the filter design under the new worst-case scenario (i.e. light load) may consider a reduced number of phases. Nonetheless, recalling the design conditions of Sect. 5.1, the high frequency decoupling capacitor count may largely be defined by the load current slew rate rather than the output voltage ripple. When this condition is met, the impact of phase shedding operation on the filter capacitor size may be low.

Furthermore, any moderate increase of the output voltage ripple under phase shedding operation might be effectively canceled out by a corresponding increase of the switching frequency at light load provided that ZVS is achieved in the LE transient (see case example of Sect. 5.4.2). Thus, the combination of phase shedding and switching frequency as function of the load current may allow maximizing the efficiency throughout the entire load range without compromising the benefits of multiphase interleaving with regard to the output filter size.

Comparing the efficiency curves of Figs. 6.25 and 6.26 clearly evidences the advantages of phase shedding to enhance the converter performance practically in the entire load range (except in deep light load). In the illustration, this is achieved in detriment of the number of required output filter ceramic capacitors, which increases from 55 to 70 (since the switching frequency is kept constant and independent on the load).

In view of this outcome, phase shedding control is revealed as an additional control feature of future IPM developments, which may need to be combined with nonlinear hysteretic control schemes in order to produce the above estimated results and further be complaint with the specifications of the load.

6.6 Layout Arrangement

As seen thus far, next generation MOSFET and IPM technologies will enable meeting the target VR specifications of Table 6.1. As a remarkable result, the power density of the proposed VR solution will be such that the hardware implementation may not require additional motherboard area than the CPU load and the associated heatsink already impose.

Figures 6.27–6.34 illustrate one possible PCB layout arrangement of the presented solution where all VR power elements (except the input filter inductor) are

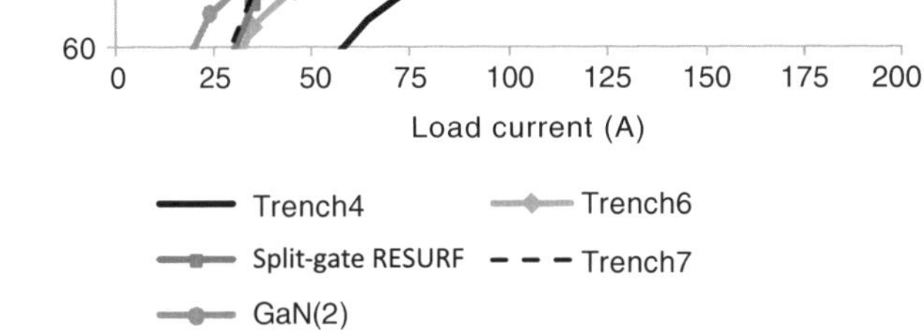

Fig. 6.25 Efficiency curves corresponding to various power device technologies under the conditions of Table 6.1

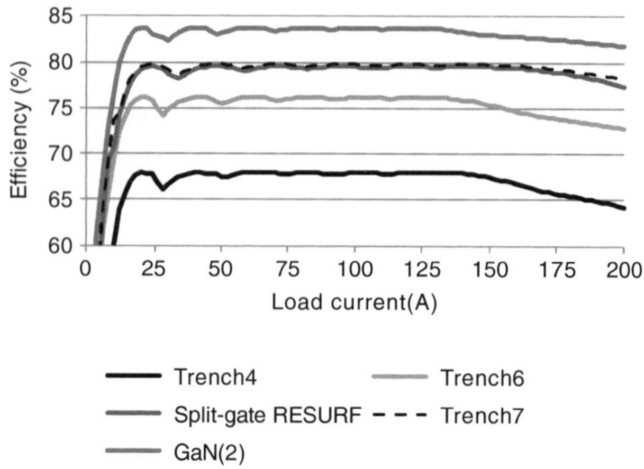

Fig. 6.26 Efficiency curves corresponding to various power device technologies under the conditions of Table 6.1 (except that the MLCC count increases from 55 to 70) and with phase shedding control

fitted underneath the microprocessor heatsink and within the permissible boundary limits defined by its fixture design according to Intel's specifications (see [1], Chap. 1) [21]. Each VR phase consists out of one IPM IC with same footprint as the PIP212-12M from NXP Semiconductors [22], and a 40 nH power coil from the SCL7649 series of Coilcraft [23].

Four out of the six phases lay along one of the longest sides of the microprocessor socket, where the available PCB area is the largest, with $\sim$15 cm^2. The remaining two phases are placed right next to one of the shortest sides of the CPU socket with an available area of 9 cm^2 and a maximum height of 10 mm. Without exceeding this maximum vertical distance, two dedicated low profile heatsinks may be designed for the groups of four and two IPM ICs according to the depicted sketch.

Besides the advantage in PCB area reduction compared to existing solutions (see Sect. 1.2.9), the close proximity and distributed arrangement of the VR phases relative to the CPU socket may notably enhance the heat dissipation of both power coils and IPM ICs as they benefit from a more direct airflow produced by the CPU fan.

These gains are conditioned by a suitable exploitation of the mainboard's bottom layer, where all bulk capacitors corresponding to the output filter are orderly located right underneath the VR phases. In line with the specifications, the bulk capacitors must be low profile and have a maximum height of $\sim$2.5 mm. The polymer electrolytic capacitors discussed previously are potential candidates for their compliance with both the physical and the electrical requirements.

Both top and bottom layers are used to accommodate the high frequency decoupling capacitors in the PCB. For the purpose of reducing the ESL, the most

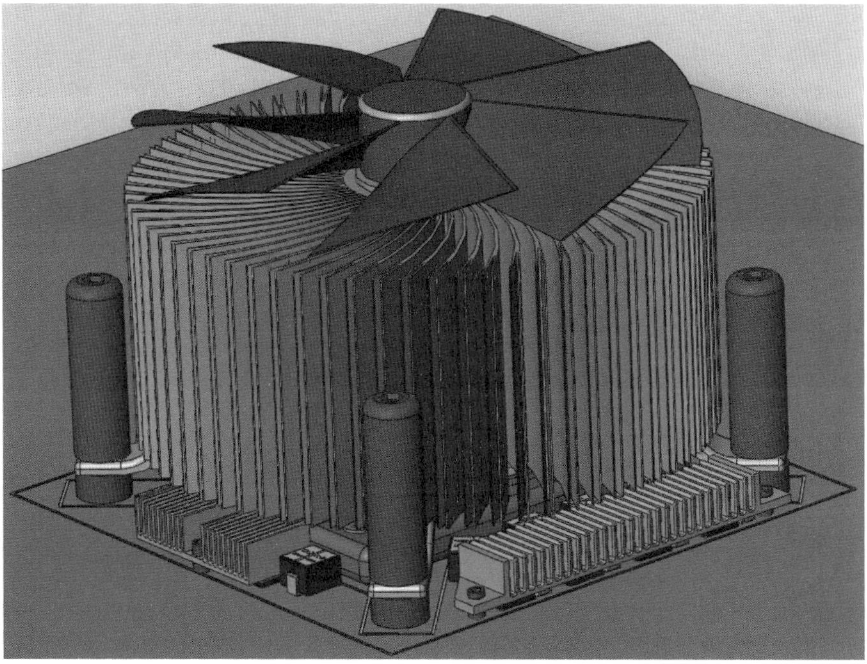

Fig. 6.27 3D view of a 6 phase VR layout arrangement underneath the microprocessor fan

Fig. 6.28 Top view of a 6 phase VR layout arrangement underneath the microprocessor fan

Fig. 6.29 Lateral view of a 6 phase VR layout arrangement underneath the microprocessor fan

Fig. 6.30 Top view of a 4 + 2 phase VR layout arrangement. Low profile heatsinks used to enhance heat dissipation of the IPM ICs

Fig. 6.31 Top view of a 4 + 2 phase VR layout arrangement without heatsinks

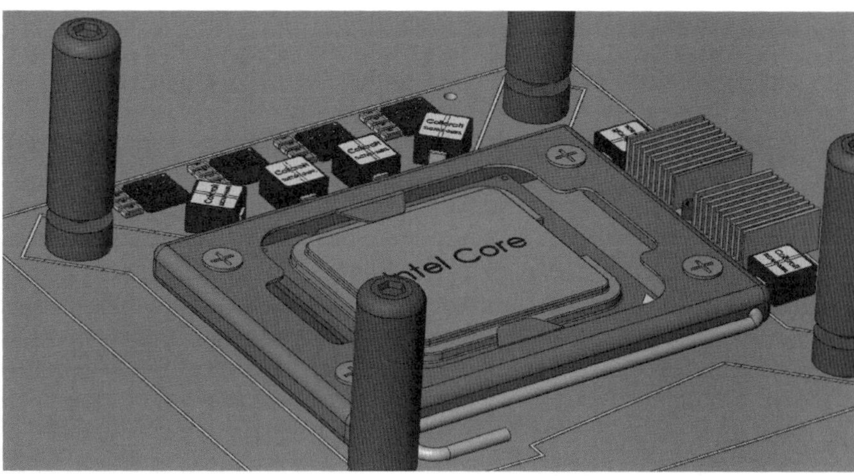

Fig. 6.32 3D view of a 4 + 2 VR layout arrangement without microprocessor heatsink

suitable location is the center socket window (the top view is not shown in the drawings) with an area of ~2.6 cm^2 per layer. Up to 28 ceramic capacitors of the size 1,206 can fit in each layer.

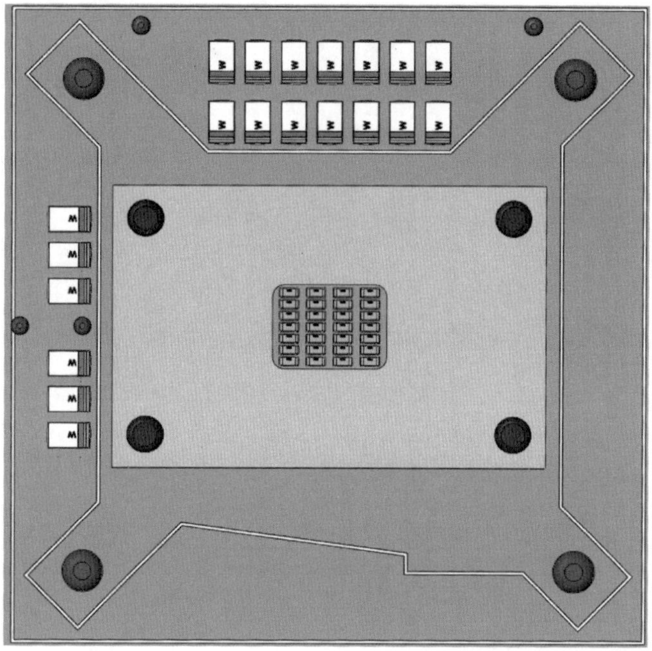

Fig. 6.33 Bottom layer top view of a 4 + 2 phase VR layout arrangement. Ceramic capacitors located in the PCB heatsink window for ESL reduction. Low profile polymer electrolytic capacitors located on the sides of the PCB heatsink

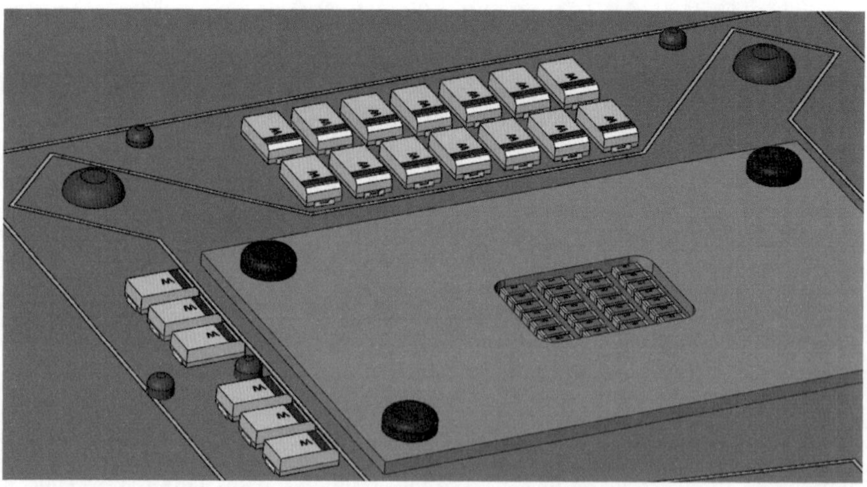

Fig. 6.34 Bottom layer 3D view of a 4 + 2 VR layout arrangements. Distribution of output filter capacitors

6.7 Mobile Laptop Applications

The technology demands for next generation laptop applications may differ to some degree from the ones previously identified for desktop and workstations. The following study establishes dedicated improvement options for mobile computer VRs. Unless otherwise specified, the analysis is based upon MOSFETs mounted in LFPak. This discrete solution employs two parallel devices as SyncFET and one as CtrlFET. The gate driver and half-bridge parameters are configured according to Table 2.4. The specifications of the application correspond to the year 2014 from Table 1.2.

Figure 6.35 shows the loss breakdown of a candidate solution employing three phases switching at 700 kHz. Since the MOSFET devices are 30 V rated and the maximum input voltage is 19 V, avalanche breakdown can only be avoided by significantly slowing down the CtrlFET switching speed, particularly that of the turn-off transient. This is done by adding a resistance of 0.5 Ω in the gate path, thus resulting in a total gate resistance of 2 Ω at turn-on and 2.5 Ω at turn-off. The moderate switching speed explains the dominating contribution of $W_{\text{ioHS(c)}}$, which is slightly higher than the ON conduction losses.

Despite of the low Q_{oss}, another significant contributor due to the high-switching frequency and increased input voltage is $W_{\text{ioHBC(s)}}$. This portion can get even higher than the CtrlFET turn-on hard-switching transient.

Largely attributed to L_{HB}, the CtrlFET turn-off hard-switching loss is the largest contribution from the overall switching losses. Consequently, L_{HB} represents, together with $R_{\text{DS(on)sp}}$, the most important improvement target parameters. Along

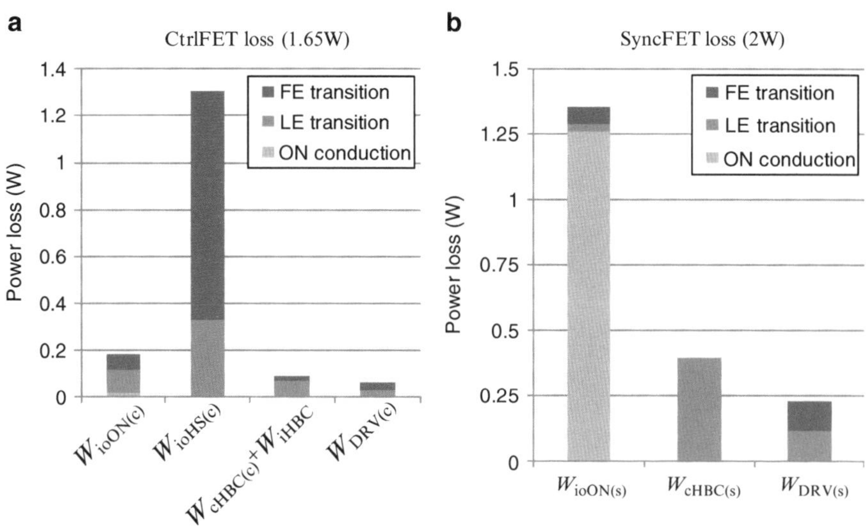

Fig. 6.35 Trench6 power MOSFET loss breakdown. $F_s = 700$ kHz, $N_p = 3$, $L_o = 150$ nH (a) CtrlFET loss breakdown, (b) SyncFET loss breakdown

with these measures, the reduction of $C_{oss(s)}$ and its strong non-linear voltage dependence may be considered to both reduce charging-up losses and voltage stress.

The significance of hard-switching losses renders the gate drive loss the least relevant contribution, as it can be deduced from the loss breakdown of Fig. 6.36. In contrast to desktop applications, hard-switching loss represents the largest portion of all loss contributions.

Simulations indicate that a 60% reduction in L_{HB} (from 2.8 to 1.1 nH) can reduce hard-switching loss by almost 50%, which represents about 25% reduction of the total MOSFET losses. Combining this measure with a reduction in $R_{DS(on)sp}$ by 50% with respect to Trench6 yields an efficiency improvement of approximately 7%.

Table 6.6 overviews the general requirements for the next generation laptop applications based on two different solutions. The first approach is implemented with currently available discrete devices, which results in a VR featuring 83% efficiency and a maximum power density of 7 W/cm^3 with three interleaved phases. The main limitation of this solution is the high number of bulk capacitors required to fulfill the load regulations. In order to minimize the capacitor count and increase the power density, both output filter inductance and switching period may be reduced. Accordingly, the proposed solution aims at halving the output inductance and increasing the switching frequency to 700 kHz, thereby achieving 30% higher power density and a drastic reduction of the bulk capacitors. The converter efficiency at heavy load lowers only slightly with the replacement of the Trench6 by the Trench7 technology and by enabling the use of the PIP212-12M package for 20 V input voltage applications. This would effectively reduce L_{HB} from 2.8 nH to around 1.6 nH.

Additional gains in efficiency and power density may imply the use of 40 V devices to allow for higher switching speeds. Similar measures as proposed in the previous sections may further be adopted in laptop applications as well (e.g., phase shedding).

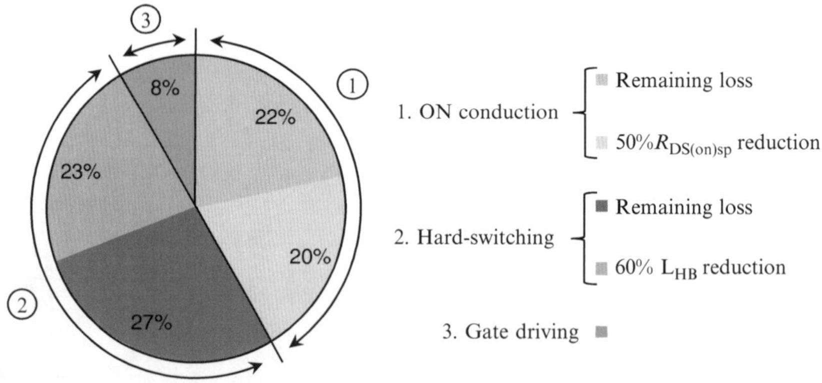

Fig. 6.36 Improvement options on the case-example of Fig. 6.35 to achieve ~85% efficiency in next generation VR laptop solutions

Table 6.6 Proposed roadmap for next generation laptop applications.

	Existing technology	Proposed solution
Package	Discrete LFPak	SOT684-4
MOSFET technology (NXP Semicon.)	Trench6	Trench7
Output inductor (nH)	300	150
Number of bulk capacitors	18	7
Number of phases	3	3
Switching frequency (kHz)	300	700
Efficiency at full load (%)	83	82
Power density (W/cm^3)	7	10

References

1. Intel® (2009) Voltage regulator-down (VRD) 11.1, processor power delivery design guidelines, Document number: 322171-001, Sept 2009
2. N-channel TrenchMOS logic level FET PH3330L, Trench4 technology NXP Semiconductors, Feb 2006
3. Peake ST, Rutter P, Hodgskiss S, Gadja M, Irwin N (2008) A fully realized "field balanced" trenchMOS technology. In: Proceedings of the 20th international symposium on power semiconductor devices & ICs, Orlando, 18–22 May 2008, pp 28–31
4. N-channel TrenchMOS logic level FET PSMN1R7-30YL, Trench6 technology NXP, Sept 2008
5. Goarin P, Koops GEJ, van Dalen R, Le Cam C, Saby J (2007) Split-gate resurf stepped oxide (RSO) MOSFETs for 25V applications with record low gate-to-drain charge. In: Proceedings of the 19th international symposium on power semiconductor devices & ICs, Jeju Island, 27–31 May 2007, pp 61–64
6. N-channel PowerTrench® (2009) MOSFET FDMS7650, Fairchild Semiconductors, Aug 2009
7. Rutter P, Peak ST (2010) Low voltage TrenchMOS combining low specific R_{DSon} and Q_{G} FOM. In: Proceedings of the 22nd international symposium on power semiconductor devices and ICs, ISPSD, Hiroshima, 6–10 June 2010, pp 325–328
8. http://www.irf.com/product-info/ganpowir/. Accessed 12 Oct 2010
9. Briere MA (2009) GaN-based power device platform. In: Power systems design north America, Jan 2009, http://www.powersystemsdesign.com
10. Briere MA (2008) GaN based power devices: cost-effective revolutionary performance. Power Electron Europe 7:29–31
11. Briere MA (2008) High frequency GaN-based power conversion stages. In: Invited talk at the first international workshop on power supply on chip, Cork, Ireland, 22–24 Sept 2008
12. Chin R, Chia TS, Wan K, Houng T, Peels W (2008) Development of flex-based embedded actives packages. In: Electronic circuit world convention, vol 11, Shanghai, March 2008
13. Peels W, Heyes D, Kengen M (2009) Embedded die technology, next generation packaging for discrete semiconductors. In: SEMICON Europa, Dresden, Germany, Oct 2009
14. Bencuya I, Estacio MCB, Sapp SP, Tangpuz CN, Baje GS, Maligro RD (2002) Low resistance package for semiconductor devices. US patent US6423623B1, 23 July 2002
15. Coilcraft. http://www.coilcraft.com/smpower.cfm. Accessed 25 Oct 2010
16. Ngo KDT, Kuo MH (1988) Effects of air gaps on winding loss in high-frequency planar magnetics. In: 19th power electronics specialists conference, PESC 1988, vol 2, Kyoto, Japan, 11–14 Apr (1988), pp 1112–1119

17. Severns R (1992) Additional losses in high frequency magnetic due to non ideal field distributions. In: 7th applied power electronics conference and exposition, APEC 1992, Boston, pp 333–338
18. Kutkut NH, Divan DM (1998) Optimal air-gap design in high-frequency foil windings. IEEE Trans Power Electron 13(5):942–949
19. Chew WM, Evans PD (1991) High frequency inductor design concepts. In: 22nd power electronics specialists conference, Cambridge, 24–27 June 1991, pp 673–678
20. Hu J, Sullivan CR (1998) Optimization of shapes for round wire, high frequency gapped inductor windings. In: IEEE industry applications society annual meeting, Oct 1998, pp 907–911
21. Intel® (2009) Intel Core i7-900 desktop processor extreme edition series and Intel Core i7-900 desktop processor series and LGA1366 Socket, thermal and mechanical design guide. Document number: 320837-003, Oct 2009
22. NXP Semiconductors (2006) DC-to-DC converter powertrain. Datasheet PIP212-12M, 2006. www.nxp.com
23. Coilcraft, SMT power inductors – SLC7649 series. Document 481-1. www.coilcraft.com

Chapter 7
Conclusions and Future Work

The contributions of this thesis work can be summarized as follows:

- Development of practical power MOSFET and synchronous rectifier buck converter (SRBC) models for investigating switching phenomena in high-switching frequency VR applications.
- Implementation of a generalized design methodology to aid the development of power conversion technologies, in particular, modern power MOSFETs and integrated power modules for high current, low voltage, and high-switching frequency operation.
- Comprehensive description of switching phenomena in high-switching frequency VRs, from which loss mechanisms can be identified, understood, and their effects quantified.
- Guidelines for the design optimization of VRs.
- Identification of both power density and efficiency limitations of conventional VR topologies to power next generation computer microprocessors.
- Establishment of roadmap targets to fulfill the power demands of next generation computer microprocessors.

The SRBC has been studied as a power solution to fulfill the energy demands of next generation computer microprocessors. A number of relevant technology performance requirements have been identified that, once available, will make the SRBC-based multiphase architecture a potential candidate to power modern desktop, workstation, and server CPUs in the years to come.

Predictions indicate that efficiencies over 75% at full load can be achieved in next generation desktop CPUs. With this minimum acceptable efficiency (according to Intel® specifications), the occupied mainboard area may be such that the entire VR can fit underneath the processor's heatsink. This is provided that

- At least six VR phases are interleaved (i.e., up to 33 A per phase), each of which comprises one IPM IC.
- $R_{DS(on)sp}$ is further reduced by at least 30% w.r.t. the recently developed 30 V Trench6 MOSFETs (from NXP Semiconductors) without compromising the switching performance of this technology. This may imply the following figures to be met: $R_{DS(on)sp} < 10$ mΩ mm^2, $Q_{GD}R_{DSon} < 15$ nC mΩ, and $Q_{Gt}R_{DSon}$

T. López et al., *Voltage Regulators for Next Generation Microprocessors*, 279
DOI 10.1007/978-1-4419-7560-7_7, © Springer Science+Business Media, LLC 2011

< 60 nC mΩ. This allows keeping the maximum die area within the limits of existing IPM footprints (e.g., <17 mm^2).
- The gate driving and input impedance of power MOSFETs enable switching frequencies of $\sim$2 MHz with extreme low duty cycles (conversion from 12 to 0.5 V).
- Dedicated efficient gate drive voltage supply (i.e., 5 V in both MOSFETs) can be arranged.
- Package inductance of IPM reduces further to meet $L_{HB} < 1$ nH.
- Reliable adaptive deadtime control at $F_s \leq 2$ MHz is enabled to eliminate body diode conduction and shoot-through.
- Gate bounce susceptibility is low enough to avoid SyncFET spurious turn-on losses.
- Nonlinear hysteretic control (and additionally phase shedding control for improved performance at light load) is chosen as main regulation scheme.
- Use of low profile bulk capacitors such as polymer electrolytic mounted on the bottom layer of the mainboard.

Forecasted MOSFET and IPM technologies such as GaN power switches and embedded active packages may potentially enable these requirements and thus represent imminent alternatives to presently exploited mature silicon technologies.

These findings are predictions primarily based on simulations. Experimental results (once the technology is available) should be the most obvious follow-up to this thesis work.

Essential to achieving the demanded targets are the integrated powertrains, the performance advantages of which have been clearly identified along the discussions. These IPM solutions represent the potential pathways to enable faster switching, reduced parasitic inductances, smaller footprint as well as the effective exploitation of power saving functionalities such as automatic deadtime reduction.

Additional identified improvement options include the implementation of a resonant gate driver topology for the SyncFET capable of reducing driver losses by as much as 40% with a 6% penalty in ON conduction losses. These results are possible provided that the resistance of the gate driver path can be minimized with respect to present typical values. One of the most relevant gate resistance contributors is that of the polysilicon gate in current MOSFET devices, which should be reduced down to at least 0.2 Ω so as to make resonant gate driver topologies effective in this application.

Regarding optimization guidelines, it has been argued that the switching frequency required to achieve a given target converter size (and/or dynamic response) can be effectively minimized by allowing higher ripple current operation. This is particularly accentuated for extreme duty cycle operation and in those cases where the inductor size dominates the VR volume. Within certain limits, the output current ripple increase happens to be generally more efficient than the otherwise increase of the switching frequency needed to achieve the same intended performance. The output filter capacitor count is not generally affected by a moderate increase of the output current ripple as long as it is the load transient response (rather than the steady-state ripple) what mainly determines the size of the capacitor bank. For the conditions considered herein, optimum coil inductances can effectively reach values down to 40 nH at 2 MHz, and 8 nH at 9 MHz with an optimum current ripple of ~7 A (i.e., ~20% of the maximum current per phase). Accordingly, in order to achieve the maximum effectiveness, low inductance power coils should be optimized for these high frequency and high-current ripple waveforms.

Of significant importance in the optimization process is the CtrlFET gate drive current. The benefits of arranging individual gate drive paths to enable different turn-on and turn-off speeds have been identified. The limited switching speed capabilities of both driver and MOSFET can make the existing parasitic half-bridge loop inductance and output capacitance advantageous to minimize the switching losses. In order to achieve the intended snubber effects, the optimum turn-on and turn-off speeds must be adjusted according to the value of these parasitic elements. The lower these become (by means, for instance, of improvements in device and integration technologies), the higher will be the demands for shorter switching times.

Upon such optimum operating conditions, a loss breakdown of commercial low inductance airgaped power coils for VR shows that winding losses may dominate over core losses at heavy load. The significance of the copper losses despite of the extremely low DC resistance may be attributed to current crowding effects in the conductor region near the airgap. Distributed airgap cores may emerge as preferred alternatives to enhance current uniformity, hence reducing conduction loss. Further investigations on power losses and heat transfer in magnetic components may be of relevant interest as switching frequencies enable the use of miniaturized power coils that may be integrated within the semiconductor module.

Achieving significant efficiency gains with the use of existing technologies may only be possible by way of alternative ZVS transformer-based topologies, such as resonant converters. This is particularly the case in laptop applications, as the large voltage conversion ratios represent one of the main barriers against high-switching frequency operation. This is supported by the following conclusive switching analysis results of IPM-based solutions:

- The half-bridge loop inductance triggers avalanche breakdown in both switches. In order to avoid it, the switching speed, and hence the switching frequency, may need to be limited. The optimum switching speed may primarily depend on the avalanche voltage level, the load current (mainly at turn-off), the input voltage, and the resonant circuit formed by the output capacitance and the loop inductance. The nonlinearity of the output capacitance further increases the overshoot voltage stress.
- Reducing the parasitic half-bridge loop inductance is the most effective measure to mitigate turn-off switching losses at high currents. In most cases, lowering L_{HB} further represents the most relevant improvement option as switching off losses frequently dominate.
- Because of the overshoot spikes, the maximum voltage rating of the power MOSFETs should be limited to >30 V in most cases. In laptop applications the voltage rating may be even higher (>45 V).

- Overshoot spikes can be mitigated by reducing the nonlinearity of the MOSFETs output capacitances (for instance, with the parallel connexion of a linear capacitance).
- The SyncFET output charge may produce (due to the large chip size) significant switching losses in the leading edge transient, particularly at high input voltages. The nonlinearity of the SyncFET output capacitance further increases these charging losses.

For their intrinsic features, all of the above identified issues of the SRBC can be avoided with ZVS topologies, which may justify an in-depth exploration of resonant converters in future VR-related research.

Proper VR operation relies on measures to shun body diode conduction and gate bounce: The first measure is accomplished by means of IPM equipped with integrated adaptive dead time control that effectively allows avoiding body diode ON conduction and reverse recovery losses. The latter loss mechanism can vanish without enforcing shoot-through. Namely, according to the *body effect*, it is sufficient that the gate voltage stays above zero volts and within the subthreshold region prior to the CtrlFET turn-on. Therefore, the body effect relaxes the required accuracy of the dead time control to avoid reverse recovery.

On the other hand, automatic deadtime control increases the susceptibility of the SyncFET to gate bounce shoot-through since the gate-source capacitance at the time of the CtrlFET turn-on is not fully discharged. The proposed definition of gate bounce susceptibility takes into account the initial conditions in the device and the nonlinearity of the transfer capacitance. Of primary importance is to insure that the gate-drain capacitance is sufficiently lower than the total input capacitance such that, under worst-case conditions, the SyncFET channel cannot be spuriously turned on.

These conclusions rely on a number of modeling tools, the development of which have been one main focus of this thesis work. These models represent the foundations of the proposed co-design approach that allows examining the behavior of power MOSFETs in the converter circuit under various control schemes. This has been performed in an iterative development process flow referred to as *virtual design loop*. Therefore, the SRBC has been comprehensively analyzed to identify and quantify power MOSFET loss mechanisms. Combined with device physics simulations, an improved power MOSFET technology (Trench6) has been developed by NXP semiconductors, which has led to the validation of both the models and the methodology. The new MOSFET structure complements the extensive experimental characterization and measurement results presented along simulated data. As predicted by the virtual loop analysis, Trench6 offers significant performance improvements primarily by way of reducing ON state resistance and improved gate bounce susceptibility without significantly degrading any switching characteristics.

The *virtual design loop* is a general methodological concept that may be applied to other engineering disciplines. Its effectiveness lies on the accuracy and tractability of the models employed, which constitute the main advantages of the approach.

In this study, four different models have been developed and exploited for different purposes.

Model level 0 is presented as a behavioral circuit simulator model featuring high accuracy and allowing a detailed description of switching phenomena. The model has been crucial to understand switching behavior, identify switching loss mechanisms, and validate simplified models.

Model level 1 is analytical-based and describes switching waveforms by means of piecewise linear intervals. The model effectively represents the fundamentals of switching phenomena by ignoring second-order effects and sets the basis for the derivation of compact switching loss expressions. The accuracy of the model can be adapted to the simulation demands by increasing the number of concatenated linear intervals. Simulations results demonstrate that model level 1 can be accurate and significantly faster than the SPICE simulations of model level 0.

Model level 2 is based on simplified loss expressions mostly derived from model level 1. The equations, though being in some cases less accurate than the numerical computations of model level 0, allow a compact description of the converter performance from which parameter dependencies can be readily assessed both quantitatively and qualitatively.

Model level 3 results into FoM for circuit and device optimization aiming at a particular application. Thus, all relevant circuit and device parameters are included in the expressions. FoM units are expressed in joules so that different technologies may be compared in terms of losses under optimum conditions. These FoM derive from simplifications of the equations from model level 2. Expressions approximate the optimum area of both SyncFET and CtrlFET for a set of electrical conditions such as the driver voltages.

Combining these FoM with the loss estimations of both the PCB layout and filter components sets out a series of VR design guidelines that allow a rigorous exploration of the performance limitations of existing and future technologies. Based on various design criteria, these guidelines offer a systematic procedure to optimize converter circuit. This includes the estimation of parameters such as the number of phases, the switching frequency, and the filter size.

Appendix A
Third Quadrant DC Output Characteristics of Low Voltage Trench MOSFETs

The advantages of higher switching frequencies in SMPS cannot often be realized due to switching losses. Thus, a fundamental understanding of switching transients has become a key goal to future optimal high frequency SMPS designs.

Today, device modeling and circuit simulators are fundamental tools to carry out transient analysis and power loss calculations of rather complex nonlinear networks. For the synchronous buck converter used in VR applications, major concerns are the power losses generated in the SyncFET, particularly, reverse recovery and conduction losses in the body diode. Several studies on these loss contributions have been published, stating the importance of reducing them and proposing techniques to achieve it. However, in order to come up with effective measures to mitigate SyncFET losses, further insights of the third quadrant operation are required.

This section deals with large die size trench MOSFETs for high current, high frequency applications, aiming at a comprehensive study of the device static behavior in the third quadrant. The investigations demonstrate that body diode conduction is strongly influenced by significant sub-threshold reverse channel conduction. This intrinsic effect is fundamental and allows explaining some switching transient behavior, as it shall be seen in Appendix B. Therefore, the inclusion of the third quadrant output characteristics in any circuital MOSFET model leads to a more accurate device description and improved power loss predictions.

A.1 DC Output Characteristics

Semiconductor manufacturers typically specify the DC output characteristics of power MOSFETs in datasheets, i.e., drain current i_D vs. voltage v_{DS} at different v_{GS} voltages. Yet, such specification only refers to the operation in the first quadrant, i.e., positive v_{DS} voltage. Regarding the third quadrant, i.e., voltage v_{DS} is negative, only the body diode forward characteristic is usually specified for zero volts v_{GS}. No further information about the channel current in the third quadrant and its v_{GS} dependence is provided.

T. López et al., *Voltage Regulators for Next Generation Microprocessors*,
DOI 10.1007/978-1-4419-7560-7, © Springer Science+Business Media, LLC 2011

In circuit simulations where this information may be relevant, it is usually assumed that the first and third quadrants are symmetric and that the body diode forward characteristic is independent on v_{GS} [1, 2]. As shown in Fig. A.1, such assumption is not always valid. The plot depicts experimental results corresponding to the output characteristics of an N-channel power trench MOSFET (PHB96NQ03LT) for VR applications. The main characteristics of the measured device are:

- Maximum drain current $i_D = 75$ A.
- Maximum drain-to-source voltage $v_{DS} = 25$ V.
- 5 mΩ R_{DSon} at gate-to-source voltage $v_{GS} = 10$ V and junction temperature $T_j = 25°C$.
- Typical gate-to-source threshold voltage ($i_D = 1$ mA, $v_{DS} = v_{GS}$) $V_{GS(th)} = 1.5$ V at $T_j = 25°C$.

Note that the measurements are performed with a curve tracer featuring fast measurement acquisition times (<150 μs for each v_{GS} trace) and therefore self-heating effects in the device under test are minimized.

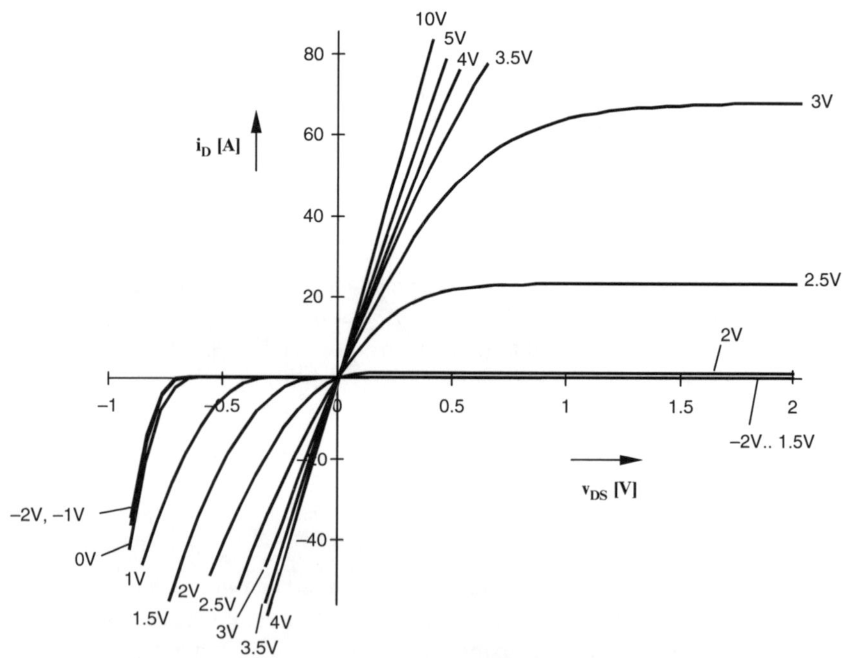

Fig. A.1 Measurement results of the output characteristics of a trench MOSFET device (PHB96NQ03LT from NXP Semiconductors). First and third quadrants are depicted for various v_{GS} at 25°C

From the measurement results it can be observed that:

- The body diode forward characteristic seems to be modulated by voltage v_{GS} in the sub-threshold region, even at negative v_{GS} values down to -1 V.
- At a given v_{GS} in the threshold region (i.e., voltage v_{GS} close to $V_{GS(th)}$), the drain current magnitude in the third quadrant is much larger than that in the first quadrant, also at low v_{DS}. For instance, at $v_{GS} = 2$ V and $v_{DS} = -0.5$ V the drain current reaches -40 A. In the first quadrant however, the maximum drain current at the same v_{GS} is about only a few amps.
- A symmetric characteristic between the first and third quadrant appears to be at high v_{GS}.

This characteristic behavior is not exclusive of trench MOS technologies as it may also occur in other types of vertical MOSFETs such as DMOS, CoolMOS^{TM}, and planar structures.

In the following subsections the third quadrant output characteristics is described in more detail by looking at the internal structure of the device and analyzing the origin of this significant reverse current conduction.

A.2 Device Physics Simulations vs. Measurement Results

MEDICI simulations are carried out to get further insights into the device's fundamental behavior in the third quadrant output characteristics. Current flow lines and carrier concentrations are analyzed under specific boundary conditions. MEDICI is based on a method approach to simulate complex semiconductor structures. A wide variety of device technologies like vertical MOSFETs can be analyzed. In this study, a trench MOS cell geometry corresponding to the PHB96NQ03LT device is modeled. Due to the symmetry of the structure, only half of the trench is modeled. The width of the channel (i.e., the third dimension) is supposed to be very large.

Figure A.2 compares measurement and simulation results of the DC output characteristics of the trench MOSFET. Note that the simulation data are scaled to the corresponding size of the MOSFET die. Package resistance is not included. The good matching between simulation and measurement demonstrates the high accuracy that the device physics simulator can offer with the simplified 2D geometrical representation.

A.3 Comparison of the First and 3Q DC Output Characteristics

Figures A.3 and A.4 represents 2D numerical simulated results in 3D graphs of the first and third quadrant DC output characteristics, respectively. The drain current difference between the first and third quadrant is depicted in Fig. A.5. For the

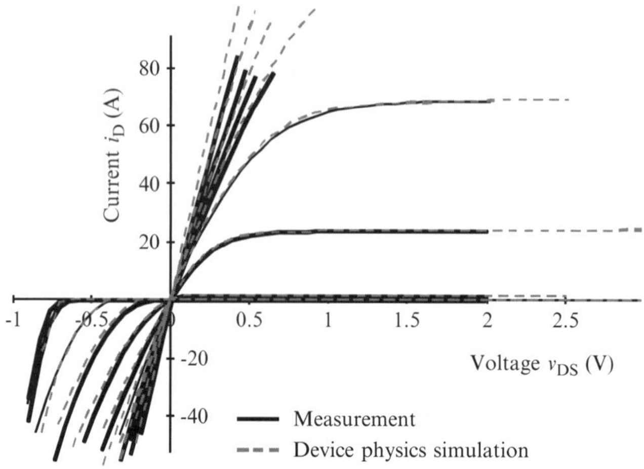

Fig. A.2 Comparison between the measurement and simulation of the output characteristics of a trench MOSFET device (PHB96NQ03L from NXP Semiconductors). First and third quadrants are depicted for various v_{GS}. Both measurement and simulation refer to 25°C

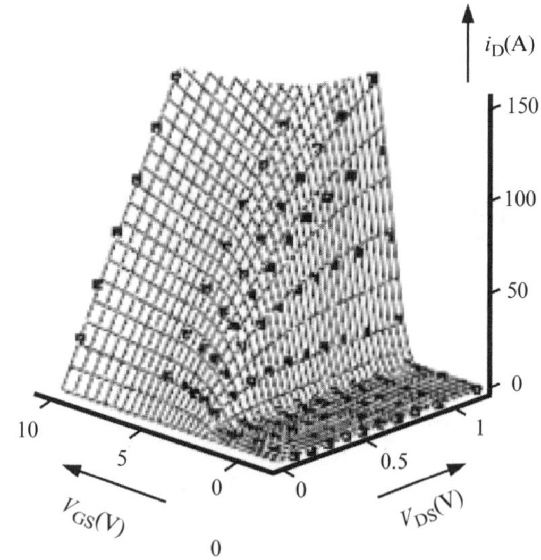

Fig. A.3 Simulated 3D representation of the first quadrant DC output characteristics of a trench MOSFET device (PHB96NQ03LT from NXP Semiconductors) at 25°C

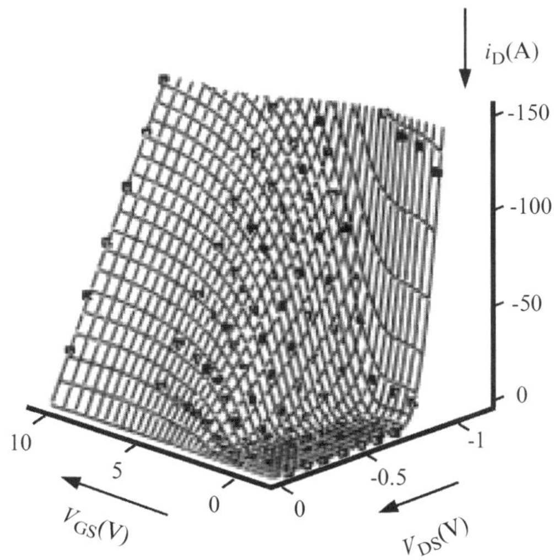

Fig. A.4 Simulated 3D representation of the third quadrant DC output characteristics of a trench MOSFET device (PHB96NQ03LT from NXP Semiconductors) at 25°C

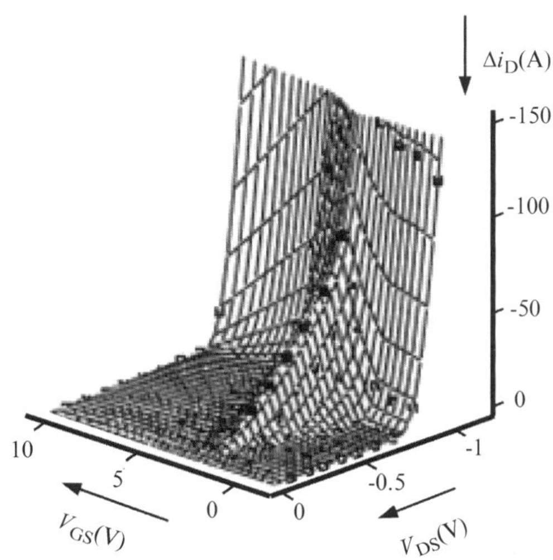

Fig. A.5 Simulated 3D representation of the current difference between third and first quadrant of the DC output characteristics of a trench MOSFET device (PHB96NQ03LT from NXP Semiconductors) at 25°C. Note that for the representation, the first quadrant is mirrored into the third quadrant

representation, the first quadrant current is mirrored onto the third one. Current difference Δi_D is therefore obtained by applying,

$$\Delta i_D = i_{D3Q} - i_{D1Q}|_{3Q}. \quad (A.1)$$

Current i_{D3Q} is the third quadrant drain current and $i_{D1Q}|_{3Q}$ the first quadrant drain current mirrored into the third quadrant.

Figure A.5 shows that the difference in current conduction between the first and third quadrants is not only caused by the body diode conduction above 0.7 V, but also due to a significant current modulated by v_{GS} in the threshold and subthreshold regions.

A further comparative study of the first and third quadrant can be made by representing the measurement results of Fig. A.1 in the way shown in Fig. A.6. Every curve in the graph corresponds to a specific drain current value and shows the required v_{GS} level to keep the drain current constant as function of v_{DS}. Therefore, the curves represent the switching trajectories of an inductive load application where the current is impressed into the devices rather than the voltage.

According to the trajectories, in the first quadrant every curve is expected to flatten out as the device enters the active region and v_{DS} rises. Each v_{GS} value where the curve saturates corresponds to a specific point in the MOSFETs transfer curve. Though, in the third quadrant, the curves do not flatten but continue to decrease even further. Thus, reverse current conduction occurs below the first quadrant threshold voltage and below the conventional body diode forward

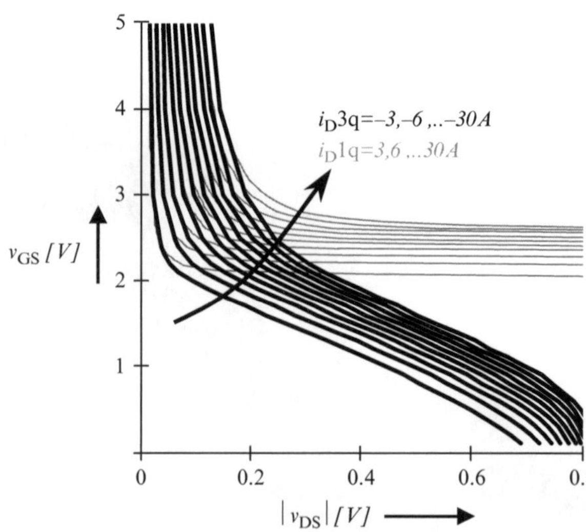

Fig. A.6 Representation of the gate-to-source voltage required to keep a certain drain current constant at different drain-to-source voltages. Comparison between the first (i_D1Q) and third quadrant (i_D3Q) operation. Measurement results at 30°C. MOSFET: PHB96NQ03LT from NXP Semiconductors

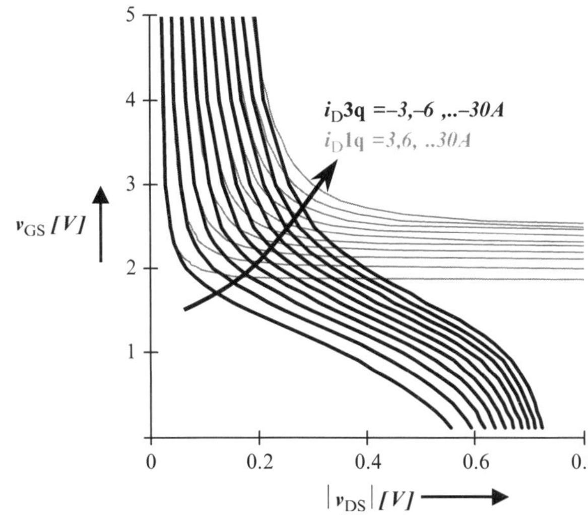

Fig. A.7 Representation of the gate-to-source voltage required to keep a certain drain current constant at different drain-to-source voltages. Comparison between the first (i_D1Q) and third quadrant (i_D3Q) operation. Measurement results at 125°C. MOSFET: PHB96NQ03LT from NXP Semiconductors

voltage conduction. A drain current value in the active region of the first quadrant at a given v_{GS} is therefore always lower than that in the third quadrant for the same v_{GS}.

At low v_{DS} the MOSFET enters the ohmic region, leading to a symmetric behavior between the first and the third quadrant.

Figure A.7 shows measurement results at 125°C. Temperature affects MOSFET parameters such as R_{DSon}, threshold voltage, and forward diode conduction. The positive thermal coefficient of the R_{DSon} makes the MOSFET to be in the ohmic region at higher v_{DS} with increased temperature. The threshold voltage reduction as temperature rises is reflected in the third quadrant curves by a reduction of v_{GS} in the active region. The negative thermal coefficient of the body diode forward voltage curve makes the diode current to be relevant at lower v_{DS} (absolute values) in the third quadrant.

A.4 3Q Current Flow Through a Trench Cell

The aim of the following simulations is to study the current flow in the third quadrant of the output characteristics. For this purpose, two important aspects are considered. First, the amount of electrons and holes that carry the current, and second the trench cell regions from which carriers flow. Figure A.8 depicts the ratio of electron to hole current at the drain ($q_{eh,\ drain}$) and source ($q_{eh,\ source}$) electrodes as function of $-v_{DS}$ (i.e., source-to-drain voltage) and for various v_{GS}. The results

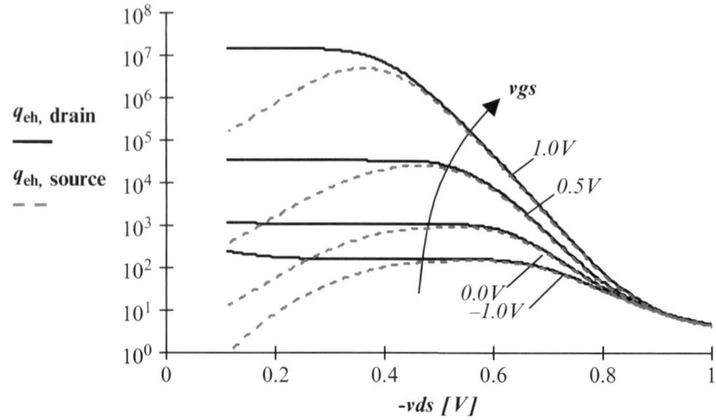

Fig. A.8 Ratio of electron to hole current at the drain and source contacts. Device: PHB96NQ03LT from NXP Semiconductors. Temperature: 25°C

Table A.1 Simulated electrical conditions for the current flow line graphs

Voltage v_{DS} (V)	Voltage v_{GS} (V)	Current i_D (A)
−0.1	0	-400×10^{-12}
	1	-32×10^{-6}
−0.4	0	-30×10^{-6}
	1	−1.2
−0.8	0	−13.8
	1	−42

demonstrate that the major contributors to current flow are electrons rather than holes, not only at the drain but also at the source electrode at moderated $-v_{DS}$ and with increasing v_{GS}. At high $-v_{DS}$, i.e., the body diode fully conducts, holes play an important role in current conduction, both at the drain (as minority carriers) and source electrodes. The electron to hole current ratio decreases at low $-v_{DS}$.

The current distribution through the trench cell is investigated under the six different electrical conditions of Table A.1. The drain current refers to Fig. A.2. The v_{GS} values are within the subthreshold region. Three different v_{DS} voltages are applied. Corresponding current flow lines and electron concentration are depicted in Figs. A.9–A.14.

Figure A.9 indicates the location of the gate, drain, and source electrodes as well as body region in the MOSFET cell. The *Y*-axis corresponds to the trench cell depth direction, whereas the *X*-axis refers to the cell length direction. The length of half trench cell is 2 μm. The trench cell depth including substrate and metallization is 6 μm. The trench depth is 1.5 μm. Source and body are shortened at the surface with a metal contact (i.e., source electrode).

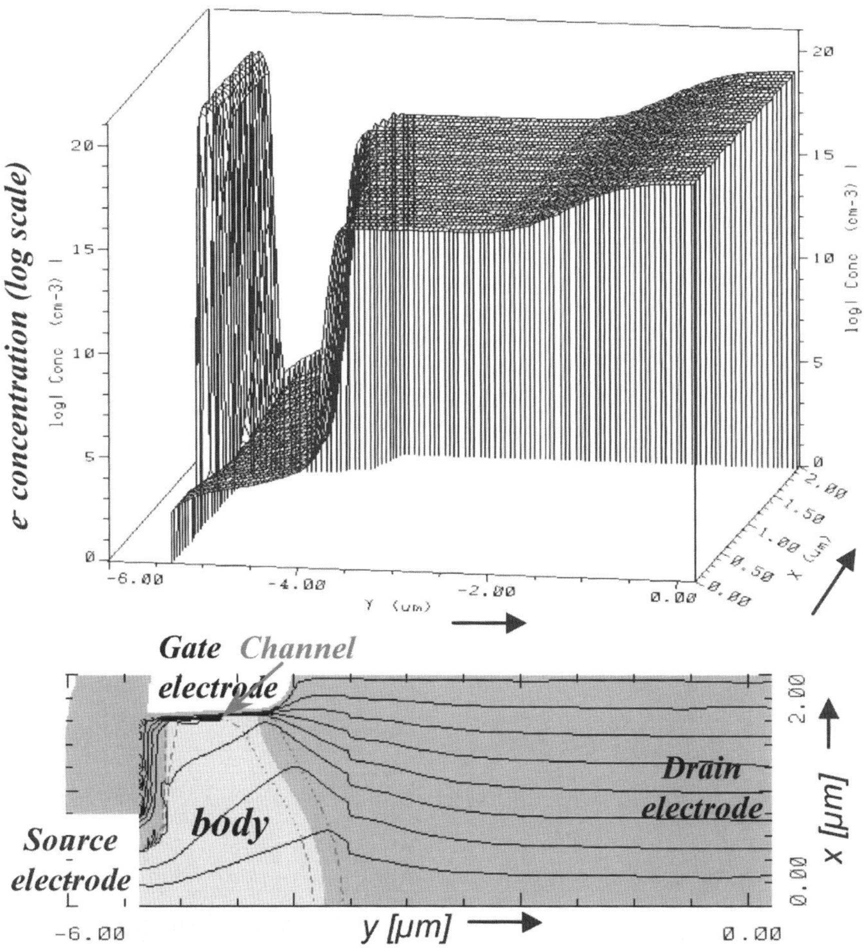

Fig. A.9 2D numerical simulation of a half trench MOSFET cell. Subthreshold electron concentration and current flow lines graphs at $v_{GS} = 0$ V and $v_{DS} = -0.1$ V. Note that simulated currents are specified in A/mm^2

The electron concentration and current flow lines graph show for every electrical condition how current distributes through the cell structure, i.e., the current distribution through body diode and channel regions.

In the simulation of Fig. A.9 ($v_{DS} = -0.1$ V, $v_{GS} = 0$ V), only a few hundred picoamps flow. Current flow lines indicate that this small amount of current is distributed through both body diode and channel regions even though the inversion channel layer is not yet formed. Although the channel region is much smaller than the body region, most of the flow lines are concentrated at the surface, showing that an important percentage of the total current is due to channel conduction. This is in agreement with the electron concentration graph showing

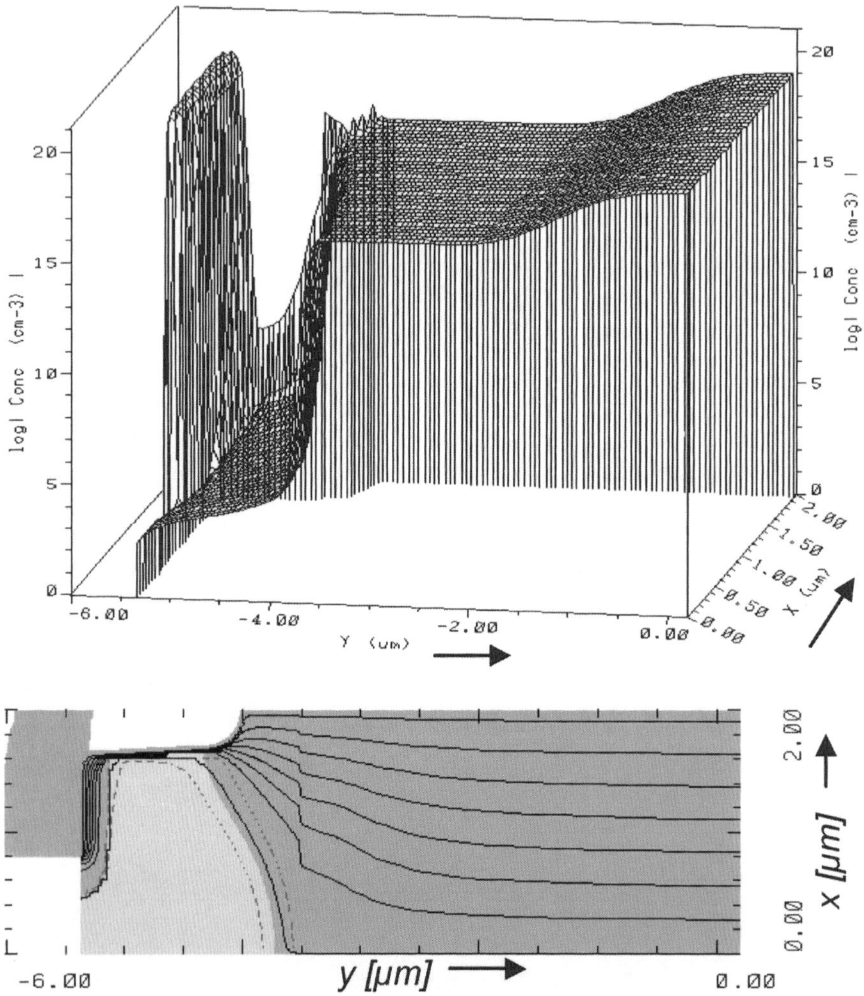

Fig. A.10 2D numerical simulation of a half trench MOSFET cell. Subthreshold electron concentration and current flow lines graphs at $v_{GS} = 1$ V and $v_{DS} = -0.1$ V

the presence of a higher electron concentration in the channel than in the body region.

Figure A.10 shows the result of increasing voltage v_{GS} from 0 to 1 V. The total current is still low but has increased to some tens of microamps. The electron concentration graph shows that there has been an increase of electrons in the channel region, whereas the electron concentration in the body region remains virtually constant. This indicates that the total current increase is caused by channel conduction. Since the positive v_{GS} depletes the channel of holes, the majority current carriers are conduction electrons in that region.

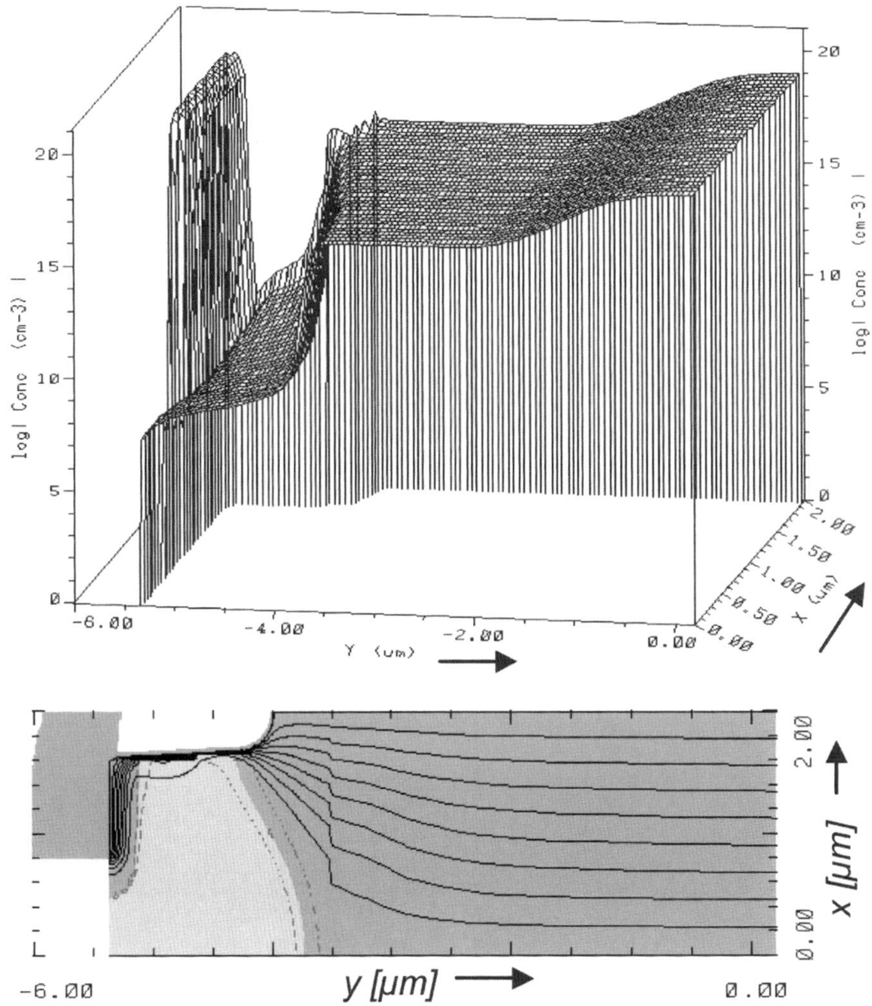

Fig. A.11 2D numerical simulation of a half trench MOSFET cell. Subthreshold electron concentration and current flow lines graphs at $v_{GS} = 0$ V and $v_{DS} = -0.4$ V

A similar behavior can also be observed at $v_{DS} = -0.4$ V from Figs. A.11 and A.12. A higher electron concentration in the channel than in the body region already occurs at $v_{GS} = 0$ V, indicating that the drain current mainly flows through the channel. The difference in work functions between the body and gate materials (semiconductor and metal, respectively) generates a partial depletion of holes in the channel. This is the reason why free electrons are likely concentrated in the channel region rather than in the body even without any external positive gate-to-source potential [3]. The drain current increases dramatically above 1 amp as soon as v_{GS} is increased from 0 to 1 V since a large number of free electrons from the drain are

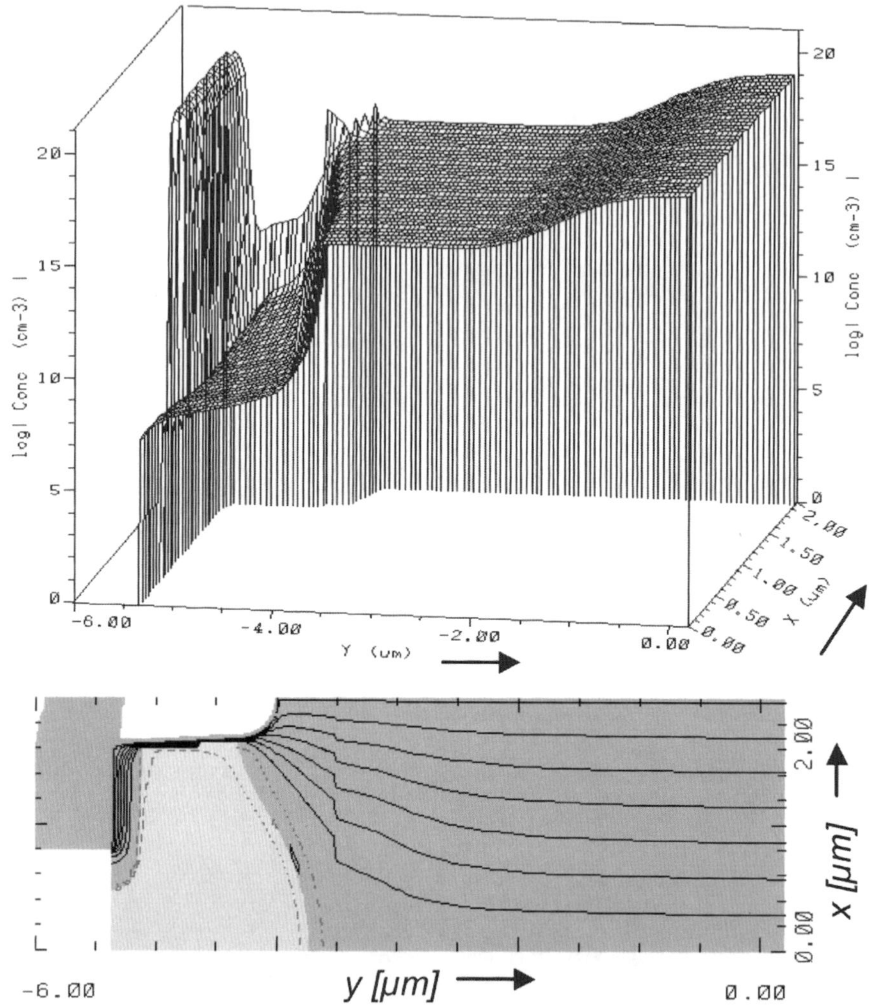

Fig. A.12 2D numerical simulation of a half trench MOSFET cell. Subthreshold electron concentration and current flow lines graphs at $v_{GS} = 1$ V and $v_{DS} = -0.4$ V

attracted to the channel. Again, the drain current rise is essentially caused by an increase of electron concentration in the channel.

High drain current results from applying a forward voltage of 0.8 V across the body diode. This is depicted in Fig. A.13. Voltage v_{GS} is kept at zero volts during the simulation. Figure A.14 shows the results of applying 1 V v_{GS}. The electron concentration in the body region is approximately equal to that for $v_{GS} = 0$ V, whereas the electron concentration in the channel region has increased. This results in a total current of 42 A (28.2 A higher than in case of $v_{GS} = 0$ V). The drain current has been amplified by 3 due to channel electron conduction.

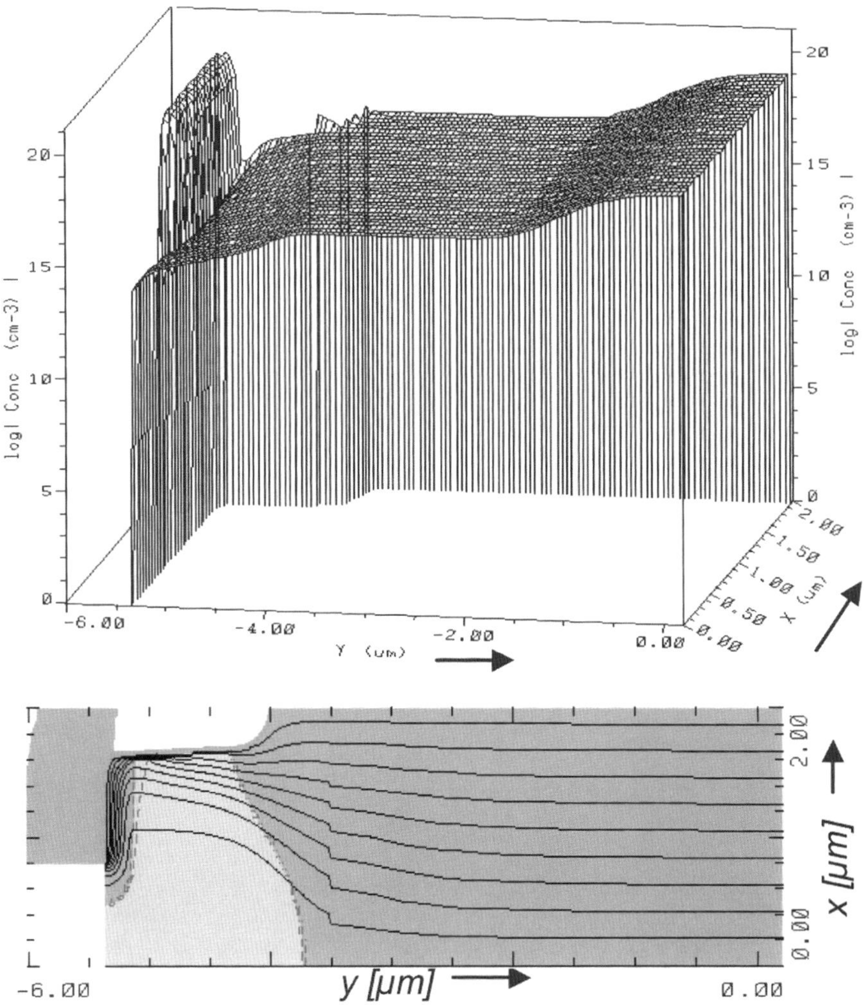

Fig. A.13 2D numerical simulation of a half trench MOSFET cell. Subthreshold electron concentration and current flow lines graphs at $v_{GS} = 0$ V and $v_{DS} = -0.8$ V

These simulation results demonstrate that the drain current in the third quadrant of the output characteristics is characterized by electron conduction in the channel region, which is modulated by the gate-to-source voltage at even values below threshold. This is in agreement with [4, 5]. The body diode current becomes important when it is forward biased with voltages above 0.7 V (at 25°C). Below this voltage value, the drain current is essentially caused by channel electron conduction.

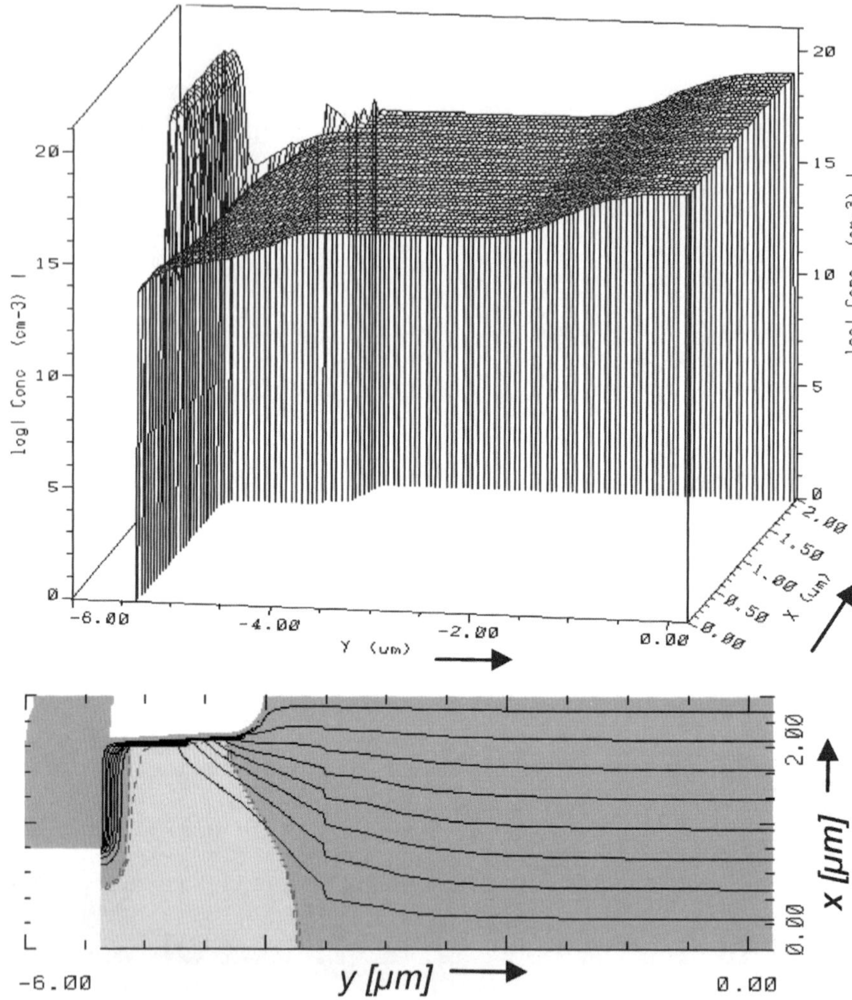

Fig. A.14 2D numerical simulation of a half trench MOSFET cell. Subthreshold electron concentration and current flow lines graphs at $v_{GS} = 1$ V and $v_{DS} = -0.8$ V

References

1. Ferreira A, Castro Simas MI (1991) Power MOSFETs reverse conduction revised. In: IEEE proceeding on power electronics specialists conference, PESC 1991, pp 416–422
2. Castro Simas MI, Costa Freire J (1994) CAD tools to optimise power MOSFET performance using channel reverse conduction. IEEE Trans Power Electron 9(5):522–531
3. Grant DA, Gowar J (1989) Power MOSFETs. Theory and applications. Wiley, New York
4. Dolny GM, Sapp S, Elbanhaway A, Wheatley CF (2004) The influence of body effect and threshold voltage reduction on Trench MOSFET body diode characteristics. In: IEEE

international symposium on power semiconductor devices and ICs, ISPSD 2004, pp 217–220

5. Sun NX, Huang AQ. The impact of sub-threshold current on ultra high density Trench MOSFET for synchronous rectifier application. In: IEEE international symposium on power semiconductor devices and ICs, ISPSD 200, pp 358–361
6. Maxim A (1998) The analog behavioural SPICE macro modelling – a novel method of power semiconductor devices modelling. IEEE Trans Power Electron
7. Lauritzen P, Ma C (1991) A simple diode model with reverse recovery. IEEE Trans Power Electron 6(2):188–191
8. Tolle T, Duerbaum T, Elferich R, López T (2003) Quantification of switching loss contributions in synchronous rectifier applications. In: European power electronics, EPE 2003
9. Djekic O, Brkovic M (1997) Synchronous rectifiers vs. Schottky diodes in a buck topology for low voltage applications. In: IEEE proceeding on power electronics specialists conference, PESC 1997, vol 2, pp 1374–1380
10. Bridge CD (2002) Control method to reduce body diode conduction and reverse recovery losses. Patent No. US 6396250 B1, 28 May 2002

Appendix B
Reverse Recovery in LV Trench MOSFETs

The optimization of power MOSFETs for high-frequency power converters needs a precise description of their dynamic behavior and the impact in the application. Circuit simulators are effective to analyze switching transients of power converters, yet they require accurate models to assess the device performance. This section presents a reverse recovery model for low voltage trench MOSFETs [1].

A simple lumped model approach is employed to represent the dynamics of diffusion and recombination processes of the MOSFETs intrinsic body diode. The parameters of the model are adjusted from reverse recovery waveforms resulting from rather accurate device physics simulations or, alternatively, from measurements. In either way, the required body diode diffusion current is extracted from the drain terminal current. This involves the identification of both channel current and MOSFET capacitance current, which are superimposed to the body diode diffusion current.

Channel conduction is manifested by the so-called "body effect" [2, 3] even though the gate voltage is well below the nominal threshold voltage of the device (see Appendix A). This channel current is proven to have a significant impact in the reverse recovery transient curves as it provides an alternative low ohmic path to the drain current. Device physics simulations reveal the existence of this channel conduction. Further, internal circulating currents between the body diode and channel regions exist that cannot be observed from the terminals. The de-embedding of the drain current forms the basis of a curve fitting procedure to derive the parameter values of the diode model. Device physics simulations validate the performance of the proposed reverse recovery representation.

B.1 Impact of the Body-Effect in Reverse Recovery Transients

According to the description of Appendix A, the effective threshold voltage reduction in the third quadrant conduction observed in the DC output characteristics of MOSFETs is consequence of the body-effect.

The DC current in the third quadrant may therefore result from two main carrier transport phenomena; namely, diffusion of minority carriers in the body diode, and

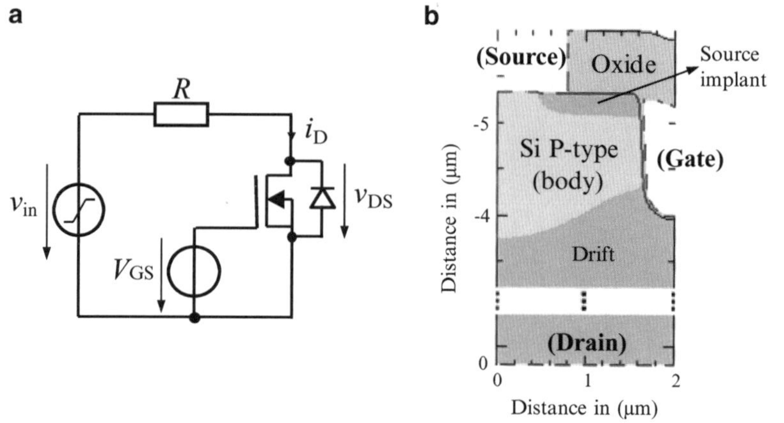

Fig. B.1 (**a**) Resistive load reverse recovery circuit diagram. (**b**) 2D representation of the simulated half-trench MOSFET cell

drift of majority carriers in the channel even though the gate voltage may be lower than the nominal threshold voltage (V_{TH0}).

This static characteristic behavior has a major impact on the dynamics of reverse recovery. First, the reverse recovery current is observed to be strongly dependent on the gate voltage in the subthreshold region because the stored charge is affected by a split of the drain current between the channel and body diode regions prior to switching. And second, since the diffusion current during reverse recovery may alternatively flow through the channel, an internal circulating current flow may arise that reduces the external drain current. Thus, the measured reverse current at the drain terminal may appear significantly lower than the actual internal body diode current.

Figure B.1 illustrates the impact of the body-effect in the reverse recovery curves. The data results from device physics simulations corresponding to the resistive reverse recovery setup and trench cell structure of Fig. B.2. Current and voltage waveforms are shown at three different gate voltage biases. Furthermore, current flow lines through the MOSFETs cell are depicted at various time instants for investigations on the current distribution.

As observed in the time dependent graphs, the reverse current conduction[1] in the three considered cases are significantly different, being the simulation at $V_{\mathrm{GS}} = -5$ V the one with the longest reverse current time, and the simulation at $V_{\mathrm{GS}}= 1$ V with the shortest. The current flow lines reveal that, in the latter case, the drain current flows almost entirely through the channel region prior to switching. Since no diffusion current is present, the reverse current during the transient is uniquely

[1]Reverse current conduction refers to the body diode. Thus, this happens when the drain current is positive.

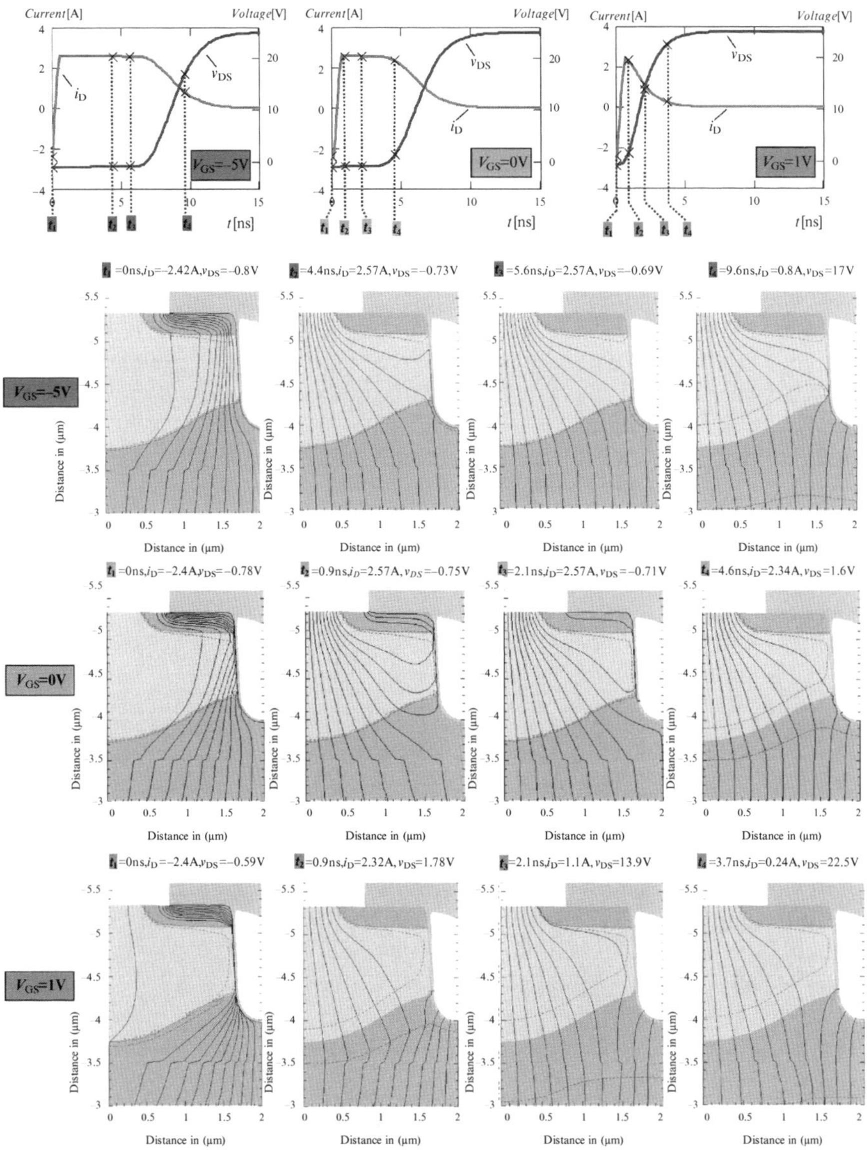

Fig. B.2 Resistive reverse recovery transient curves and current flow lines for various gate voltages. Data result from 2D device physics simulations of the trench cell of Fig. B.2b with a scale factor corresponding to a 1 mm^2 active area. Steady-state conditions prior to switching correspond to those at time t_1

caused by the output capacitance charge, i.e., the depletion capacitance of the body diode and the gate-to-drain oxide capacitance. Voltage V_{DS} rapidly rises when the current reverses, which causes the channel to block. Prior to switching at $V_{GS} = -5$ V, the drain current entirely flows through the body diode, i.e., the inversion layer is not formed. During reverse conduction, current flows from drain-to-source through the body region as the drain-to-source voltage remains virtually constant at -0.7 V, which is a clear indication of current flow due to diffusion of minority stored charges.

The current flow lines at $V_{GS} = 0$ V indicate that the drain current flows through both body diode and channel regions. Thus, the reverse current caused by charge diffusion is lower than in case of $V_{GS} = -5$ V, as the initial forward current through the body diode is lower. It is further observed that an internal circulating current flows between channel and body diode during reverse current conduction. Some of the flow lines cross the p–n junction from the body diode to the drift region and return to the source through the channel, which is conducting since V_{DS} remains negative. Other flow lines do not cross the p–n junction to get into the drift region, but trace a circulating path through the body region, channel, source implant, and source metallization. It is therefore expected that the internal diffusion current is higher than the current at the drain terminal. When the body diode blocks, V_{DS} rises and the drain current is predominantly capacitive.

B.2 Reverse Recovery Circuit Model

The main purpose of this section is to present a behavioral model that represents reverse recovery transients according to the description given in the previous section. The proposed model reproduces the MOSFET operation as the sum of the diffusion and drift carrier transport phenomena, as well as the capacitance current. The resulting circuit diagram is depicted in Fig. B.3. The three current contributors are independently modeled as described in [1].

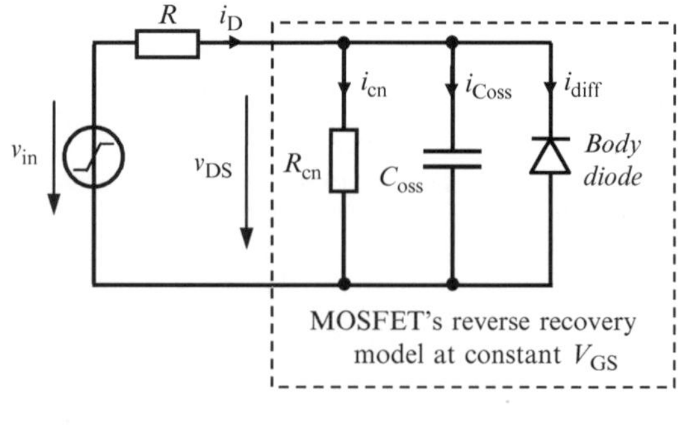

Fig. B.3 Circuit diagram of the resistive load reverse recovery and MOSFETs equivalent circuit

The output capacitance shown in Fig. B.4 is obtained from device physics simulations by means of voltage ramp transients. The resulting data are table-based represented in the circuit simulator.

For modeling the channel conduction in the third quadrant, the following assumption based on the body-effect is taken: the increase of drain current when the gate voltage rises is merely due to an increase of channel conduction. According to this, channel and body diode current can be distinguished from the output characteristics. That is, in the third quadrant there exists a gate voltage below which the drain current gets independent of. Such drain current curve corresponds to the diode current. The difference between the body diode current and the total drain current is thus the channel current. Figure B.5 shows the de-embedding of both channel and body diode currents for the present case study. As it is depicted, the channel current reaches maximum values at V_{DS} below -0.7 V, and tends to zero at strong body diode conduction. The resulting channel curve at a low current level is shown in the diagram of Fig. B.6 in comparison to the predicted curve resulting from applying the body-effect theory [4]. The theoretical curve is very

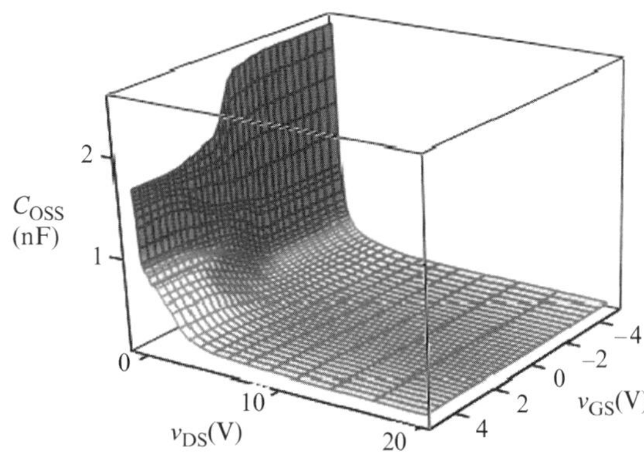

Fig. B.4 MOSFETs output capacitance. Data result from 2D device physics simulations of the trench cell of Fig. B.2b with a scale factor corresponding to a 1 mm^2 active area

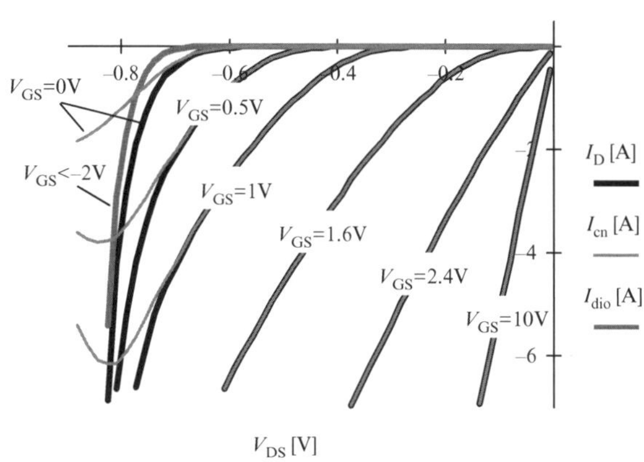

Fig. B.5 Third quadrant output characteristics. Drain current de-embedding. Data result from 2D device physics simulations of the trench cell of Fig. B.2b using a scale factor corresponding to a 1 mm^2 active area

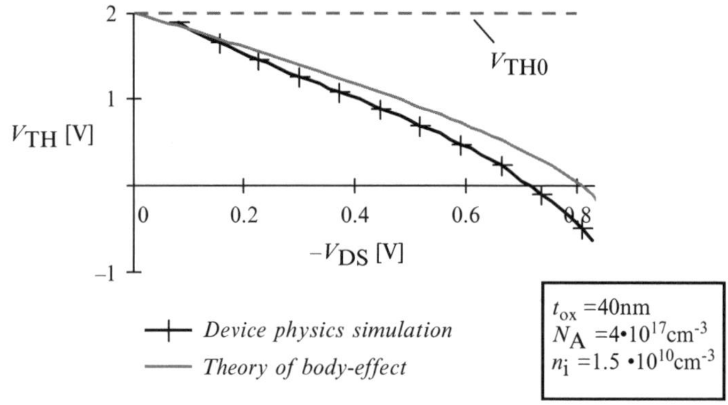

Fig. B.6 Effective third quadrant threshold voltage reduction. Comparison between device physics simulations and body-effect theory

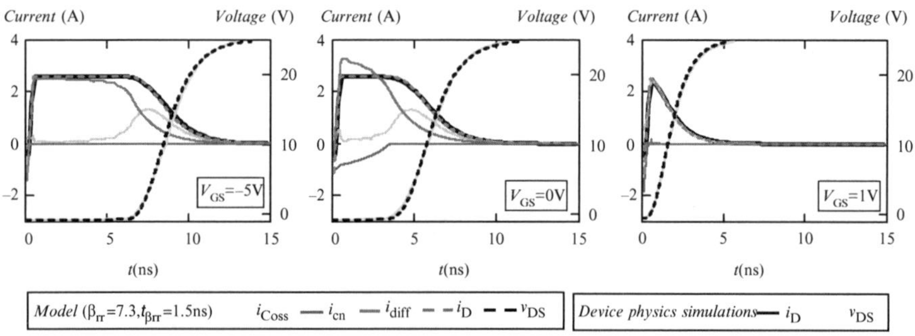

Fig. B.7 Comparison of resistive reverse recovery curves resulting from device physics simulations (Fig. B.1) and SPICE simulations of the proposed behavioral model

sensitive to the concentration of holes N_A in the body region. The high nonuniformity of this concentration in the simulated trench cell makes it difficult to assign a uniform equivalent value assumed in the theory. The channel behavior is modeled in a circuit simulator by means of a curve fitting function (Fig. B.7).

Finally, the body diode diffusion current is modeled employing the following empiric equation based on [5],

$$i_{\text{diff}}(t) = \beta_{\text{rr}} \cdot I_{\text{dio}}(v_{\text{DS}}(t)) + \frac{1}{t_{\beta\text{rr}}} \int [I_{\text{dio}}(v_{\text{DS}}(t)) - i_{\text{diff}}(t)]\, \mathrm{d}t + \cdots + I_{\text{diff0}}, \qquad \text{(B.1)}$$

where I_{dio} is the body diode static characteristics (see Fig. B.5), I_{diff0} is the initial condition of the integral, and β_{rr} and $t_{\beta\text{rr}}$ are fitting parameters. As done for the channel current, I_{dio} is modeled in the circuit simulator using a curve fitting function.

Figure B.6 shows reverse recovery transient curves resulting from both device physics- and circuit simulations. Parameters β_{rr} and $t_{\beta\text{rr}}$ are adjusted to best match the device physics-simulated curves. Regarding voltage and current waveforms, the simple body diode model together with the channel and capacitance data agree with the device physics simulation approach. For the case of $V_{GS} = 0$ V, circuit simulation predicts and quantifies the internal circulating current between channel and diode, which results in a body diode current peak higher than the drain current.

As described in [1], the reverse recovery model can also be applied for describing real devices by means of experimental characterization.

B.3 SPICE Simulations of a Synchronous Buck Converter

The effect of channel electron conduction as majority carriers in the third quadrant subthreshold region may impact on both switching and conduction losses in the application. First, because as channel conduction increases, the power loss due to the reverse recovery of the minority mobile stored charge reduces. And second, because v_{DS} voltage drop can be considerably lower than the conventional body diode forward voltage, with a consequent conduction power loss reduction.

The importance of this effect is illustrated by means of SPICE simulations of a synchronous buck converter. Figure B.8 shows the circuit diagram under consideration. For the sake of simplicity, the input and output of the converter is modeled as an ideal voltage and current source, respectively. Values are specified in Table B.1. Note that the parasitic impedance values *Lg*, *Rg*, *Ld*, *Rd*, *Ls*, and *Rs* do not correspond to any particular circuit arrangement (package device, PCB layout, etc.). A table based behavioral macro MOSFET model [1, 6] is applied for modeling in SPICE switches T1 and T2. The required data (i.e., capacitance, channel, and body diode characteristics) are gathered from MEDICI simulations corresponding to the PHB96NQ03LT device. The body diode reverse recovery effect is modeled as proposed in [5] and integrated as a part of the macro model.

The switching interval in which switch T2 turns off and switch T1 turns on is analyzed. In the usual operation, there exists a certain time delay between the T2 turn-off and the T1 turn-on that avoids cross conduction. The body diode of T2 takes over the output current during that delay time. Thus, reverse recovery occurs when T1 turns on and the diode blocks. SPICE simulation results from Figs. B.9 to B.11 illustrate this behavior in detail. In Fig. B.9 (case (a)), driving voltage v_{DRV2} of T2 switches from 10 to 0 V linearly. When v_{GS2} drops down to a sufficiently low value the body diode starts conducting and partially takes over the output current $i0$ (at v_{GS}= 0 V part of the total drain current still flows through the channel in the third quadrant). Voltage v_{GS2} may significantly bounce during current and/or voltage variation due to the gate parasitic impedance, which affects the body diode current. Meanwhile switch T1 is turned on. Driving signal[2] v_{DRV1} switches from 0 to 10 V

[2]The driver signals have not been optimized to reduce power losses in order to avoid effects that are out of the subject. The setup is arranged in a way to illustrate the impact of the third quadrant effect on the application.

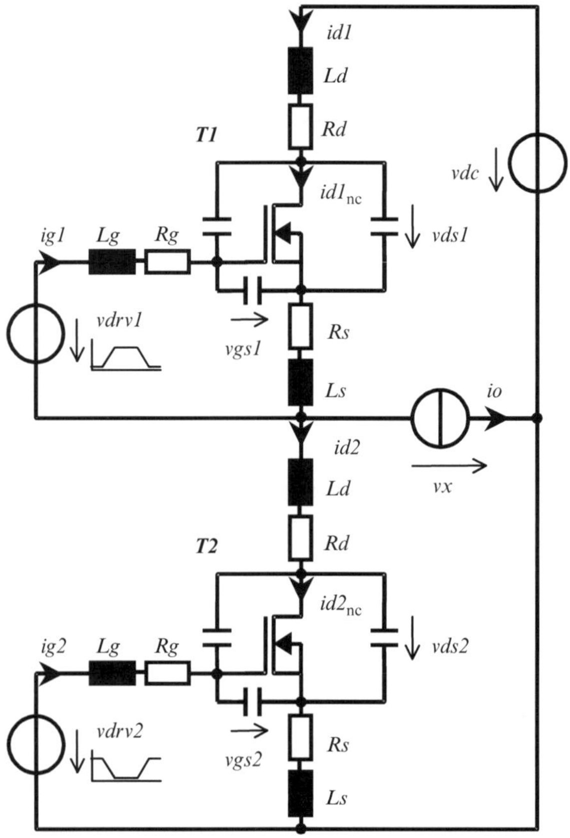

Fig. B.8 SPICE simulation circuit diagram of a synchronous buck converter

Table B.1 Specifications of the synchronous buck converter circuit of Fig. B.8

Operation DC-link voltage output current	DC-link voltage v_{DC}	10 V
	Output current I_o	10 A
Gate circuits (T1, T2)	Total gate inductance L_G	2 nH
	Total gate resistance R_G	1 Ω
Source, drain circuit (T1, T2)	Source inductance L_s	0.5 nH
	Drain inductance L_D	0.1 nH
	Source and drain resistance R_S, R_D	1 mΩ

MOSFETs T1, T2 Channel current and inter-electrode capacitances are functions of (v_{DS}, v_{GS}). The body diode current is a function of (v_{DS}, v_{GS}, t) and comprises reverse breakdown and reverse recovery

linearly in 20 ns. The feedback effect of the source inductance considerably delays the commutation, which is translated in a long quasi-plateau shape in the v_{GS1} waveform. During the Miller plateau voltage v_{DS1} rises and switch T1 conducts the output current. The oscillatory energy generated in the commutation is dissipated in

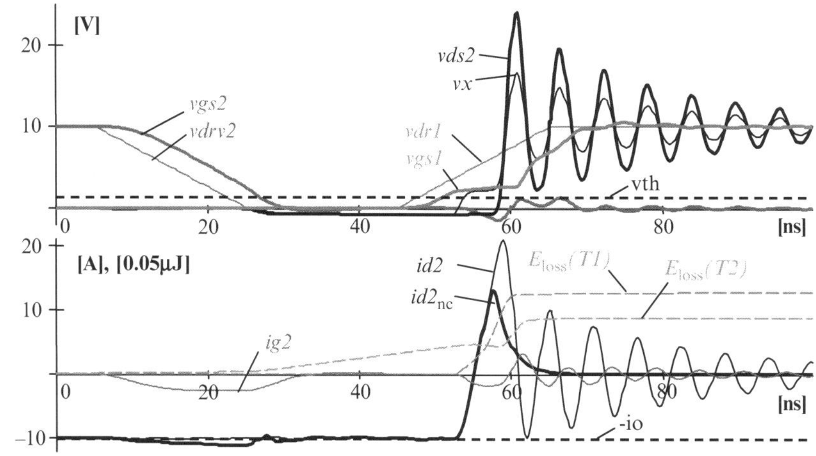

Fig. B.9 SPICE simulations case (a) of circuit diagram of Fig. B.8. Min(v_{DRV2}) = 0 V, max ($id2$) = 20.8 A, max($id2_{nc}$) = 13.4 A (reverse recovery), E_{loss}(T1, T2) = 1.08 μJ, temperature = 25°C

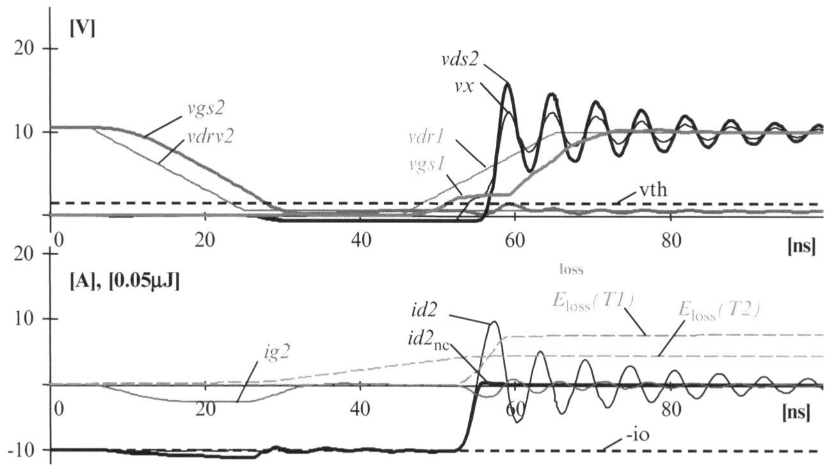

Fig. B.10 SPICE simulations case (b) of circuit diagram of Fig. B.8. Min(v_{DRV2}) = 0.5 V, max ($id2$) = 9.7 A, max($id2_{nc}$) = 0.4 A (reverse recovery), E_{loss}(T1, T2) = 0.61 μJ, temperature = 25°C

the parasitic resistance of the circuit. Note that the maximum peak of the voltage oscillation across v_{DS2} can be considerably high, even enough to cause breakthrough.

The loss energy in body diode and channel can be calculated by the subtraction of the capacitive current from the total drain current, i.e., looking at Fig. B.8 the subtraction leaves $id1_{nc}$ or $id2_{nc}$ [7], and multiplying the result by v_{DS} (i.e., v_{DS1} or v_{DS2} in the proposed example). It provides the power loss in the MOSFET

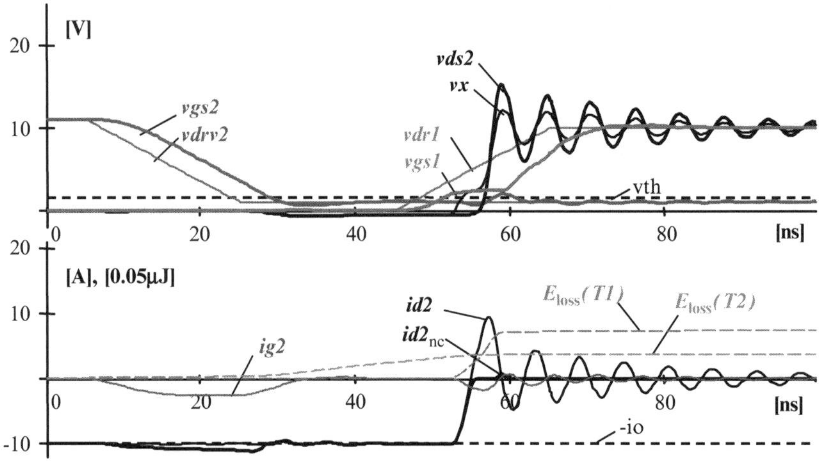

Fig. B.11 SPICE simulations case (c) of circuit diagram of Fig. B.8. Min(v_{DRV2}) = 1 V, max(*id2*) = 9.4 A, max($id2_{nc}$) = 0.9 A (gate bouncing), E_{loss}(T1, T2) = 0.56 μJ, Temperature=25°C

excluding the gate, package and pins resistance at a certain instant of time. The loss energy is then obtained by integrating this instantaneous power along the time. In the example, this operation is calculated for switches T1 and T2 at the time interval (0, 100 ns). The resulting loss energy is 1.08 μJ.

In simulation results of Fig. B.9 (case (b)) a DC voltage of 0.5 V is added to the drive signal of switch T2. Therefore the upper limit of v_{DRV2} becomes 10.5 V and the lower limit 0.5 V. This change alone significantly impacts on the switching behavior. Three relevant aspects of the resulting waveforms must be pointed out. First, the positive voltage value across v_{GS2} during the third quadrant operation of switch T2 is sufficiently high to practically eliminate body diode conduction and thus reverse recovery. Note that during the turn-on of T1, v_{GS2} bounces around 0.5 V, always below threshold and above zero volts. Second, the mitigation of reverse recovery current has the direct implication of lowering down current and voltage oscillations, which in turn reduces switching power losses in switch T1 as well as electromagnetic interference (EMI). And third, voltage v_{DS2} during the third quadrant operation is slightly reduced, and so are the conduction losses. The calculated loss energy in this case is 0.61 μJ, which means a 43% reduction in comparison to case (a).

In the simulation of Fig. B.9 (case (c)), the driving voltage offset of switch T2 is further increased to 1 V, so that the lower limit corresponds to 1 V. This value is so critically close to the threshold voltage that a spurious turn on occurs due to gate bouncing voltage [7]. The described effects taking place in case (b) are slightly pronounced. That is, reverse recovery is completely vanished, the amplitude of the oscillations is further reduced and the conduction in the third quadrant is improved. This accounts for a loss reduction of about 48% in comparison to case (a).

B.4 Summary

The third quadrant DC output characteristics of a high density trench power MOSFET is analyzed using measurements, 2D numerical simulations, SPICE, and a table base macro MOSFET model. In contrast to the first quadrant conduction, a substantial third quadrant current in the threshold and subthreshold regions is observed in measurements and analyzed in simulations. This asymmetric current is essentially caused by channel electron conduction and not by body diode conduction. The repercussions of this effect in the application are illustrated with switching transient simulations of a synchronous buck converter in SPICE. The example demonstrates that the third quadrant output characteristics may strongly influence the converter behavior, up to the point where relevant issues such as reverse recovery, hard switching losses, and EMI can be reduced or even vanished. Therefore, the inclusion of the third quadrant output characteristics in circuital MOSFET models is a step forward to better understand and predict switching transients in SMPS.

The reverse recovery transient is accurately described with a proposed behavioral model for circuit simulators. According to the physical phenomena, the drain current is de-embedded in three components: Diode, channel, and capacitance currents. The dynamics of the body diode is effectively modeled with a simple empirical equation that only requires two fitting parameters. The model approach allows assessing the impact in the reverse recovery transient curve of the different current contributors.

References

1. Elferich R, López T (2005) Accurate behavioural modelling of power MOSFETs based on device measurements and device physics simulations. In: European power electronics, EPE 2005, Sept 2005
2. Dolny GM, Sapp S, Elbanhaway A, Wheatley CF (2004) The influence of body effect and threshold voltage reduction on trench MOSFET body diode characteristics. In: IEEE international symposium on power semiconductor devices and ICs, ISPSD 2004, pp 217–220
3. López T, Elferich R, Tolle T, Koper N, Duerbaum T (2004) Third quadrant output characteristics in high density Trench MOSFETs. In: IEEE international power electronics and motion control conference, EPE-PEMC 2004, vol 2, pp 26–34
4. Mouthaan T (1999) Semiconductor devices explained using active simulation. Wiley, New York, pp 274–278
5. Lauritzen PO, Ma CL (1991) A simple diode model with reverse recovery. IEEE Trans Power Electron 6(2):188–191
6. Maxim A (1998) The analog behavioural SPICE macro modelling – a novel method of power semiconductor devices modelling. In: IEEE Transactions on PE, 1998
7. Tolle T, Duerbaum T, Elferich R, López T (2003) Quantification of switching loss contributions in synchronous rectifier applications. In: European power electronics, EPE 2003

Appendix C
Reverse Recovery Lumped Models

Lumped models for circuit simulators are widely used for approximating the distributed effects of carrier diffusion and recombination phenomena in semiconductor diodes. Over the last 50 years a great number of diode lumped models has been presented. Most of the proposals are based upon the concept of concentrated charge storage nodes, which allows for simple implementations and fast simulation times.

Among the most relevant proposals, the lumped charge node models from [1–3] are implementations of the diffusion equation employing analogous physical variables. This feature enables a direct assessment of the circuit response to changes in inherent device physics properties, such as geometry, materials, and doping levels. Furthermore, in these proposed models, the computation time can be readily improved in detriment of accuracy by reducing the number of charge nodes [2, 3].

In most circuit simulations though, the models tend to take a rather simpler form to describe the physical principles. The variables of these models, whose values are typically determined from parametric measurements [4], are usually chosen to compromise between accuracy and mathematical convenience, and thus often do not necessarily keep a direct correlation to the physical variables of the device [5].

While there are existence of many alternatives (see [6]), the most extended concentrated charge-based model, which stands about halfway between physics and circuit considerations, is perhaps the simplest in its form and implementation, i.e., the charge-control model [7]. Nonetheless, eventhough the charge-control concept can be entirely treated as an equivalent circuit tool, its prediction for transient analysis can be highly inaccurate [8].

Today's increasing demands for high performance electronic systems trigger the need of accurate semiconductor models that allow for stringent design optimizations. Particularly, the reverse recovery of diodes is one of the crucial limiting factors of switching transient performance in many power switched converters. Consequently, proposed advanced diode models that have not gained much attention in the past years may now be potential candidates as alternatives to the often-misused charge-control model.

In this section, lumped models for circuit simulators are investigated to represent the body diode reverse recovery of low voltage power trench MOSFETs in synchronous rectifier applications. Three previously presented models are compared in terms of internal constitution, circuit implementation, prediction accuracy, and required computing power.

2D device physics simulations of a trench MOSFET cell underlie the prediction accuracy study. device physics simulators of semiconductor structures can accurately model the electrical characteristics of a device from its physical properties, which makes them suitable for the development of manufacturing processes. However, circuit analysis based on device physics simulations is cumbersome, and requires high computation power. Thus their use is rather limited. The purpose of this work is to obtain an equivalent accurate lumped model of the "virtual" device to effectively analyze its performance in the circuit application before it is manufactured [5].

First, a systematic procedure is carried out to de-embed the body diode reverse recovery current from inductive switching simulations of the 2D structure. The resulting data are then used to adjust the parameters of the lumped models by means of curve fitting. It is therefore exposed how effectively these lumped models can reproduce the response from the device physics simulations under different load currents and switching speed conditions of the application.

C.1 Device Physics Simulations of Power Trench MOSFETs

Figure C.1a shows the 2D representation of a trench MOSFET cell. Due to the symmetry, only half of the structure is simulated. Note that a third dimension, z, orthogonal to x and y, is considered in order to compute the current through

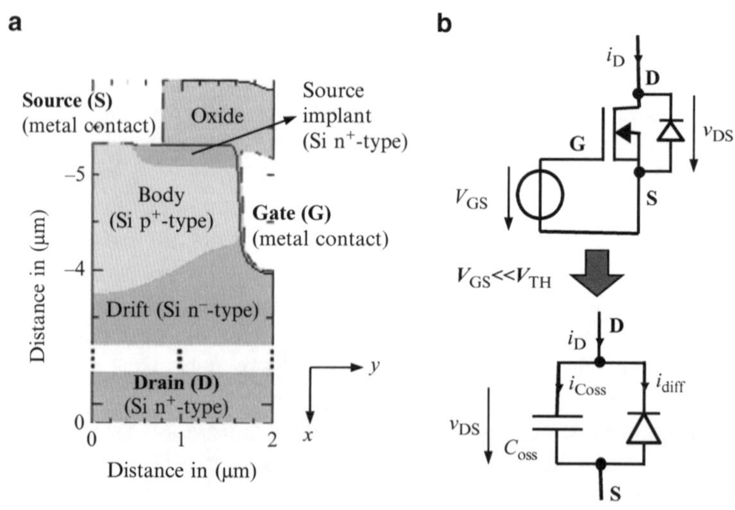

Fig. C.1 (**a**) 2D representation of a half trench MOSFET cell for low-voltage applications. (**b**) Simplified lumped model representation of a MOSFET at a strong negative gate voltage

a 1-mm^2 stripe. That is, although only two dimensions of half trench are simulated, the results shown here refer to a structure that has 4-μm length in the *y* direction, and extends 25 cm in the *z* direction.

To focus the study on the body diode behavior only, the experiments are performed at a strong negative gate voltage, thereby forcing the channel to block in both directions of the drain current, as described in [9]. Other effects of the MOSFETs are further modeled and combined in [5]. The MOSFET can therefore be represented by the body diode behavior in parallel to the output capacitance of the device, as shown in Fig. C.1b. The body diode diffusion current is obtained by,

$$i_{\text{diff}}(t) = i_{\text{D}}(t) - i_{C_{\text{oss}}}(t) = i_{\text{D}}(t) - C_{\text{oss}}\frac{\mathrm{d}v_{\text{DS}}(t)}{\mathrm{d}t}. \tag{C.1}$$

Capacitance C_{oss} is derived from small signal device physics simulations. Figure C.2a shows the resulting capacitance curve.

Table C.1 lists the parameter values used in the device physics simulations. Parameters W and p_{n0} are given as approximated averaged values.

Note that in low voltage diodes the length of the drift region is typically significantly shorter than the diffusion length, $W \ll L_{\text{p}}$, where,

$$L_{\text{p}} = \sqrt{D_{\text{p}}\tau_{\text{p}}}. \tag{C.2}$$

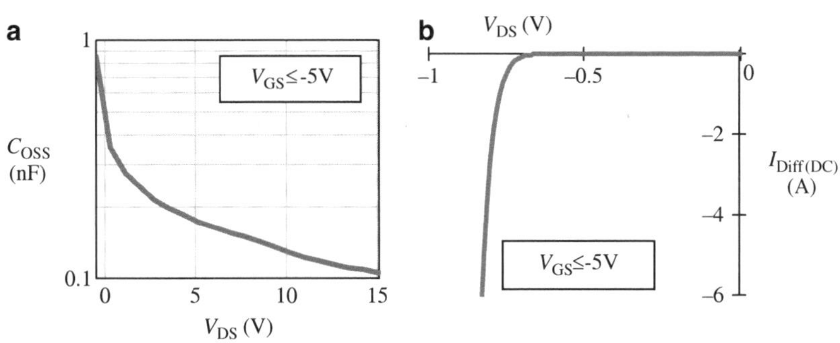

Fig. C.2 Device physics simulations of a trench MOSFET cell structure. (**a**) Output capacitance. (**b**) Static diffusion current

Table C.1 Device parameter values used in the device physics simulations

Parameter	Value
Cross-sectional area of the device, A	1 mm^2
Width of the drift region, W	3.4 μm
Temperature, T	300 K
Average equilibrium concentration of holes in the drift region, p_{n0}	10^9 cm^{-3}
Lifetime of holes in the drift region, τ_{p}	10 μs
Diffusion coefficient of holes, D_{p}	6.5 cm^2/s

Current i_{diff} is mainly caused by diffusion of holes in the drift region. In order to calculate it, the device physics simulator computes the continuity equation of excess of holes in the drift region, p'_n,

$$\frac{\partial p'_n}{\partial t} = -\frac{1}{q}\vec{\nabla}\cdot\overrightarrow{J_{\mathrm{p}}} - \frac{p'_n}{\tau_{\mathrm{p}}} \tag{C.3}$$

in combination with the diffusion-based transport equation, which is,

$$\overrightarrow{J_{\mathrm{diff}}} = \overrightarrow{J_{\mathrm{p}}} = -qD_{\mathrm{p}}\vec{\nabla}p'_n. \tag{C.4}$$

The static characteristics of the body diode, shown in Fig. C.2b as $I_{\mathrm{Diff(DC)}}(V_{\mathrm{DS}})$, derive from the above two equations when (C.3) is set to equal zero. As it will be shown later, this static curve is required as input data to the lumped models.

Figure C.3 illustrates the circuit diagram for the generation of the inductive reverse recovery curves. Table C.2 lists the parameter values of the circuit. Inductance L is varied to adjust the ramp slope K_i of the current, which is expressed in units A/s. Capacitance C and resistance R are used to avoid an excessive voltage across v_{DS} that may drive the device into avalanche. The circuit simulation starts in static equilibrium with zero current through inductance L. The initial voltage V_{iC} across capacitance C depends on current I_{o} and the static characteristics of the body diode (see Fig. C.2b).

Figure C.4 shows an example of a reverse recovery transient curve derived from device physics simulator. Current i_{diff} is obtained from (C.1). Besides current I_{o} and current slope K_i, three other parameters are defined to characterize the reverse recovery curves. These are the reverse recovery charge, Q_{rr}, the reverse current peak, I_{rp}, and time t_{rp}, defined to be the time interval between the zero current crossing and I_{rp}, as indicated in Fig. C.4.

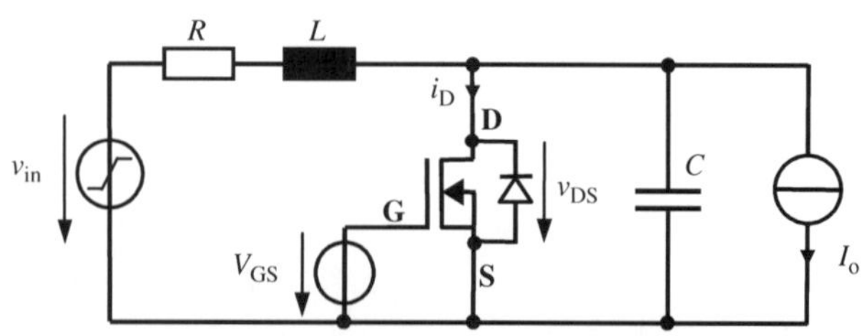

Fig. C.3 Inductive load reverse recovery circuit diagram

Table C.2 Default parameter values of circuit of Fig. C.3

R	L	C	V_{GS}	v_{in}
0.1 Ω	$\frac{10-V_{\mathrm{iC}}}{K_i}$	1 nF	−5 V	Ideal step function from V_{iC} to 10 V

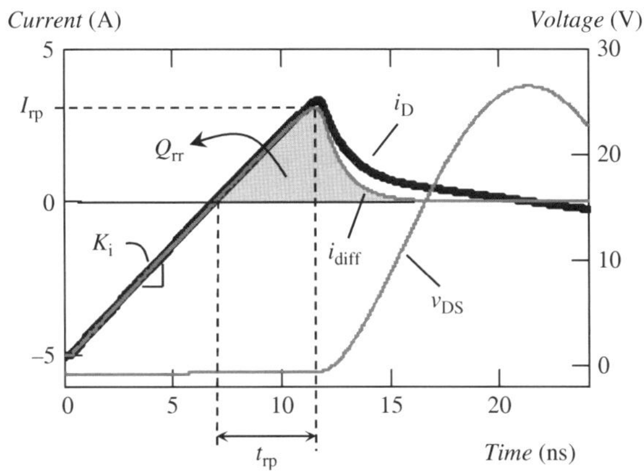

Fig. C.4 Inductive load reverse recovery waveforms corresponding to circuit diagram of Fig. C.1.3. 2D device physics simulations of the trench MOSFET cell of Fig. C.1.1(a)

C.2 Circuit Simulator Lumped Models

Three lumped models are analyzed. These are: the generalized lumped model approach based on the transmission line analogy presented in [1, 2], the model based on two charge storage nodes proposed in [4], and the empiric lumped model from [5]. The following subsections describe their internal constitution and basic circuit implementation.

C.2.1 Transmission Line Based Lumped Model

The model proposed in [1, 2] deals with a simplified 1D space problem of the diffusion equation, in the way indicated as follows,

$$p_n' = p_n'(x,t) \tag{C.5}$$

being the x direction in correspondence to the diagram of Fig. B.2a. Combining (C.3)–(C.5) yields the 1D diffusion equation of holes in the drift region,

$$D_\mathrm{p}\frac{\partial^2 p_n'(x,t)}{\partial x^2} = \frac{\partial p_n'(x,t)}{\partial t} + \frac{p_n'(x,t)}{\tau_\mathrm{p}}. \tag{C.6}$$

The diffusion current through the diode is calculated by,

$$i_{\mathrm{diff}}(t) \cong i_{\mathrm{diff_{1D}}}(t) = -qAD_\mathrm{p}\left.\frac{\mathrm{d}p_n'(x,t)}{\mathrm{d}x}\right|_{x=X_\mathrm{p}}. \tag{C.7}$$

Parameter X_p is the boundary between the depletion and quasi-neutral areas in the drift region.

The lumped models can be obtained by identifying the existing analogy between the diffusion equation in (C.6) and the electric transmission line equation (or telegrapher's equation) [3]. Figure C.5 shows the equivalent circuit of an incremental transmission line section, called basic transmission line lumped cell (TLLC).

According to the fact that two forms of energy behave identically when the basic differential equations that describe them have the same form and the initial and boundary conditions are identical, then applying the following equivalences,

$$p_n'(x,t) \Leftrightarrow v(x,t), \quad D_p \Leftrightarrow (RC)^{-1}, \quad \tau_p \Leftrightarrow \frac{C}{G}, \quad L = 0, \qquad \text{(C.8)–(C.11)}$$

where v is the voltage across the transmission line, R its resistance per unit length, C its capacitance per unit length, and G its transverse conductance per unit length.

The diffusion and recombination processes described by (C.6) can be precisely modeled with the transmission line circuit of Fig. C.6 when the number of employed TLLCs, N_c, tends to infinite and Δx tends to 0 for each individual TLLC.

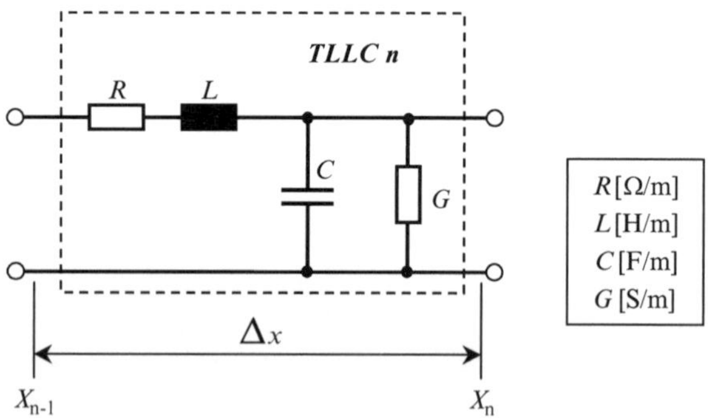

Fig. C.5 Basic transmission line lumped cell (TLLC)

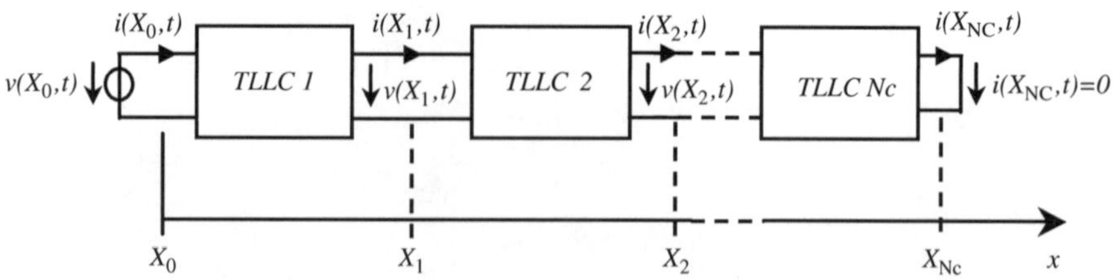

Fig. C.6 Analog circuit lumped model of diffusion of holes in the drift region of the body diode

According to the following equivalences,

$$X_{\mathrm{p}} \Leftrightarrow X_0, \quad W - X_{\mathrm{p}} \Leftrightarrow N_{\mathrm{c}}\Delta x. \tag{C.12)–(C.13}$$

It is straightforward to verify that the diode diffusion current is proportional to the current through the transmission line of Fig. C.6 at $x = X_0$, as follows,

$$\frac{i_{\mathrm{diff\_TL}}(t)}{qA} \Leftrightarrow \frac{i(X_0, t)}{C}, \tag{C.14}$$

where $i_{\mathrm{diff\_TL}}$ is approximately equal to (C.7) when N_{c} is sufficiently high. Note that the transmission line is terminated at $x = X_{N\mathrm{c}}$ with a short so as to emulate an idealized metal contact to the drain terminal. The boundary value at the other side of the transmission line varies instantaneously in proportion to the static characteristics [10], that is,

$$p_n'(X_0, t) = -\frac{K_{\mathrm{p}}}{qA} \cdot I_{\mathrm{Diff(DC)}}(v_{\mathrm{DS}}(t)), \tag{C.15}$$

where K_{p} is a parameter of the model to be determined. $I_{\mathrm{Diff(DC)}}$ in (C.15) corresponds to the simulated steady-state diffusion current from Fig. C.2b, which can be table based implemented in the circuit simulator. Parameter K_{p} must be adjusted to force the steady-state conditions in $i_{\mathrm{diff\_TL}}$ equal to $I_{\mathrm{Diff(DC)}}$.

If expression (C.14) is multiplied by qA, then $i_{\mathrm{diff\_TL}}$ is directly equivalent to $i(X_0, t)/C$. Thus, numerical operations with extreme magnitudes are mitigated.

C.2.2 Lumped Model Based on Two Lumped Storage Nodes

The proposed model from [4] implements approximately the diffusion current equation from (C.7) and a charge control continuity equation that derives from (C.6). The diffusion current from (C.7) is approximated by means of two charge storage nodes, q_{E} and q_{M}, as follows,

$$i_{\mathrm{diff}}(t) \cong i_{\mathrm{diff\_L}}(t) = \frac{q_{\mathrm{E}}(t) - q_{\mathrm{M}}(t)}{T_{\mathrm{M}}}, \tag{C.16}$$

where,

$$q_{\mathrm{E}}(t) = \tau_{\mathrm{m}} \cdot I_{\mathrm{Diff(DC)}}(v_{\mathrm{DS}}(t)), \tag{C.17}$$

$$q_{\mathrm{M}}(t) = \int_0^t q_{\mathrm{E}}(t_{\mathrm{o}}) h_{\mathrm{L}}(t_{\mathrm{o}})\, \mathrm{d}t_{\mathrm{o}} + \frac{q_{\mathrm{E}}(0^-)}{1 + (T_{\mathrm{M}}/\tau_{\mathrm{m}})} \mathrm{e}^{-((1/T_{\mathrm{M}}) + (1/\tau_{\mathrm{m}}))t}, \tag{C.18}$$

$$h_{\mathrm{L}}(t) = \frac{1}{T_{\mathrm{M}}} \mathrm{e}^{-((1/T_{\mathrm{M}})+(1/\tau_{\mathrm{m}}))t} \cdot u(t). \tag{C.19}$$

Note that $u(t)$ is the unit step function. The circuit implementation of the above equations results in a single TLLC, where the following equivalences apply,

$$T_{\mathrm{M}} \Leftrightarrow R, \quad \tau_{\mathrm{m}} \Leftrightarrow \frac{1}{G}, \quad L = 0. \qquad \text{(C.20)–(C.22)}$$

Variable q_{E} represents the stored charge at the boundary between the depletion and quasi-neutral areas of the drift region, whereas q_{M} is the concentrated lumped charge in the quasi-neutral region. Thus, τ_{m} approximates the holes' lifetime, and T_{M} the diffusion time across the quasi-static region.

Due to its arrangement, the DC static current that the model predicts differs from $I_{\mathrm{Diff(DC)}}$. The error is proportional to the following factor,

$$I_{\mathrm{Diff(DC)}} \text{ error factor} = \left(1 + \frac{T_{\mathrm{M}}}{\tau_{\mathrm{m}}}\right)^{-1}. \tag{C.23}$$

In applications where current is impressed into the device, the above error might well be negligible in case of diodes with steep static characteristics.

C.2.3 Empirical Lumped Model

The empirical model from [5] aims at representing the reverse recovery of high current power MOSFETs. As its name indicates, the model does not derive from device physics theory. Instead, its parameters are arranged to allow for an effective curve fitting procedure to match reverse recovery transient curves from device physics simulations and/or measurements.

The diode diffusion current of the body diode is expressed as,

$$\begin{aligned} i_{\mathrm{diff}}(t) \cong i_{\mathrm{diff_E}}(t) &= \beta_{\mathrm{rr}} \cdot I_{\mathrm{Diff(DC)}}(v_{\mathrm{DS}}(t)) + \cdots \\ &+ \frac{1}{t_{\beta\mathrm{rr}}} \int [I_{\mathrm{Diff(DC)}}(v_{\mathrm{DS}}(t)) - i_{\mathrm{diff\_E}}(t)]\, \mathrm{d}t + I_{qt=0}, \end{aligned} \tag{C.24}$$

where β_{rr} and $t_{\beta\mathrm{rr}}$ are the parameters of adjustment and $I_{qt=0}$ the initial conditions of the integral. Rearranging it follows that,

$$i_{\mathrm{diff_E}}(t) = \int_0^t r_{\mathrm{e}}(t_{\mathrm{o}}) h_{\mathrm{e}}(t - t_{\mathrm{o}})\, \mathrm{d}t_{\mathrm{o}} + I_{\mathrm{Diff(DC)}}(v_{\mathrm{DS}}(0)) \cdot \mathrm{e}^{-(t/t_{\beta\mathrm{rr}})}. \tag{C.25}$$

The convolution integral is composed by,

$$h_e(t) = e^{-(t/t_{\beta rr})} \cdot u(t), \tag{C.26}$$

$$r_e(t) = \beta_{rr} \frac{dI_{Diff(DC)}(v_{DS}(t))}{dt} + \frac{I_{Diff(DC)}(v_{DS}(t))}{t_{\beta rr}}. \tag{C.27}$$

Looking at (C.24) one identifies the following,

- For the DC diffusion current to be equal to $I_{Diff(DC)}$ the integral term plus the initial condition must equal to $(1-\beta_{rr})$ times the value of $I_{Diff(DC)}$.
- Parameter β_{rr} controls the contribution of $I_{Diff(DC)}$ to the total current. Increasing its value amplifies the current transient. Thus, β_{rr} allows adjusting the reverse current peak value.
- Parameter $t_{\beta rr}$ controls the integral factor, that is, the rate of change of total amount of charge being stored or removed during the transient current. Thus, $t_{\beta rr}$ allows adjusting the reverse recovery time, particularly, the decay curve time.

Unlike in [4], the steady-state diffusion current error is not present if the initial conditions of the integral are appropriately set.

The model can be realized in a circuit simulator by using (C.25). This requires though that the simulator allows for numerical expressions of mathematical formulations.

C.3 Models Performance

It is of general interest to investigate how accurately the models can predict the reverse recovery curves upon different circuit conditions, particularly, various output current levels and switching speeds. In the way described in Sect. C.2, device physics simulations are performed to produce inductive reverse recovery transients at current slopes from 100 to 1,000 A/μs in 1 mm^2 trench cell. The output current is varied from 0.5 to 5 A. These might well be usual ranges in applications such as point of load that employ multi-chip modules or advance discrete solutions. The values must be scaled according to the total active area of the device. For example, in a 10 mm^2 die size, the corresponding ranges are 1–10 A/ns and 5–50 A, respectively.

The device physics simulations curves are compared to those from the lumped models after calibrating their parameters. The latter is particularly simple in case of the transmission line based model, which only needs extracting parameter K_p. The rest of parameters are directly set according to the equivalences (C.8)–(C.13) and the physical parameter values from Table C.1. A single DC bias point calculation is sufficient to find the value of K_p for matching a point in the static curve of Fig. C.2b.

Table C.3 Simulated electrical conditions for the current flow line graphs

Voltage v_{DS} (V)	Voltage v_{GS} (V)	Current i_D (A)
−0.1	0	-400×10^{-12}
	1	-32×10^{-6}
−0.4	0	-30×10^{-6}
	1	−1.2
−0.8	0	−13.8
	1	−42

The other two lumped models require the derivation of their parameters by means of curve fitting of the reverse recovery transients. The square error of the diffusion current with $K_i = 505$ A/μs and $I_o = 5$ A is minimized. Table C.3 lists the extracted parameter values from the three models.

Regarding the lumped model from [4], the resulting value for τ_m is several orders of magnitude lower than the lifetime of holes τ_p, which suggests that no direct correlation exists between the two parameters. Parameter τ_m is though expected to have a closer value to τ_p for long drift devices, i.e. $L_p \gg W$, which is typical in high voltage diodes [11]. Anyway, parameter T_M cannot be deduced from the physical properties and thus curve fitting is always required.

Figures C.7 and C.8 compare reverse recovery transient curves predicted by the lumped models and device physics simulations. The lumped representations from [4, 5] virtually produce the same response and so the dashed curves refer to both models. They can predict the device physics simulations at the calibrated points with virtually no error. However, when K_i is varied and I_o is kept constant, i.e., upper graphs of Figs. C.7 and C.8, the accuracy error increases. The more K_i is deviated from the calibrated point, the higher the error gets. For the ranges given in the example, the maximum accuracy error is: 35, 30, and 16% in predicting Q_{rr}, I_{rp}, and t_{rp}, respectively. These errors occur when K_i is reduced by factor of 5 from the calibrated point. When I_o is varied and K_i is kept constant (lower graphs of Figs. C.7 and C.8), the maximum accuracy error is 32%, 28, and 27% for Q_{rr}, I_{rp} and t_{rp}, respectively. The latter worst-case numbers result from reducing I_o one order of magnitude from the calibrated point. When comparing the computation power it is observed that the device physics simulations are two orders of magnitude slower than the lumped models from [4, 5], which show similar simulation times.

The transmission line based model is inaccurate at low N_c. At high N_c though the accuracy rapidly increases, being comparable to the other two lumped models when N_c approaches 100. The recovery decay curves at high N_c are however slightly slower than those from device physics simulations, which result in an overestimation of Q_{rr}, as shown in Fig. C.8. Figure C.9 shows the dependency of the accuracy error and simulation time on N_c. The error rapidly reduces when increasing N_c up to 100. Above it the accuracy error practically does not change. Thus, the use of 100 TLLCs appears to be optimum, considering the rapid increase of required computation power at N_c above 100, as shown in Fig. C.9b. At $N_c = 100$, it already doubles the computation time of the other two lumped models, whose simulation times equal to those of a single TLLC (Table C.4).

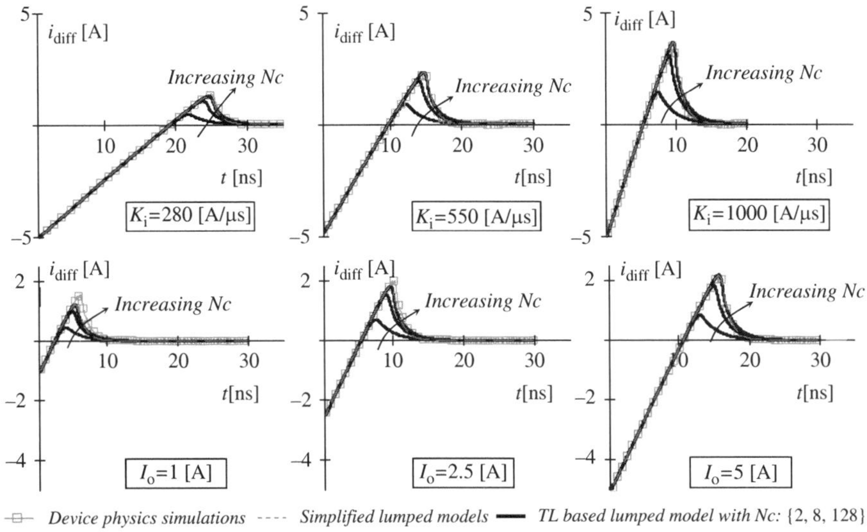

Fig. C.7 Inductive reverse recovery comparison at various current slopes and output currents

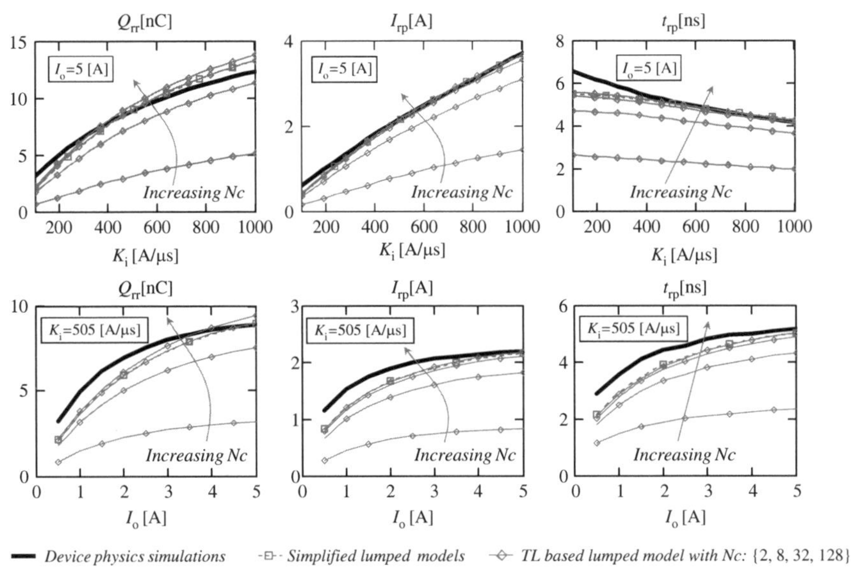

Fig. C.8 Inductive load reverse recovery parameters of various models as a function of the current slope (*upper* graphs), and as a function of the output current (*lower* graphs). From *left* to *right*: Total reverse recovery charge, Q_{rr}, maximum reverse recovery peak, time interval between zero current crossing and maximum reverse current peak, t_{rp}

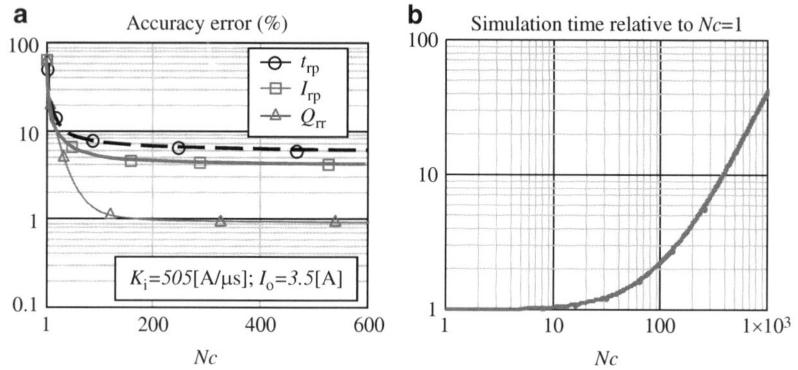

Fig. C.9 Transmission line based model performance from [1, 2] at various TLLCs. (**a**) Accuracy errors with respect to device physics simulations. (**b**) Simulation time relative to $N_c = 1$

Table C.4 Calibrated model parameters

TL model [1–3]	Single TLLC [4]	Empirical model from [5]
$K_p = 54$ µs/cm	$T_M = 1.3$ ns, $\tau_m = 6.8$ ns	$\beta_{rr} = 6.5$, $t_{\beta rr} = 1$ ns

C.4 Summary

Lumped diode models for circuit simulators are suitable for the representation of the body diode dynamics of low voltage trench MOSFETs. Here the considered lumped models require as input data the DC static characteristics of the body diode. Additionally, two parameters must be calibrated in the models from [4, 5] by means of curve fitting of a reference reverse recovery transient curve. This curve fitting involves a procedure to de-embed the diffusion current from the total drain current. device physics simulations can effectively be used to provide all the required data for the lumped models.

It is shown that the lumped models from [4, 5] have a very simple structure, are easy to implement, and furthermore predict with virtually no error the transient response of device physics simulations upon the circuit conditions of the parameters calibration. The accuracy error may though increase when these conditions are varied. Up to 35% prediction error in Q_{rr} is reached when the current ramp slope is reduced by a factor of 5 from the value used in the calibration. The parameters of these two models are not directly correlated to the physical magnitudes of the device. As example, it is shown that the parameter from [4] that is related to the lifetime of holes takes a value that is three orders of magnitude lower than the real value.

The transmission line based model from [1, 2] does not necessarily require the curve fitting of a transient curve for parameter calibration. Instead, only a very simple adjustment is needed. All but one model parameters can be directly deduced from the physical magnitudes of the device. The transmission line representation requires of at least 100 TLLCs and twice computation time to approach the accuracy of the lumped models from [4, 5].

References

1. Linvill JG (1958) Lumped models of transistors and diodes. In: IEEE Proceedings of the IRE, vol 46, pp 1141–1152, June 1958
2. Wang PP, Branin FH Jr (1973) Multi-section network modeling of junction diodes. In: Proceedings of the 4th annual Pittsburgh conference modeling and simulation, pp 470–474
3. Chang F-Y (1996) Transient analysis of diode switching circuits including charge storage effect. In: IEEE transactions on circuits and systems, I: fundamental theory and applications, vol 43, no. 3, pp 177–190, March 1996
4. Lauritzen PO (1991) A simple diode model with reverse recovery. IEEE Trans Power Electron 6(2):188–191
5. Elferich R, López T, Koper N (2005) Accurate behavioural modelling of power MOSFETs based on device measurements and device physics simulations. In: IEEE 11th European conference on power electronics and applications, Sept 2005
6. Tan CM, Tseng K-J (1999) Using power diode models for circuit simulations – a comprehensive review. IEEE Trans Ind Electron 46(3):637–645
7. Koehler D (1967) The charge-control concept in the form of equivalent circuits, representing a link between the classic large signal diode and transistor models. Bell Syst Tech J 46:523–576, ISSN:0005-8580
8. Darling RB (1995) A full dynamic model for pn-junction diode switching transients. IEEE Trans Electron Dev 42(5):969–976
9. López T, Elferich R, Koper N (2006) Reverse recovery in high density trench MOSFETs with regard to the body effect. In: IEEE international symposium on power semiconductor devices and ICs, ISPSD 2006, pp 1–4
10. Sze SM (2002) Semiconductor devices physics and technology, 2nd edn New York, Wiley
11. López T, Alarcón E (2006) Performance comparison of pn-junction diode lumped models for circuit simulators. In: IEEE international symposium on circuit and systems, ISCAS 2006

Appendix D
Loss Quantification

As seen in Chap. 2, the proposed circuit models level 0 and 1 allow breaking down the overall power losses into the individual lossy elements of the converter circuit in the predefined switching time intervals of Sect. 2.5.1.1. This however does not provide a direct correlation between the separated quantities and the loss mechanisms defined in Sect. 2.5.1.2.

In order to quantify loss mechanisms, the dissipated power in every loss by element is systematically regarded as the sum of individual contributions from various current loops in the circuit, each of which can be mathematically formulated and related to a loss mechanism.

D.1 Conceptual Approach

The diagram of Fig. D.1 represents N different current loops formed by the path of resistor R and, paralleled to it, N additional current branches. The instantaneous power contribution to heat dissipation in R caused by the loop of the ith current branch is defined as,

$$p_i = R \cdot i_R^2 \cdot \Gamma_t^N. \tag{D.1}$$

Function Γ expresses the weight of loss contribution of a given loop as follows,

$$\Gamma_i^N = t_{r0}(i_i, i_R) \cdot \left(\sum_{m=1}^{N} t_{r0}(i_m, i_R) \right)^{-1}. \tag{D.2}$$

Function t_{r0} is defined as,

$$t_{r0}(x, y) = \begin{cases} x, & \text{sign}(x) = \text{sign}(y), \\ 0, & \text{otherwise.} \end{cases} \tag{D.3}$$

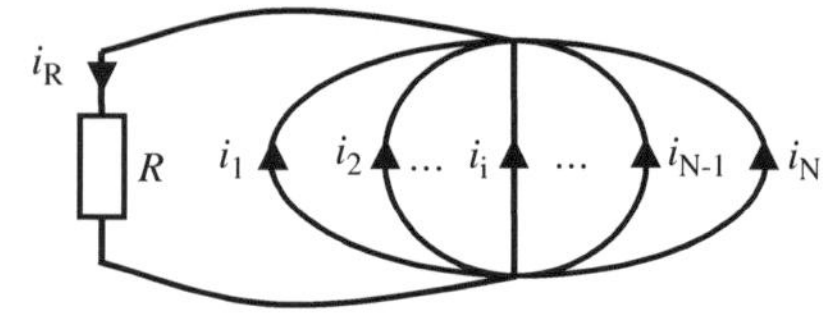

Fig. D.1 Basic current loops diagram of parallel current branches

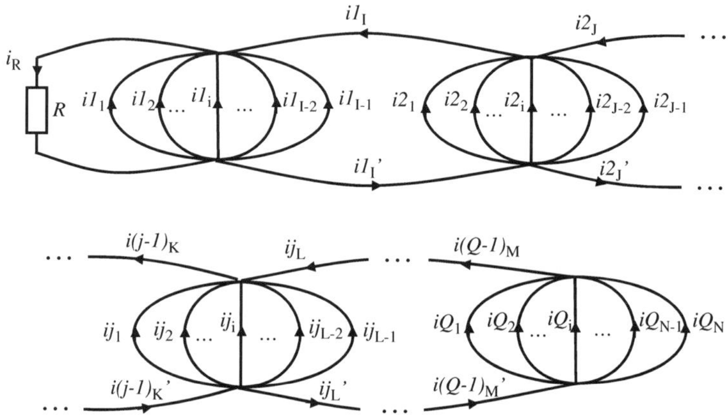

Fig. D.2 Diagram of concatenated groups of parallel current branches

According to (D.1)–(D.3), the loops of those current branches with opposite current direction to that of R have a null contribution to losses. The sum of all current loop contributions must be equal to the total power dissipation in R, that is,

$$\sum_{k=1}^{N} \Gamma_k^N = 1. \tag{D.4}$$

The above definition also applies in case the current through R is formed by a single loop of several series interconnected circuit elements. In such case, the currents in (D.2) must simply be replaced by the voltages across the series elements.

Figure D.2 depicts an extension of the diagram of Fig. D.1. Each current loop may be any closed path formed by the concatenation of two or more current branches. However, as in the simple diagram of Fig. D.1, only those loops containing the current branch of R are considered.

In the extended diagram, series branches (i.e. $i1_I$, $i2_J$, $i(j-1)_K, \ldots$) concatenate groups of parallel branches in a cascade configuration. For instance, group 2 contains $J-1$ parallel branches and 2 series concatenating branches, i.e. $i2_J$ and $i2_{J'}$.

Note that $i2_J = i2_{J'}$. In such arrangement, the power contribution in R of the loop containing the ith parallel branch of the jth group is defined to be,

$$p_{j_i} = \begin{cases} R \cdot i_R^2 \cdot \Gamma_{j_i}^{n(j)} \cdot \prod_{k=1}^{j-1} \Gamma_{k_{n(k)}}^{n(k)}, & j > 1, \\ R \cdot i_R^2 \cdot \Gamma_{j_i}^{n(j)}, & j = 1. \end{cases} \tag{D.5}$$

Function $n(k)$ provides the number of branches of a group minus 1, e.g., $n(k) = L$ in Fig. D.2. For the particular case of the last group of the chain, $n(k)$ equals to the total number of parallel branches of the group since it does not contain series branches. The defined weight of contribution Γ of a current branch is given by,

$$\Gamma_{j_i}^{L} = t_{r0}\left(ij_i, \sum_{k=1}^{L} ij_k\right) \cdot \left(\sum_{m=1}^{L} t_{r0}(ij_m, ij_k)\right)^{-1}. \tag{D.6}$$

Note from (D.5) that the loss contribution of a current loop may be dependent on the product of multiple Γ functions. The total number of different current loops that include R is,

$$\text{Number of loops} = 1 - Q + \sum_{k=1}^{Q} n(k). \tag{D.7}$$

Thus, (D.7) provides the total number of loss contributors in R, each of which corresponds to the current loop of a particular parallel current branch.

D.2 Implementation

The formulation of the previous section is now put in practice by deriving the corresponding current loops diagram of each lossy element of the switched circuit. Figure D.3 illustrates the case of the CtrlFET channel. It is straightforward to verify

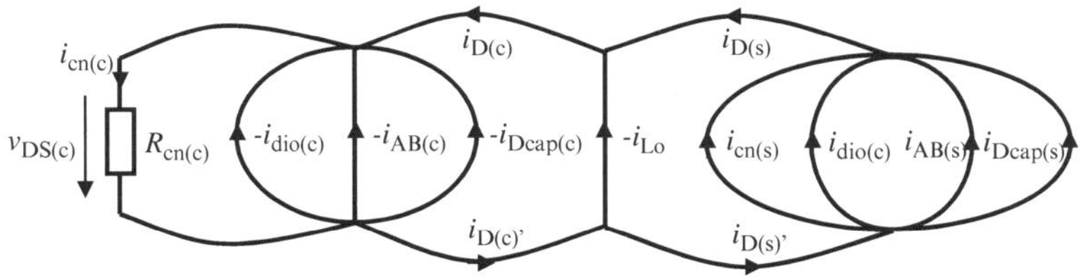

Fig. D.3 Diagram of current loops corresponding to the CtrlFET channel

that all current loops of the diagram that include $i_{\mathrm{cn(c)}}$ have a correspondence to the circuit model of Fig. D.3.

Expressions (D.5) and (D.6) are applied to assess the loss contribution of the various current loops in the form of a power quantity. Thus, for instance, the instantaneous power loss contribution of the output current loop in the CtrlFET channel is,

$$p_{\mathrm{cn(c)_{io}}} = v_{\mathrm{DS(c)}} \cdot i_{\mathrm{cn(c)}} \cdot \Gamma_{\mathrm{cn(c)_{dC}}} \cdot \Gamma_{\mathrm{cn(c)_{io}}}. \tag{D.8}$$

The product $\Gamma_{\mathrm{cn(c)_{dC}}}$ by $\Gamma_{\mathrm{cn(c)_{io}}}$ from (D.8) is the weight of the contribution of the output current loop. Their individual quantities are calculated as follows,

$$\Gamma_{\mathrm{cn(c)_{dC}}} = \frac{t_{r0}\left(i_{\mathrm{D(c)}}, i_{\mathrm{cn(c)}}\right)}{\mathrm{den}\ \Gamma_{\mathrm{cn(c)_{CtrlF}}}}, \tag{D.9}$$

$$\Gamma_{\mathrm{cn(c)_{io}}} = \frac{t_{r0}\left(i_{\mathrm{Lo}}, i_{\mathrm{D(c)}}\right)}{t_{r0}\left(i_{\mathrm{Lo}}, i_{\mathrm{D(c)}}\right) + t_{r0}\left(i_{\mathrm{D(s)}}, i_{\mathrm{D(c)}}\right)}, \tag{D.10}$$

$$\begin{aligned} \mathrm{den}\ \Gamma_{\mathrm{cn(c)_{CtrlF}}} &= t_{r0}\left(-i_{\mathrm{dio(c)}}, i_{\mathrm{cn(c)}}\right) + t_{r0}\left(-i_{\mathrm{AB(c)}}, i_{\mathrm{cn(c)}}\right) + \cdots \\ &\quad + t_{r0}\left(-i_{\mathrm{Dcap(c)}}, i_{\mathrm{cn(c)}}\right) + t_{r0}\left(i_{\mathrm{D(c)}}, i_{\mathrm{cn(c)}}\right). \end{aligned} \tag{D.11}$$

Likewise, the loss contribution of the loops corresponding to the body diode and avalanche breakdown branches can be calculated from their equivalent current loops diagrams.

The current loops diagrams corresponding to the drain and source resistances of the MOSFETs differ according to the state of the switches. Namely, the current flow through these resistances may be attributed to the output load, the gate circuit path, and the paths of the corresponding off-state device. Regarding the OFF state device paths, current flow may be produced by gate bounce, capacitive charge variation, reverse recovery and avalanche breakdown. Thus, the considered current loops in the diagram may be those containing as parallel branches the output coil, the gate circuit and the internal branches of the OFF state device.

Figure D.4 shows the diagram for $R_{\mathrm{S(c)}}$ during the ON conduction time of the CtrlFET. In the example of Sect. 2.5.1, this time may correspond to any that is contained in the interval [$t1$, $t7$]. Similarly, Fig. D.5 shows the loops diagram for $R_{\mathrm{S(c)}}$ during the OFF state of the CtrlFET.

The deduction of the loss contribution quantities corresponding to the six current loops of Figs. D.4 and D.5 is immediate employing (D.5) and (D.6). Likewise, the calculations of the loss contributions in resistances $R_{\mathrm{D(c)}}$, $R_{\mathrm{D(s)}}$, $R_{\mathrm{S(s)}}$ result from similar current loop diagrams.

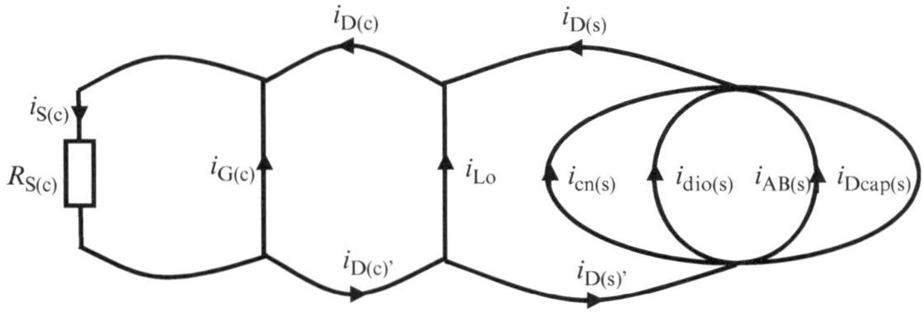

Fig. D.4 Diagram of current loops corresponding to the CtrlFET source resistance during the ON state of the device

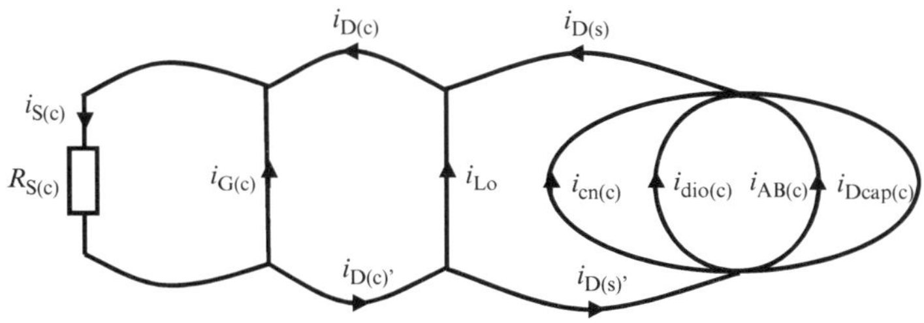

Fig. D.5 Diagram of current loops corresponding to the CtrlFET source resistance during the OFF state of the device

D.3 Loss Mechanisms Formulation

This section formulates the relation between the voltage and current contributions to power losses and the different loss mechanisms defined in Chap. 2.

D.3.1 Load Current Hard-Switching

According to its definition, snubbed hard-switching is related to the loss contribution of the output current loop and the CtrlFET channel voltage without the contribution of the half-bridge inductance voltage. This can be expressed as follows,

$$p_{\mathrm{SHS}} = v_{\mathrm{DS(c)}} \cdot \left[\Gamma_{\mathrm{cn(c)}V_{\mathrm{in}}} + \Gamma_{\mathrm{cn(c)}v_{\mathrm{DS(s)}}}\right] \cdot i_{\mathrm{cn(c)}} \cdot \Gamma_{\mathrm{cn(c)dC}} \cdot \Gamma_{\mathrm{cn(c)}_{\mathrm{io}}}. \tag{D.12}$$

Functions $\Gamma_{\mathrm{cn(c)}V_{\mathrm{in}}}$ and $\Gamma_{\mathrm{cn(c)}_{v_{\mathrm{DS(s)}}}}$ are the weights of loss contributions of voltages V_{in} and $v_{\mathrm{DS(s)}}$. These quantities are obtained by replacing currents for voltages in (D.12) according to the following expression,

$$v_{\mathrm{DS(c)}} = V_{\mathrm{in}} - V_{\mathrm{L}} - v_{\mathrm{DS(s)}}. \tag{D.13}$$

Voltage V_L is the voltage induced across the half-bridge inductances. Note from (D.13) that the voltages across R_S and R_D from both MOSFETs are, for this case, neglected.

D.3.2 Half-Bridge Charging

Half-bridge charging is related to the loss contributions of the MOSFET output capacitive currents, and the over voltage across the CtrlFET channel caused by the half-bridge inductances during hard-switching. The latter is calculated as follows,

$$p_{\mathrm{HBC(L)}} = v_{\mathrm{DS(c)}} \cdot \Gamma_{\mathrm{cn(c)}V_L} \cdot i_{\mathrm{cn(c)}} \cdot \Gamma_{\mathrm{cn(c)dC}} \cdot \Gamma_{\mathrm{cn(c)_{io}}}. \tag{D.14}$$

Function $\Gamma_{\mathrm{cn(c)}_{V_L}}$ is the weight of loss contribution of the half-bridge inductances. Like $\Gamma_{\mathrm{cn(c)}_{V_{in}}}$ and $\Gamma_{\mathrm{cn(c)}_{v_{\mathrm{DS(s)}}}}$ in (D.12), function $\Gamma_{\mathrm{cn(c)}_{V_L}}$ results from (D.2) and (D.13).

The output capacitance current contributions to heat dissipation are obtained from (D.5) to (D.6) for each lossy element of the half-bridge, i.e., from their corresponding current loops diagrams. For instance, the loss contribution of the SyncFET output capacitance current loop in the CtrlFET channel is,

$$p_{\mathrm{cn(c)_{coS}}} = v_{\mathrm{DS(c)}} \cdot i_{\mathrm{cn(c)}} \cdot \Gamma_{\mathrm{cn(c)_{dC}}} \cdot \Gamma_{\mathrm{cn(c)_{dS}}} \cdot \Gamma_{\mathrm{cn(c)_{coS}}}. \tag{D.15}$$

The product of $\Gamma_{\mathrm{cn(c)_{dC}}}$, $\Gamma_{\mathrm{cn(c)_{dS}}}$ and $\Gamma_{\mathrm{cn(c)_{coS}}}$ is the weight of loss contribution of the SyncFET output capacitance current loop corresponding to the diagram of Fig. D.3.

D.3.3 Gate Charging

Gate charging is associated to the power loss in the gate drive circuit. It includes the total power dissipation in the drivers and gate resistances, and also part of the dissipation in the source resistances. The latter requires discrimination of the loss contributions from the gate and drain currents. Thus, the following expression is applied for the case of the CtrlFET,

$$p_{\mathrm{Rs(c)_{igC}}} = R_{\mathrm{S(c)}} \cdot i^2_{\mathrm{S(c)}} \cdot \Gamma_{\mathrm{Rs(c)_{igC}}}, \tag{D.16}$$

where $i_{\mathrm{S(c)}}$ is the source current and $\Gamma_{\mathrm{Rs(c)_{igC}}}$ the weight of loss contribution of the gate current. Similarly, for the case of the SyncFET,

$$p_{\mathrm{Rs(s)_{igS}}} = R_{\mathrm{S(s)}} \cdot i^2_{\mathrm{S(s)}} \cdot \Gamma_{\mathrm{Rs(s)_{igS}}}. \tag{D.17}$$

D.3.4 Reverse Recovery

Reverse recovery is related to the loss occurring in the half-bridge path during the reverse recovery time intervals. Outside this interval, the reverse recovery tail generates further loss in the half-bridge. For instance, the loss in the CtrlFET channel is calculated as follows,

$$p_{\mathrm{cn(c)}_{\mathrm{dioS}}} = v_{\mathrm{DS(c)}} \cdot i_{\mathrm{cn(c)}} \cdot \Gamma_{\mathrm{cn(c)}_{\mathrm{dC}}} \cdot \Gamma_{\mathrm{cn(c)}_{\mathrm{dS}}} \cdot \Gamma_{\mathrm{cn(c)}_{\mathrm{dioS}}}. \tag{D.18}$$

The product of $\Gamma_{\mathrm{cn(c)}_{\mathrm{dC}}}$, $\Gamma_{\mathrm{cn(c)}_{\mathrm{dS}}}$, and $\Gamma_{\mathrm{cn(c)}_{\mathrm{dioS}}}$ is the weight of contribution of the SyncFET body diode current loop corresponding to the diagram of Fig. D.3. The loss contribution in the other lossy elements of the half-bridge can be likewise calculated.

D.3.5 Gate Bouncing

Gate bounce related loss is derived from the current contribution of the MOSFETs channel that is spuriously turned on. For instance, the loss contribution of the SyncFET channel loop in $R_{\mathrm{S(c)}}$ due to gate bounce is,

$$p_{\mathrm{Rs(c)}_{\mathrm{cnS}}} = R_{\mathrm{S(c)}} \cdot i^2_{\mathrm{S(c)}} \cdot \Gamma_{\mathrm{Rs(c)}_{\mathrm{dC}}} \cdot \Gamma_{\mathrm{Rs(c)}_{\mathrm{dS}}} \cdot \Gamma_{\mathrm{Rs(c)}_{\mathrm{cnS}}}. \tag{D.19}$$

The product of $\Gamma_{\mathrm{Rs(c)}_{\mathrm{dC}}}$, $\Gamma_{\mathrm{Rs(c)}_{\mathrm{dS}}}$ and $\Gamma_{\mathrm{Rs(c)}_{\mathrm{cnS}}}$ is the weight of contribution of the SyncFET channel loop according to the diagram of Fig. D.4. The contributions of this loss mechanism in the other half-bridge lossy elements are similarly calculated.

D.3.6 Avalanche Breakdown

Most of the avalanche breakdown related loss occurs in the avalanching device. It is readily calculated as,

$$p_{\mathrm{AB}} = v_{\mathrm{DS}} \cdot i_{\mathrm{AB}}. \tag{D.20}$$

The avalanche current i_{AB} may further generate heat dissipation in the half-bridge. For instance, the instantaneous power loss in $R_{\mathrm{S(c)}}$ due to CtrlFET avalanche is quantified as,

$$p_{\mathrm{Rs(c)}_{\mathrm{abC}}} = R_{\mathrm{S(c)}} \cdot i^2_{\mathrm{S(c)}} \cdot \Gamma_{\mathrm{Rs(c)}_{\mathrm{dC}}} \cdot \Gamma_{\mathrm{Rs(c)}_{\mathrm{dS}}} \cdot \Gamma_{\mathrm{Rs(c)}_{\mathrm{abC}}}. \tag{D.21}$$

The product of $\Gamma_{\mathrm{Rs(c)}_{\mathrm{dC}}}$, $\Gamma_{\mathrm{Rs(c)}_{\mathrm{dS}}}$ and $\Gamma_{\mathrm{Rs(c)}_{\mathrm{abC}}}$ is the weight of contribution of the CtrlFET avalanche current loop corresponding to the diagram of Fig. D.5.

D.3.7 Output Current ON Conduction

It accounts for the loss contribution of the output current loop in the MOSFETs throughout the ohmic and dead time intervals. Expressions (D.8)–(D.11) provide the instantaneous loss contribution of i_{Lo} in the CtrlFET channel. The loss contributions from the other half-bridge lossy elements result from similar calculations.

Appendix E
Magnetic Losses

This section presents illustrative calculations on magnetic losses and parasitic impedances in a PCB track section, a power MOSFET package and an output filter coil for high switching frequency VR converters. Magnetic fields created by time dependent excitation currents are numerically solved by means of an FEM software [1, 2]. Eddy current losses are determined in the time domain by applying typical excitation waveforms of the converter. Alternatively, harmonic calculations leading to estimations on ESR will be used to analytically approximate transient Eddy current losses. For simple cases, the ESR will be analytically expressed as a function of physical parameters such as the skin depth.

Simulations in the frequency domain will also be applied to estimate the ESL of a multilayer PCB track section and the leadframe of the power LFPak. The latter is modeled in 3D in combination with a multilayer PCB layout from which the source inductance will be estimated and compared to experimental results.

Simulations of the power coil include the nonlinear behavior of the ferrite material, in particular, the characteristics of the B–H curves, for which the hysteresis model from Jiles–Atherton is implemented in the device physics simulator. Besides Eddy current and hysteresis losses, excess loss will be derived from specific power loss curves of manufacturer data in combination with the field calculations.

E.1 Quasi-Static Formulation

The quasi-static formulation of time varying electromagnetic fields is suitable for the analysis of structures with a size much smaller than the characteristic wavelength. The physical assumption of this situation is that the current and charges generating the electromagnetic fields vary so slowly in time that are practically the same everywhere within the domain structure at every instant as if they had been generated by stationary sources.

This approximation implies that the equation of continuity can be written as,

$$\vec{\nabla} \cdot \vec{J} = 0. \tag{E.1}$$

Also, the electric displacement can be disregarded in the Maxwell-Ampère's law, thus yielding,

$$\vec{\nabla} \times \vec{H} = \vec{J}. \tag{E.2}$$

Vector $\vec{J}$ is the current density, which may be formed by an external supply and the time varying magnetic field, i.e.,

$$\vec{J} = \sigma \vec{E} = \sigma \left(\vec{J_e} - \frac{\partial \vec{A}}{\partial t} \right), \tag{E.3}$$

where parameter σ is the material's conductivity, current vector $\vec{J_e}$ the current density created by an external source and $\vec{A}$ the magnetic vector potential, which is related to the magnetic field $\vec{B}$ through the curl operator. Furthermore, from Faraday's law, it follows that,

$$\vec{\nabla} \times \vec{E} = -\frac{\partial \vec{B}}{\partial t}. \tag{E.4}$$

Combining (E.1)–(E.4) yields the following partial differential equation,

$$\sigma \frac{\partial \vec{A}}{\partial t} + \vec{\nabla} \times (\mu^{-1} \vec{\nabla} \times \vec{A}) = \vec{J_e}. \tag{E.5}$$

Parameter μ is the material's permeability that relates $\vec{B}$ and $\vec{H}$. Equation (E.5) is solved numerically by the FE method in order to estimate the effects of Eddy currents in parts of the converter circuit.

E.2 PCB Layout

It is frequently advantageous to base the PCB layout design on multilayer structures where the inner layer is used as return ground path in such a way that the loop inductance is minimized. Figure E.1 depicts a cross-sectional view of such layout arrangement, where the currents in the top and bottom traces are equal in magnitude but flow in opposite directions. For simplicity, only the section to the right of the line of symmetry is represented and simulated.

Harmonic device physics simulations of the PCB cross section allow calculating the associated ESR and ESL contribution as a function of frequency. The results of Fig. E.2 show that the skin effect becomes significant at frequencies around 1 MHz and above, reaching 10 times higher ESR at 25 MHz with respect to DC operation.

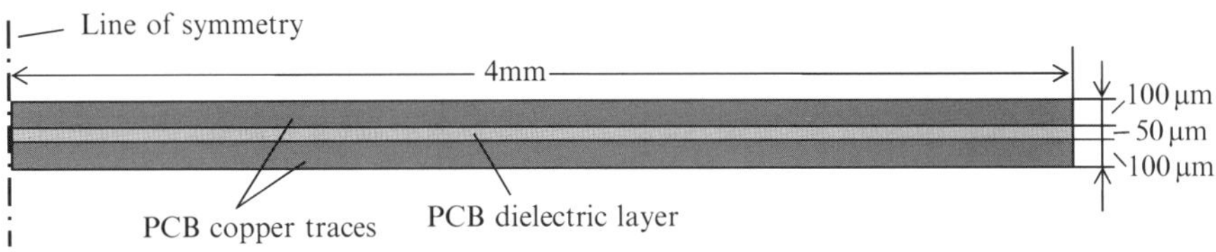

Fig. E.1 Simulated 2D cross-sectional geometry corresponding to a double-layer PCB

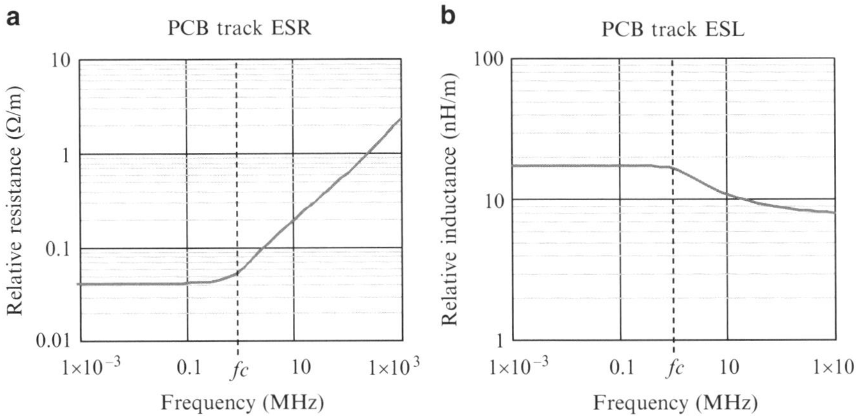

Fig. E.2 Harmonic simulations of the double-layer PCB layout of Fig. E.1

The influence of the dielectric thickness on the ESL is reflected in the curves of Fig. E.3, showing that the loop area increase boosts the parasitic ESL in a nonlinear fashion. This contrasts with the ESR, which is fairly independent of the separation distance between the two parallel conductive layers.

A relevant physical quantity related to magnetic losses is the *skin depth*, which defines the penetration distance of the magnetic field into a conductive material. The skin depth is defined as,

$$\delta(f) = \frac{1}{\sqrt{\pi \cdot \mu_0}} \sqrt{\frac{\rho}{\mu_r \cdot f}} \approx 503 \sqrt{\frac{\rho}{\mu_r \cdot f}}. \tag{E.6}$$

Figure E.4 shows an example of the skin depth dependence on the frequency in the case of copper material.

The AC resistance estimated from harmonic calculations can be analytically calculated for simple geometries like that shown in Fig. E.1. Provided that the current density mainly crowds in the conductor sides in contact to the dielectric layer, the ESR may be approximately expressed as follows,

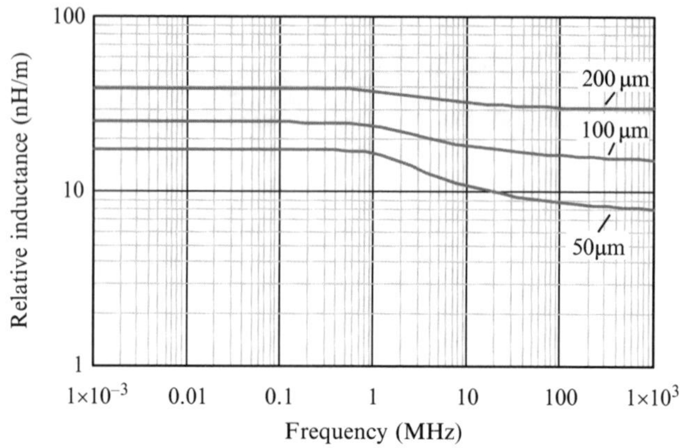

Fig. E.3 Harmonic simulations of the double-layer PCB layout of Fig. E.1. Inductance extraction for various gap thickness

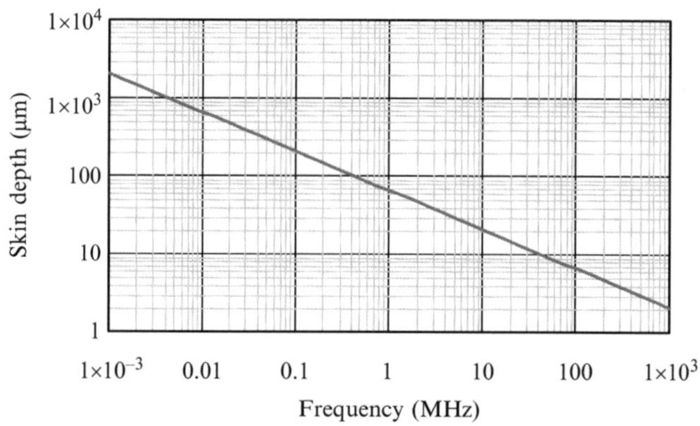

Fig. E.4 Skin depth in copper material at 25°C. ρ = 16.67 nm, $\mu_r = 1$

$$R_{\mathrm{pcb}}(f) = 2\rho \begin{bmatrix} 2 \cdot \left(H_x - 2 \cdot \min\left(\frac{H_x}{2}, \delta(f)\right)\right) \cdot \min\left(\frac{L_x}{2}, \delta(f)\right) + \cdots \\ 2 \cdot \left(L_x - 2 \cdot \min\left(\frac{L_x}{2}, \delta(f)\right)\right) \cdot \min\left(\frac{H_x}{2}, \delta(f)\right) + \cdots \\ 4 \cdot \min\left(\frac{H_x}{2}, \delta(f)\right) \cdot \min\left(\frac{L_x}{2}, \delta(f)\right) \end{bmatrix}^{-1} . \quad \text{(E.7)}$$

Parameters H_x and L_x are the thickness and width of the conductive layers, respectively (e.g. 100 μm and 8 mm in the example of Fig. E.1). Comparing device

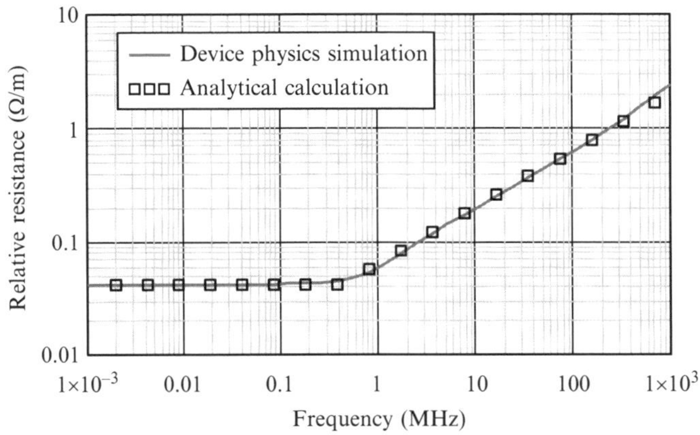

Fig. E.5 AC resistance of the double layer PCB track section of Fig. E.1. Comparison between device physics simulations and analytical calculations, i.e., (E.7) with $H_x = 100$ μm and $L_x = 8$ mm

physics simulations with (E.7) as in Fig. E.5 shows that the approximations adopted are adequate for this particular case.

In the application of interest, however, PCB losses may result from trapezoidal current waveforms rather than sinusoidal excitations. The following establishes a relation between the above calculated AC resistance and the power losses associated to ramp current waveforms.

The transient skin effect simulations of Figs. E.6–E.8 show current crowding plots resulting from a current ramp excitation of 10 A/μs. Regardless of the magnitude of the slope, however, the current density distribution varies as a function of time from a boundary conduction at $t = 0$, to a quasi uniform distribution across the cross-sectional area of the copper layers in steady-state, i.e., as $t \to \infty$.

The transient time between the two states strongly depends on the geometry and is proportional to a time constant defined by the cut-off frequency in the harmonic domain (i.e., f_c in Fig. E.2). Steady-state operation may thus be considered for $t > 5\tau_c$, where,

$$\tau_c = \frac{1}{2 \cdot \pi \cdot f_c}. \tag{E.8}$$

The same expression can be used to relate the energy loss at any time instant to its corresponding resistance in the frequency domain as follows,

$$w_{rx}(t_x) = R_x(t_x) \int_0^{t_x} i^2 \, dt \cong R_{pcb}\left(\frac{1}{2 \cdot \pi \cdot t_x}\right) \int_0^{t_x} i^2 \, dt, \tag{E.9}$$

where w_{rx} represents the energy loss per unit length.

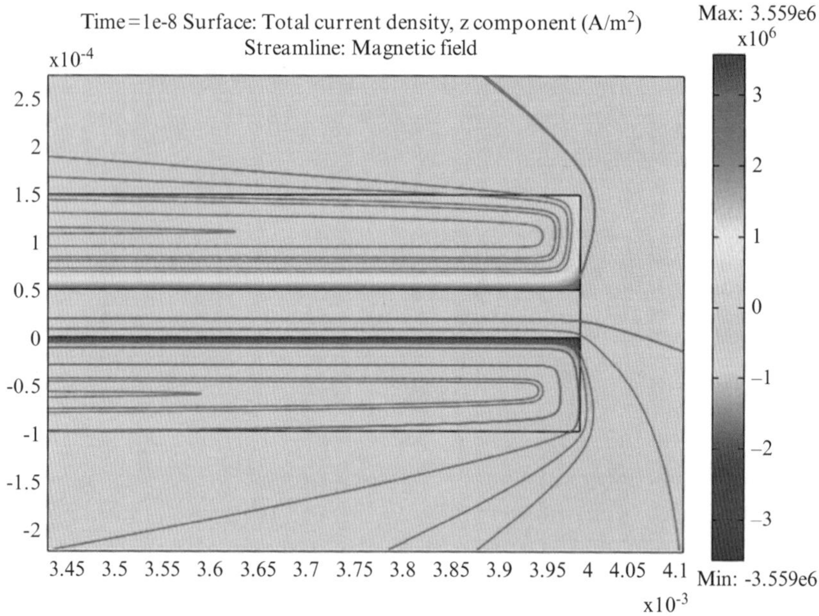

Fig. E.6 2D transient skin effect simulation of the double-layer PCB track section of Fig. E.1. Cross-sectional view of the right edge. Ramp excitation of 10 A/μs starting with zero initial conditions. Total current density distribution and magnetic field lines at $t = 10$ ns

Similarly, the equivalent ESL may be calculated from the stored magnetic energy at any given time,

$$L_x(t_x) = \frac{2 \cdot w_{mx}(t_x)}{\int_0^{t_x} i^2 \, \mathrm{d}t}. \tag{E.10}$$

Figure E.9 depicts the close correlation between the estimated ESR and ESL in both frequency and time domains.

Expression (E.9) can be applied to approximate the PCB power loss corresponding to a trapezoidal current waveform as shown in Fig. E.10a. The approximation consists of dividing the waveform into three sections defined by the ramp slopes and computing (E.9) in each of the resulting intervals. This is, for the case example of Fig. E.10, as follows,

$$P_\mathrm{T} \cong \frac{1}{T_\mathrm{s}} \left[R_x(10\,\mathrm{ns}) \cdot \left(\int_0^{10\,\mathrm{ns}} i^2(t) \cdot \mathrm{d}t + \int_{50\,\mathrm{ns}}^{60\,\mathrm{ns}} i^2(t) \cdot \mathrm{d}t \right) + R_x(40\,\mathrm{ns}) \int_{10\,\mathrm{ns}}^{50\,\mathrm{ns}} i^2(t) \cdot \mathrm{d}t \right]$$
$$= 3.3\,\mathrm{W/m}. \tag{E.11}$$

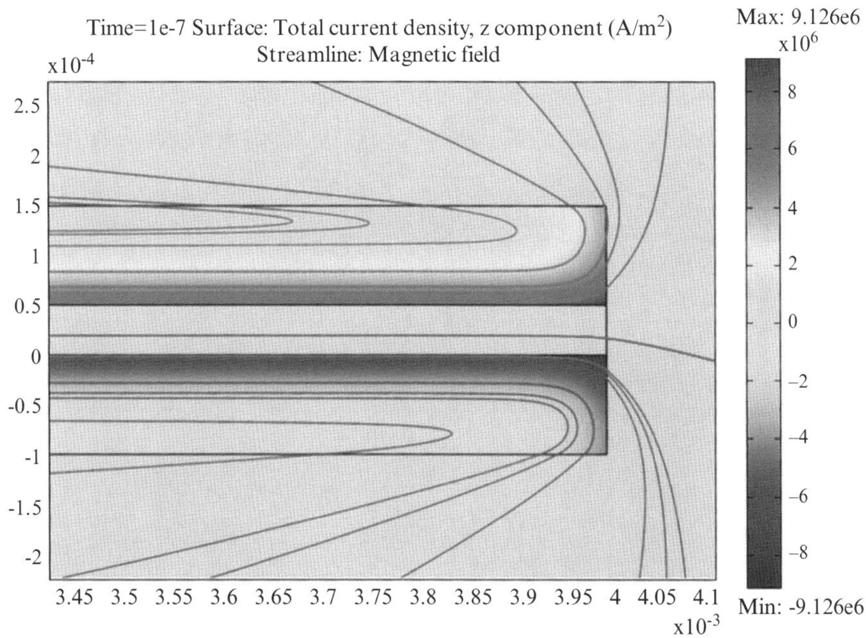

Fig. E.7 2D transient skin effect simulation of the double-layer PCB track section of Fig. E.1. Cross-sectional view of the right edge. Ramp excitation of 10 A/μs starting with zero initial conditions. Total current density distribution and magnetic field lines at $t = 100$ ns

This result matches the average power loss estimated from calculations, as shown in the instantaneous power loss plot of Fig. E.10b.

The trapezoidal signal of Fig. E.10a resembles the typical current waveform through the CtrlFET, where the ON time is generally much shorter than the switching period. While adequate for short ON times, expression (E.11) becomes inaccurate for calculating the resistive losses of the SyncFET path since the ON time of the switch is closer to the switching period. In this case the following expression appears to be more accurate,

$$w_{rs} = \left[R_x(0) \cdot I_{\mathrm{RMS}}^2 + R_x(T_{\mathrm{S}}) \cdot (I_{\mathrm{RMS}} - |I_{\mathrm{AVG}}|)^2\right] \cdot T_{\mathrm{s}}^{-1}. \qquad \text{(E.12)}$$

Currents I_{RMS} and I_{AVG} in (E.12) are the RMS and average quantities of the impressed waveform, respectively. The second term accounts for the AC component and vanishes as the duty cycle approaches unity.

Figure E.11 shows transient skin effect simulations of the PCB track section of Fig. E.1 with an impressed current that resembles that of the SyncFET path. Applying (E.12) with the use of Fig. E.9 yields an estimated track section loss of 6.5 W/m, which is close to the average value of the instantaneous power curve of Fig. E.11b.

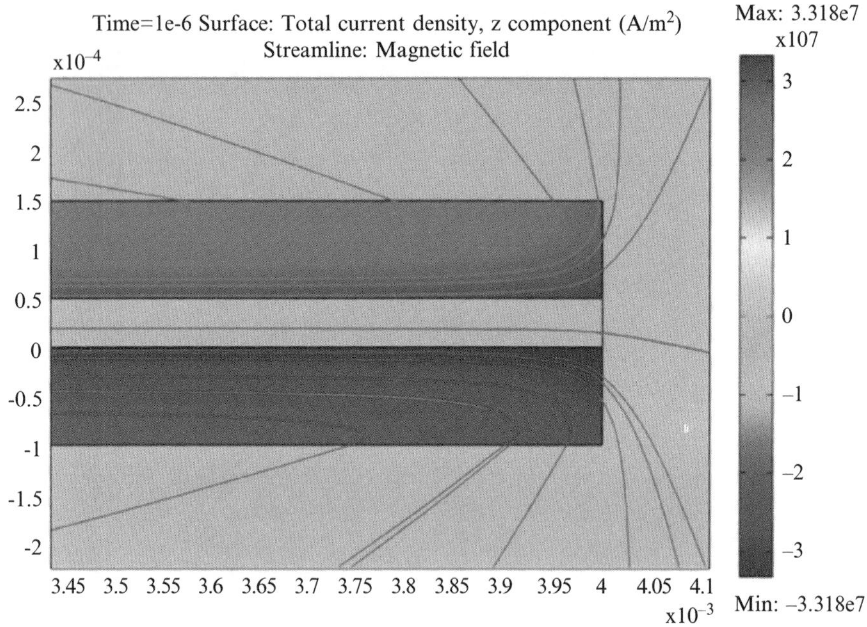

Fig. E.8 2D transient skin effect simulation of the double-layer PCB track section of Fig. E.1. Cross-sectional view of the right edge. Ramp excitation of 10 A/μs starting with zero initial conditions. Total current density distribution and magnetic field lines at $t = 1{,}000$ ns

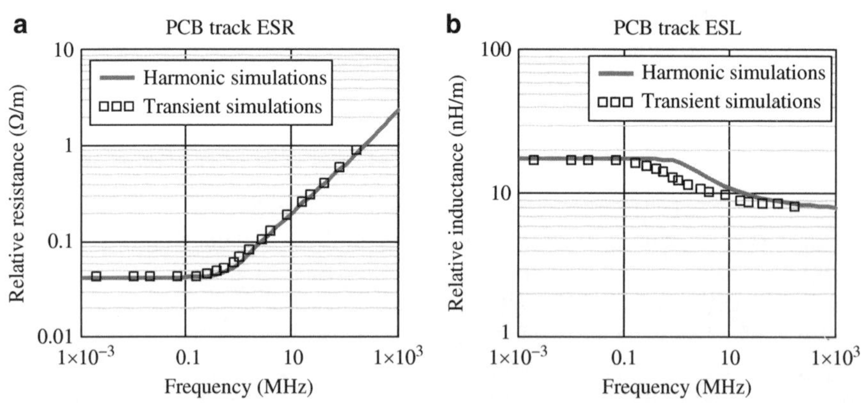

Fig. E.9 Relative effective ESR and ESL of the double-layer PCB track section of Fig. E.1. Comparison between harmonic and transient simulations

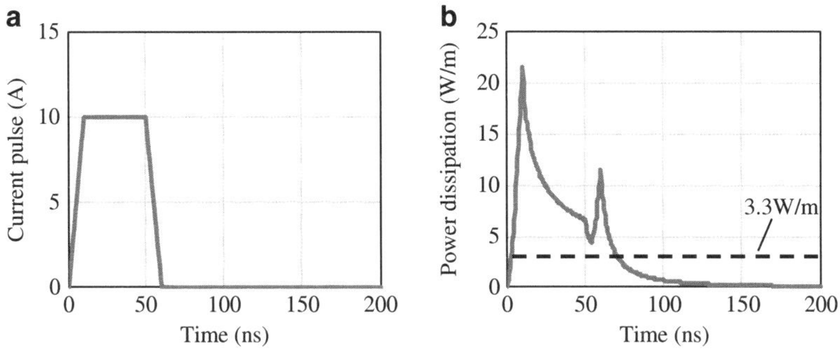

Fig. E.10 2D transient skin effect simulation of the double-layer PCB track section of Fig. E.1. (**a**) Excitation current waveform of CtrlFET drain current path. (**b**) Instantaneous and average power dissipation

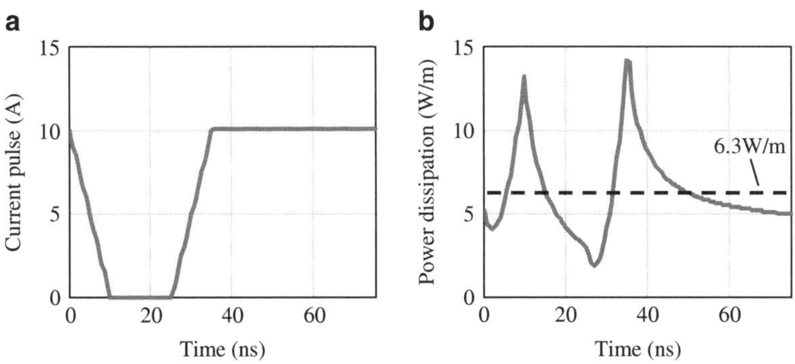

Fig. E.11 2D transient skin effect simulation of the double-layer PCB track section of Fig. E.1. (**a**) Excitation current waveform of SyncFET drain current path. (**b**) Instantaneous and average power dissipation

For low and moderate ripple currents, empirical equation (E.12) may also be adequately applied to approximate the resistive losses of the PCB track section corresponding to the output filter inductor path.

E.3 Power MOSFET Package: LFPak

This section presents simulation results on the LFPak impedance, in particular, the leadframe of the source contact. This electrode contributes predominantly to the overall package resistance as well as the source inductance. The goal is to estimate these two parameters in a wide frequency range. In order to emulate operating

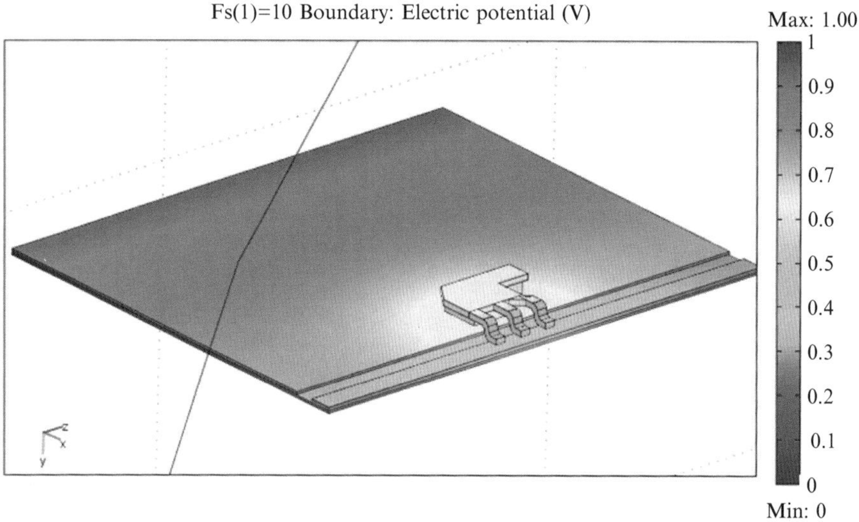

Fig. E.12 Simulated LFpak leadframe mounted on a double layer PCB. View of the voltage distribution in DC operation

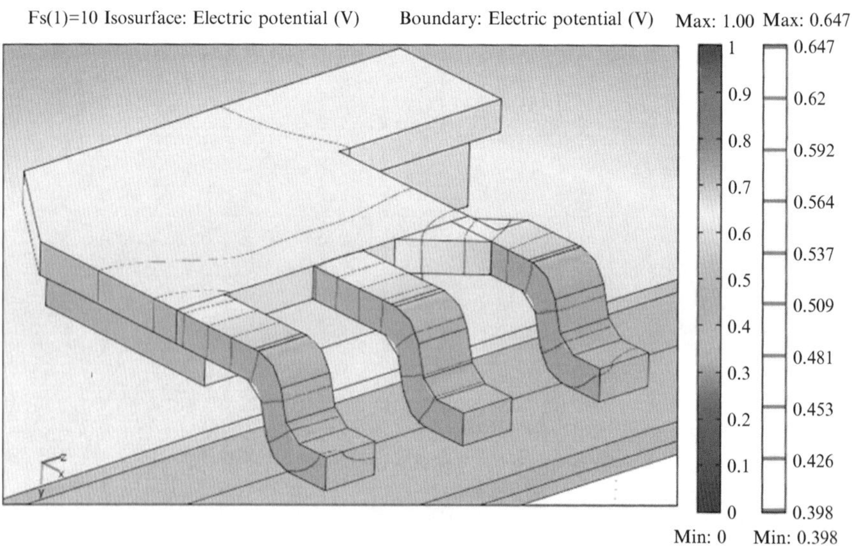

Fig. E.13 Zoom in view of the LFpak leadframe corresponding to the simulation of Fig. E.12

conditions similar to the application at hand, the simulations include both package leadframe and PCB board with a ground layer that confines the magnetic field and thus minimizes the loop inductance.

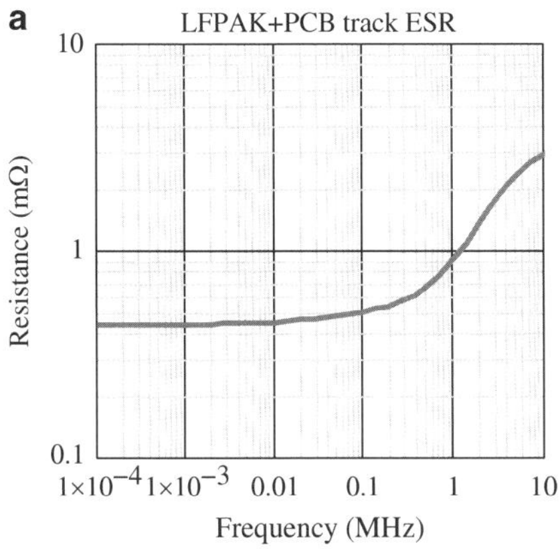

Fig. E.14 3D harmonic simulations of the LFpak leadframe impedance mounted on a double layer PCB (see Fig. E.12)

Figures E.12 and E.13 show a 3D view of the simulated leadframe plus PCB board arrangement comprising two metal layers (top and bottom) separated by a 50 μm dielectric layer. The excitation signal is impressed across the conductive layers in the y–z plane at $x = X_o$, X_o being at the PCB edge furthest away from the leadframe.

The isolines of Fig. E.13 describe the voltage distribution across the leadframe, which corresponds to DC operation. The total voltage across the leadframe divided by the total current yields a source resistance of ~110 μΩ. This is 25% of the total estimated resistance, as it can be derived from Fig. E.14a.

According to the simulations of Fig. E.14, the ESL at 1 MHz is around 500 pH, whereas the ESR increases to 0.9 mΩ from the 0.42 mΩ of DC operation. From these total 0.9 mΩ AC resistance, approximately 0.4 mΩ belong to the leadframe, which suffers a dramatic resistance increase due to Eddy currents. As to the ESL, the PCB tracks contribute with ~150 pH according to 2D simulations like shown in Sect. E.2. This leaves a ~350 pH of source inductance in the proposed geometrical arrangement.

The ESR frequency dependence is a consequence of Eddy currents in the conductors that force the current flow to confine at the conductor boundaries as the frequency increases. This is illustrated in Figs. E.15–E.17 at various harmonic frequencies. The plots show that Eddy currents are low at 10 kHz and below, whereas in the MHz range, strong current crowding is manifested in the corner leads on the leadframe as shown in Fig. E.17.

E.4 Power Coil

The purpose of this section is to illustrate the performance of a VR power inductor for high-switching frequency operation by means of device physics simulations and the theory of magnetism in ferromagnetic materials. The results provide

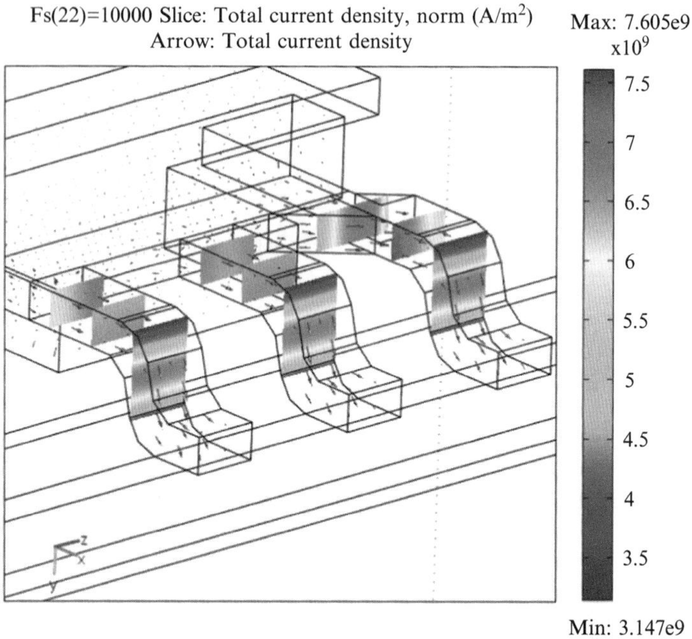

Fig. E.15 3D harmonic simulation of the LFpak leadframe mounted on a double layer PCB as shown in Fig. E.14. Current distribution through the source leads at 10 kHz

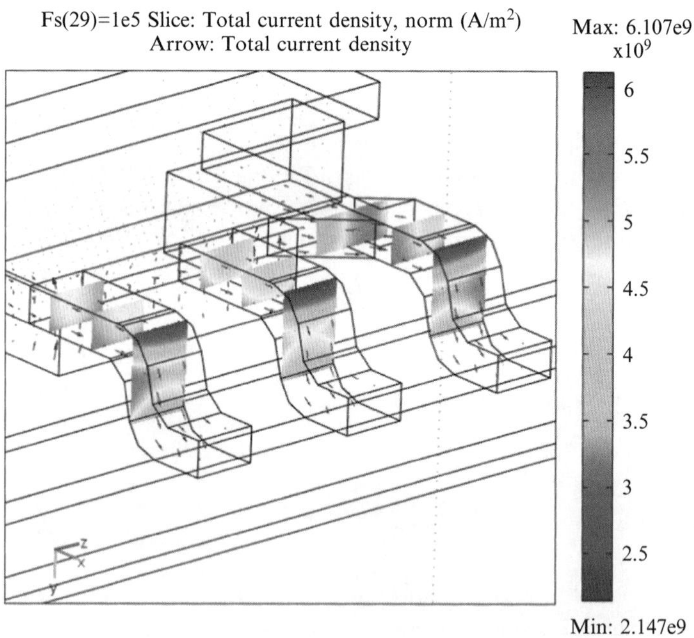

Fig. E.16 3D harmonic simulation of the LFpak leadframe mounted on a double layer PCB as shown in Fig. E.14. Current distribution through the source leads at 100 kHz

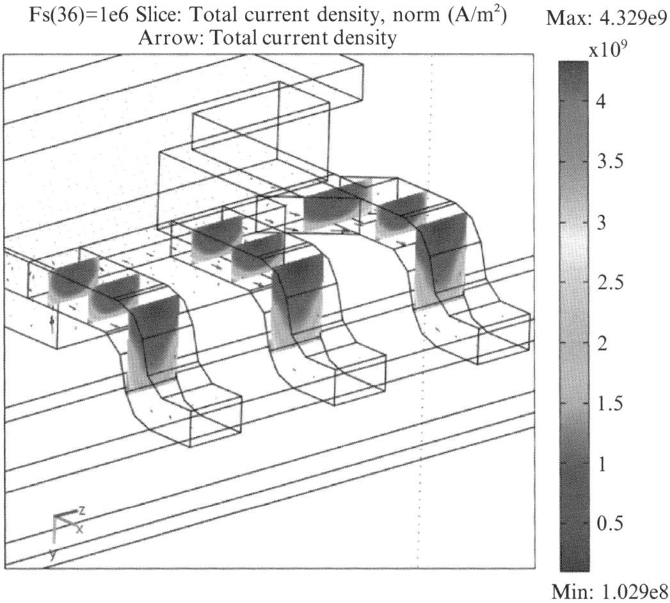

Fig. E.17 3D harmonic simulation of the LFpak leadframe mounted on a double layer PCB as shown in Fig. E.14. Current distribution through the source leads at 1 MHz

estimations on inductor losses in a wide range of operating conditions in the VR circuit as described in Chaps. 5 and 6. Furthermore, insights on the magnetic field distribution and loss breakdown allow identifying potential improvement options in the coil design.

The magnetic loss calculations presented herein are founded upon the assumption that the overall coil losses can be decomposed into the sum of three contributions, namely the *classical Eddy current loss*, *hysteresis loss*, and *excess loss*. While the first contributor is present in all materials and governed according to the classical theory of electromagnetism, the other two portions particularly manifest in ferromagnetic materials due to the existence of multiple magnetic domains that may divide the medium into regions with different magnetic moment alignments called magnetic domains. The reason of magnetic domains within the crystal structure is that their formation reduces the magnetic free energy. The notion of magnetic domains was initially introduced by Weiss in 1906 as a hypothesis, and later in 1949, Williams, Borzot, and Shockley demonstrated experimentally their existence. Providing mathematical treatment to stochastic processes and other existing mechanisms in the microscopic scale, it can be consistently shown that the movement, rotation, and interaction between the magnetic domains at their interfaces (also called magnetic walls) are fundamentally responsible for significant losses in the ferrite materials. The theory of magnetic domains and domain wall motion is comprehensibly described in [3].

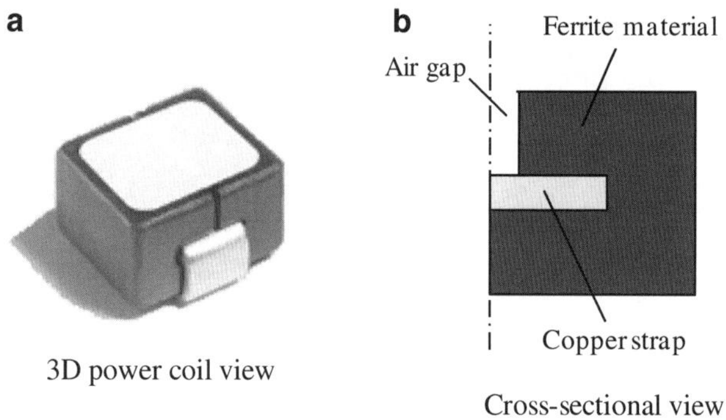

Fig. E.18 Commercial power coil inductor for high-current, high-switching frequency operation from Coilcraft (part number SCL7649S-360KL). 3D view and 2D cross-sectional representation. Total airgap = 850 μm

The three loss components derive from macroscopic models of the magnetic medium can be combined with the field calculations provided that the field sources are known. The latter is not a generalization given the fact that magnetic losses can influence the current waveforms, in which case the direct coupling existing between the loss and field equations may need to be taken into account.

The following subsections briefly describe the underlying physics of magnetic losses and the model implementation with the software package Comsol Multiphysics [1]. The description is illustrated with modeling results referring to the commercial power inductor of Fig. E.18a [4], from which specific material data are available. The geometry is simplified and described in 2D as depicted in Fig. E.18b. Likewise, other coil structures may be analyzed based on the same proposed methodological approach.

E.4.1 Classical Eddy Current Loss Model

Classical Eddy current losses are calculated from Maxwell's equations as described in Sect. E.1 and illustrated for the cases of the PCB and package impedance calculations of Sects. E.2 and E.3. From a macroscopic perspective, the magnetic material is assumed to be homogeneous where the effects of magnetic domains are averaged out.

E.4.2 Hysteresis Loss Model

The survey of magnetic core material models provided in [5] shows the great diversity of approaches to describe magnetization phenomena. Among the most widespread hysteresis models is that of Jiles–Atherton (J–A) [6], which can predict the associated hysteresis losses by a nontrivial differential function dependent on the magnetic field. One of the advantageous features of the J–A model is that such differential equation is computationally efficient, simple to use, and employs few and physically related material parameters.

In the J–A model, magnetization M is represented as the sum of the irreversible magnetization M_{irr} due to domain wall displacement and the reversible magnetization M_{rev} due to domain wall bending. In the ideal case, where the solid is free of impedances to changes in magnetization, the relation between M and magnetic field H follows a single-valued curve referred to as the anhysteretic curve M_{an} (i.e., curve without hysteresis), which is represented with a modified Langevin expression. Along this curve, both M_{rev} and M_{irr} take place at low H, whereas M_{irr} dominates at higher H. In the region of saturation, where all domain moments line up, domain rotation processes may dominate and be responsible for additional quasi-static effects, which are not considered by the presented approach.

The model's fundamental assumption for the cause of hysteresis in ferromagnets is the presence of pinning sites in the solid that locally oppose the motion of domain walls and trigger domain wall bulging. The pinning mechanism arises as a consequence of imperfections in the material such as dislocations, nonmagnetic inclusions within the grain, and inhomogeneous strains, all of which are identified without distinction as pinning sites or impedances to domain wall motion.

In terms of energy balance, the magnetization energy is the difference between the energy produced by the bulk magnetization in the lossless case and the energy associated to the frictional type of force produced by the pinning sites. This yields the following differential expression for the irreversible magnetization,

$$\frac{\mathrm{d}M_{\mathrm{irr}}}{\mathrm{d}H_{\mathrm{e}}} = \frac{M_{\mathrm{an}} - M_{\mathrm{irr}}}{k\delta}. \tag{E.13}$$

Parameter k is the pinning coefficient and is related to the density of pinning defects in the medium. Also, $\delta \equiv \mathrm{sign}(\mathrm{d}H_{\mathrm{e}}/\mathrm{d}t)$, where $H_{\mathrm{e}} = H + \alpha M$ is an effective field, and is analogous to the Weiss mean field experienced by the individual magnetic moments within a domain.

The total magnetization M is expressed as the sum of M_{rev} due to domain wall bending and M_{irr} due to wall displacement, i.e.,

$$M = M_{\mathrm{irr}} + M_{\mathrm{rev}}. \tag{E.14}$$

The reversible component is proportional to the difference between M_{irr} and M_{an}, giving,

$$M_{\mathrm{rev}} = c(M_{\mathrm{an}} - M_{\mathrm{irr}}). \tag{E.15}$$

The anhysteretic curve, which is the locus of global equilibrium states, is empirically defined with a modified Langevin expression,

$$M_{an}(H_e) = M_s\left(\coth\left(\frac{H_e}{a}\right) - \frac{a}{H_e}\right). \tag{E.16}$$

Parameter a has dimensions of magnetic field and characterizes the shape of M_{an}, whereas M_s defines the saturation magnetization.

Figure E.19 illustrates the use of the presented model to closely reproduce the characteristic B–H curves of a commercial ferrite material. A representation of the decomposed magnetization field is shown in Fig. E.20.

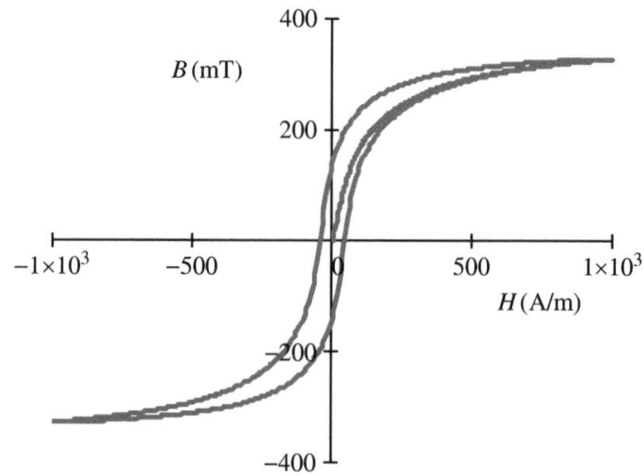

Fig. E.19 Jiles–Atherton model based representation of typical $B(H)$ loop of the 3F5 material from Ferroxcube at 100°C. Parameter values: $M_s = 2.9 \times 10^5$ (A/m), $a = 115$ (A/m), $k = 280$ (A/m), $c = 0.77$, $\alpha = 0.001$

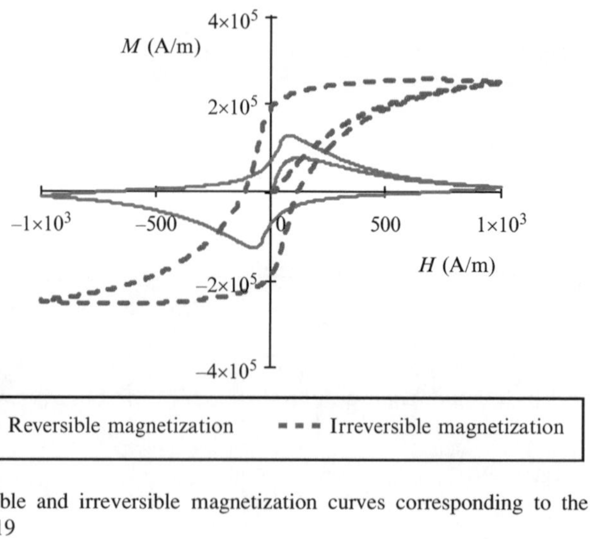

Fig. E.20 Reversible and irreversible magnetization curves corresponding to the $B(H)$ loop example of Fig. E.19

The hysteresis model is implemented in the device physics simulator. After computing the field distribution, the hysteresis energy losses can be numerically calculated from,

$$w_{\text{hys}} = \int_0^B H \,\mathrm{d}B. \qquad \text{(E.17)}$$

Equation (E.17) is expressed in J/m^3. The total power loss thus results from integrating the energy over the entire volume and multiplying the Joules quantity by the switching frequency.

E.4.3 Excess Loss Model

Thus far, Eddy current losses formulated under the basis of the classical electromagnetic field theory have been calculated in homogeneous materials such as PCB traces and package leads. In ferromagnetic materials, however, additional Eddy current losses arise due to the existence of magnetic domains and the dynamics of their associated domain walls. Namely, when the excitation field varies, the domain walls move accordingly producing changes in the magnetization only in or near the domain walls, leaving the magnetization inside the bulk practically unchanged. Since domain walls occupy only a small fraction of the total volume, the change in the magnetization in and around the walls has to be much larger than the average magnetization over the entire specimen. As a consequence, the locally induced Eddy currents by fluctuations of the flux density caused by domain wall motion are generally higher than those calculated on the basis of uniform magnetization. The formed *excess field* H_{ex}, which is part of the external field needed to compensate the fields originated by the local Eddy currents, gives rise to additional losses termed *anomalous* or *excess losses*.

The concept of magnetic object MO, corresponding to a group of neighbor walls evolving in a highly correlated fashion, is introduced in [3] in order to take into proper consideration the role of these short-range internal correlation fields. It is then assumed that H_{ex} is proportional to the velocity of variation of the local flux induced by MO moving. This proportionality is given in (E.18), where G expresses the MO friction coefficient and σ the material's conductivity.

$$H_{\text{ex}} = G\sigma \frac{\mathrm{d}\phi}{\mathrm{d}t}. \qquad \text{(E.18)}$$

The velocity of variation of the global flux $S(\mathrm{d}B/\mathrm{d}t)$ results from the contribution of the number n_{om} of MO taking place in the magnetization process. This is expressed as,

$$S\frac{\mathrm{d}B}{\mathrm{d}t} = n_{\text{om}}(t)\frac{\mathrm{d}\phi}{\mathrm{d}t}. \qquad \text{(E.19)}$$

Parameter S is the cross-sectional area of the magnetic material with assumed sheet shape. Experimental results obtained by Bertotti for several crystalline magnetic materials show that there exists a linear relationship between H_{ex} and n_{om} as,

$$n_{om} = \frac{H_{ex}}{V_0}. \tag{E.20}$$

The energy loss per unit volume originated by the n_{om} MO is given by,

$$w_{ex} = \int_0^T H_{ex} \frac{dB}{dt}\, dt. \tag{E.21}$$

Combining expressions yields the following expression for the mean value of excess loss in units J/m^3,

$$w_{ex} = \sqrt{\sigma G V_0 S} \int_0^T \left|\frac{dB}{dt}\right|^{\frac{3}{2}} dt. \tag{E.22}$$

The term outside the integral sign can be adjusted to best match the specific power loss plots from the manufacturer datasheets [7, 8].

E.4.4 Simulation Results

Figure E.21 depicts the inductor current resulting from applying a zero average square voltage as in the buck converter operation. From the ripple current, the total inductance is estimated to be 33 nH, a value within the specified range according to the specifications of the miniaturized coil. From the total inductance, 27 nH are actually resulting from the simulated cross-sectional area of the power

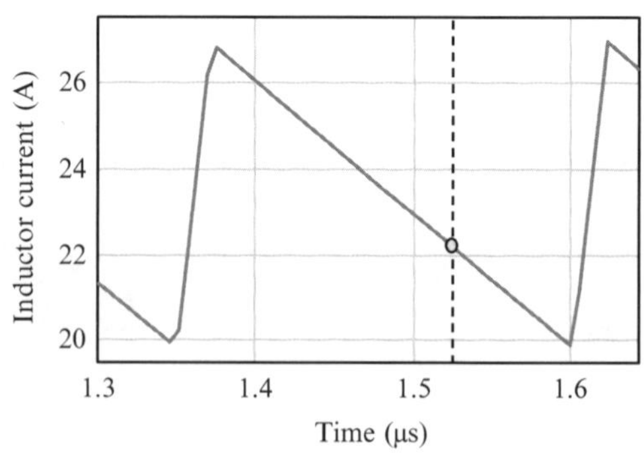

Fig. E.21 Example of simulated steady-state current resulting from an applied squared voltage as in a buck converter with $V_{in} = 12$ V, $V_o = 1$ V, $F_s = 4$ MHz and 23 A load current

coil. The other 6 nH are added to account for the total portion of the external leads.

The magnetic flux density plot of Fig. E.22 reveals high stress in the interior corners and airgap boundaries of the ferrite core that may induce local magnetic saturation with a consequent loss increase and development of hot spot areas.

The current density distribution of Fig. E.23 describes high current crowding in the region nearby the airgap, thus suggesting a dramatic increase of the conductor resistance in the MHz frequency range. The graph further indicates that the low conductivity of the ferrite appears very effective in this frequency range as no signs of Eddy current across the core are present.

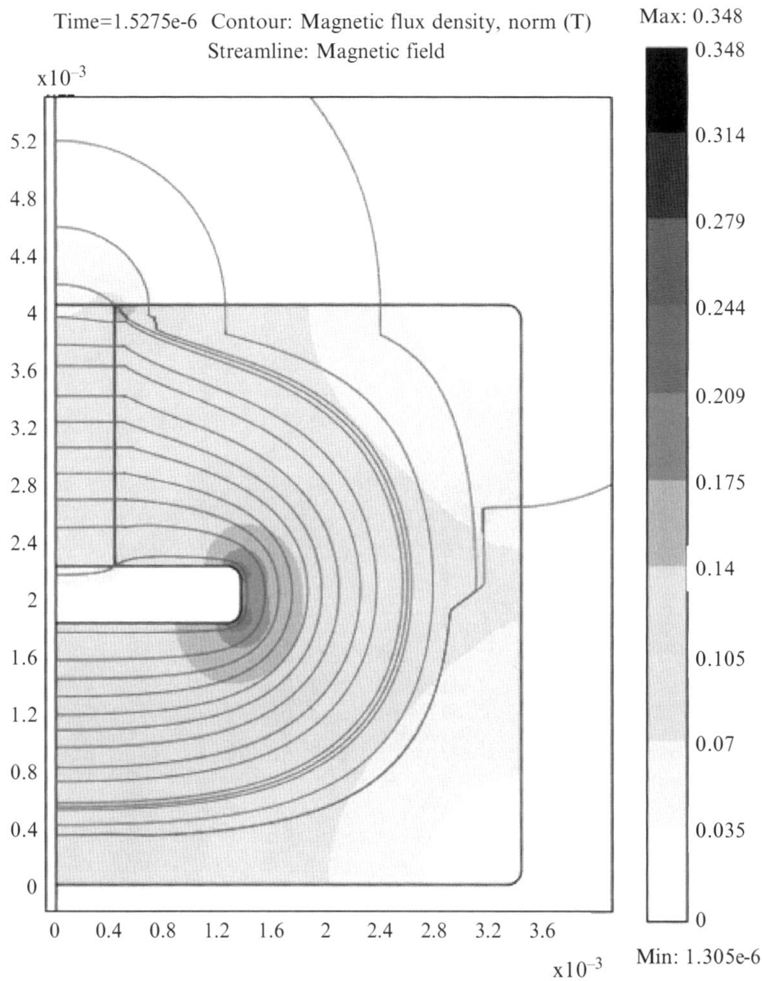

Fig. E.22 Magnetic flux density and field lines corresponding to the transient simulation of Fig. E.21 at time $t = 1.53\ \mu s$

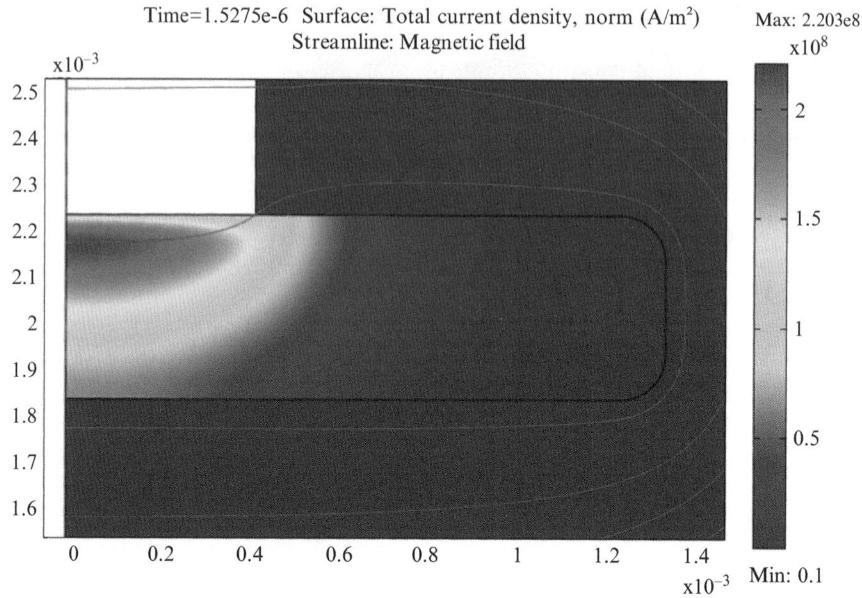

Fig. E.23 Total current density distribution and magnetic field lines corresponding to the transient simulation of Fig. E.21 at time $t = 1.53$ μs

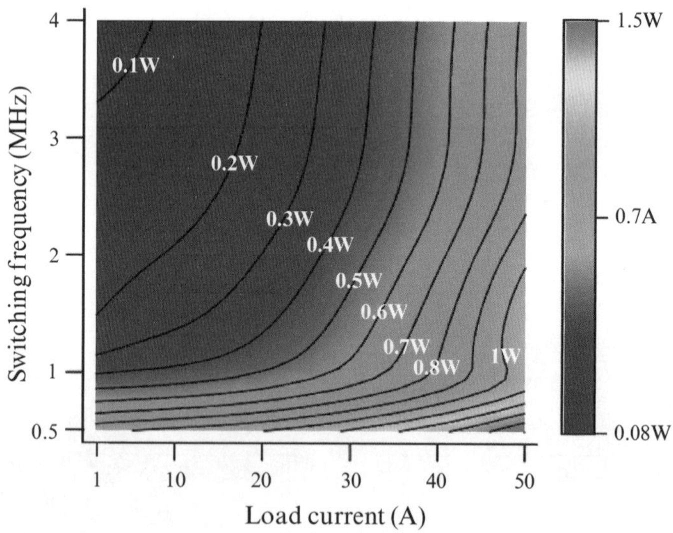

Fig. E.24 Simulated losses of a buck converter coil as a function of the switching frequency and load current at $V_{in} = 12$ V, $V_o = 1$ V, and 100°C. Reference part: SCL7649S-360 from Coilcraft

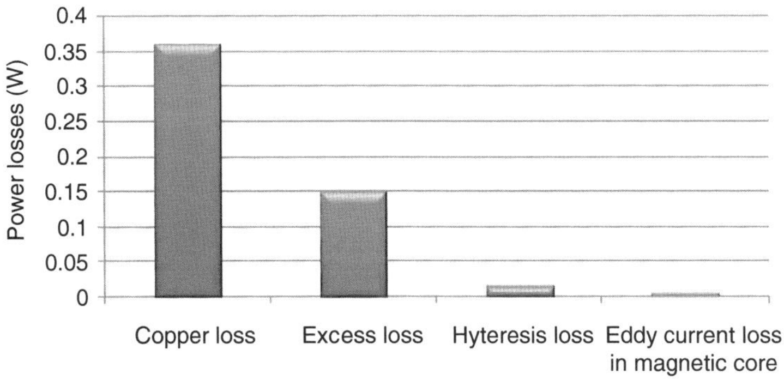

Fig. E.25 Estimated buck converter coil loss breakdown. Simulated conditions: $V_{in} = 12$ V, $V_o = 0.7$ V, $F_s = 2$ MHz, 33.3 A average coil current and 100°C coil temperature

Figure E.24 shows the simulated coil loss as function of the switching frequency and load current under typical VR operating conditions. Due to thermal limitations, the power consumption may typically be limited to a maximum of ~0.5 W, which prevents the use of the inductor in the kHz range to avoid excessive current ripple.

As the contour plot indicates, losses are weakly dependent on the switching frequency above 3 MHz since the ripple current amplitude becomes low. Nonetheless, the maximum load current with negligible ripple current might be lower than 40 A so as to maintain the coil temperature within its specified limits. Higher current operation may be enabled with the used of heat dissipation enhancement measures (e.g., fans).

The loss breakdown of Fig. E.25 shows that in a potential operating target regime heat generation is mainly dominated by copper losses, while hysteresis and Eddy current loss in the magnetic material may be neglected. Excess loss appears as a significant loss mechanism with a ~30% contribution. This is consistent with [8] and references therein.

References

1. Comsol Multiphysics version 3.5a, www.comsol.com
2. Bastos JPA, Sadowski N (2003) Electromagnetic modeling by finite element methods. Marcel Dekker, New York, ISBN: 0-8247-4269-9
3. Bertotti G (1998) Hysteresis in magnetism for physicists, materials scientists and engineers, Academic, New York, ISBN-10: 0-12-093270-9
4. SMT Power Inductors – SLC7649 Series. Datasheet document 481-1, April 2008, www.coilcraft.com
5. Takach MD, Lauritzen PO (1995) Survey of magnetic core models. In: Applied power electronics conference and exposition, APEC, March 1995, vol 2, pp 560–566

6. Jiles DC, Atherton DL (1986) Theory of ferromagnetic hysteresis. J Magn Magn Mater 61: 48–60
7. Albach M, Duerbaum T, Brockmeyer A (1996) Calculating core losses in transformers for arbitrary magnetizing currents a comparison of different approaches. In: 27th annual IEEE power electronics conference, PESC 1996, vol 2, pp 1463–1468
8. Li J, Abdallah T, Sullivan CR (2001) Improved calculation of core loss with nonsinusoidal waveforms. In: IEEE industry applications society annual meeting, October 2001, pp 2203–2210

Appendix F
Quality Factor in Resonant Gate Drivers

For years, resonant gate drivers have been presented as alternatives to conventional solutions to drive power MOSFETs with reduced energy loss. The conceptual idea of resonant gate drivers consists of replacing the resistance of the path through which gate charge flows in the conventional approach by an inductance. With the absence of dissipative elements, the full energy required to change the state of the power device can be controllably recovered and returned to the power supply, or alternatively stored in auxiliary tanks for reuse. Several basic circuit topologies have been proposed to most effectively implement this concept [1, 2].

Commonly, the inductance of the resonant gate drivers determines the switching speed of the power MOSFET as the gate resistance does in the conventional way. Thus, in high-switching frequency applications, the use of low lossy resonant gate drivers featuring low inductance may be of high interest, for size and switching times need also be reduced.

On the other hand, reducing the gate inductance negatively impacts the gate losses as soon as parasitic resistances are no longer negligible. In fact, the gate path circuit is not exempt from dissipative elements. Parasitic resistances exist in the polysilicon gate of the power MOSFET, in the gate switches and as equivalent series to the gate inductance and other storage elements that may be present in the gate circuit. Attempts in reducing the resistance may result in a loss increase due to the parasitic charge of the gate switches. Therefore, whenever the inductance of the resonant gate driver is to be minimized, both the gate parasitic charge and resistance may have to be considered for optimization. This strong loss dependence on the switching speed in resonant gate drivers is in contrast to the behavior of conventional solutions.

Quality factor Q is a measure of the rate of energy dissipation in a resonant system. Q is defined as the ratio of energy stored in the resonant circuit and the energy dissipated per oscillation cycle. Its use is frequent for characterizing the quality of oscillators.

This section makes use of Q to analyze the performance of the resonant gate driver topologies of Fig. F.1 in terms of energy loss, resonant inductor size, and switching speed capability. Simplified resonant gate driver models are expressed as a function of the fundamental variable Q. It results into compact and general analytical close-form expressions that provide easy and direct assessment as to how the different topologies compare. The magnitudes of comparison are given in

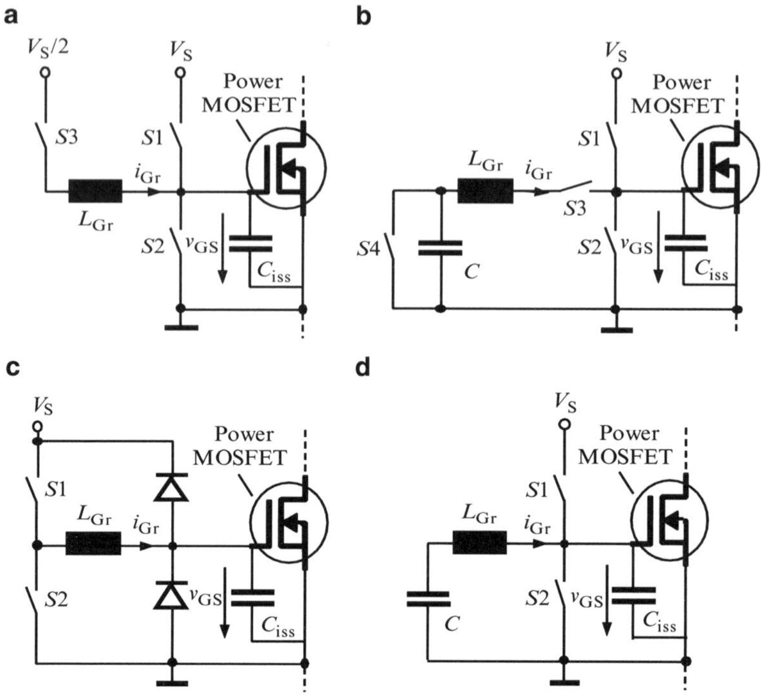

Fig. F.1 Investigated resonant gate driver topologies Topology (a) corresponds to [7], whereas (b) is proposed by 1261 [8], (c) by [9] and (d) by [3].

terms of normalized quantities against performance figures of conventional gate drivers. This further allows highlighting the differences between resonant and conventional solutions, thereby providing further value, generalization, and compactness to the information contained in the equations.

Seeking for the circuit topology that offers maximum switching speed capability at minimum energy loss and inductor size, conclusions reveal that such topology does not exist among those considered. Instead, the topology of choice must result from compromising driving requirements according to a particular application. This is illustrated with an example and demonstrated with circuit simulations employing a rather accurate representation of the gate driver and power MOSFET. The model allows parameters optimization, such as power MOSFET chip size and driving voltage, that minimizes the overall power loss in zero voltage switching applications.

F.1 Gate Driver Requirements

Further loss reduction in resonant gate drivers is possible in detriment of the switching speed, e.g. [3, 4], which usually translates into an increase of switching losses in the power train, i.e., energy dissipation in the channel and/or body diode

Table F.1 Case example. Parameter values of resonant gate driver topology (a)

Basic specifications of synchronous buck converter	Input voltage = 12 V Output voltage = 1.5 V Average output current = 15 A Output ripple current = 6 A Switching frequency = 3 MHz
Power MOSFET (SyncFET)	$C_{iss} = 0.61$ nF/mm^2 Specific ON resistance = 25 mΩ/mm^2 Chip area = 11 mm^2 Polysilicon gate resistance = $R_{GM} = 0.2\ \Omega$
MOSFET switches $S1$ and $S2$	ON resistance = $R_{Sc} = 0.7\Omega$ $C_{iss} = C_{iSc} = 57$ pF $C_{oss} = C_{oSc} = 57$ pF
Two anti-series MOSFET switch $S3$	ON resistance = $R_{Sr} = 0.2\Omega$ $C_{iss} = C_{iSr} = 0.4$ nF $C_{oss} = C_{oSr} = 0.2$ nF (values of one MOSFET)
Gate driving voltages	Gate switches = $V_{drvS} = 3.5$ V, power MOSFET = 7 V ($V_S = 3.5$ V)
Dead times	4 ns for both leading and falling edges
Gate inductor L_{Gr}	25 nH, negligible ESR

of the power MOSFET. In some cases it may even occur that the loss reduction achieved in the gate does not compensate the loss increase in the power train. An optimum switching time must therefore exist that minimizes the overall losses.

Hard-switching operation of the power device usually involves faster switching times than zero voltage switching, ZVS. Since the benefits can thus be greater, this work focuses on the study of resonant gate drivers for ZVS applications.

In the following example, the resonant gate driver (a) from Fig. F.1 is designed to illustrate the demands for driving a synchronous rectifier MOSFET of a step down converter for high current applications. Table F.1 summarizes the most relevant parameters of the proposed application example.

A circuit model of both driver and power MOSFET is employed for carrying out parameter sweeps that yield identification of minimum energy loss conditions. Optimized parameters include the chip area of the power device, inductance L_{Gr}, and driving voltage V_S. Figure F.2 shows simulated waveforms of the steady-state switching cycle operation upon optimized conditions.[3] Both drain current id and drain-to-source voltage v_{DS} are accurately predicted during third quadrant operation, when the losses in the channel and body diode are mainly generated. Note that the third quadrant behavior is modeled according to [5, 6]. Furthermore, the simulation incorporates dead time control functionality to minimize body diode conduction. An algorithm controls the beginning of the turn-on and turn-off times such that the minimum voltage across v_{DS} equals to a predefined value, V_{thd}. In the example of Fig. F.2, $V_{thd} = -0.3$ V, which corresponds to a dead time (t_d) of approximately 4 ns for both leading and falling edge transitions of v_{GS}. Dead time t_d

[3]The loss model is described in the following sections.

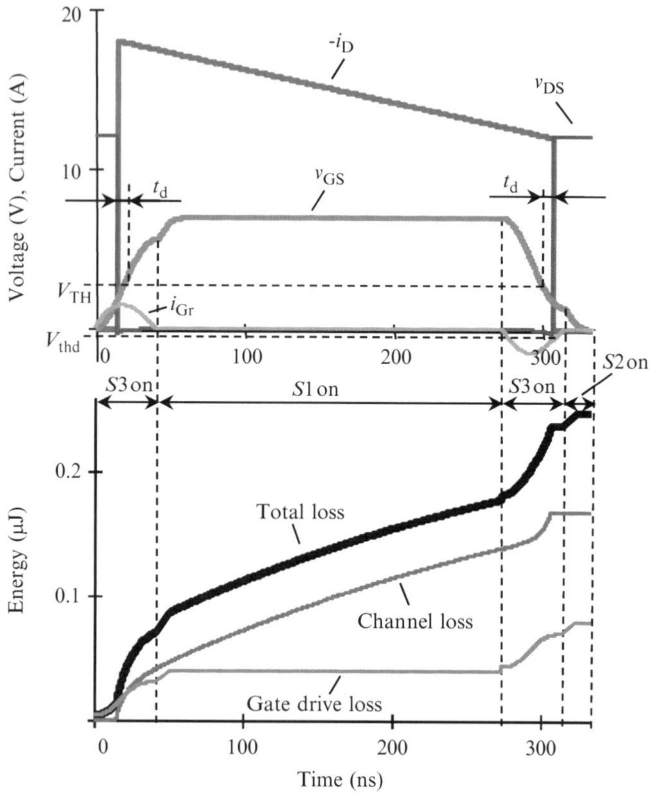

Fig. F.2 Switching cycle operation of topology (a) for a case example upon optimized power loss conditions (see Table C.1)

is defined as the interval between the time instant for v_{DS} to reach the threshold voltage V_{TH} of the power MOSFET and the time of reversion of v_{DS}.

As depicted in the lowest plot of Fig. F.2, energy loss in gate results from current through the parasitic gate resistances during the switching times and from the stored charge of parasitic capacitances from switches $S1$–$S3$. Note that all parasitic elements of the gate circuit are assumed linear. Significant energy loss also occurs across the terminals drain and source during the switching times as a consequence of the channel resistance and current id. This loss contribution can be reduced by further shortening the dead times, or increasing the switching speed. The latter is possible by reducing L_{Gr}, as illustrated in Fig. F.3, which depicts the power loss dependence on resonant inductor L_{Gr} and for various dead times.

It becomes obvious that the dead time increase significantly worsens the overall power loss. Thus, faster switching speed is required, which limits the size of L_{Gr}. In the given example, the optimum size of L_{Gr} varies from 25 nH to just a few nH for a dead time increase of 7 ns. It indicates that the loss reduction achieved with the use

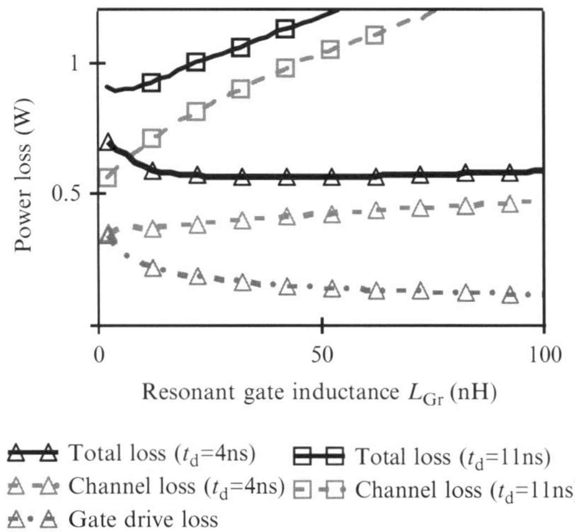

Fig. F.3 Power loss in both power MOSFET channel and gate circuit as a function of L_{Gr} and for various dead times

of resonant gate drivers may be extremely conditioned by the switching speed demands.

F.2 Formulation of Basic Resonant Gate Driver Topologies

This section describes various fundamental resonant gate drive topologies as a function of Q, and compares their performance in terms of switching speed, energy loss, and inductor size.

F.2.1 Basic Resonant Gate Driver Topologies

Figure F.1 shows the four different fundamental resonant gate driver topologies under investigation. Topology (a) corresponds to [7], whereas (b) is proposed by [8], (c) by [9] and (d) by [3].

F.2.2 Assumptions and Definitions

Along the formulation it is assumed that the driven power MOSFET operates in ZVS. Thus, its input capacitance may be assumed fairly linear. A capacitance C_{iss} is defined as the average input capacitance over the operating gate voltage range. Likewise, the output and input capacitance of the gate switches C_{oS} and C_{iS} are defined as average quantities. The switches of the gate driver are MOSFETs. Their parasitic ON resistance and capacitances are related by the figure of merits $\mathrm{FoM}_i = C_{\mathrm{iS}}R_{\mathrm{S}}$ and $\mathrm{FoM}_o = C_{\mathrm{oS}}R_{\mathrm{S}}$. ON resistance R_{S} is assumed linear and independent on the current flow. Further parasitic elements may also be considered, like the equivalent series resistance (ESR) of the resonant inductor and the gate resistance of the power MOSFET. The turn-on and turn-off times of the gate switches are assumed ideally zero.

Topologies (a) and (b) require switch $S3$ to be bidirectional. Such feature is realized by means of two anti-series MOSFETs.

In the next sections, the energy loss and switching speed of the resonant gate drivers are expressed as normalized quantities against those of conventional gate drivers, which are given by,

$$E_{\mathrm{CGD}} = \frac{1}{2}(C_{\mathrm{iss}} + 2C_{\mathrm{oSc}})V_{\mathrm{s}}^2 + C_{\mathrm{iSc}}V_{\mathrm{DRV(s)}}^2, \tag{F.1}$$

$$\tau_{\mathrm{CGD}} = 5C_{\mathrm{iss}}(R_{\mathrm{GM}} + R_{\mathrm{Sc}}). \tag{F.2}$$

Parameters C_{iSc}, C_{oSc}, and R_{Sc} are the input capacitance, the output capacitance and the ON resistance of the gate switches, respectively. Voltage V_{drvS} is the driving voltage of the gate switches. The energy loss of (F.1) corresponds to one switching transient. The switching time of (F.2) is defined as the time required to reach 99.3% the final value after beginning of commutation. Note that the impact of the parasitic inductances on the switching time is neglected.

F.2.3 Resonant Gate Driver (a)

The principle of operation of this resonant gate driver is described in the example of Fig. F.2. Power loss occurs due to three different causes: the stored charge in the gate switches $S1$–$S3$, the resonant current through the parasitic resistances during the charge of C_{iss}, and the voltage clamping across C_{iss} at the end of the resonant transitions. Based on the aforementioned assumptions the following normalized energy loss expression for one switching transition can be obtained,

$$e_{\mathrm{n(a)}} = \frac{e_{\mathrm{nrc(a)}}\left(1+\frac{\alpha}{2}\right)\beta^2 + \delta + \frac{1}{2}\phi_{\mathrm{a}}\left(\frac{\beta}{2}\right)^2 + 2\phi_{\mathrm{a}}\delta}{\left(1+\frac{\alpha}{2}\right)\beta^2 + \delta}, \tag{F.3}$$

$$\alpha = \frac{C_{\mathrm{iss}}}{C_{\mathrm{oSc}}}, \quad \beta = \frac{V_{\mathrm{s}}}{V_{\mathrm{DRV(s)}}}, \quad \delta = \frac{\mathrm{FoM}_i}{\mathrm{FoMo}}, \quad \phi_{\mathrm{a}} = \frac{R_{\mathrm{Sc}}}{R_{\mathrm{Sr}}}. \qquad \text{(F.4 – F.7)}$$

As indicated in Table C.1, R_{Sc} and C_{oSc} refer to clamping switches $S1$ and $S2$. R_{Sr} is the ON resistance of any of the two MOSFETs constituting bidirectional switch $S3$. V_{drvS} is the driving voltage of gate switches $S1$–$S3$. As α increases, expression (F.3) tends to,

$$e_{\mathrm{nrc(a)}} = \frac{1}{2}\left(1 - \mathrm{e}^{-\frac{\pi}{2\sqrt{Q_{\mathrm{a}}^2 - \frac{1}{4}}}}\right). \tag{F.8}$$

The series equivalent resonant capacitance is approximated as the sum of C_{iss} and C_{oSc}, which yields the following quality factor,

$$Q \equiv Q_{\mathrm{a}} = \frac{1}{R_{\mathrm{L}} + R_{\mathrm{GM}} + 2R_{\mathrm{Sr}}}\sqrt{\frac{L_{\mathrm{Gr}}}{C_{\mathrm{iss}} + 2C_{\mathrm{oSc}}}}. \tag{F.9}$$

Resistances R_{L} and R_{GM} are the parasitic ESRs of L_{Gr} and the power MOSFET, respectively.

The normalized switching time can also be analytically calculated as a function of Q as follows,

$$\tau_{\mathrm{n(a)}} = \frac{\alpha + 2}{5\alpha}\frac{\rho_{\mathrm{a}}\pi Q_{\mathrm{a}}^2}{\sqrt{Q_{\mathrm{a}}^2 - \frac{1}{4}}} + \frac{1}{5}\ln\left(\frac{1 - \mathrm{e}^{-\frac{\pi}{2\sqrt{Q_{\mathrm{a}}^2 - \frac{1}{4}}}}}{2\mathrm{e}^{-5}}\right), \tag{F.10}$$

$$\rho_{\mathrm{a}} = \frac{R_{\mathrm{GM}} + R_{\mathrm{Sc}}}{R_{\mathrm{L}} + R_{\mathrm{GM}} + 2R_{\mathrm{Sr}}}. \tag{F.11}$$

The first term in (F.10) refers to the time during the resonant transition, whereas the second one accounts for the time after the activation of the clamping switch, i.e., $S1$ at turn-on, and $S2$ at turn-off. This latter term is valid as long as the voltage across C_{iss} at the end of the resonant transition has not yet arrived at the 99.3% of its final value. It should otherwise be ignored, that is, whenever the term within the logarithm becomes lower than one.

Due to the assumed linearity at the transitions, (F.3)–(F.11) apply to both turn-on and turn-off transients.

F.2.4 Resonant Gate Driver (b)

This topology is conceptually equal to (a) with respect to the resonant charge duration, i.e., half resonant period, as well as the clamping means. The main difference lays in its realization, which allows the use of a single power supply. For an optimum operation, capacitance C is set to equal to C_{iss}, whereas the output capacitance of switches $S1$ and $S2$ should equal half value from that of switch $S4$.

The principle of operation is as follows: Prior to turn-off, switch $S4$ resets voltage across C. At turn-off, the stored energy in C_{iss} is resonantly transferred to

C through switch $S3$. Switch $S2$ clamps v_{GS} to zero at the end of the resonant semi-cycle. The stored energy in C is sent back to C_{iss} at turn-on. Switch $S1$ clamps the ON state to voltage V_S as soon as current i_{Gr} reaches zero at the end of the resonant semi-cycle.

Similarly to topology (a), the normalized energy loss against (F.1) can be calculated by,

$$e_{n(b)} = \frac{e_{nrc(b)}\left(1+\frac{\alpha}{2}\right)\beta^2 + 2\delta(1+\phi_b) + \frac{1}{16}\beta^2\phi_b[4+(1+e^{-\chi_b})^2]}{\left(1+\frac{\alpha}{2}\right)\beta^2+\delta}, \tag{F.12}$$

$$\chi_b = \frac{\frac{\pi}{2\sqrt{2}}}{\sqrt{Q_a^2 - \left(\frac{1}{2\sqrt{2}}\right)^2}}, \quad \phi_a = \phi_b = \frac{R_{Sc}}{R_{Sr}}, \tag{F.13, F.14}$$

$$Q \equiv Q_b \frac{1}{R_C + R_L + R_{GM} + 2R_{Sr}}\sqrt{\frac{L_{Gr}}{C_{iss}+2C_{oSc}}}, \tag{F.15}$$

$$e_{nrc(b)} = \frac{1}{4}[3 - e^{-\chi_b}(2+e^{-\chi_b})]. \tag{F.16}$$

Expression (F.12) simplifies to (F.16) for large α. The switching time results, after normalization, in,

$$\tau_{n(b)} = 2\frac{\alpha+2}{5\alpha}\rho_b Q_b^2\chi_b + \frac{1}{5}\ln\left(\frac{1 - e^{-(\pi/(2\sqrt{2}))\Big/\sqrt{Q_b^2-\left(\frac{1}{2\sqrt{2}}\right)^2}}}{2e^{-5}}\right), \tag{F.17}$$

$$\rho_b = \frac{R_{GM}+R_{Sc}}{R_C+R_L+R_{GM}+2R_{Sr}}. \tag{F.18}$$

As in topology (a), the second term in (F.17) should be ignored whenever it becomes negative.

The symmetry of the turn-on and turn-off transitions allows the use of (F.12)–(F.18) to obtain the total power loss in gate.

F.2.5 Resonant Gate Driver (c)

As in drivers (a) and (b), topology (c) resonantly charges C_{iss}. The duration of the transition is though approximately one quarter of the resonant period, just when v_{GS}

is clamped by the action of the diodes. The remaining stored energy in L_{Gr} is delivered to V_S after the resonant interval. The normalized energy loss within one switching interval can be derived from the following expression,

$$e_{n(c)} = \frac{0.5(e_{nr(c)} + e_{nc(c)} + e_{nd(c)})\alpha\beta^2 + \delta + 2\beta^2}{0.5\alpha\beta^2 + \delta + \beta^2}, \tag{F.19}$$

where,

$$e_{nr(c)} = 1 - e^{-2\chi_c}, \tag{F.20}$$

$$e_{nc(c)} = 2\left(\frac{1+\varepsilon_c}{\phi_c}Q_c\right)^2 \ln\left(1 + \frac{\phi_c}{1+\varepsilon_c}\frac{e^{-\chi_c}}{Q_c}\right) + \left(e^{-\chi_c} - 2\frac{1+\varepsilon_c}{\phi_c}Q_c\right)e^{-\chi_c}, \tag{F.21}$$

$$e_{nd(c)} = 2\varepsilon_c\left(\frac{Q_c}{\phi_c}\right)^2\left[\frac{\phi_c}{Q_c}e^{-\chi_c} - (1+\beta)\ln\left(1 + \frac{\phi_c}{1+\varepsilon_c}\frac{e^{-\chi_c}}{Q_c}\right)\right], \tag{F.22}$$

$$\chi_c = \frac{\pi - \tan^{-1}\left(2\sqrt{Q_c^2 - \left(\frac{1}{2}\right)^2}\right)}{2\sqrt{Q_c^2 - \left(\frac{1}{2}\right)^2}}, \tag{F.23}$$

$$\phi_c = \frac{R_{Sc} + R_L}{R_{Sc} + R_L + R_{GM}}, \quad \varepsilon_c = 2\frac{V_d}{V_S}. \tag{F.24, F.25}$$

Voltage V_d in (F.25) is the forward voltage of the diodes. Their ESR is neglected. The quality factor differs from topologies (a) and (b) as,

$$Q \equiv Q_c = \frac{1}{R_L + R_{GM} + R_{Sc}}\sqrt{\frac{L_{Gr}}{C_{iss}}}. \tag{F.26}$$

The normalized energy losses of (F.20)–(F.22) are related to the resonant transition, the time interval of energy recovery of L_{Gr}, and the contribution of the clamping diodes, respectively.

Finally, the normalized switching time is,

$$\tau_{n(b)} = \frac{1}{5}\rho_c Q_c^2 \chi_c, \tag{F.27}$$

$$\rho_c = \frac{R_{GM} + R_{Sc}}{R_{Sc} + R_L + R_{GM}}. \tag{F.28}$$

Equations (F.19)–(F.28) are applicable to both turn-on and turn-off switching times. For large α, (F.19) simplifies to the sum of (F.20)–(F.22).

F.2.6 Resonant Gate Driver (d)

Unlike topologies (a)–(c), this resonant gate driver employs L_{Gr} as a constant current means to charge C_{iss}. One of the conditions to guarantee the principle of operation is that the time constant formed by L_{Gr} and the parasitic resistances of the charging path has to be much larger than the converter's switching period. For, say, a factor 5 higher, the following minimum Q is required,

$$Q_{\mathrm{min}} = \frac{5\rho_{\mathrm{d}}\sqrt{2}}{\sqrt{d(1-d)p}}, \tag{F.29}$$

where ρ_{d} is defined as,

$$\rho_{\mathrm{d}} = \frac{6p \cdot R_{\mathrm{GM}} + R_{\mathrm{Sc}}(1-6p) + R_{\mathrm{C}} + R_{\mathrm{L}}}{R_{\mathrm{C}} + R_{\mathrm{L}} + R_{\mathrm{GM}}}. \tag{F.30}$$

Furthermore, d is the duty cycle and p is the ratio switching time to the switching period. Resistance R_{C} is the ESR of blocking capacitor C. The quality factor is, for this topology,

$$Q \equiv Q_{\mathrm{d}} = \frac{1}{R_{\mathrm{L}} + R_{\mathrm{GM}} + R_{\mathrm{C}}}\sqrt{\frac{L_{\mathrm{Gr}}}{C_{\mathrm{iss}} + 2C_{\mathrm{oSc}}}}. \tag{F.31}$$

Approximating the input capacitance of the power MOSFET as the sum of C_{iss} and $2C_{\mathrm{oSc}}$, and following the assumptions from [3] yield the following expressions for the normalized energy loss,

$$e_{\mathrm{n(d)}} = \frac{e_{\mathrm{nr(d)}}(\alpha+2)\beta^2 + 2\delta}{(\alpha+2)\beta^2 + 2\delta}, \tag{F.32}$$

$$e_{\mathrm{nr(d)}} = \frac{1+\frac{2}{\alpha}}{6\rho_{\mathrm{d}}pQ_{\mathrm{d}}}\sqrt{\frac{2d(1-d)}{p}}. \tag{F.33}$$

The normalized switching time is,

$$\tau_{\mathrm{n(d)}} = \frac{Q_{\mathrm{d}}}{5}\,\frac{1+\frac{2}{\alpha}}{\phi_{\mathrm{d}}}\sqrt{\frac{2p}{d(1-d)}}, \tag{F.34}$$

$$\phi_{\mathrm{d}} = \frac{R_{\mathrm{GM}} + R_{\mathrm{Sc}}}{R_{\mathrm{C}} + R_{\mathrm{L}} + R_{\mathrm{GM}}}. \tag{F.35}$$

Finally, combining (F.29) and (F.35) yields the minimum normalized switching time capability of the driver,

$$\tau_{\mathrm{n(d)min}} = 2\frac{\rho_{\mathrm{d}}}{\phi_{\mathrm{d}}}\frac{1 + (2/\alpha)}{d(1-d)}. \tag{F.36}$$

F.3 Topology Comparison

The formulation provided previously is used in this section to compare the proposed topologies. Table F.2 summarizes the values assigned to the gate driver parameters, which may correspond to an application example similar to that of Sect. F.3. Note that parameter α may rather be higher in topologies (a) and (b) than in (c) and (d) as the ON resistance of clamping switches $S1$ and $S2$ of the first ones do not affect Q (see (F.19), (F.15), (F.26), and (F.31)).

Figure F.4 compares the required quality factor among the four topologies. As already indicated, Q increases monotonically with the switching time in all cases, that is to say, the switching time increases with L_{Gr}.

However, there exist large differences in the required Q depending on the topology of choice. Topology (a) features the lowest required Q, whereas topology (d) may need as much as a factor 10 higher than (a) for low p. This translates into an inductance 100 times larger for equal values of C and R in Q.

Somewhere in between is found topology (c), which offers the shortest switching times, even shorter than conventional gate drivers at very low Q. Topology (d) has the lowest switching speed capability, limited to a factor 7 slower than conventional gate drivers in the given example. It is followed by topology (a), with a factor 1.5 slower at $Q = 0.5$.

Figure F.5 compares the loss reduction as function of the switching speed. Requiring the largest Q, topology (d) features the highest energy reduction at low switching times. The overall performance of this driver in terms of loss reduction and Q improves as p increases. Its use is though limited to rather large switching times.

Topology (a) provides the second best energy loss reduction capability at moderate and high-switching times.

Table F.2 Parameter values of the analyzed resonant gate drivers

Topology (a)	$\alpha = 100$, $\beta = 1$, $\delta = 1$, $\phi_{\mathrm{a}} = 2$, $\rho_{\mathrm{a}} = 1$
Topology (b)	$\alpha = 100$, $\beta = 1$, $\delta = 1$, $\phi_{\mathrm{b}} = 2$, $\rho_{\mathrm{b}} = 1$
Topology (c)	$\alpha = 50$, $\beta = 1$, $\delta = 1$, $\phi_{\mathrm{c}} = 0.5$, $\rho_{\mathrm{c}} = 0.95$, $\varepsilon_{\mathrm{c}} = 0.1$
Topology (d)	$\alpha = 50$, $\beta = 1$, $\delta = 1$, $\phi_{\mathrm{d}} = 1.2$, $\rho_{\mathrm{d}}(p = 5\%) = 1$, $d = 50\%$

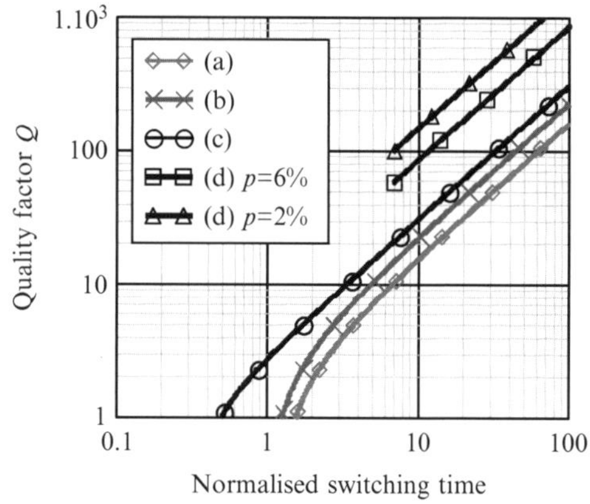

Fig. F.4 Quality factor vs. normalized switching time of the proposed resonant gate driver topologies

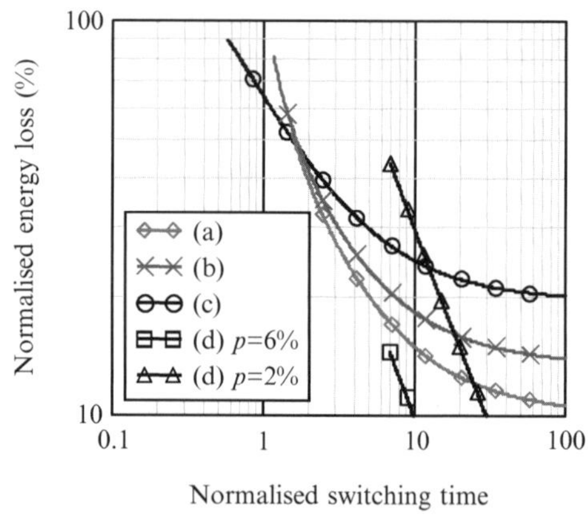

Fig. F.5 Normalized energy loss vs. normalized switching speed of the proposed resonant gate driver topologies

Table F.3 Summary of the topology comparison study

	Switching speed	Inductance size	Loss savings
Topology (a)		✓	
Topology (b)		✓	☒
Topology (c)	✓		☒
Topology (d)	☒	☒	✓

✓ pros; ☒ cons

On the other hand, topologies (b) and (c) show the best performance in switching speed, though in detriment of rather high energy losses, e.g. close to 70% for topology (c) at normalized switching time equal to one.

In general terms, Table F.3 summarizes the information contained in Figs. F.4 and F.5.

F.4 Application Example

The above analysis suggests that the topology of choice must be dependent on the application. This implies to know a priori the characteristics of the power MOSFET, its mode of operation as well as the particular electrical conditions in the power train.

This section compares the performance of a conventional gate driver with topologies (a) and (c) for the application example given in Sect. F.2. In order to ensure minimum power losses, circuit parameters are optimized accordingly, as described in Sect. F.2. Results are shown in Tables F.4 and F.5 for topology (c) and a conventional gate driver, respectively. The optimization results of topology (a) are given in Table C.1.

Power losses are calculated from simulated waveforms as shown in Figs. F.2, F.6 and F.7. The gate drive losses result in the magnitudes predicted by the equations of Sect. F.3. Gate charge losses are obtained considering the assumptions of Sect. F.3.

Figure F.8 shows the overall power loss for each case upon optimized conditions. Topology (c) improves the overall losses (i.e., channel and driver) by 23%, whereas topology (a) does it by 34%. The main improvement with respect to the conventional solution lays on the gate driver part, though channel losses are also reduced. The latter is achieved in topology (a) with a smaller chip size power MOSFET due to a significantly higher V_S value.

The optimum switching time of the resonant gate drivers is relatively high, e.g., 60 ns in case of topology (a). As illustrated in Sect. F.2, this relaxed switching speed requirements are influenced by the soft-switching operation of the power MOSFET as well as the short dead times. Thus, according to Fig. F.5, topologies (a) and (d) provide higher loss reduction than (b) and (c). This agrees with the results of Fig. F.8. The optimized inductance results are also consistent with the predictions of Fig. F.4, i.e., topology (a) offers the lowest inductance, just 25 nH, which might

Table F.4 Optimized circuit parameters of topology (c)

Clamping switches $S1$ and $S2$	Input $R_{DSon} = R_{Sc} = 0.3\ \Omega$, $FoM_i = FoM_o = 40$ ps
Power MOSFET (SyncFET)	Chip area $= 13\ mm^2$
Gate driving voltage	Gate switches $V_{drvS} = 5.5$ V, power MOSFET $V_S = 5.5$ V
Resonant gate inductor L_{Gr}	50 nH, negligible ESR

Table F.5 Optimized circuit parameters of a conventional gate driver

Clamping switches $S1$ and $S2$	Input $R_{DSon} = R_{Sc} = 0.4\ \Omega$, $FoM_i = FoM_o = 40$ ps
Power MOSFET (SyncFET)	Chip area $= 13.5\ mm^2$
Gate driving voltage	Gate switches $V_{drvS} = 4.5$ V, power MOSFET $V_S = 4.5$ V

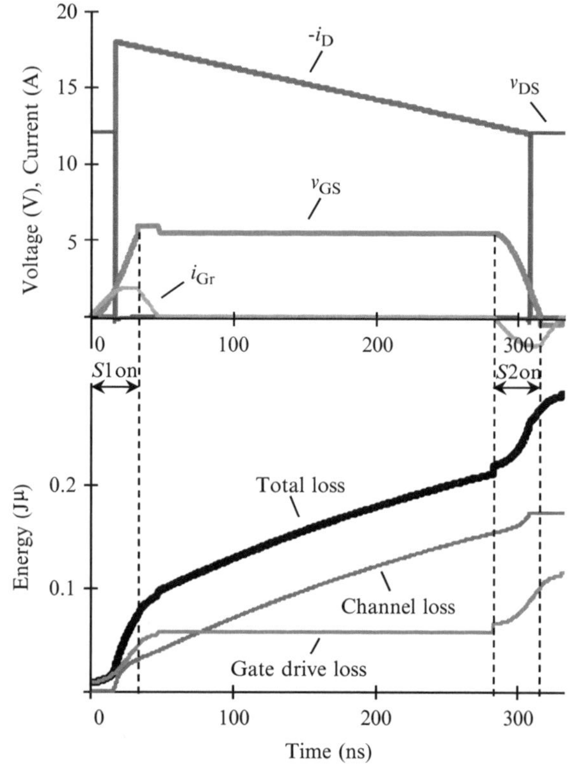

Fig. F.6 Switching cycle operation of topology (c) for a case example upon optimized power loss conditions

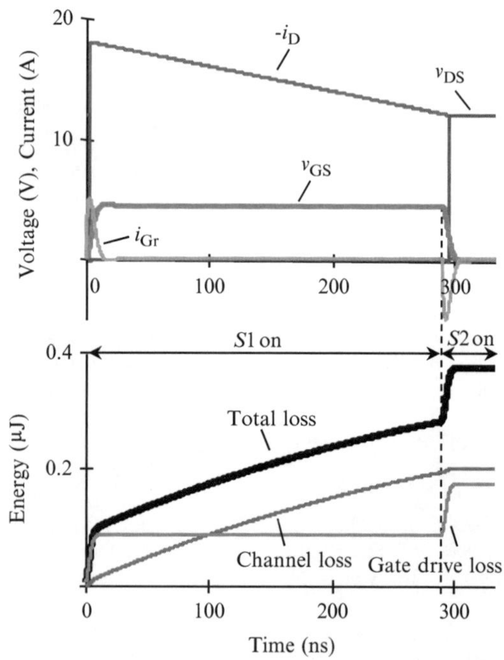

Fig. F.7 Switching cycle operation of a conventional gate driver for a case example upon optimized power loss conditions. Switches $S1$ and $S2$ refer to the upper and lower switch, respectively

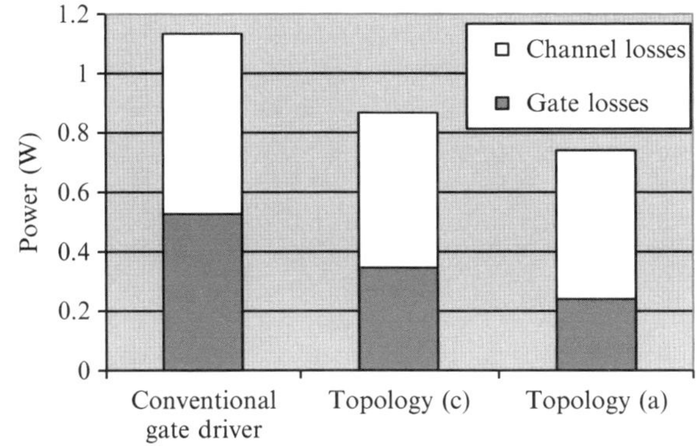

Fig. F.8 Power loss comparison of various optimized gate drivers

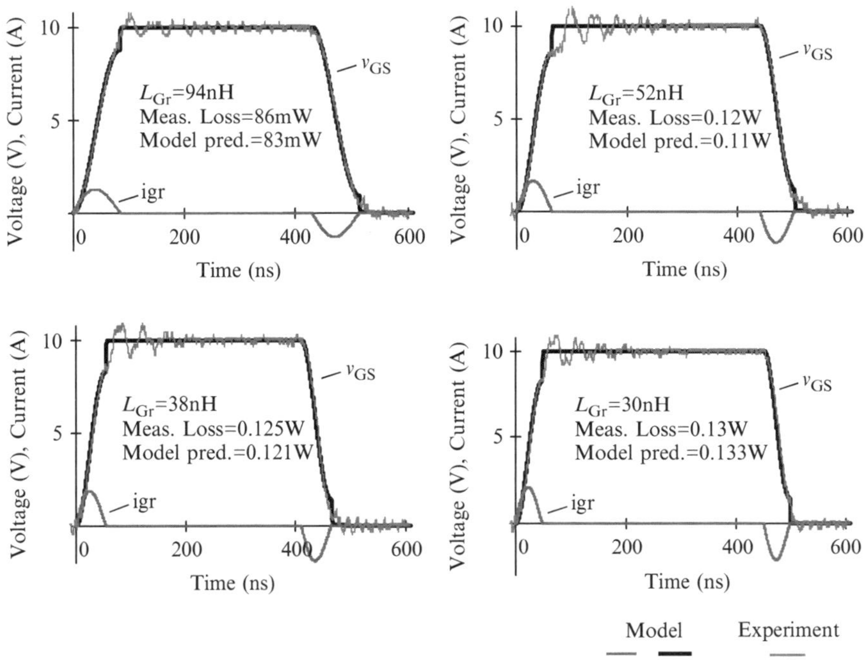

Fig. F.9 Measured and simulated data of topology (a) for various L_{Gr}

well be suitable for integration. This contrasts with typical inductances for topology (d), which may provide even higher loss reduction, yet in detriment of a significantly higher inductance.

Fig. F.10 Conventional gate driver losses as function of switching frequency

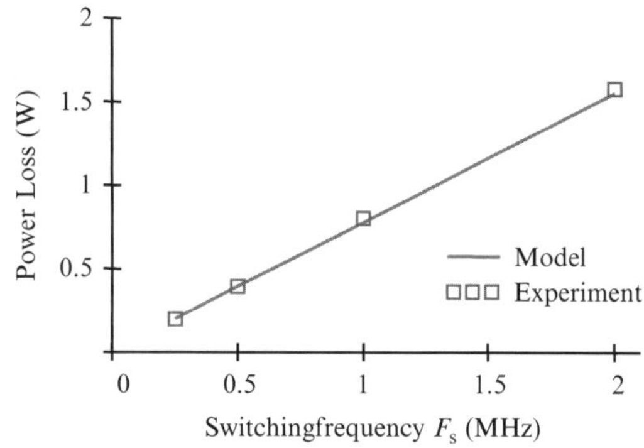

Table F.6 Experimental setup data

Clamping switches $S1$ and $S2$ (used for conventional gate driver as well)	2xN-channel MOSFETs TN0200T from Vishay, $R_{DSon} = 0.25\ \Omega$, $C_{oss} = 160$ pF
Bidirectional switch S3 (two anti-series MOSFETs)	2xN-channel MOSFETs PHN110 from NXP Semiconductors, $R_{DSon} = 0.1\ \Omega$, $C_{oss} = 400$ pF
Power MOSFET (SyncFET)	Replaced by a ceramic capacitor: negligible R_{GM}, $C_{iss} = 6.9$ nF
Gate driving voltage V_s	10 V
Switching frequency F_s	1 MHz (Unless otherwise specified)
Resonant gate inductor L_{Gr}	See Fig. F.9, negligible ESR

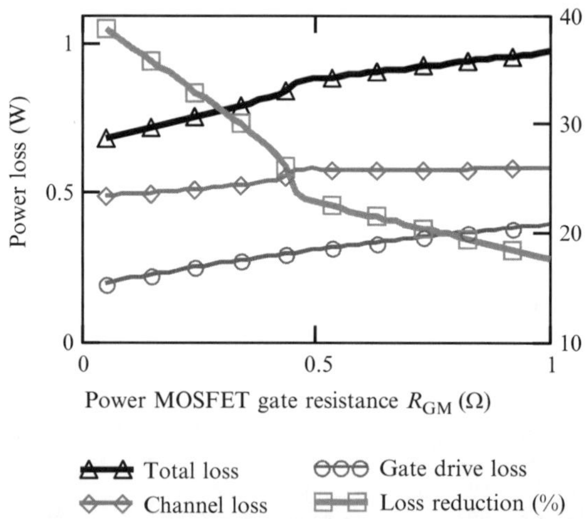

Fig. F.11 Influence of power MOSFET gate resistance on power losses. Results refer to application example of section F.2 employing topology (a)

F.5 Experimental Results

The circuit model and equations of Sect. F.3 corresponding to the conventional gate driver and topology (a) are experimentally validated. Figure F.9 shows the good agreement between experimental and simulated waveforms as well as measured and predicted power losses upon various operating conditions. Note that the gate drive losses of the gate switches are not considered.

In order to validate (F.1), Fig. F.10 compares the measured and predicted power loss of a conventional gate driver. Table F.6 provide information of the measurement setup.

F.6 Discussion and Conclusions

Several aspects shall be considered: First, the polysilicon gate resistance R_{GM} strongly impacts power losses and thus it should be reduced. Current typical values are near to 1 Ω, which may yield to a poor loss reduction, as illustrated in Fig. F.11. Note that the sudden gradient change in the channel loss curve at around 0.45 Ω is entirely attributed to artifacts of the dead time controller. Second, the switching speed of conventional gate drivers is typically limited by the saturation of the gate switches due to the high current peaks. This may be mitigated in resonant gate drivers since current peaks are lower. Third, slowing down the switching speed may limit the use of extreme duty cycles. Four, miniaturization is typically the main motivation for going into high-switching frequency operation. Topologies (a) and (b) may offer the lowest inductance, i.e., the smallest size. Yet, both circuit and mode of operation are in principle more complex than topologies (c) and (d). When size and switching speed are less demanding, then topology (d) may be the most appropriate choice, for its simplicity and reduced power loss. Five, considerations of the power MOSFETs operating conditions are crucial to identify the topology of choice and optimize design parameters.

References

1. Strydom JT, de Rooij MA, van Wyk JD (2004) A comparison of fundamental gate driver topologies for high frequency applications. In: IEEE applied power electronics conference, APEC 2004, pp 1045–1052
2. Dwane P, O' Sullivan D, Egan MG (2005) An assessment of resonant gate drive techniques for use in modern low power DC-DC converters. In: IEEE applied power electronics conference, APEC 2005, pp 1572–1580
3. Maksimovic D (1991) A MOS gate drive with resonant transitions. In: IEEE power electronics specialists conference, PESC 1991, pp 527–532

4. López T, Sauerlaender G, Duerbaum T, Tolle T (2003) A detailed analysis of a resonant gate driver for PWM applications. In: IEEE applied power electronics conference, APEC 2003 pp 873–878
5. Dolny GM, Sapp S, Elbanhaway A, Wheatley CF (2004) The influence of body effect and threshold voltage reduction on trench MOSFET body diode characteristics. In: International symposium in power semiconductor devices, ISPSD 2004, pp 217–220
6. López T, Elferich R, Koper N (2006) Reverse recovery in high density trench MOSFETs with regard to the body-effect. In: IEEE international symposium in power semiconductor devices, ISPSD 2006, pp 1–4
7. Van Der Broeck H, Wendt M, Steinbush H (2006) Method of controlling a circuit arrangement for the AC power supply of a plasma display panel. US7064732B2, 20 June 2006
8. Steigerwald RL. Lossless gate driver circuit for a high frequency converter. Patent US5010261
9. Chen Y, Lee FC, Amoroso L (2004) A resonant MOSFET gate driver with efficient energy recovery. IEEE Trans Power Electron 470–477

Appendix G
Experimental Prototypes

This section presents the circuit diagrams and PCB layouts of the prototype boards employed throughout this thesis work. Three different converter circuits are experimented with for different purposes:

- Synchronous buck converter prototype for switching measurements
- Point-of-load converter based on discrete LFPak based MOSFETs for efficiency measurements
- Multi-phase buck converter based on the IPM PIP212-12M for efficiency measurements

G.1 Synchronous Buck Converter Board

Experimental switching waveforms presented in the model validation section of Chap. 2 are based on the following circuit board. Furthermore, MOSFET characterization parameters such as reverse recovery are also performed with this PCB as it allows both current and voltage measurements in practically all relevant loops and nodes of the switched converter. The board is designed for 30 V MOSFET devices and it allows high-current, high-frequency switching operation. MOSFETs temperature control is enabled by means of thermal pads on the bottom layer, right underneath the switch devices. External pulse generator for the individual control of the switches is required (Figs. G.1–G.6).

Fig. G.1 Synchronous buck converter prototype board. Designed by Dr. Tobias Tolle from Philips Research

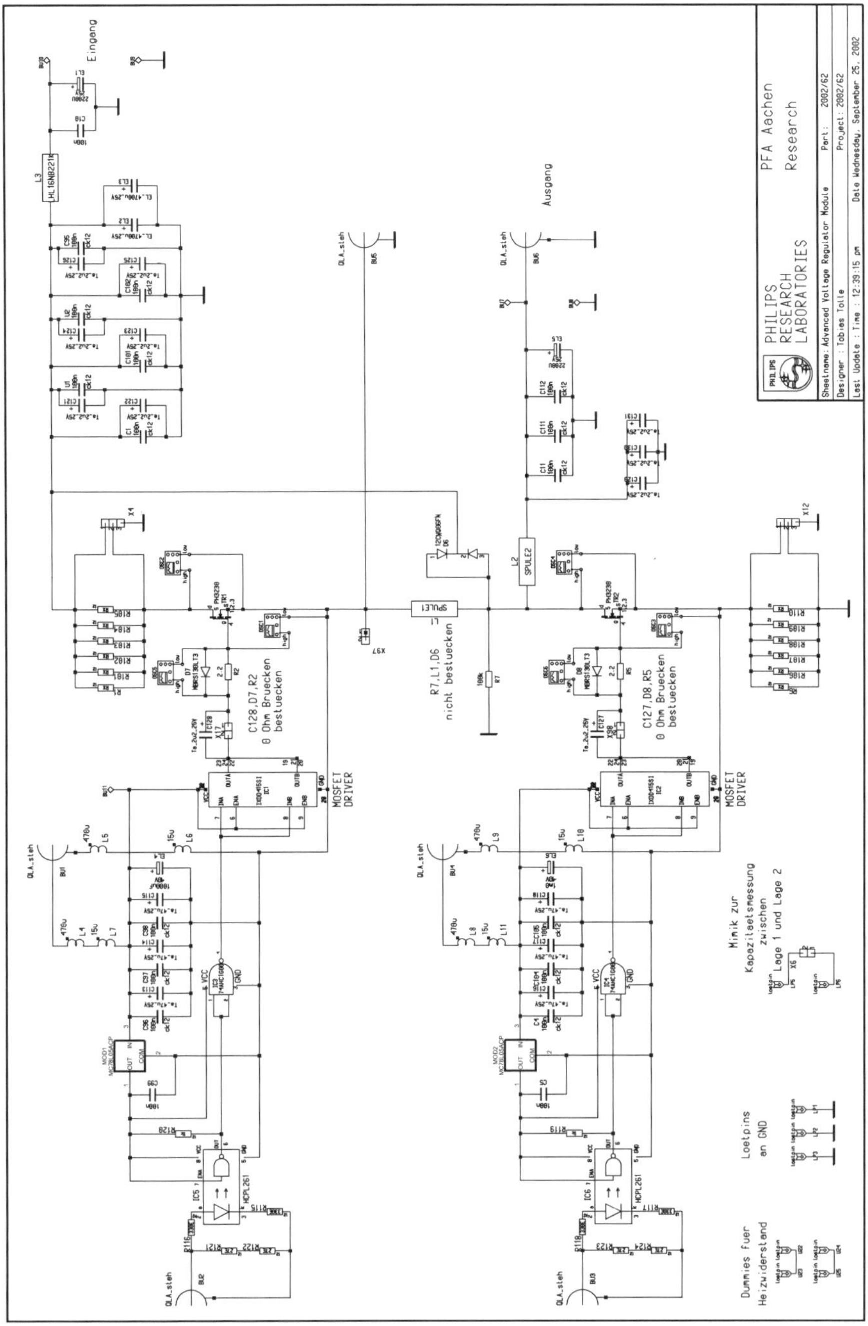

Fig. G.2 Circuit diagram of a synchronous buck converter prototype board of Fig. G.1

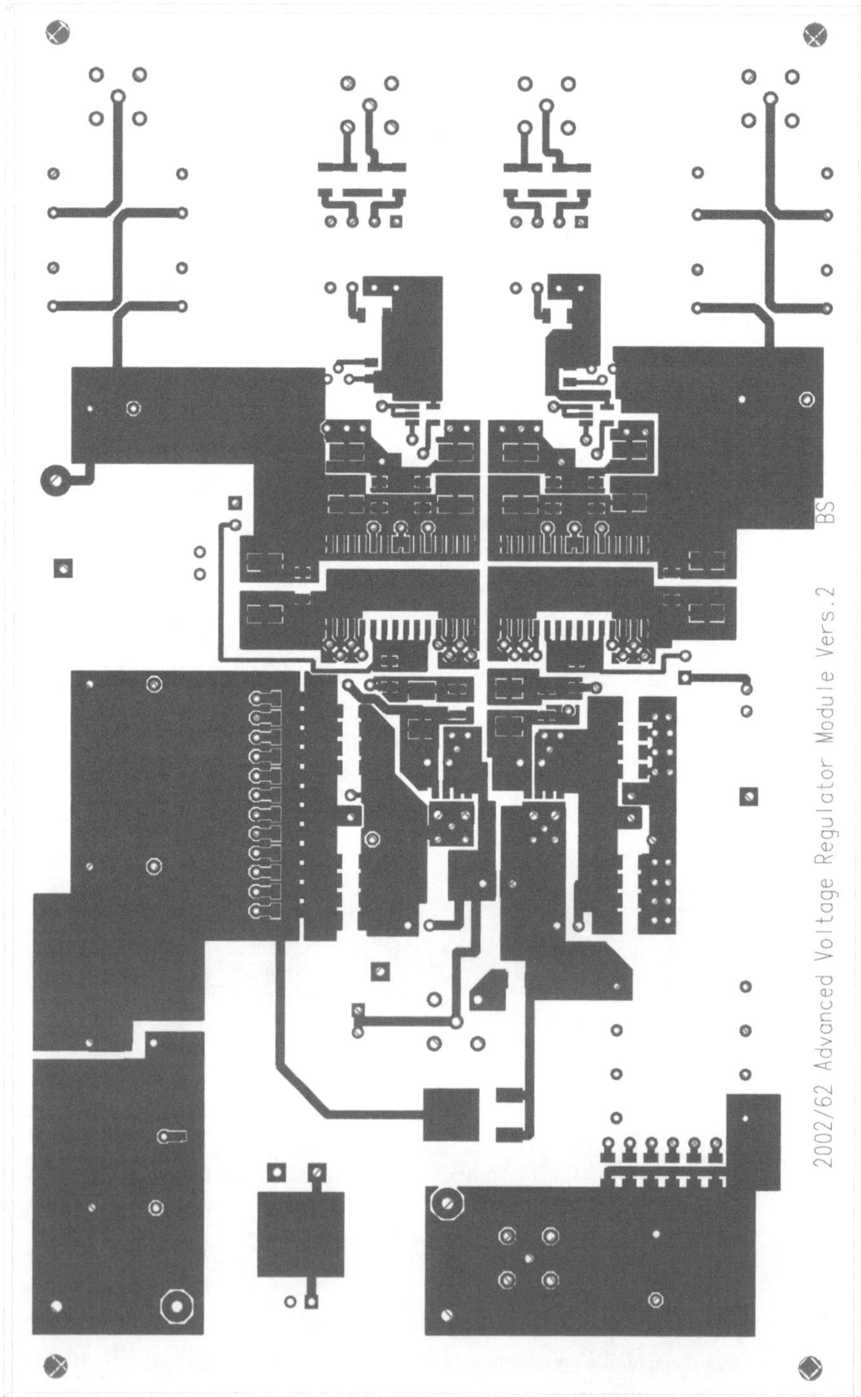

Fig. G.3 PCB design of top layer corresponding to the test board of Fig. G.1

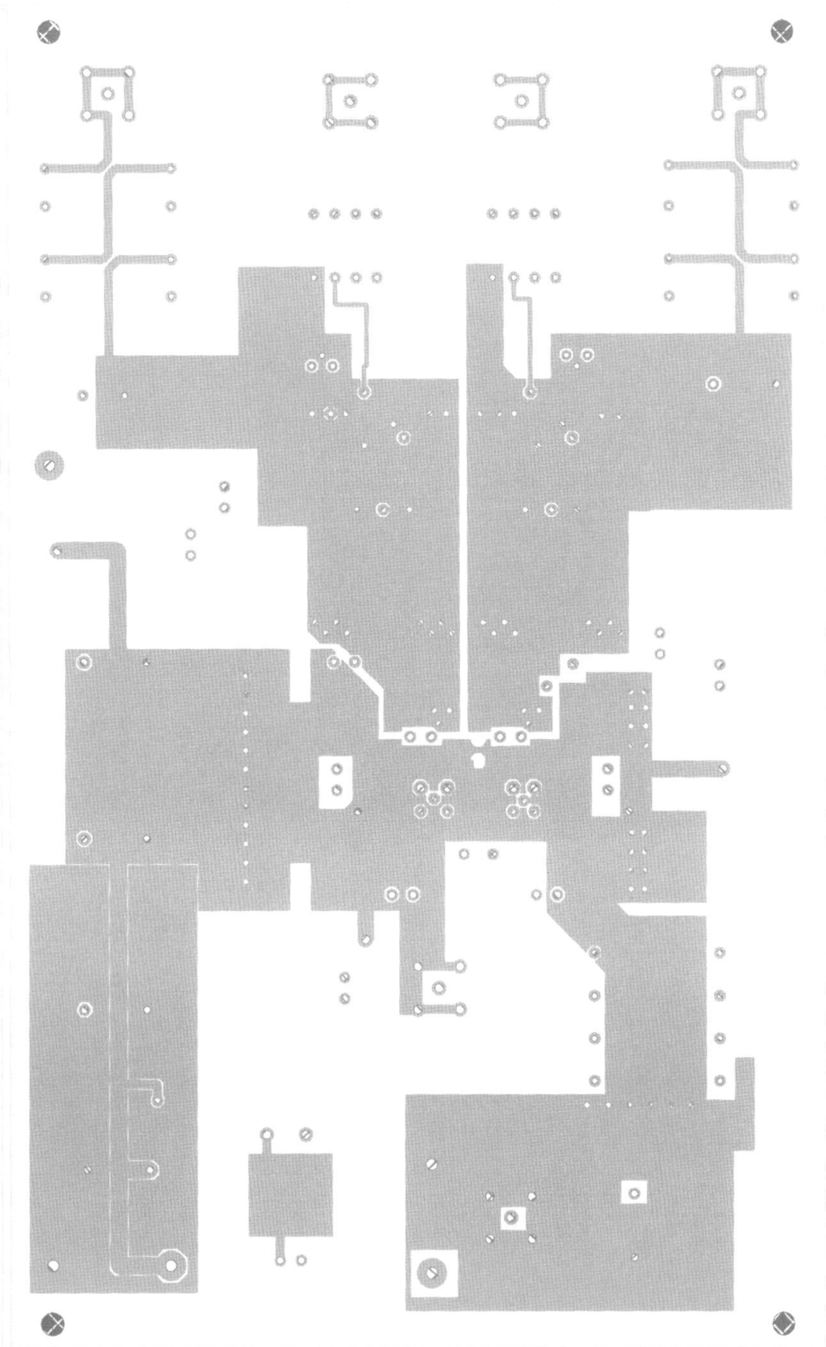

Fig. G.4 PCB design of inner layer corresponding to the test board of Fig. G.1

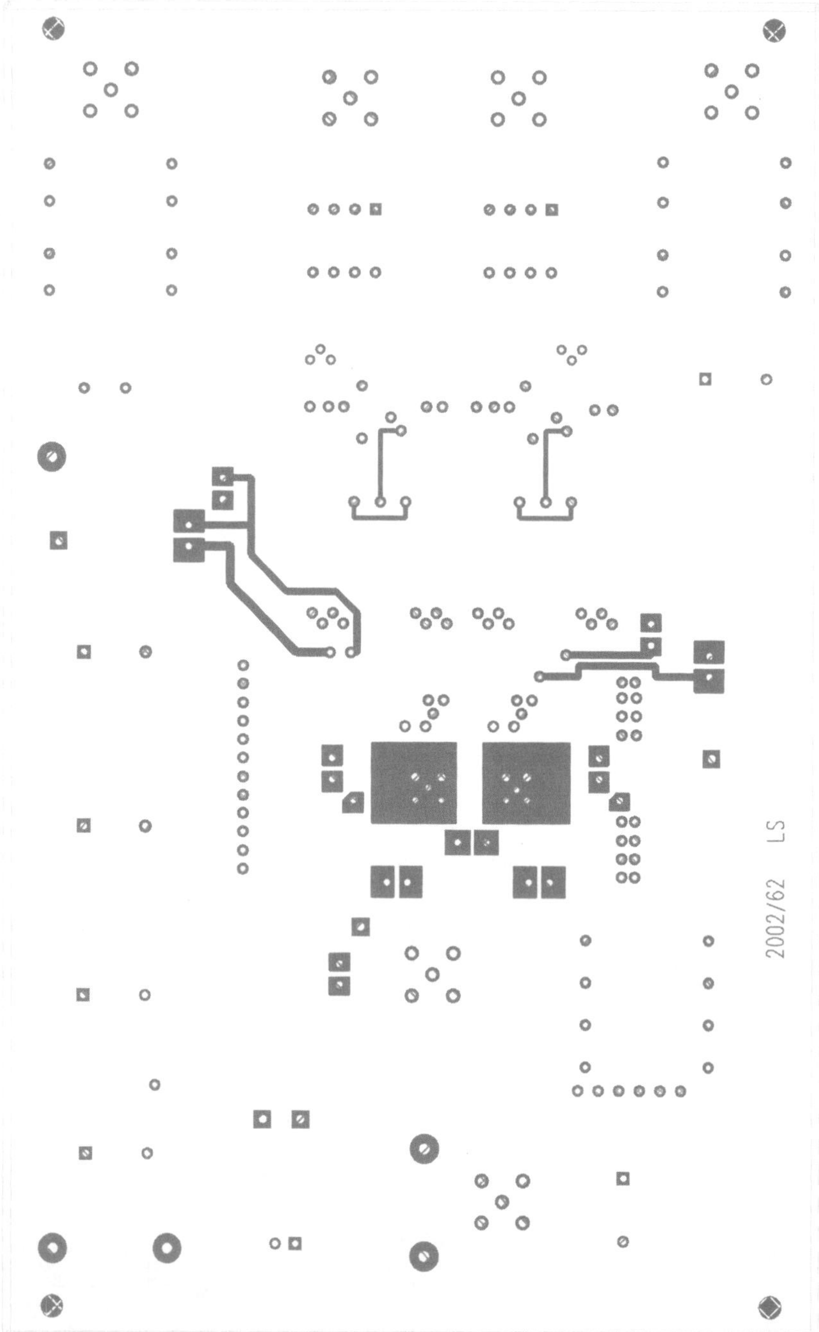

Fig. G.5 PCB design of bottom layer corresponding to the test board of Fig. G.1

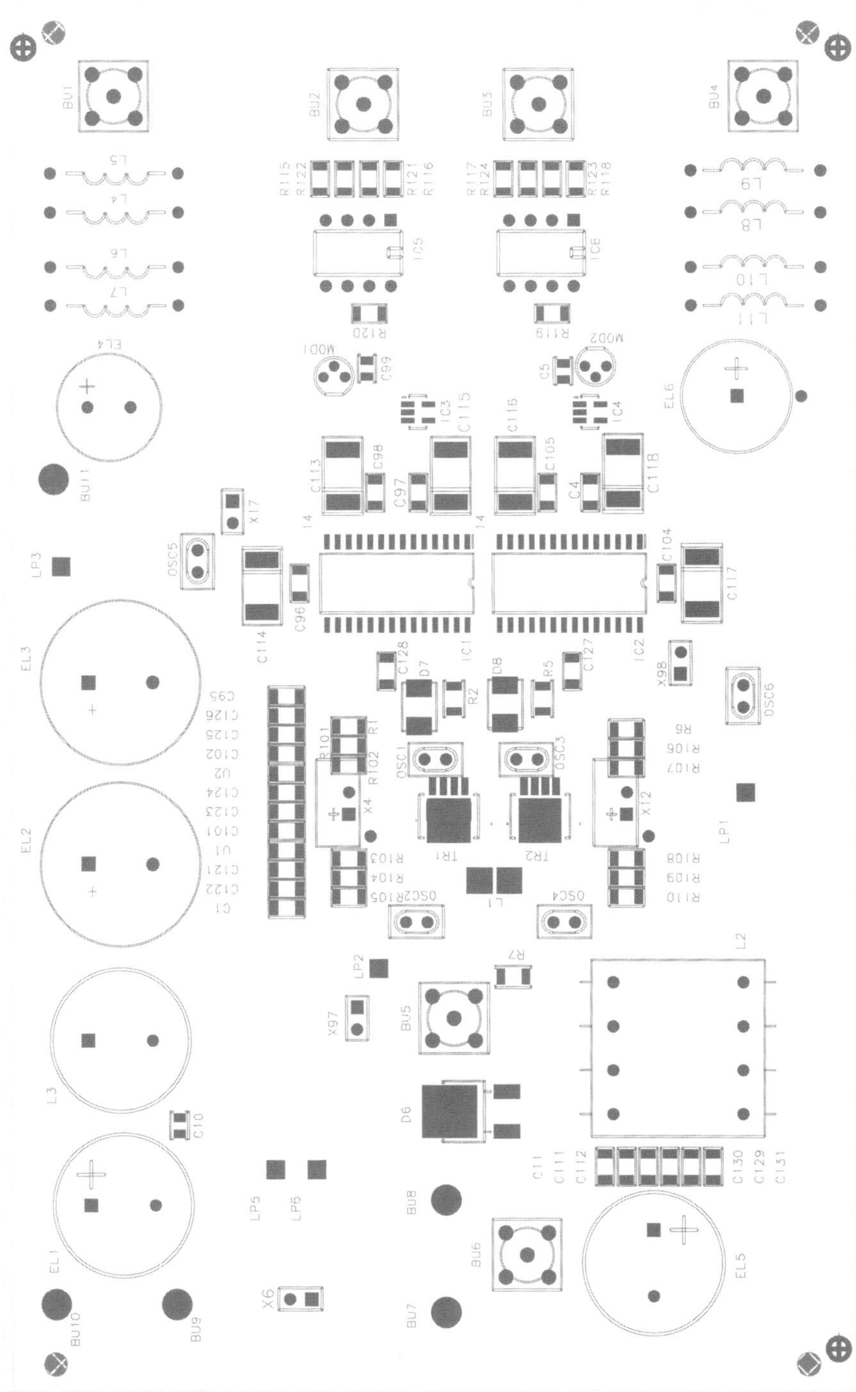

Fig. G.6 PCB design of top overlay corresponding to the test board of Fig. G.1

G.2 Point-of-Load Demo Board

The main purpose of this demo board for LFPak MOSFETs is to demonstrate the highest possible efficiency in an industry standard module size of 33 × 14.2 mm with the existing device technologies. Simulation results comparing Trench4 with Trench6 in Chap. 6 are performed on this test board, which is designed to feature low parasitic inductances and allow load currents up to 20 A. The PoL is equipped with a commercial synchronous buck converter regulator for high-switching frequency operation with automatic dead time reduction and other loss saving features (Figs. G.7– G.11).

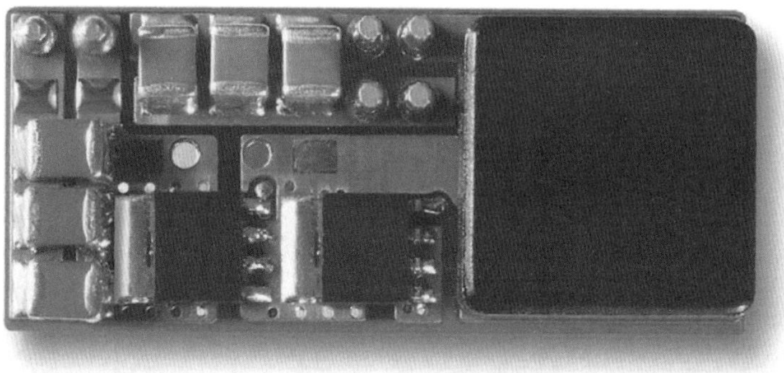

Fig. G.7 Point-of-load demo board. Designed by Victor Guijarro from Philips Semiconductors

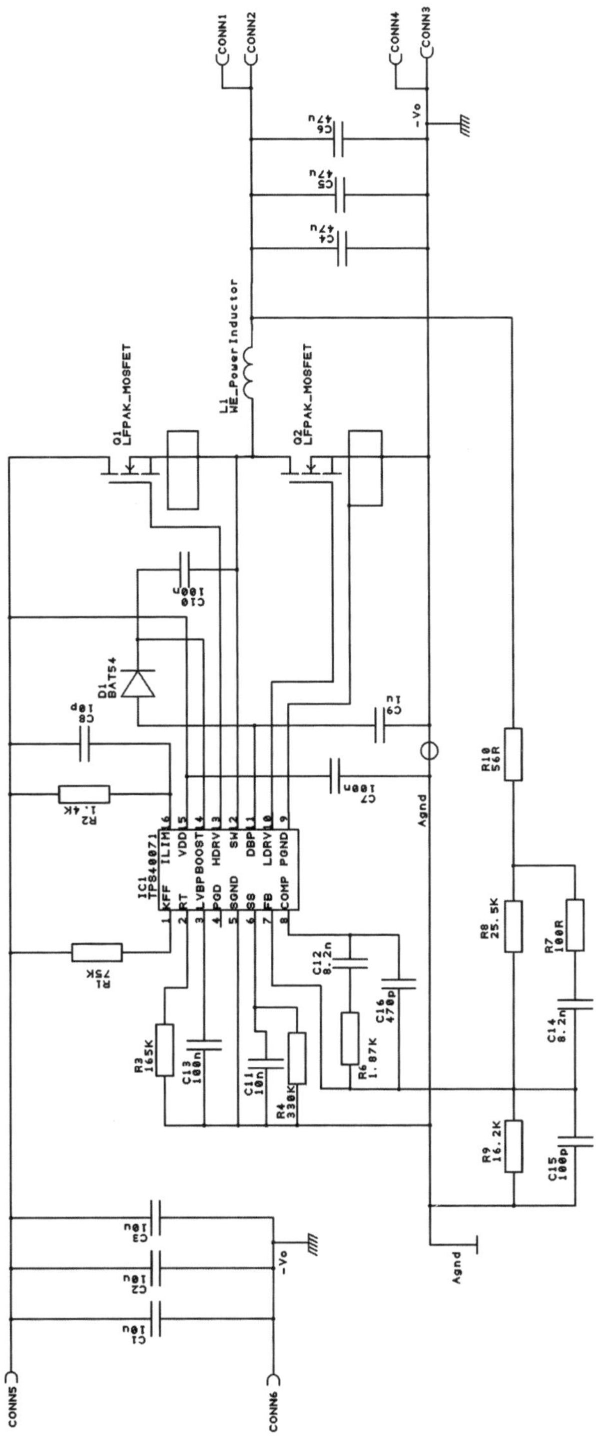

Fig. G.8 Circuit diagram of a synchronous buck converter prototype board of Fig. G.7

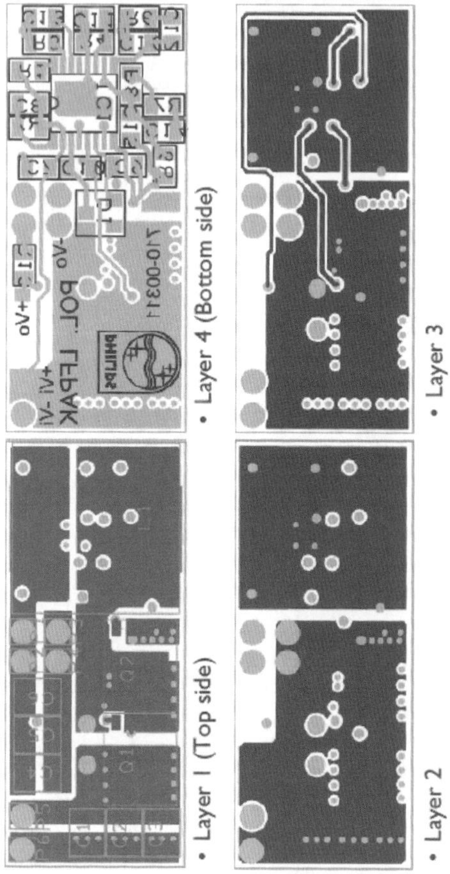

Fig. G.9 PCB layer design corresponding to the demo board of Fig. G.7

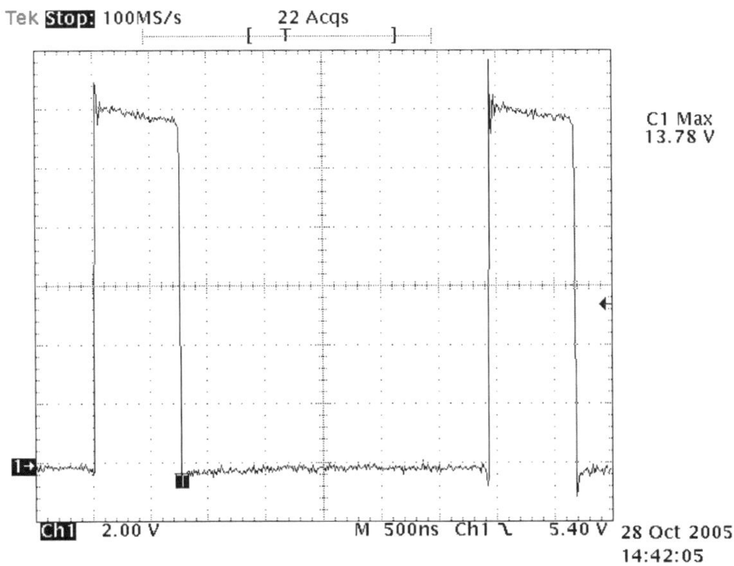

Fig. G.10 Measured switched-node waveform from PoL demo board of Fig. G.7. Employed MOSFETs: PH3330L (SyncFET) and PH8030L (CtrlFET). Conditions: $V_{in} = 12$ V, $V_o = 2.5$ V, $I_o = 17$ A. Scope settings: 500 ns/div

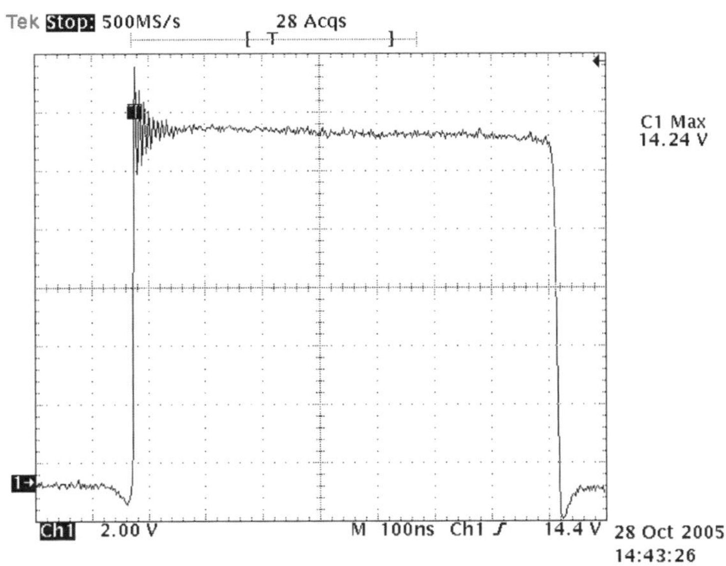

Fig. G.11 Measured switched-node waveform from PoL demo board of Fig. G.7. Employed MOSFETs: PH3330L (SyncFET) and PH8030L (CtrlFET). Conditions: $V_{in} = 12$ V, $V_o = 2.5$ V, $I_o = 17$ A. Scope settings: 100 ns/div

G.3 Multiphase VR Demo Board

This VR demo board allows performance evaluations of the IPM PIP212-12M in multiphase operation driving load currents up to 140 A. The nominal source voltage is 12 V and can be converted down to 0.8 V. Phases are controlled by a central unit and can be individually disabled. The demonstration board contains a low power onboard switching regulator to provide the gate drive voltage of the SyncFET and drive a series of LED indicators, which are turned-off for efficiency measurements. The experimental curves of the PIP212-12M from Chap. 3 are based on this test board (Figs. G.12–G.21).

Fig. G.12 Multiphase voltage regulator test board. Designed by Dr. Phil Good, from Philips Semiconductors

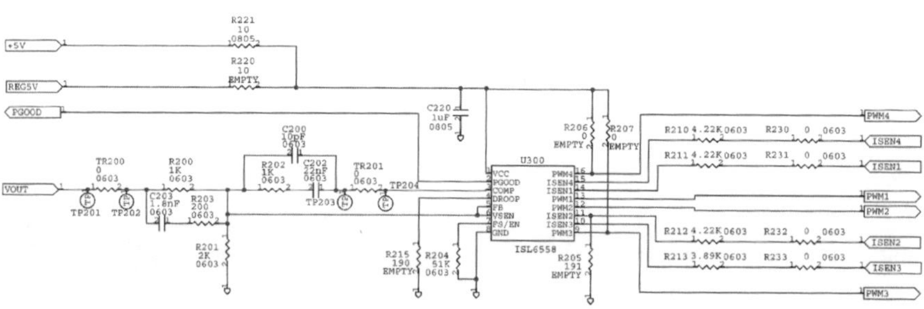

Fig. G.13 Circuit diagram of the multiphase VR test board of Fig. G.12. Multiphase regulator section

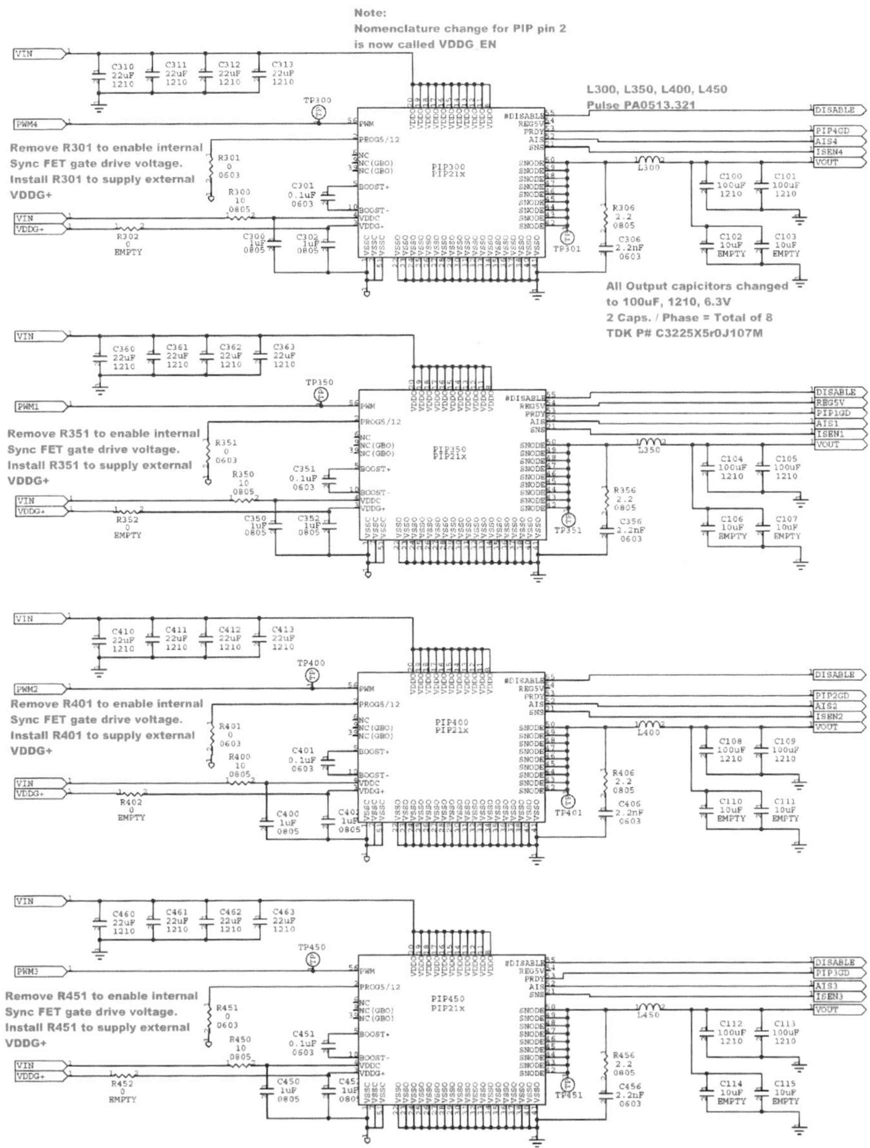

Fig. G.14 Circuit diagram of the multiphase VR test board of Fig. G.12. Converter phases

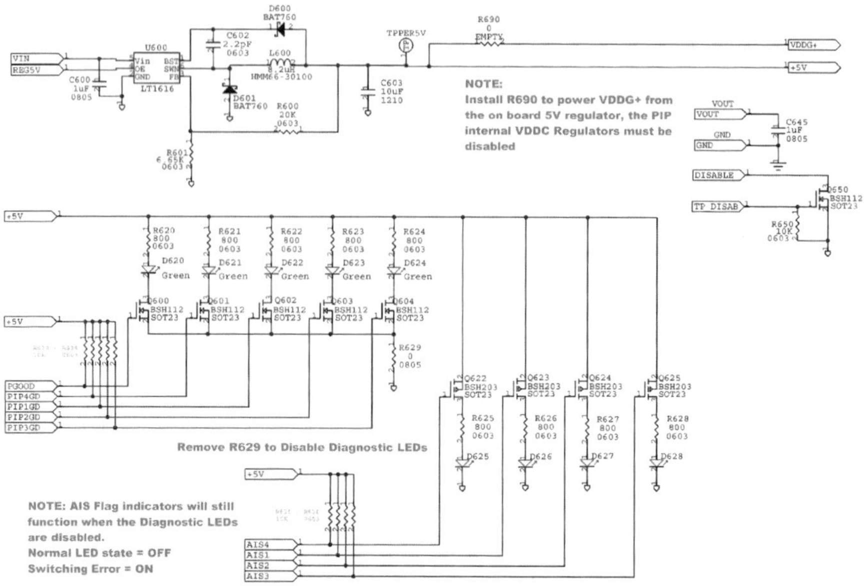

Fig. G.15 Circuit diagram of the multiphase VR test board of Fig. G.12. Low power regulator and signal LED circuitry

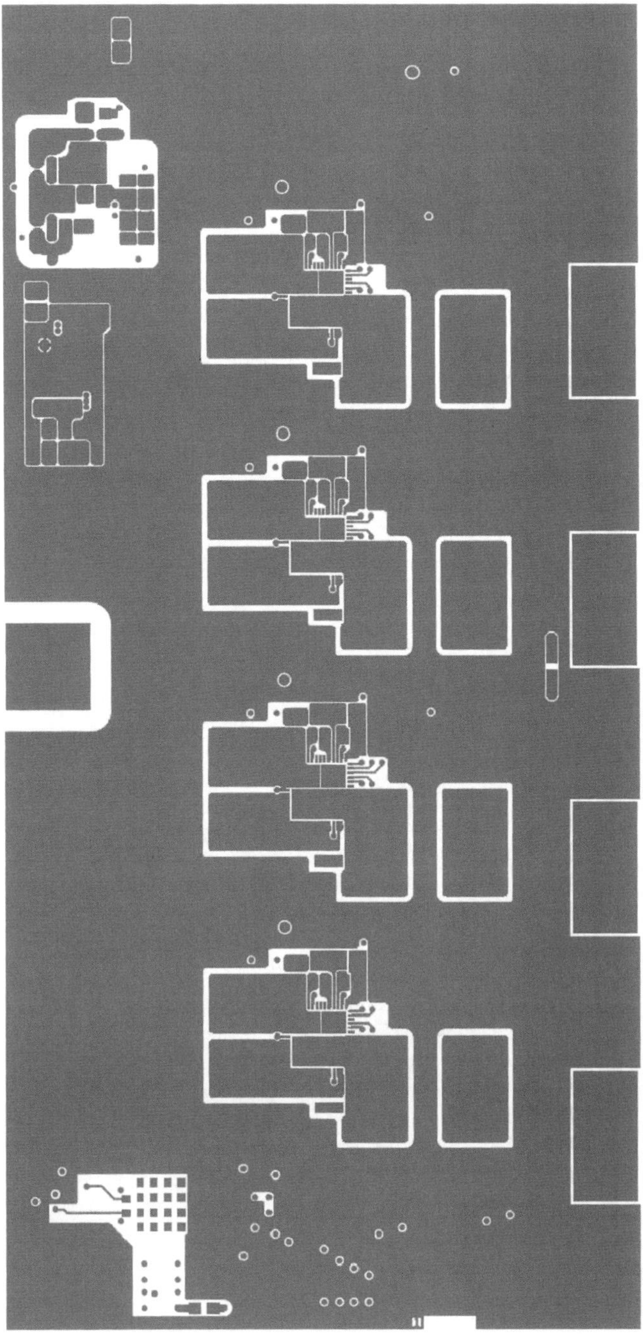

Fig. G.16 PCB design of top layer corresponding to the test board of Fig. G.12

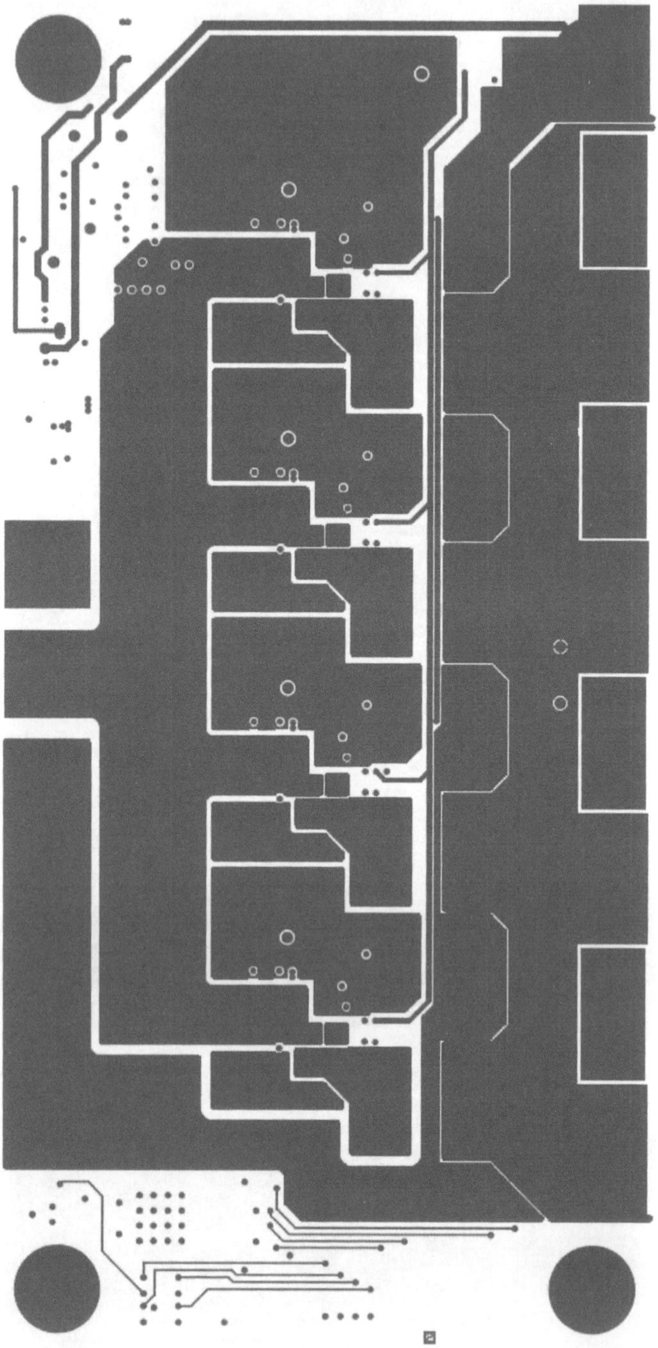

Fig. G.17 PCB design of inner layer 1 corresponding to the test board of Fig. G.12

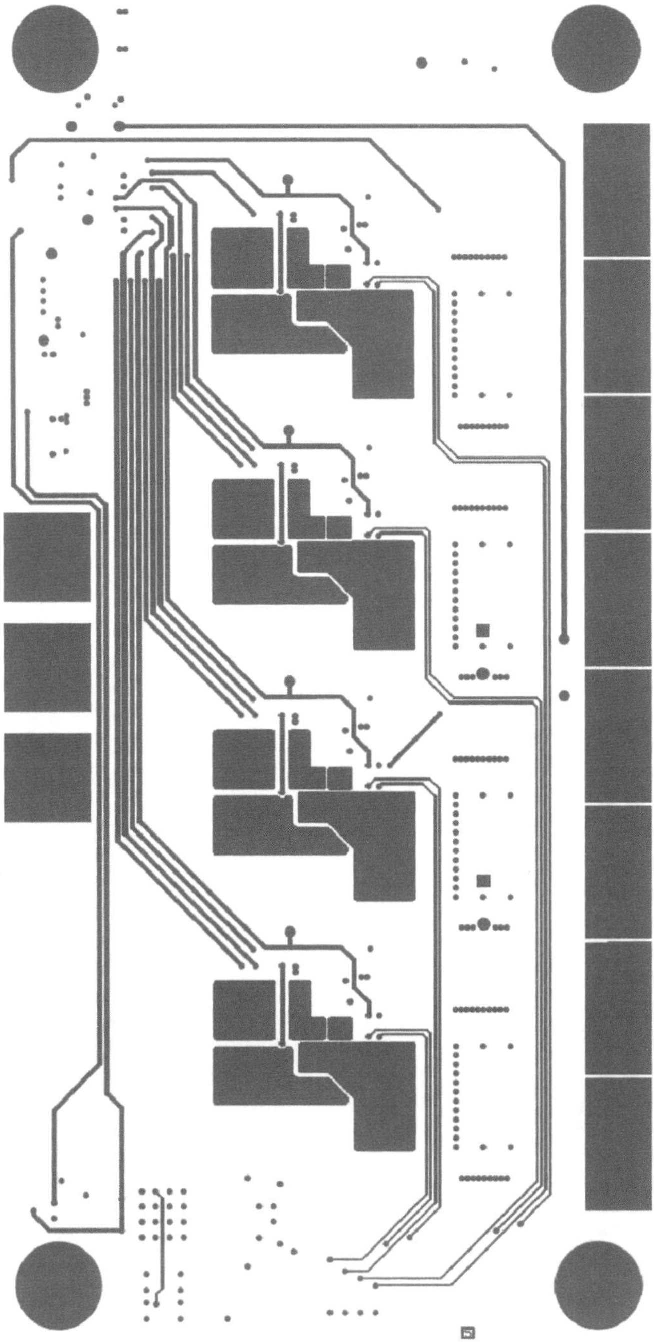

Fig. G.18 PCB design of inner layer 2 corresponding to the test board of Fig. G.12

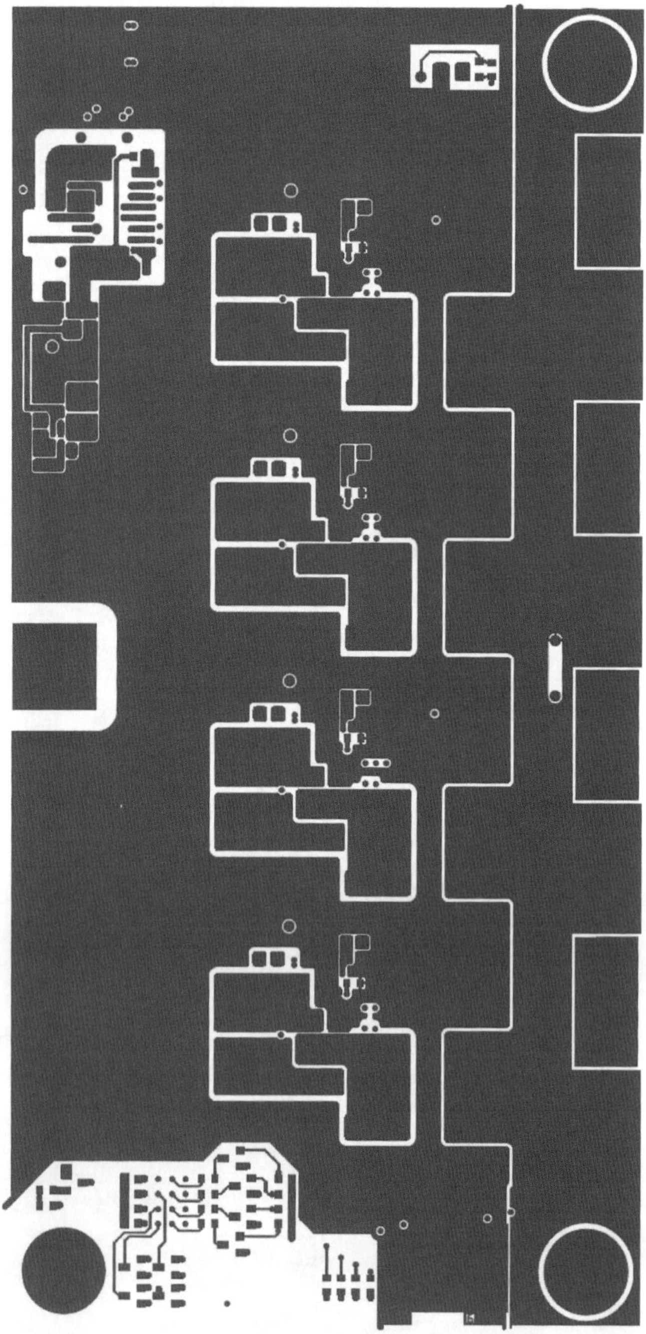

Fig. G.19 PCB design of bottom layer corresponding to the test board of Fig. G.12

Fig. G.20 PCB design of top overlay corresponding to the test board of Fig. G.12

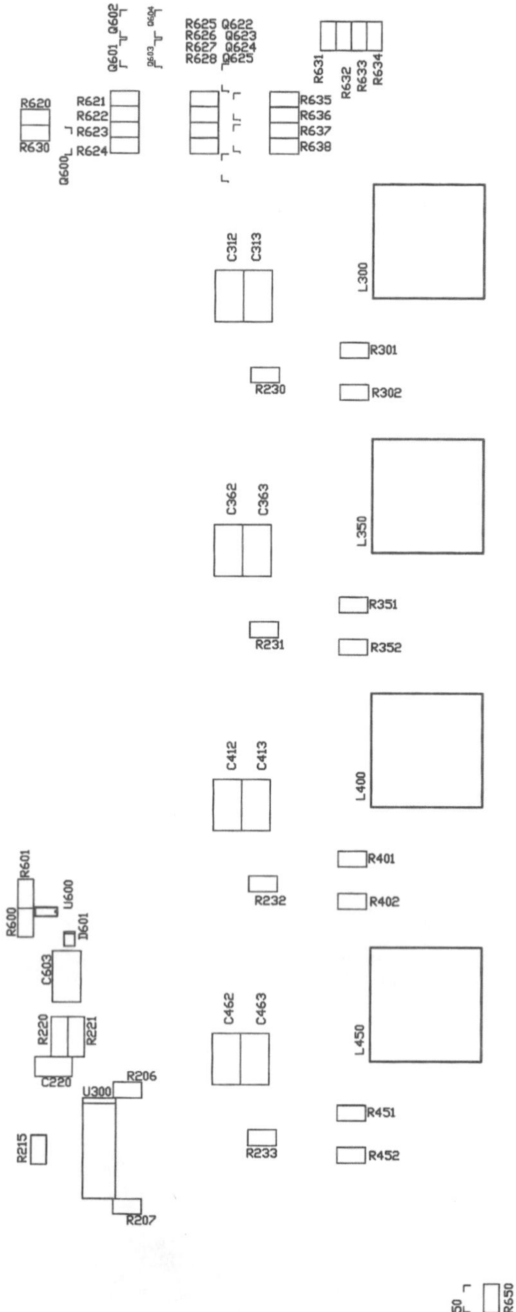

Fig. G.21 PCB design of bottom overlay corresponding to the test board of Fig. G.12

Index